Proceedings in Life Sciences

ADP-Ribosylation of Proteins

Edited by
F. R. Althaus, H. Hilz, and S. Shall

With 272 Figures

Springer-Verlag
Berlin Heidelberg New York Tokyo

Felix R. Althaus
Institut für Pharmakologie und Biochemie
Veterinärmedizinische Fakultät der Universität
CH-8057 Zürich, Switzerland

Helmuth Hilz
Physiologisch-chemisches Institut
Universität Hamburg
2000 Hamburg 20–40, FRG

Sydney Shall
The University of Sussex
Biology Building
Brighton, Sussex BN1 9QG, England

ISBN 3-540-15598-8 Springer-Verlag Berlin Heidelberg New York Tokyo
ISBN 0-387-15598-8 Springer-Verlag New York Heidelberg Berlin Tokyo

Library of Congress Cataloging in Publication Data. Main entry under title: ADP-ribosylation of proteins. (Proceedings in life sciences) "Synopses of contributions which were presented at the Seventh International Symposium on ADP-Ribosylation Reactions, held in Vitznau, Switzerland, from September 23 to 27, 1984" – P. Bibliography: p. Includes indexes. 1. Adenosine diphosphate ribose – Congresses. 2. Nucleoproteins – Congresses. I. Althaus, F.R. (Felix R.), 1949– . II. Hilz, H. (Helmuth), 1924– . III. Shall, S. (Sydney), 1932– . IV. International Symposium on ADP-Ribosylation Reactions (7th: 1984: Vitznau, Switzerland). V. Series. QP625.A29A35   1985   574.19′256   85-22219

Offsetprinting and bookbinding: Brühlsche Universitätsdruckerei, Giessen
2131/3130-543210

This book is dedicated to
the pioneers of the field of ADP-ribosylation

Paul Mandel, Osamu Hayaishi, and
Takashi Sugimura

Preface

This book provides an update on recent advances in the field of ADP-ribosylation reactions. The individual chapters represent the synopses of contributions which were presented at the Seventh International Symposium on ADP-Ribosylation Reactions, held in Vitznau, Switzerland, from September 23 to 27, 1984. This volume covers new developments in the field since the last meeting was held on this topic in 1982, in Tokyo. Therefore, the present text is not meant to form a comprehensive account of a specialized research area, but encompasses a collection of state-of-the-art reports from the vast majority of laboratories currently involved in ADP-ribosylation work. For the sake of rapid publication, the editorial policy was to ensure easy accessibility of information contained in individual articles rather than to provide elaborate cross references or reference to work published prior to 1982. However, a detailed subject index will help the reader find complementary information.

The enzymes of ADP-ribose metabolism have not yet acquired universally acceptable trivial names and the Enzyme Commission has not yet definitely decided on formal appellations. Consequently, a variety of names for the nuclear enzyme appear in this book, including nuclear(ADP-ribosyl)transferase, poly(ADP-ribose) polymerase, or synthetase or synthase. Hopefully, a common convention will soon be established.

The Seventh International Symposium on ADP-Ribosylation Reactions was only possible because of the generous support which we have been given by our sponsors, listed below. It is only by such far-sighted support of apparently purely academic research that subsequently we shall have the basic knowledge to advance medical science or to lay the technical foundations of new industries. Thus, our thanks go to the following institutions:

The Commission of the European Community in the frame of
    European Cooperation in the field of Scientific and
    Technical Research

The European Society of Toxicology

Schweizerische Naturforschende Gesellschaft

Schweizerischer Nationalfonds zur Förderung der wissenschaftlichen
    Forschung

Schweizerische Akademie der Medizinischen Wissenschaften

Union Schweizerischer Gesellschaften für Experimentelle Biologie

Schweizerische Krebsliga

Krebsliga des Kantons Zürich, Zürich, Switzerland

Fonds für Medizinische Forschungen der Universität Zürich,
    Zürich, Switzerland

Prof. Bruno Bloch Stiftung, Zürich, Switzerland

Interpharma, Basel, Switzerland

Pharmaton SA, Lugano, Switzerland

Jubiläumsstiftung der Schweizerischen Lebensversicherungs- und
    Rentenanstalt, Zürich, Switzerland.

We wish to thank Dr. Marianne Althaus-Salzmann for her contributions in various stages of preparation for the Seventh International Symposium on ADP-Ribosylation Reactions. The excellent secretarial work of Mrs. Jeanne-Marie Burger is gratefully acknowledged.

August 1985   Felix R. Althaus
                                                     Helmuth Hilz
                                                     Sydney Shall

Contents

ADP-Ribosylation and Chromatin Structure

Mono(ADP-Ribosylation) Reactions

Contributors

You will find the addresses at the beginning of the respective contributions

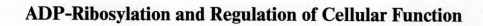

ADP-Ribosylation and Regulation of Cellular Function

Multiple Aspects of ADP-Ribosylation Reactions

PAUL MANDEL[1]

1. Introduction

It has been extensively reported that the amount of NAD in proliferating tumor cells is much lower than that of nonproliferating cells (see for review [1]). Considering this observation we asked the question (1961) whether an increase of NAD, induced by administration of nicotinamide to an animal would block normal cell proliferation in an adult animal: the so-called compensatory kidney hypertrophy and hyperplasia which occur after ablation of one kidney. In the experiments performed we found in parallel with a block of the hyperplasia and hypertrophy an increase of kidney NAD, a decrease of RNA and DNA synthesis, and an appearance of a polyadenylic compound [2]. Taking into account the increase of NAD, the alteration of RNA and DNA synthesis, and the appearance of the polyadenylic compound we incubated nuclear extracts in the presence of $[\alpha\text{-}^{32}\text{P}]$-ATP and nicotinamide mononucleotides (NMN), the two moieties of NAD. Under these conditions a 600—1500-fold stimulation of incorporation of AMP into a polyadenylic compound was observed. The reaction was DNA dependent [3]. Very rapidly, however, we discovered that the subunit of the compound synthesized was not AMP and that an adenosine fraction from ATP and phosphoribosyl from NMN was incorporated into the polymer in equimolar amounts. We also obtained evidence that NAD is the immediate precursor of the polyadenylate compound. From enzymatic degradation experiments we were able to conclude that the polymer synthesized from NAD was polyadenosine diphosphate ribose and not polyadenylate [4]. Confirmation of our structure came from full methylation of the polymer. We could conclude that poly(ADP-ribose) is a homopolymer in which the ADPR groups are linked glycosidically (C1-C2 of ribose) and that the homopolymer is synthesized from NAD by a nuclear enzyme with concomitant release of nicotinamide. The enzymatic activity appeared to be DNA dependent [4, 5]. The structure was confirmed the next year by the laboratories of Hayaishi and Sugimura [6, 7]. Several comprehensive reviews cover the work on the subject in detail [37—41].

The enzyme poly(ADP-ribose) synthetase and the polymer poly(ADP-ribose) were born. We demonstrated the existence of the polymer in vivo with J. Doly [8]. Evidence for a branched structure of poly(ADP-ribose) was reported by Miwa et al.

---

1 Centre de Neurochimie du CNRS, 5, rue Blaise Pascal, 67084 Strasbourg Cedex, France

ADP-Ribosylation of Proteins
(ed. by F.R. Althaus, H. Hilz, and S. Shall)

[9] and we could provide electron microscopic evidence for the existence of this branched structure [10].

The next step in ADP-ribosylation reactions was the demonstration of a transfer of ADP-ribose to proteins by Nishizuka et al. [11] and Otake et al. [12].

2. Nuclear ADP-Ribosylation Reactions

In 1976 we developed a new procedure for poly(ADPR) polymerase purification from thymus gland with less than 1% protein contamination [13].

Using the protein blotting technique with polyclonal antibodies we could provide evidence which indicated species-spore (cow, rat, chicken) or tissue-specific differences in molecular weight of poly(ADPR) polymerase [14]. This finding suggests that the global enzyme structure/function relationship has been well preserved during evolution. However, no immunoprecipitation line with poly(ADPR) polymerase derived from rabbit or rat brain and liver nuclei could be obtained, showing the existence of species-specific antigenic determinants on the poly(ADPR) polymerase molecule [15]. Using microcomplement fixation we could approximately evaluate the poly(ADPR) level per mg DNA in calf thymus, chick neurons, and rat liver nuclei: 0.16 μg, 0.55 μg, and 0.01 μg, respectively. It could also be demonstrated by the microcomplement fixation technique that the DNA fraction which was copurified with the enzyme must occupy a specific site on the poly(ADPR) polymerase molecule and that the ADP-ribosylation site of the enzyme differs from the antigenic site [15].

After incubation of our poly(ADPR) polymerase with tritiated NAD we obtained poly(ADP-ribosylated) poly(ADPR) polymerase with an average chain length of 20–30 ADPR moieties [16–18]. The automodification reaction of poly(ADPR) polymerase could be visualized by electron microscopy [19]. The dark field micrographs demonstrate that as the reaction proceeds the apparent size of the enzyme increases and is surrounded by numerous points of uranyl acetate which also stains the products of the reaction. With increasing times the proportion and the size of this material increases, whereas the quantity of enzyme bound to DNA and of product decreases. A number of branches, up to 100 nanometers long, increases the size of the synthesized polymer. These results fully confirm the proposed branched structure of poly(ADP ribose) reported by Miwa et al. [9]. A detachment of the poly(ADP-ribosylated) enzyme from DNA was also observed. It may be a part of an autoregulation mechanism linked to the enzymatic activity. The branched structure may contribute more efficiently to locally modify a greater number of histone H1 molecules participating in the solenoidal structure of the chromosome. Moreover, the negatively charged polymer which can bind histone cores at physiological ionic strength may serve as an histone acceptor in the replication fork region as suggested by De Murcia et al. [10].

The poly(ADPR) polymerase which we purified to homogeneity was shown to contain about 10% DNA on a weight basis. After removing this fragment of DNA which we call sDNA, the enzyme became DNA dependent. The activity was, however, completely restored by adding calf thymus DNA or sDNA. Nevertheless, the calf thymus DNA concentration needed was 100 times higher than that of sDNA [16].

The poly(ADPR) polymerase was stimulated by histone H1 when the H1 enzyme ratio was of 2; histones H2A, H2B, H3, and H4 had little effect on the sDNA-linked enzyme [16—18]. The specific sDNA fraction also protects some enzyme sites against inhibition. Ultraviolet spectroscopic properties of sDNA suggested the presence of single stranded or denatured DNA portions in sDNA. The Km for NAD was shown to vary with the DNA concentration and two apparent Km could be defined, depending on the DNA concentration. It may correspond to some conformational changes in the enzyme or in the enzyme-DNA complex. The ratio of the apparent K_m for sDNA to the enzyme concentration is constant at any enzyme level. Moreover, the minimum estimation of the number of base pairs of sDNA required for maximal activation of one enzyme molecule was 16, while for calf thymus DNA this estimation was 640 [16].

The electron microscopic observations of the enzyme sDNA complex at high magnification by the dark field technique strongly suggests that sDNA is located at the periphery of the poly(ADPR) polymerase molecule forming a nucleosomal-like structure [10]. The characteristics of the structure will be reported by Ittel et al. in this symposium.

When rat pancreatic polynucleosomes were poly(ADP-ribosylated) with purified calf thymus poly(ADPR) polymerase and examined by electron microscopy a relaxation of their native zigzag structure was observed, even at high ionic strengths; they showed a close resemblance to chromatin depleted of histones H1. The relaxed state of poly(ADP-ribosylated) polynucleosomes was also confirmed by sedimentation velocity analysis [19, 20]. Locally relaxed regions can also be generated within polynucleosome chains by the activity of their intrinsic poly(ADPR) polymerase and appeared to be correlated with the formation of hyper(ADP-ribosylated) forms of histone H1 and an increase of DNA polymerase activity [21]. The posttranslational transitory modifications of histones are potential modulators of chromatin structure. This may be involved in DNA transcription, replication, and repair.

In eukaryotes, DNA supercoiling is an essential step in forming the DNA histone complexes of chromatin, and thus, in organizing the chromosome. DNA supercoiling is controlled by a class of enzymes called topoisomerases [22]. We could demonstrate that the topoisomerase I molecule may be a target for ADP-ribosylation and that this covalent modification inactivates this enzyme involved in chromosome organization [23].

Moreover, a reverse copurification of the two enzymes [poly(ADPR) polymerase and topoisomerase I] occurred in two independent enzyme preparations; it may suggest a close localization and a possible spatial and functional correlation of the two enzymes.

In view of the data indicating the involvement of poly(ADPR) polymerase and of topoisomerase I in DNA repair it is not unlikely that the two enzymes also interact during this process. ADP-ribosylation may be a regulatory mechanism of topoisomerase I activity. Obviously, this hypothesis remains to be confirmed in vivo.

3. Cytoplasmic ADP-Ribosylation Reactions

3.1 ADP-Ribosylation in Brain Mitochondria

The enzymatic transfer of ADPR to an acceptor protein in mitochondria was reported in rat liver and in testis (for review see [24, 25]). Recently, Hilz described a non-enzymic ADP-ribosylation of specific mitochondrial acceptors [26].

Recently, a method which provides highly purified mitochondria from rat brain was developed in our laboratory [27]. We became interested in the ADP-ribosylation reaction in brain mitochondria since these organelles possess a very high rate of phosphorylation, and since there are differences between the metabolic activity of mitochondria located in nerve endings and those located in the cell soma. It turned out that per protein unit the ADPR transferase activity of permeabilized mitochondria or mitoplasts was similar for both synaptic and cell-body mitochondria. At least four ADP-ribosylated proteins could be detected in the auto(ADP-ribosylated) mitochondria [28]. Finally, accumulating data suggest that only about 25% of the ADP-ribosylation activity in mitochondria was DNA sensitive and apparently inhibited by nicotinamide and thymidine to a rather small extent. The poly(ADPR) attached to the proteins generally does not exceed two ADPR units. The activity of the ADP-ribosylation reaction appeared to be higher in brain mitochondria than in liver mitochondria [28]. It also appeared that ADP-ribosylation is not only a multifunctional reaction, but may also be a result of strikingly different mechanisms. Whether the mitochondrial ADP-ribosylation corresponds to a system appearing early in evolution which is taken over later by specific proteins in nuclei or other cytoplasmic particles remains an open question, which seems to us of great interest.

3.2 Cytoplasmic Particle-Associated Poly(ADPR) Polymerase Activity

A ribosome- and polysome-associated poly(ADPR) polymerase activity was reported by Roberts et al. [29] and Burzio et al. [30]. We have demonstrated that ADP-ribosyl transferase activity is associated with free ribonucleoprotein particles carrying messenger RNA, the so-called free mRNP particles [31].

It was shown that the polyribosomal form of mRNP complexes is actively translated, whereas the free form is not. One might expect that a covalent chemical modification of some of the mRNA proteins, such as ADP-ribosylation, will render the mRNA available for translation. We characterized the mRNA-associated ADP-ribosyl transferase in plasmacytoma, in Krebs II, ascite tumor cells, and in liver. Several auto(ADP-ribosylated) proteins could be obtained when mRNP particles were incubated with NAD. It is unlikely that we are dealing with a contamination of chromatin since in plasmocytoma the enzymatic activity in mRNP represent 34% of the total cellular activity, while the maximum DNA contamination is only 4%. Moreover, after DNAse hydrolysis the enzymatic activity remains unchanged and addition of DNA is without effect [31]. More information on these mRNP particles will be given by Thomassin et al. (this volume).

4. ADP-Ribosylation at the Cellular Level

4.1 ADP-Ribosylation Reaction in Cultured Neurons and Glial Cells

Having investigated at the molecular and subcellular level there is now a need to place the ADP-ribosylation reaction in the whole of cell biology. It seemed to us that pure neuronal and/or glial cell cultures are well adapted to explore the multiple correlations of ADP-ribosylations. Mature neurons do not proliferate and are highly differentiated. Glial cells do proliferate and still possess several well explored structural and dynamic markers. It appeared that in nonproliferating chick brain neurons the poly(ADPR) polymerase reaction is as high as in less differentiated rapidly proliferating glioma cells. In a spontaneously transformed and proliferating, but poorly differentiated, astroblast line recently established in our laboratory, the poly(ADPR) polymerase activity is much lower than in a normal astroblast primary culture of proliferating and differentiated cells [32]. The problem is now to identify those proteins which act as acceptors in neurons and astroblasts in order to establish their role.

4.2 ADP-Ribosylation Reaction in Eye Lens

Much effort has been devoted to the effect of aging on the control of cell division and DNA integrity (see for review [33]). Moreover, it has been shown that induction of DNA repair function parallels an increase of poly(ADPR) polymerase, or ADP-ribosyl transferase activity. There is a good deal of evidence that alteration of DNA repair is involved in the lens aging processes.

Thus, we investigated poly(ADPR) polymerase activity in bovine lens epithelial cells during aging. A fourfold increase of this enzymatic activity could be recorded when lenses of old and young animals were compared [34].

5. ADP-Ribosylation and Cell Proliferation

There are several reports concerning a reduction or blockage of cell proliferation by inhibitors of ADP-ribosylation reaction. We obtained similar results using hepatoma, glioma, and astrocyte cultures. Moreover, it appeared that in phytohemagglutin-induced human lymphocyte proliferation ADP-ribosylation is involved in the early stage of the induction [35]. Finally, in addition to the already observed synergistic effect of inhibitors of ADP-ribosylation and some antimitotics [36] we could observe an agonistic effect with some others (unpublished data). The mechanism remains to be explained.

ADP-ribosylation discovered 18 years ago in Strasbourg became the subject of collaborative work in several countries. In addition to the demonstrated involvement in cell proliferation, differentiation, and DNA repair, a hope for a new approach in cancer therapy may not be an illusion.

References

1. Wintzerith M, Klein N, Mandel L, Mandel P (1961) Comparison of pyridine nucleotides in the liver and in an ascitic hepatoma. Nature (London) 191:467–649
2. Revel M, Mandel P (1962) Effect of an induced synthesis of pyridine nucleotides in vivo on the metabolism of ribonucleic acid. Cancer Res 22:456–462
3. Chambon P, Weill JD, Mandel P (1963) Nicotinamide mononucleotide activation of a new DNA-dependent polyadenylic acid synthesizing nuclear enzyme. Biochem Biophys Res Commun 11:39–43
4. Chambon P, Weill JD, Doly J, Strosser MT, Mandel P (1966) On the formation of a novel adenylic compound by enzymatic extracts of liver nuclei. Biochem Biophys Res Commun 25:638–643
5. Doly J, Petek F (1966) Etude de la structure d'un composé "polyADP-ribose" synthétisé par des extraits nucléaires de foie du poulet. C R Acad Sci 263:1341–1344
6. Reeder RH, Ueda K, Honjo T, Nishizuka Y, Hayaishi O (1967) Studies on the polymer of adenosine diphosphate ribose. II. Characterisation of the polymer. J Biol Chem 242:3172–3179
7. Hasegawa S, Fujimura S, Shimizu Y, Sugimura T (1967) The polymerization of adenosine 5'-diphosphate-ribose moiety of NAD by nuclear enzyme. II. Properties of the reaction product. Biochim Biophys Acta 149:369–376
8. Doly J, Mandel P (1967) Mise en évidence de la biosynthèse in vivo d'un polymère composé, le polyadénosine diphosphoribose dans les noyaux de foie de poulet. C R Acad Sci 264:2687–2690
9. Miwa M, Saikawa N, Yamaizumi Z, Nishimura S, Sugimura T (1979) Structure of poly(adenosine diphosphate ribose): Identification of 2'-[1'''-ribosyl-2''-(or 3''-) (1'' '-ribosyl)]adenosine-5',5'',5'' '-tris(phosphate) as a branch linkage. Proc Natl Acad Sci USA 76:595–599
10. De Murcia G, Jongstra-Bilen J, Ittel ME, Mandel P, Delain E (1983) Poly(ADP-ribose) polymerase auto-modification and interaction with DNA: electron microscopic visualization. EMBO J 2:543–548
11. Nishizuka Y, Ueda K, Honjo T, Hayaishi O (1968) Enzymic adenosine diphosphate ribosylation of histone and poly(adenosine diphosphate ribose) synthesis in rat liver nuclei. J Biol Chem 243:3765–3767
12. Otake H, Miwa M, Fujimura S, Sugimura T (1969) Binding of ADP-ribose polymer with histone. J Biochem (Tokyo) 64:145–146
13. Okazaki H, Niedergang C, Mandel P (1976) Purification and properties of calf thymus poly-adenosine diphosphate ribose polymerase. FEBS Lett 62:255–258
14. Jongstra-Bilen J, Ittel ME, Jongstra J, Mandel P (1981) Similarities in the molecular weight of poly(ADPR) polymerase. Biochem Biophys Res Commun 103:383–390
15. Okazaki H, Delaunoy JP, Hog F, Bilen J, Niedergang C, Creppy EE, Ittel M, Mandel P (1980) Studies on polyADPR polymerase using the specific antibody. Biochem Biophys Res Commun 97:1512–1520
16. Niedergang C, Okazaki H, Mandel P (1979) Properties of purified calf thymus poly(adenosine diphosphate ribose) polymerase. Comparison of the DNA-independent and the DNA-dependent enzyme. Eur J Biochem 102:43–57
17. Okazaki H, Niedergang C, Mandel P (1980) Adenosine diphosphate ribosylation of histone H1 by purified calf thymus polyadenosine diphosphate ribose polymerase. Biochimie 62:147–157
18. Okazaki H, Niedergang C, Couppez M, Martinage A, Sautière A, Mandel P (1980) In vitro ADP-ribosylation of histones by purified calf thymus poly(adenosine diphosphate ribose) polymerase. FEBS Lett 110:227–229
19. Poirier GG, De Murcia G, Jongstra-Bilen J, Niedergang C, Mandel P (1982) Poly(ADP-ribosylation) of polynucleosomes causes relaxation of chromatin structure. Proc Natl Acad Sci USA 79:3423–3427

20. Aubin R, Fréchette A, De Murcia G, Malouin F, Lord A, Mandel P, Poirier GG (1983) Nucleosome poly(ADP-ribose) polymerase. In: Miwa et al. (eds) ADP-ribosylation, DNA repair and cancer. Jpn Sci Soc Press, Tokyo/VNU Science Press, Utrecht, p 83

21. Poirier GG, Niedergang C, Champagne M, Mazen A, Mandel P (1982) Adenosine diphosphate ribosylation of chicken-erythrocyte histones H1, H5 and high-mobility group proteins by purified calf thymus polyadenosine diphosphate-ribose polymerase. Eur J Biochem 127:437–442

22. Gellert M (1981) DNA topoisomerases. Annu Rev Biochem 50:879–910

23. Jongstra-Bilen J, Ittel ME, Niedergang C, Vosberg HP, Nadle P (1983) DNA topoisomerase I from calf thymus is inhibited in vitro by poly(ADP-ribosylation). Eur J Biochem 136:391–396

24. Kun E, Kirsten E (1982) Mitochondrial ADP-ribosyltransferase system. In: Hayaishi, Ueda (eds) ADP-ribosylation reactions. Academic Press, London New York, p 193

25. Burzio LO, Saez L, Cornejo R (1981) Poly(ADP-ribose) synthetase activity in rat testis mitochondria. Biochem Biophys Res Commun 103:369–375

26. Hilz H, Koch R, Fanick W, Klapproth K, Adamietz P (1984) Nonenzymic ADP-ribosylation of specific mitochondrial polypeptides. Proc Natl Acad Sci USA 81:3929–3933

27. Rendon A, Masmoudi A (to be published) Purification of nonsynaptic and synaptic mitochondria and plasma membranes from rat brain by a rapid percoll gradient procedure.

28. Masmoudi A, Thomassin H, Rendon A, Mandel P (in preparation) ADP-ribosylation in mitochondria

29. Roberts JH, Stark P, Giri CP, Smulson M (1975) Cytoplasmic poly(ADPR) polymerase during the HeLa cell cycle. Arch Biochem Biophys 171:305–315

30. Burzio LO, Concha II, Figueroa J, Concha M (1983) Poly(ADP-ribose) synthetase associated to cytoplasmic structures of meiotic cells. In: Miwa M, Hayaishi O, Shall S, Smulson M, Sugimura T (eds) ADP-ribosylation, DNA repair and cancer. Jpn Sci Soc Press, Tokyo/VNU Science Press, Utrecht, p 141

31. Elkaim R, Thomassin H, Niedergang C, Egly JM, Kempf J, Mandel P (1983) Adenosine diphosphate ribosyltransferase and protein acceptors associated with cytoplasmic free messenger ribonucleoprotein particles. Biochimie 65:653–659

32. Mandel P, Niedergang C, Mersel M (in preparation)

33. Generoso WM, Shelby MD, De Serres FJ (eds) (1980) DNA repair and mutagenesis in eukaryotes. Plenum Press, New York

34. Bizec JC, Mandel P, Klethi J (1985) PolyADP-ribosyl polymérase (adénosine diphosphate-ribosyl-transférase) du cristallin de bovidés; modulations au cours du vieillissement. C R Acad Sci 300:37–41

35. Ittel ME, Jongstra-Bilen J, Rochette-Egly C, Mandel P (1983) Involvement of poly(ADP-ribose) polymerase in the initiation of phytohemagglutinin-induced human lymphocyte proliveration. Bicohim Biophys Res Commun 116:428–434

36. Mandel P, Bergerat JP, Dufour P, Herbrecht R (1983) Inhibition de la polyADP-ribose synthétase, voie nouvelle dans la recherche d'une réduction de la prolifération cellulaire. Bull Acad Nat Med Paris 167:327–332

37. Mandel P, Okazaki H, Niedergang C (1982) Poly(adenosine diphosphate ribose). In: Progress in nucleic acid research and molecular biology, vol 27. Academic Press, London New York

38. Purnell MR, Stone PR, Whish WJD (1980) ADP-ribosylation reactions. Biochem Soc Trans 8:215–227

39. Hayaishi O, Ueda K (eds) (1982) ADP-ribosylation reactions. Academic Press, London New York

40. Smulson ME, Sugimura T (eds) (1980) Novel ADP-ribosylations of regulatory enzymes and proteins. In: Developments in cell biology, vol VI. Elsevier/North Holland, Amsterdam New York

41. Hilz H (1981) ADP-ribosylation of proteins – a multifunctional process. Hoppe-Seyler's Z Physiol Chem 362:1415–1425

ADP-Ribosylation as a Cellular Control Mechanism

SYDNEY SHALL[1]

Introduction

It is now about 21 years since the discovery of poly(ADP-ribose). This coming of age is reflected also in the interest and importance of ADP-ribosylation reactions in cell and molecular biology.

The title of this talk gives the conclusion I wish to draw, namely that ADP-ribosylation reactions are a ubiquitous cellular control mechanism. The available evidence is reasonable grounds for inferring that ADP-ribosylation reactions are cellular control mechanisms in a variety of different, independent situations. An instructive analogy is the phosphorylation of a large variety of enzymes and proteins that are part of regulatory cascades. But this analogy drives us immediately to ask why has ADP-ribosylation and specifically poly(ADP-ribose) evolved? Why did cells not evolve the ability to phosphorylate the appropriate proteins? What is so special about ADP-ribosylation and particularly poly(ADP-ribosylation)?

Up to the present time, ADP-ribosylation reactions have been identified as control mechanisms in four well-defined physiological processes in eukaryotic cells; (1) in the control of adenyl cyclase, (2) in the control of protein biosynthesis, (3) in the regulation of certain cases of cell differentiation and (4) in DNA repair. It is not yet obvious that there is an underlying unity in all these examples. Of course, there need not be a common thread. But past biochemical experience suggests that it is well worthwhile looking for underlying coherence. With the evidence available at the present time, it does seem to me as if we could at least talk of poly(ADP-ribosylation) and of mono(ADP-ribosylation) as two rather separate categories. It is probably worthwhile also to wonder whether we can legitimately assign polymers to the cell nucleus and monomers to the cytoplasm, although there is some evidence for a nuclear mono(ADP-ribosyl) transferase. This view implies, as first suggested I think, by H. Hilz, that some or perhaps all of nuclear mono(ADP-ribose) is a metabolic product of the polymer. This distinction between the monomer and the polymer clearly remains to be resolved.

ADP-ribosylation reactions were apparently independently discovered in the laboratories of P. Mandel, O. Hayaishi and T. Sugimura about 21 years ago, at about the

---

1 Cell and Molecular Biology Laboratory, Biology Building, University of Sussex, Brighton BN1 9QG, Great Britain

ADP-Ribosylation of Proteins
(ed. by F.R. Althaus, H. Hilz, and S. Shall)
© Springer-Verlag Berlin Heidelberg 1985

same time that phosphorylation and acetylation of chromatin proteins were described. Since then, we have collectively acquired a reasonable understanding of the structure, enzymology and of some of the cellular reactions which are regulated by ADP-ribo-sylation reactions. ADP-ribosyl adducts may be monomers or polymers and the latter may be linear or branched. It seems that ADP-ribosylation reactions are quite ubi-quitous; they occur in bacteria, bacterial viruses, animals, animal viruses and plants; although DNA-dependent poly(ADP-ribosylation) seems to be confined to true eukaryotic cells.

Poly(ADP-ribosylation) of chromatin proteins seems to be an integral part of DNA repair in eukaryotic cells. It is now clear that these reactions are required for an effi-cient cellular response to some DNA-damaging agents such as (monofunctional) alkylating agents and ionizing radiation. It would appear that the nuclear enzyme ADP-ribosyl transferase (nuclear ADPRT) is sensitive to DNA breaks and may modu-late the rate of their rejoining, possibly by regulating the activity of DNA ligase II.

DNA Repair

The role of nuclear ADPRT in DNA repair has been extensively reviewed [1—4] and I will only very briefly summarise the observations and arguments here.

It has been known for a long time that DNA-damaging agents cause a fall in cellular NAD^+ level [1, 5—7]. DNA-damaging agents (both chemicals and ionizing radiation) activate ADPRT [7], probably in proportion to the steady-state level of DNA strand-breaks that are associated with these several DNA-damaging agents. Some 9 years ago we suggested that the loss of NAD^+ was due to the biosynthesis of poly(ADP-ribose) [5]. This accelerated biosynthesis of poly(ADP-ribose) from NAD^+ in intact cells exposed to DNA-damaging agents has been directly shown several times. These elegant experiments of the Jacobsons and of Hilz and colleagues have confirmed that there is a very rapid turnover of poly(ADP-ribose) in intact cells.

All the inhibitors of ADPRT prevent the depletion of cellular NAD^+ that is caused by DNA-damaging agents [7]. This adds further weight to the notion that DNA damage activates ADPRT resulting in increased synthesis of poly(ADP-ribose) and a loss of cellular NAD^+. In general, the kinetics of NAD^+ disappearance and of ADPRT activa-tion correlate rather well.

The requirement of ADP-ribosylation for efficient DNA excision repair was directly demonstrated some years ago [8]; inhibitors of ADPRT retard DNA repair as measured on alkaline sucrose gradients. 3-Aminobenzamide (and a variety of other ADPRT inhibitors including 5-methyl nicotinamide and thymidine) will retard DNA strand-rejoining in L1210 cells exposed to dimethyl sulphate (DMS). The enzyme inhibitors will also retard strand-rejoining after exposure to methyl-nitroso-urea, N-methyl-N^1-nitro-N-nitrosoguanidine, methyl methane sulphonate, bleomycin and a variety of other DNA-damaging agents. All the inhibitors tested retard DNA strand-rejoining in L1210 cells, while *none* of the non-inhibitory acid analogues have this inhibitory effect [9]. Consequently, we can reasonably conclude that the inhibition is mediated by ADPRT and not by other processes such as changes in nucleotide pools. This inhibition of DNA repair is also evident when tested by following the sedimentation behaviour of nucleoids [10]. This technique is very sensitive and is performed at neutral pH. Taking the results from alkaline sucrose gradients and the nucleoids together, one can exclude any relevance of alkali-labile sites included in the alkaline gradient assays and DNA-supercoiling and -conformation which is included in the nucleoid measurements, and we can identify the common element as measuring the rejoining of preformed DNA strand-breaks.

Efficient DNA repair requires ADP-ribosylation not only in rodent cells but in human cells as well; no significant species differences have yet been reported. It seems to be a general property of nucleated cells.

ADPRT participates in DNA repair in a wide variety of cell types. However, there are significant differences in the responses of different cell types. The ability of the enzyme inhibitors to retard DNA repair varies in different cell types. In addition, there are clear differences in the responses to the various DNA-damaging agents. Ultraviolet radiation is least efficient at activating ADPRT; I would associate many bulky adducts and bifunctional reagents with ultraviolet radiation in this respect. Ionizing radiation is somewhat more efficient in activating ADPRT. By contrast, small monofunctional alkylating- and especially methylating-agents are extremely efficient at activating the nuclear ADPRT. The efficacy of ADPRT inhibitors quite predictably follows the same pattern. These responses are probably in proportion to the number of DNA strand-breaks present after each treatment. I shall describe a plausible explanation for the differences after I have outlined the probable molecular mechanism of ADP-ribosylation in DNA repair.

I want now to consider the possible molecular mechanism by which nuclear ADPRT participates in DNA repair.

The published data indicate that ADPRT inhibitors do not block early steps in DNA repair, because the single-strand breaks appear in the presence of the enzyme inhibitor, as they do in its absence. DNA repair synthesis is not inhibited by the ADPRT inhibitors [11], as was first reported from the laboratories of N. Berger and M. Miwa. These observations suggest that ADPRT activity is not needed for the early steps in DNA excision repair. We therefore turn our attention to the DNA ligase reaction. The available published data has led us to propose that ADP-ribosylation regulates the rejoining or the ligation step of DNA repair.

Inhibitors of nuclear ADPRT or nicotinamide deprivation interfere only very slightly with DNA biosynthesis or with cell growth of undamaged cells, by elongating the cell cycle slightly. I would like to digress for a moment to suggest a possible explanation for this observation. During DNA biosynthesis DNA ligation occurs at two distinct steps. At an early stage, Okazaki fragments, of about 200 bp length in eukaryotic cells, are joined together to make sub-replicon fragments. At a later time, these fragments are joined to replicon size pieces, then clusters of replicons are ligated together until finally a chromosome is complete. I suggest that perhaps different ligases operate during these DNA biosynthetic processes. Perhaps DNA ligase I is mostly used in the synthesis of the smallest fragments; by contrast, I suggest that perhaps DNA ligase II is the main enzyme catalyzing the later fragment condensation reactions. This predicts that condensation of intermediate fragments to high molecular weight DNA would be inhibited by 3-aminobenzamide. This hypothesis would give the nuclear

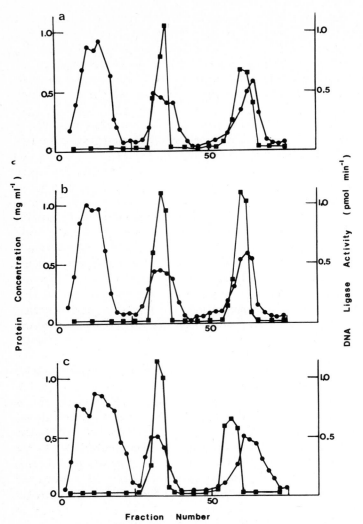

Fig. 1a–c. DNA ligase activities in mammalian cells. **a** Separation of DNA ligase activities from wild-type L1210 cells by hydroxylapatite column chromatography. **b** Increase in DNA ligase II activity after exposure of L1210 cells for 20 min to 100 μM dimethylsulphate. **c** 3-aminobenzamide prevents the increase in DNA ligase II activity after treatment with dimethylsulphate. ● Protein concentration; ■ DNA ligase activity

ADPRT a marginal role in DNA biosynthesis, and hence in the cell cycle of nucleated cells. Partial inhibition of ligase II would delay the condensation of the newly-formed DNA fragments, this in turn may elongate the cell cycle slightly.

It is necessary to explain that, in contrast to bacteria and yeast, in animal cells there are two forms or species of DNA ligase which have been separated chromatographically, serologically and on sucrose gradients (Fig. 1a). It has been assumed that

DNA ligase I is involved mainly in semi-conservative DNA replication; its activity fluctuates following physiological or experimental changes in growth behaviour, in a similar way to DNA polymerase α. It has been suggested that the second ligase activity, DNA ligase II, may be a "repair enzyme". The existence of two ligase activities makes it reasonable to consider whether DNA ligase II, the "repair enzyme activity" is specifically regulated by ADP-ribosylation.

Two years ago, we reported direct evidence that ADP-ribosylation regulates DNA ligase II activity in the course of DNA repair [12]. We assayed partially purified DNA ligase activity from mouse leukaemia cells with purified exogenous DNA. When the cells were exposed to the alkylating agent, dimethyl sulphate, the total DNA ligase activity in the cell extracts increased about 2.3-fold. The increase in ligase activity is probably a consequence of ADPRT activity because all classes of ADPRT inhibitors prevented this increase, and non-inhibitory acid analogues did not. The ADPRT-dependent increase in ligase activity is predominantly due to DNA ligase II. While the DNA ligase I activity is essentially constant, the DNA ligase II actually increases three- to five-fold after damage (Fig. 1b). The ADPRT inhibitors that block the increase in total DNA ligase activity induced by dimethyl sulphate, work by specifically preventing the increase in DNA ligase II activity (Fig. 1c).

It seems then that alkylation damage increases DNA ligase activity, and predominantly DNA ligase II activity which is presumed to be the major repair enzyme. This increase in DNA ligase II activity is totally prevented by inhibiting poly(ADP-ribose) biosynthesis, implying that the increase in DNA ligase activity requires ADPRT activity.

The molecular basis for the activation of DNA ligase II is not yet known. There are many potential mechanisms available. For example, a DNA-binding protein which is ADP-ribosylated may regulate ligase activity. A specific case of this general hypothesis is directly testable. It is known that the ADPRT enzyme can modify other ADPRT enzyme molecules; moreover, ADPRT molecules have a very high affinity for nicks in DNA. Thus, ADPRT molecules may be bound to nicks produced during DNA repair, thus making them inaccessible both to nuclease and to ligase action. Automodification of enzyme molecules by ADPRT would probably weaken the affinity of the enzyme for the DNA; the enzyme would then dissociate from the break, thus allowing access for the DNA ligase. The increase in DNA ligase activity would then be attributed to the increased availability of substrate.

Further evidence that DNA ligase and ADPRT activities are interrelated has been derived from genetic studies [13]. We have isolated variants of mouse leukaemia L1210 cells in which cytotoxicity to dimethylsulphate is not fully potentiated by the ADPRT inhibitor 3-aminobenzamide, as occurs in normal L1210 cells (Fig. 2). This variant cell can repair low doses of DNA damage in the presence of ADPRT inhibitors as efficiently as in the absence of the inhibitors. These cells show changes in their DNA ligase activities (Table 1). The basal DNA ligase I activity is increased 66% above wild-type, whereas DNA ligase II appears to be unchanged (Fig. 3). The most striking observation, however, is that the DNA ligase II activity is not increased after dimethyl sulphate treatment as occurs in wild-type L1210 cells (Table 1). It seems that by increasing DNA ligase I levels these cells can survive damage in the presence of 3-aminobenzamide.

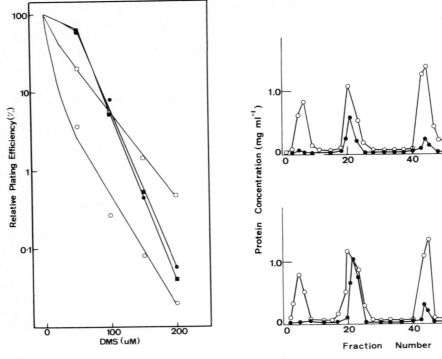

Fig. 2   Fig. 3

Fig. 2. The effect of 2 m*M* 3-aminobenzamide on the cytotoxicity of DMS in variant (mutant) 3 and in wild-type L1210 cells. ● Parental L1210 cells; ○ L1210 with 2 m*M* 3AB; ■ variant 3 cells; □ variant 3 with 2 m*M* 3AB

Fig. 3. DNA ligase I and II activity in wild-type L1210 cells (*upper panel*) and in variant (mutant) 3 (*lower panel*) cells. The ligases were separated on an hydroxylapatite column as described in [12]. *Upper panel:* wild-type cells; *lower panel:* variant (mutant) 3 cells. ● DNA ligase activity, ○ protein concentration

Table 1. DNA ligase I and II activities in parental and variant 3 cells[a]

	Controls		Exposed to DMS	
DNA Ligase	I	II	I	II
L1210	1.0 ± 0.2 (6)	0.29 ± 0.09 (6)	1.07 ± 0.05 (3)	0.91 ± 0.07 (3)
Variant 3	1.0 ± 0.13 (6)	0.16 ± 0.07 (6)	1.17 ± 0.09 (2)	0.19 ± 0.05 (2)
Variant 3[b]	1.66 ± 0.22 (6)	0.26 ± 0.12 (6)	1.96 ± 0.14 (2)	0.32 ± 0.08 (2)

[a] The enzyme activities are expressed relative to the activity in cells not exposed to DMS ± SD (n)
[b] The relative enzyme activites of variant 3 comared to wild-type cells

This variant (mutant) provides genetic evidence for our hypothesis that ADP-ribosylation is required for efficient DNA repair after DNA damage by monofunctional alkylating agents, because nuclear ADPRT activity regulates DNA ligase activity. The present mutant confirms that in mammalian cells DNA ligase II activity is regulated by nuclear ADPRT activity.

If ADPRT activity is really required for efficient cellular recovery from DNA damage, then inhibition of ADPRT activity would be expected to increase the lethality of DNA-damaging agents. This prediction has been amply confirmed [8, 14]. All the inhibitors of ADPRT activity synergistically potentiate the cytotoxicity of monofunctional DNA-damaging agents. Other DNA-damaging agents such as ionizing radiation and bleomycin also display potentiation of lethality by ADPRT inhibitors.

I should like to digress again to repeat a suggestion I have made before, which is that this synergistic potentiation of lethality will be of some clinical use, especially in relation to drugs like streptozotocin, bleomycin and with ionizing radiation. Even without any evidence of tumour specificity yet, the advantage of using lower doses of the cytotoxic agent is significant. I await the results of experiments in animals being performed in several laboratories with some optimism.

Another example of a biological consequence of ADPRT inhibitors is the increased frequency of sister chromatid exchanges and of chromosomal aberrations. Sister chromatid exchanges involve recombination between sister duplexes and therefore require breakage and rejoining of DNA. I should hasten to add, however, that so far, there is unfortunately no compelling evidence that ADPRT inhibitors induce sister chromatid exchanges by virtue of their effect on the ADPRT enzyme. This evidence is very much needed; in particular, it is important to exclude the possibility that the enzyme inhibitors induce sister chromatid exchanges by their effects on nucleotide metabolism. We have confirmed that 3-aminobenzamide at relatively high concentrations and for long times will slightly inhibit the incorporation of radioactivity from both methionine and glucose into nucleic acid. This side effect is common to the amides and the acids; this is a crucial point which I wish to emphasise. Workers in this field have always been conscious of the dangers inherent in the use of enzyme inhibitors. We have emphasised this point in our publications over the years. Careful readers (and careful experimenters) will have noticed that we always check for toxicity and that we persistently use the non-inhibitory acid analogues as controls. Thus, we can be confident that the data to which I have referred so far, probably does not involve perturbations of nucleotide metabolism.

The conclusions that I draw from the published work is that aminobenzamide does not by itself induce sister chromatid exchanges; presumably, it can only induce them when chromatid breaks are engendered by some other process. The increase in sister chromatid exchanges may well be explained by a longer lifetime of gaps in the DNA when ADPRT is inhibited. In the presence of two nearby single-stranded breaks, inhibition of ADPRT activity may prolong the life of this double-strand break and therefore increase the probability of a sister chromatid exchange. This notion presupposes that ADPRT is required not only for DNA repair but for other DNA nicking and rejoining phenomena. These results suggest that both NAD^+ and ADPRT activity are needed to maintain chromosomal integrity. The earlier suggestion that I made concerning the role of DNA ligase II in DNA replicative biosynthesis may also be relevant to the induction of sister chromatid exchanges. Inhibition of DNA fragment condensation in growing cells provides another potential source of DNA single-strand breaks. Combined with breaks from bromodeoxyuridine or other DNA repair, we may in this way produce conditions conducive to chromatid exchange reactions.

I have earlier commented on differences observed between different DNA-damaging agents; simple monofunctional alkylating agents seem to require a large and important involvement of ADP-ribosylation in their repair processes. Ionizing radiation such as X-rays and γ-irradiation, seem to require much less. For example, it was not easy to observe inhibition of DNA repair after γ-irradiation using alkaline sucrose gradients; it was necessary to resort to the nucleoid technique for an unequivocal demonstration that 3-aminobenzamide inhibits repair after γ-radiation. This observation can also be made with DNA unwinding techniques. Finally, ultraviolet radiation, bifunctional alkylating agents (such as nitrogen mustards) and very bulky substituents (such as bromomethyl-benzanthracene) do not seem to require ADPRT activity for their repair.

One plausible explanation for these differences is that they reflect varying kinetic behaviour [2, 3, 14]; that is, they reflect in each case the steady-state level of single-strand DNA breaks associated with a particular damaging agent. Ultraviolet irradiation does not break DNA directly at physiologically significant doses. DNA strand-breaks are produced only during excision repair of ultraviolet damage. But it is known that such repair is slow and that the incision step is rate-limiting. Therefore, only a very small number of breaks are present at any given time. For example, after 10 J m^{-2} which gives 10 to 20% survival, there are only about 0.3 breaks per 10^9 base pairs of DNA.

At biological doses of ionizing radiation of say, 300 rad which gives again 10 to 20% survival, the number of direct breaks is about 0.6 breaks per 10^9 base pairs of DNA. After exposure to alkylating agents the enzymatic incision steps are not rate-limiting, and a relatively large number of breaks are seen after exposure to dimethyl sulphate. $100 \ \mu M$ dimethyl sulphate gives about 10 to 20% survival in human fibroblasts; this dose generates about three breaks per 10^9 base pairs of DNA. Of course, this number of breaks is a dynamic equilibrium between the rate of production of the breaks and the rate of their repair by polymerase and ligase activity. Moreover, these breaks will continue to be produced enzymically and by spontaneous hydrolysis for several hours. ADP-ribosylation, if induced by the steady-state level of DNA breaks, will enhance DNA ligase II activity and will therefore shift the equilibrium to a position with fewer DNA breaks. This new dynamic equilibrium will activate ADPRT less and consequently DNA ligase II will be activated to a lesser degree. Gradually, therefore, the equilibrium will return to that of the undamaged state.

This quantitative discussion shows that at equitoxic doses, the ratio of steady-state DNA breaks is 1:2:10 for UV irradiation, ionizing radiation and monofunctional methylating agent, respectively. The differences in the responses after these different agents may well be found in these different kinetics. In the light of this discussion, quantitative differences in the ADPRT response are to be expected between different cell types and between different damaging agents.

This review has shown that there is clear evidence that ADP-ribosylation is required for efficient DNA repair. Furthermore, current evidence indicates that ADP-ribosylation regulates the ligation step in DNA repair. The data which I have reviewed does not by itself exclude the possibility that ADP-ribosylation also modifies chromatin structure during repair, in addition to its apparent participation in activating DNA ligase. Indeed, it may be suggested that ADP-ribosylation leads to relaxation of chromatin structure and hence favours the ligase reaction. It may also be suggested that a more local change in chromatin conformation may favour the ligase reaction. In support of this view we could adduce the elegant work of M. Smulson, G. Poirier and P. Mandel, and O. Hayaishi and their respective coworkers. However, I also note that these results are derived from subcellular systems and it is still unclear whether these reactions occur in intact cells and thus contribute in fact to DNA repair. I for one, would be most surprised to discover that ADP-ribosylation does not modulate chromatin conformation, considering the enzymology of nuclear ADPRT and the properties of poly(ADP-ribose). However, it does still remain to be formally demonstrated whether ADP-ribosylation modulates DNA repair by activating DNA ligase II at the level of enzyme activity or by changing chromatin conformation or of course, by doing both, either as related or as independent effects.

I would argue that nuclear ADPRT participates in DNA repair and that the evidence in favour of this conclusion is totally compelling. Furthermore, I would suggest that ADPRT is involved in DNA repair because it modulates DNA ligase activity. Moreover, I would argue that there is clear evidence against a requirement for ADPRT in the incision event, despite the evidence of in vitro reversible regulation of an endonuclease. However, I cannot formally exclude the possibility that nuclear ADPRT activity *suppresses* an endonuclease activity; in which case, inhibitors of ADPRT would liberate endonuclease activity and thus generate DNA breaks. Such a proposal was made by K. Wielckens a year and a half ago at the European ADP-Ribose meeting in Berlin. In order to answer this proposition we have devised what we call the 'Taxi' experiment. For a variety of reasons we have not yet done this experiment, but I am sufficiently confident that the evidence says that nuclear ADPRT does *not* regulate DNA incision events in DNA repair, that I offer the experiment as a prediction if someone would wish to do it. The formal problem is to separate incision events from ligation events and measure the effect of 3-aminobenzamide separately on the two steps. One way in which this may be done is to incorporate controlled proportions of BrUdr into cells in the dark; then by exposure to experimentally defined intensities and durations of near UV-light it should be possible to experimentally manipulate both the rate of appearance and the number of DNA breaks. In a subsequent dark period, these breaks will be repaired and no new breaks will be formed. Now one can add ADPRT inhibitors and I predict that they *would* retard the rejoining of the breaks. This experiment should provide unequivocal evidence that ADPRT activity modulates the rate of rejoining of DNA breaks. It should be possible also to use this protocol to test whether ADPRT regulates incision events, because ADPRT inhibitors can be combined with a known number of DNA breaks and one can ask whether one can metabolically produce more breaks which the Wielcken's proposal predicts. I invite anyone who firmly disbelieves our hypothesis to try this experiment.

So, I assert that ADPRT modulates DNA rejoining during DNA repair. Does ADPRT have this function in any other biological situation? I have already suggested that ADPRT activity is involved in fragment condensation during the later stages of DNA biosynthesis. I would also suggest that ADPRT activity is required in other

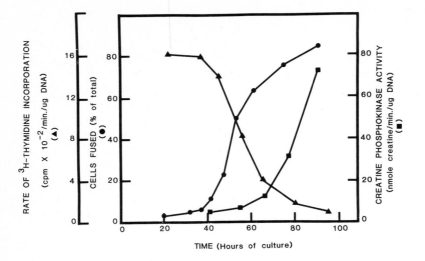

Fig. 4. Chicken muscle cell differentiation in vitro. ● Percentate of muscle cells fused; ■ creatine phosphokinase activity; ▲ the rate of DNA synthesis. (Experimental details in [16])

examples of DNA strand-breakage and rejoining, such as the gene rearrangements that accompany maturation of the immuno-globulin genes and probably also of the major histocompatibility complex, and of the T-cell receptor genes. In addition, I would propose that ADPRT is involved also in both mitotic and meiotic recombination, and finally in the integration of proviruses, exogeneous DNA sequences and amplified genes.

Cell Differentiation

I want to turn now to a second identified ADP-ribosylation function; namely, certain cases of cell differentiation. This possibility was first suggested nearly 10 years ago by A. Caplan on the basis of his extensive experimental embryological studies. There are now numerous examples which we may quote indicating that ADP-ribosylation participates in cell differentiation.

   Our first investigation of the possible participation of ADPRT activity in cell differentiation used a chick muscle primary cell culture system. When embryonic chick myoblasts are seeded onto collagen-coated petri dishes and grown in a medium containing horse-serum and embryo extract, the majority of the cells pass through an initial phase of proliferation (Fig. 4). The onset of terminal differentiation is marked by the fusion of myoblast cell membranes to form multi-nucleate syncytia of muscle fibres, the cessation of DNA biosynthesis and of cell division. There is also substantial increase in the activity of creatine phosphokinase, largely due to the de novo synthesis of a muscle specific isozyme of this protein. In this work we have made use of

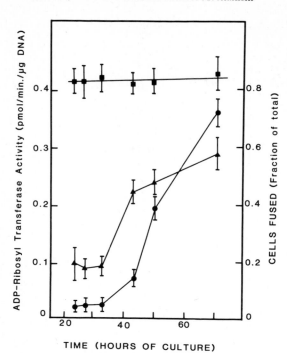

Fig. 5. Changes in ADPRT activity during chicken muscle cell differentiation in vitro. ▲ Physiological ADPRT activity; ■ total, potential ADPRT activity; ● fraction of myoblasts fused

both cell fusion and creatine phosphokinase activity as markers of terminal cell differentiation (Fig. 4).

During the cyto-differentiation of primary chick myoblasts there is a threefold increase in the endogenous physiological ADPRT activity, but there is no change in the total, potential enzyme activity (Fig. 5). This suggests that the increase in the physiological activity is due to enzyme activation, and not to an increase in the number of enzyme molecules. The ADPRT activation precedes both cell fusion and the rise in creatine phosphokinase activity.

Inhibitors of ADPRT activity reversibly block both muscle cell fusion and the rise in creatine phosphokinase activity associated with this cell differentiation. Myoblast fusion is inhibited in a concentration-dependent manner in the continuous presence of either 3-aminobenzamide or 3-methoxybenzamide (Fig. 6a). These inhibitors do not inhibit proliferation of myoblasts (Fig. 6b). The non-inhibitory analogues, 3-amino-benzoate and 3-methoxybenzoate caused no significant inhibition of fusion at corresponding concentrations (Fig. 6a). Inhibitors of nuclear ADPRT not only block myoblast fusion (Fig. 7a), but they also prevent the usual increase in creatine phosphokinase activity (Fig. 7b). Both the inhibition of cell fusion and of the increase in creatine phosphokinase activity by ADPRT inhibitors, is reversible (Fig. 7a,b). Inhibitors of ADPRT activity also inhibit the appearance of acetylcholine receptors on the muscle cell surface.

We have also experimentally manipulated ADP-ribosylation by partially depleting the cells of their NAD^+ content, by nicotinamide starvation. The NAD^+-depleted

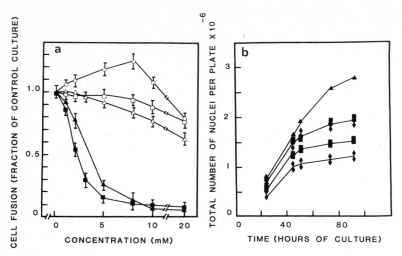

Fig. 6a,b. ADPRT inhibitors block chick myoblast differentiation but not proliferation. **a** ▲ 3-amino-benzamide; △ 3-aminobenzoate; ■ 3-methoxybenzamide; □ 3-methoxybenzoate; ○ nicotinamide. **b** ▲ 8 mM 3-aminobenzamide; ◆ 10 mM 3-aminobenzamide; ♦ 20 mM 3-aminobenzamide; ■ 5 mM 3-methoxybenzamide; ■ 8 mM 3-methoxybenzamide; ● no inhibitor

myoblasts do not fuse to form multi-nucleated fibres (Fig. 7c), nor do they show the usual rise in creatine phosphokinase activity (Fig. 7c,d).

We have determined the time during which the cells are sensitive to the effects of ADPRT inhibitors. The cells become resistant to these enzyme inhibitors between 24 and 40 h in culture (Fig. 8); cell fusion is only beginning at 40 h, while the appearance of creatine phosphokinase starts at about 60 h. Thus, the ADPRT inhibitors are blocking a relatively early event in this example of cell differentiation.

We conclude from these observations that there is an obligatory requirement for ADPRT activity in the terminal differentiation of primary chick myoblast cultures [15, 16].

Because we had previously shown that the involvement of ADPRT in DNA repair is associated with the breaking and rejoining of DNA strands, and because it seems that DNA strand-breaks are an *essential* activator of the enzyme, we were interested to investigate whether a related mechanism might also be involved in some categories of cell differentiation. There are a number of examples now in which specific gene expression is associated with gene rearrangement or DNA transposition. We have therefore examined the DNA of developing chick muscle.

We found that between 100 and 300 single-strand DNA breaks appear during the cyto-differentiation of primary chick myoblasts. These DNA strand-breaks appear in the presence of a proficient DNA repair mechanism, and are therefore not due to a general deficiency in DNA repair.

In alkaline sucrose gradients which estimate the size of single-stranded DNA, the DNA from proliferating cultures has a high molecular weight (Fig. 9a). At 44 h when only 15% of the cells have fused (see Fig. 4), there is a significant reduction in the molecular weight of the cullar DNA (Fig. 9b). By 53 h the maximum number of

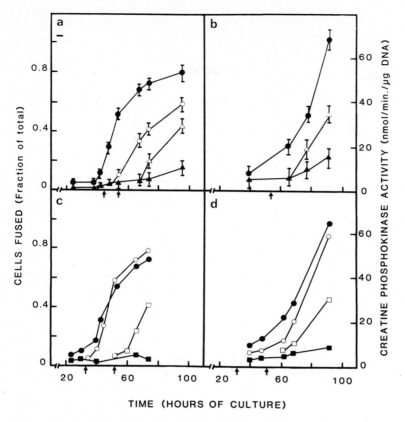

Fig. 7a–d. Reversible inhibition of chick myoblast differentiation by ADPRT inhibitors and by nicotinamide deprivation. Reversible inhibition by 8 m*M* 3-aminobenzamide (a,b), and by depletion of NAD levels (c,d). a and c show myoblast cell fusion; b and d show creatine phosphokinase activity. ● Control; ▲ continuous presence of 8 m*M* 3-aminobenzamide; the medium containing inhibitor was removed and replaced with medium from parallel control cultures of the same age at 44 h (○) or at 54 h (△). In c and d myoblasts were cultured in nicotinamide-free medium. ■ No nicotinamide; ● 1 m*M* nicotinamide added at 0 h; ○ 1 m*M* nicotinamide added at 32–34 h of culture; □ 1 m*M* nicotinamide added at 52 h of culture

single-strand breaks are present (Fig. 9c), although only 50–60% of the myoblasts have fused. The molecular weight then remains constant up to 92 h of culture (Fig. 9c–f). Only single-strand breaks appear in the DNA; there is no decrease in the molecular weight of the DNA in neutral sucrose gradients containing ionic detergents between 25 and 92 h of culture. We have been unable to establish what happens after 92 h of culture, so we do not yet know whether the DNA breaks persist or are subsequently rejoined. The experiments displayed in Fig. 9 are also 'mixing' experiments, and show that in our experimental conditions, mixing of early (without strand-breaks) and late (with strand-breaks) cells does not induce strand-breaks in the DNA of early cells (see Fig. 9b–f).

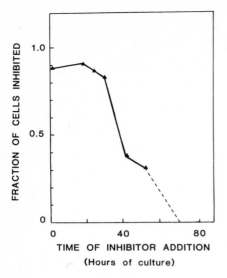

Fig. 8. Time during which myoblasts are sensitive to inhibition by 3-aminobenzamide

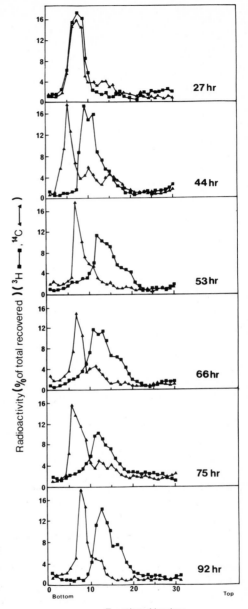

Fig. 9. Molecular weight analysis of muscle cell DNA by alkaline sucrose gradients. 27 h cultures labelled with [^{14}C] thymidine; ■ 44, 53, 66, 75 and 92 h cultures labelled with [^{3}H] thymidine. ■ [^{3}H], ▲ [^{14}C]

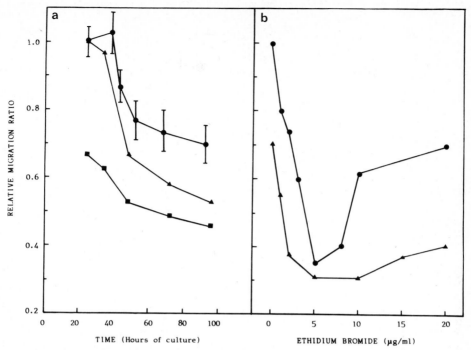

Fig. 10a,b. The sedimentation of nucleoids from differentiating chick muscle cells. **a** The lysis mixture contained 1.0 *M* NaCl (●) or 1.96 *M* NaCl (▲), or 1.95 *M* NaCl plus 5 μg/ml ethidium bromide (■). **b** The effect of ethidium bromide. The lysis mixture contained 1.95 *M* NaCl and the indicated concentration of ethidium bromide. ● 24 h cultures; ■ 72 h cultures

Alkaline sucrose gradients detect single-strand breaks that are induced by the alkaline conditions of the gradients (for example, apurinic sites), as well as true single-strand breaks which had been generated enzymatically. We have distinguished between these two phenomena by examining the sedimentation rate of nucleoids. In neutral sucrose gradients containing 1.0 *M* NaCl, nucleoids from 24 and 38 h cultures sediment rapidly; in contrast, 6 h later, at 44 h the nucleoids sediment more slowly, and by 68 h a considerable number of metabolically induced single-strand DNA breaks are evident (Fig. 10a). Increasing the salt concentration of the gradients to 1.95 *M* NaCl, and thus stripping the nuclei of yet more of their protein, does not alter the results significantly (Fig. 10a). Thus the reduction in the sedimentation of the nucleoids is not due to changes in the protein composition of the cromatin.

This inference is also drawn from several experiments with ethidium bromide. In the presence of 5 μg ml^{-1} of ethidium bromide the sedimentation of nucleoids from 24 h cultures is reduced from 1.0 to 0.67 (Fig. 10a). However, the same concentration of ethidium bromide has a much smaller effect on the sedimentation at 96 h, which is only reduced from 0.56 to 0.46. In nucleoids from 24 h cultures, increasing concentrations of ethidium bromide induce the biphasic effect characteristic of DNA molecules under topological constraints (Fig. 10b), whereas nucleoids from 72 h cultures are

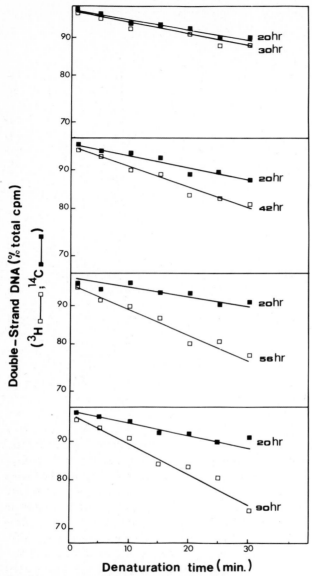

Fig. 11. DNA strand-breaks in differentiating muscle cells. ■ 20 h culture labelled with [^{14}C] thymidine; □ 30, 42, 56 or 90 h cultures labelled with [^{3}H] thymidine. □ [^{3}H]; ■ [^{14}C]. Analysis was by the alkaline unwinding technique

affected less (Fig. 10b), probably due to the presence of single-strand breaks. We have estimated the number of single-strand DNA breaks in differentiating muscle cell DNA, from 20–90 h (Fig. 11a–d); few single-strand breaks were detectable in the DNA at either 20 or 30 h, but by 42 h approx. 50–100 single-strand breaks per 10^{12}

daltons (per diploid genome) had appeared. By 56 h they had increased to 100—300 per genome (Fig. 11c), and did not increase thereafter (Fig. 11d).

The presence of DNA strand-breaks in differentiating cells may be attributed to a general deficiency of DNA repair. However, these differentiating muscle cells are proficient in repair and can readily rejoin single-strand breaks. There is no detectable decline in the ability to rejoin DNA strand-breaks caused by γ-irradiation (Fig. 12). In the 92 h cultures (Fig. 12) a significant number of DNA strand-breaks exist before irradiation; exposure to 300 rad of γ-radiation would be expected to introduce a further 600 breaks, about twice as many as were metabolically engendered. The rate at which these breaks were rejoined was the same as that in 24 h myoblasts — but rejoining stopped when the number of breaks had dropped to the endogenous level. This implies that these endogenous breaks are retained in the face of a proficient DNA strand-rejoining capacity, perhaps because they are of a special character or occur at specific nucleotide sequences; alternatively, there may be a shift in the steady-state level of breaks.

I am suggesting then, that in certain categories of cell differentiation there are rearrangements or transpositions within the mammalian genome. The single-strand breaks which we observe are possibly only indicators of these DNA movements.

The involvement of ADPRT in cell differentiation has been confirmed and very much extended by a number of workers. A. Johnstone and G. Williams and subsequently W. Greer and G. Kaplan have described the blocking of lymphocyte activation by ADPRT inhibitors. Our splendid conference organiser, F. Althaus, has demonstrated that ADPRT activity is required for the expression of some foetal proteins in adult hepatocytes in culture. G. Williams and A. Johnstone have described how ADPRT inhibitors will block the differentiation, but not the proliferation of the parasite *Trypanosoma cruzi*. We have shown that the differentiation of a human promyelocyte cell line (HL60) can be blocked by ADPRT inhibitors.

R. Ben-Ishai will describe later at this meeting interesting experiments with a myoblast cell line in which ADPRT inhibitors block the formation of muscle from these progenitor cells. G. Francis and coworkers have described the inhibition of maturation of human granulocyte-macrophage progenitor cells, and L. Sachs has also observed inhibition of differentiation in maturing mouse white blood cells. E. Uyeki and coworkers have reported influencing chondrocyte differentiation with ADPRT inhibitors. Finally, erythroleukaemia cell differentiation is inhibited by ADPRT inhibitors.

Naturally, we are all anxious to know why ADPRT is required for these varied examples of cell differentiation. At this stage we simply do not know. There are a number of possibilities to consider. Perhaps the nuclear ADPRT enzyme is working as it does in DNA repair; this implies that DNA strand-breakage and rejoining is occurring during cell differentiation. In support of this possibility I draw your attention to the enzymological results which indicate that DNA strand-breaks are an *essential* activator of the enzyme; all other effects on enzyme activity seem to be modulators of this basic enzyme activity. We have tested this possibility in both the chick muscle and the HL60 systems. With the chick muscle-differentiation system we described the appearance during cyto-differentiation of several hundred single-strand breaks which are *not* due to a general deficiency in DNA repair [15—17]. Similar observations had

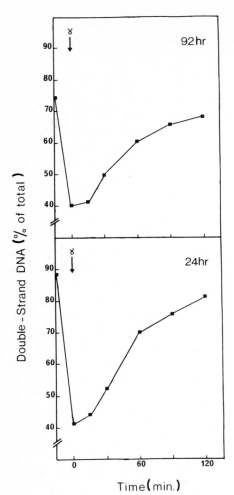

Fig. 12. Rate of DNA repair in developing muscle cells following 300 rad of γ-irradiation. Repair was analysed by measuring the rate of DNA strand separation in alkali

previously been made in developing erythroleukaemia cells and have also been made in mitogen-activated lymphocytes. We report at this meeting similar results with HL60 cells, in which the DNA strand-breaks are again not due to a general deficiency in DNA repair [18].

These observations regarding DNA strand-breaks are at most, suggestive that perhaps the molecular mechanism of ADPRT action in cell differentiation involves breakage and rejoining of DNA. However, other mechanisms are very plausible. For example, ADP-ribosylation may be required because of its effect on chromatin conformation, or specific regulatory chromatin proteins may be ADP-ribosylated as a signal for the induction of a previously non-expressed gene.

In addition to these possibilities, there is another quite different and very interesting possibility. It is known that cholera toxin can ADP-ribosylate a component of the adenyl cyclase. J. Moss has described cytoplasmic ADPRT enzymes with a similar propensity to modify guanidino groups in proteins. In addition, the endogeneous enzyme modifies the same 42kD polypeptide as does cholera toxin. Therefore, we must consider the possibility that an ADPRT is involved in cell differentiation because it regulates adenyl cyclase activity. This is a complex issue because it is conceivable that cell differentiation may involve both adenyl cyclase at an early stage of signal transduction and may also subsequently involve nuclear ADPRT in achieving the expression of new genes. In general, the adenyl cyclase seems to be more usually associated with proliferation than with differentiation, but it is certainly involved in myoblast differentiation.

Cytoplasmic and Membrane ADPRT Activities

We can thus pass on to consider very briefly the cytoplasmic and membrane ADPRT activities. There are clearly a number of separate enzymes; but so far only two clear functions have been identified: the regulation of protein synthesis by the ADP-ribosylation of elongation factor 2 (EF2) and the activation of adenyl cyclase. The physiological role of regulating elongation factor 2 is obscure because it is generally believed that regulation of protein biosynthesis at the translational level occurs at the initiation step; but a universal decrease in protein synthesis does occur in resting lymphocytes and in quiescent fibroblasts.

I have talked mostly about eukaryotic cells, so I should mention here the regulation of RNA polymerase of *E. coli* by an ADPRT of phage T4. This example, and the presence of ADPRT enzymes in protozoa such as *Physarum polycephalum, Tetrahymena pyriformis, Dictyosteleum discoideum, Trypanosoma cruzi, Trypanosoma brucei* and probably in *Plasmodium spp.* show that ADPRT enzymes evolved at a fairly early stage in evolution.

Conclusion

In summary then, it has been demonstrated that ADP-ribosylation reactions regulate diverse cellular processes. Poly(ADP-ribosylation) participates in DNA repair in eukaryotic organisms, possibly by regulating the ligation step (Fig. 13). In addition, ADP-ribosylation reactions modulate cell differentiation, protein synthesis and membrane adenyl cyclase reactions. No doubt, as time passes, the list of cellular processes modulated by ADP-ribosylation reactions will grow.

Acknowledgements. The work in my laboratory has been supported by the Cancer Research Campaign, the Medical Research Council and the Science and Engineering Research Council. I would especially like to thank all my colleagues and students together with whom, we have helped to elucidate the mysteries of ADP-ribose.

ADP-RIBOSYL TRANSFERASE

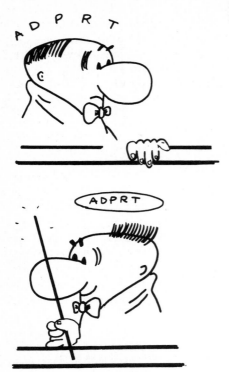

Fig. 13. Cartoon indicating our general view of nuclear ADPRT activity. The enzyme is sensitive to breaks in the DNA and modulates their rejoining

References

1. Shall S. Durkacz BW, Gray DA, Irwin J, Lewis PJ, Perera M, Tavassoli M (1982) (ADP-ribose)$_n$ participates in DNA excision repair. In: Harris CC, Cerutti PA (eds) Mechanisms of chemical carcinogenesis. Liss, New York, pp 389–407
2. Shall S (1983) ADP-ribosylation, DNA repair, cell differentiation and cancer. In: Miwa M, Hayaishi O, Shall S, Smulson M, Sugimura T (eds) ADP-ribosylation, DNA repair and cancer. Jpn Sci Soc Press, Tokyo, pp 3–25
3. Shall S (1984a) Inhibition of DNA repair by inhibitors of nuclear ADP-ribosyl transferase. Nucleic Acids Res Symp Ser 13:143–191
4. Shall S (1984b) ADP-ribose in DNA repair: A new component of DNA excision repair. Adv Radiat Biol 11:1–69
5. Whish WJD, Davies MI, Shall S (1975) Stimulation of poly(ADP-ribose) polymerase activity by the antitumour antibiotic, streptozotocin. Biochem Biophys Res Commun 65:722–730
6. Goodwin PM, Lewis PJ, Davies MI, Skidmore CJ, Shall S (1978) The effect of gamma-radiation and neo-carzinostatin on NAD and ATP levels in mouse leukaemia cells. Biochem Biophys Acta 543:576–582
7. Skidmore CJ, Davies MI, Goodwin PM, Halldorsson H, Lewis PJ, Shall S, Zia'ee A-A (1979) The involvement of poly(ADP-ribose) polymerase in the degradation of NAD caused by γ-radiation and N-methyl-N-nitroso-urea. Eur J Biochem 101:135–142

8. Durkacz BW, Omidiji O, Gray DA, Shall S (1980) (ADP-ribose)$_n$ participates in DNA excision repair. Nature (London) 283:593—596
9. Gray DA, Durkacz BW, Shall S (1981) Inhibitors of nuclear ADP-ribosyl transferase retard DNA repair after N-methyl-N-nitrosourea. FEBS Lett 131:173—177
10. Durkacz BW, Irwin J, Shall S (1981a) The effect of inhibition of (ADP-ribose)$_n$ biosynthesis on DNA repair assayed by the nucleoid technique. Eur J Biochem 121:65—69
11. Durkacz BW, Irwin J, Shall S (1981b) Inhibition of (ADP-ribose)$_n$ biosynthesis retards DNA repair but does not inhibit DNA repair synthesis. Biochem Biophys Res Commun 101:1433—1441
12. Creissen D, Shall S (1982) Regultion of DNA ligase activity by poly(ADP-ribose). Nature (London) 296:271—272
13. Murray B, Irwin J, Creissen D, Tavassoli M, Durkacz BW, Shall S (to be published) Isolation and characterisation of cell variants altered in DNA ligase activity
14. Nduka N, Skodmore CJ, Shall S (1980) The enhancement of cytotoxicity of N-methyl-N-nitrosourea and of γ-radiation by inhibitors of poly(ADP-ribose) polymerase. Eur J Biochem 105:515—530
15. Farzaneh F, Shall S, Zalin R (1980) DNA strand breaks and poly(ADP-ribose) polymerase activity during chick muscle differentiation. In: Smulson M, Sugimura T (eds) Novel ADP-ribosylation of regulatory enzymes and proteins. Elsevier/North Holland, Amsterdam New York, pp 217—225
16. Farzaneh F, Zalin R, Brill D, Shall S (1982) DNA strand breaks and ADP-ribosyl transferase activation during cell differentiation. Nature (London) 300:362—366
17. Farzaneh F, Zalin R, Shall S (1964) Single-strand DNA breaks are formed during muscle cell differentiation. Exp Biol Med 9:260—268

Poly(ADP-Ribose): Structure and Enzymology

Size Distribution of Branched Polymers of ADP-Ribose Generated in Vitro and in Vivo

RAFAEL ALVAREZ-GONZALES and MYRON K. JACOBSON[1]

Abbreviations:

HPLC	High performance liquid chromatography
$(PR)_2 AMP$	$2'$-[$1''$-ribosyl$2''$($1'''$-ribosyl)]adenosine $5',5'',5'''$ Tris-(phosphate)
PRAMP	$2'$-($1''$-ribosyl)adenosine $5',5''$ bis(phosphate)
AMP	adenosine $5'$monophosphate
EDTA	Ethylenediaminotetraacetic acid
temed	N,N,N',N' tetramethylethylenediamine

Introduction

Poly(ADP-ribose) metabolism is a chromatin associated event. The function of poly-(ADP-ribose) may be mediated by activation or inactivation of enzymatic activities by the covalent binding of the polymer and/or by noncovalent interactions of this highly negatively charged polymer with other chromatin components. We have studied the size distribution and branching frequency of polymers in order to evaluate the potential importance on noncovalent interactions with other components of chromatin.

Size Distribution of Branched Polymers Generated in Vitro

Figure 1 shows the distribution of radiolabeled poly(ADP-ribose) synthesized in nucleotide permeable SVT2 cells [1], when fractionated by HPLC size exclusion chromatography using a Bio-Sil TSK-125 column. Fractions that contained radiolabeled material were subjected to polyacrylamide gel electrophoresis using a 20% polyacrylamide gel to examine the resolution obtained. Figure 2 shows that each fraction contained poly(ADP-ribose) of a different average size. Fraction 13 through 17 con-

---

1 Department of Biochemistry, North Texas State University/Texas College of Osteopathic Medicine, Denton, TX 76203, USA

ADP-Ribosylation of Proteins
(ed. by F.R. Althaus, H. Hilz, and S. Shall)
© Springer-Verlag Berlin Heidelberg 1985

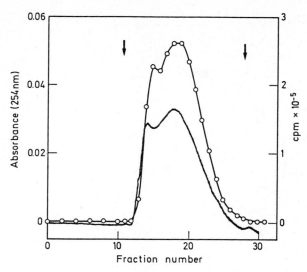

Fig. 1. Size exclusion chromatography of ADP-ribose made in nucleotide permeable SVT2 cells [1] and purified as described somewhere else [4], using one Bio Sil TSK-125 column. The running buffer used was 0.1 M sodium phosphate, pH 6.8 at a flow rate of 1.0 ml min^{-1}. 0.5 ml fractions were collected. The *arrows* indicate void volume and included volume, respectively. ●—● Radioactivity profiles; ——— absorbance profile

tained primarily material which did not migrate into the gels. In fact, this material did not effectively migrate in gels as low as 4%. We have termed this material "complex polymer".

The size distribution of polymers was directly determined by digestion to nucleotides using snake venom phosphodiesterase. The unique nucleotides PRAMP and $(PR)_2$ AMP are generated from linear internal and branched residues, respectively, and AMP is generated from terminal residues. The separation of AMP, PRAMP, and $(PR)_2$ AMP was achieved by strong anion exchange HPLC. The amount of each nucleotide derived from poly(ADP-ribose) in each of the fractions is shown in Table 1. It is important to note that the conventional method of calculation of average chain length does not account for the presence of branching in poly(ADP-ribose) [2]. This can result in an underestimation of polymer size by several fold. We have calculated the true average polymer size using a formula described elsewhere [2], as well as the average number of points of branching per molecule using the formula shown in Table 1. A very broad distribution of polymer sizes was obtained with the largest being polymers of an average size of 190 residues and containing between five and six points of branching per molecule. Since very complex polymers are generated in vitro, we have also examined for the presence of these polymers in vivo.

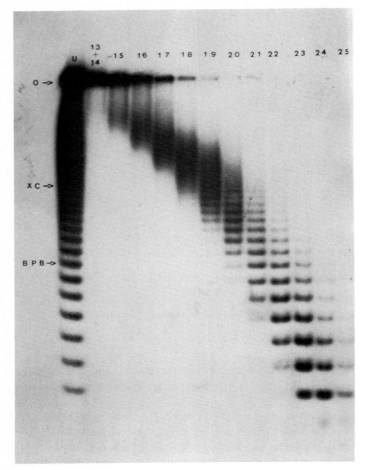

Fig. 2. Polyacrylamide gel electrophoresis of unfractionated and fractionated [^{32}P]poly(ADP-ribose). A 20% polyacrylamide gel containing a ratio of 19:1 of acrylamide:bis acrylamide, 4.4 mM ammonium persulfate, 3.4 mM temed, 1 mM EDTA, and 50 mM Tris-borate pH 8.3 was used. The gel was run for 3 h at 400 V, after a 2 h prerun at the same voltage

Size Distribution of Branched Polymers Generated in Vivo

Cells in culture were subjected to a combination of hyperthermia and DNA damage, conditions under which it has been shown that the intracellular levels of poly(ADP-ribose) are increased [3]. Cells were labeled with [^{3}H]adenine prior to treatment to label the NAD pool and thus to generate [^{3}H] labeled poly(ADP-ribose) which was purified [4] and analyzed as described above. The results are shown in Table 2. Polymers as large as 244 residues were observed containing between six and seven points of branching per molecule. Thus, highly complex polymers of ADP-ribose also exist in vivo.

Table 1. Determination of polymer size and number of branching points per molecule

Fraction	cpm			Average size	Points of branching[a] per molecule
	AMP	PRAMP	$(PR)_2 AMP$		
13	437	11,775	371	190	5.62
14	980	26,981	793	153	4.24
15	1,511	40,581	1,091	102	2.59
16	1,715	43,661	866	54	1.02
17	2,297	51,433	891	38	0.63
18	5,603	111,010	1,386	28	0.32
19	6,967	109,273	1,103	20	0.18

[a] Points of branching per molecule = $\dfrac{(PR)_2 AMP}{AMP - (PR)_2 AMP}$

Table 2. Size distribution of branched polymers of ADP-ribose from C3H10T1/2 cells treated with heat shock and MNNG

Fraction[a]	cpm			Average size	Points of branching per molecule
	AMP	PRAMP	$(PR)_2 AMP$		
24	524	13,931	463	244	6.6
25	2,664	76,391	2,097	143	3.7
26	3,449	81,687	2,689	115	3.5
27	3,211	74,685	2,073	70	1.8
28	2,691	56,296	1,520	52	1.3
29	2,280	45,534	1,140	43	1.0
30	2,420	45,440	998	34	0.7
31	2,427	42,165	844	29	0.5
32	2,649	44,120	718	25	0.4
33	3,583	49,925	763	19	0.2

[a] Two BioSil TSK-125 columns in series were used to fractionate the polymer by HPLC size exclusion chromatography

Discussion

These studies support the possibility that noncovalent interactions between highly complex polymers either free or protein-bound and other chromatin components may affect changes in chromatin structure (e.g., condensation or relaxation). It is interesting that the average number of branching points per molecule approaches one at an approximate size of 40 residues. This is also the size at which highly complex polymers are generated. We suggest that the branching event is required for the formation of these highly complex polymers.

It has been proposed that nucleosomes are arranged in chromatin in higher ordered structures termed solenoids [5]. The solenoid has been shown to have about six to eight nucleosomes per turn [6]. Since highly complex polymers have about six to eight branches per molecule, it is possible that highly complex polymers may stabilize such higher order structures of chromatin.

Acknowledgments. This work was supported by Grant B-633 from the Robert A. Welch Foundation, Grants CA23994 and CA29357 from the National Institutes of Health. RAG was supported by predoctoral fellowships from the Samuel Roberts Noble Foundation and Conacyt. We thank Hilda Mendoza-Alvarez for technical assistance and Lynne Gracy for editorial assistance during preparation of the manuscript.

References

1. Alvarez-Gonzalez R, Juarez-Salinas H, Jacobson EL, Jacobson MK (1983) Evaluation of immobilized boronates for studies of adenine and pyridine nucleotide metabolism. Anal Biochem 135:69–77
2. Sugimura T, Miwa M (1982) Structure and properties of poly(ADP-ribose). In: Hayaishi O, Ueda K (eds) ADP-ribosylation reactions. Academic Press, London New York, p 43
3. Juarez-Salinas H, Duran-Torres G, Jacobson MK (1984) Alteration of poly(ADP-ribose) by hyperthermia. Biochem Biophys Res Commun 122:1381–1388
4. Jacobson MK, Payne DM, Alvarez-Gonzalez R, Juarez-Salinas H, Sims JL, Jacobson EL (1984) Determination of in vivo levels of polymeric and monomeric ADP-ribose by fluorescence methods. Methods Enzymol 106:483–494
5. Butler PJG (1983) The folding of chromatin. CRC Crit Rev Biochem 15:57–91
6. Sau P, Bradbury EM, Baldwin JP (1979) Higher ordered structures of chromatin in solution. Eur J Biochem 97:593–599

Studies on the Structure of Poly(ADP-Ribose) by HPLC. Separation of in Vitro Generated Polymer Chains, Analysis of Chain Length, and Branching

ALAEDDIN HAKAM and ERNEST KUN[1]

Introduction

Intact chains of in vitro synthesized poly(ADP-ribose) were first isolated by reversed phase HPLC [1], a method which separated total polymeric chains from ADP-R, NAD, and AMP. Ion exchange HPLC was developed for the separation of individual intact polymeric chains of various size and this experimental approach was presently employed to explore the incidence of branching [2, 3]. The HPLC method disclosed the existence of trace amounts of as yet unidentified snake venom phosphodiesterase degradation products of poly(ADP-ribose), which could indicate end-pieces or possibly new structural components.

Experimental Procedures

Preparative isolation of poly(ADP-ribose) was done as reported [1]. Intact poly(ADP-ribose) was applied to a Sephadex G50 column and fractionated into short n_{av} = 6.96, medium n_{av} = 18.6, and long n_{av} = 44 chain lengths. HPLC solvent gradients involved at least two of the following buffer systems: buffer A, 0.05 M potassium phosphate (pH 4.25); buffer B, 0.05 M potassium phosphate (pH 4.25), 1 M ammonium sulfate, and 30% MeOH (v/v); buffer C, same as B but containing 0.2% (v/v) trifluoracetic acid, (pH 2). Flow rate was 0.6 ml min^{-1} in all cases. Chain length analysis of poly(ADP-ribose) were based on spectral analysis of emerging fractions from the AX300 column with the following solvent gradient: 100% A to 100% B in 90 min (linear) and from 100% B to 100% C (linear) in 45 min. Chain length determination of degradation products of the oligomers by phosphodiesterase were analyzed on the AX100 column with the linear gradient from 100% A to 50% B, 50% A in 45 min. The chain length of polymers (n) was calculated by a modification of a published formula [4] as follows:

1 The Cardiovascular Research Institute and Department of Pharmacology, Biochemistry and Biophysics, The University of California, San Francisco, CA 94143, USA

ADP-Ribosylation of Proteins
(ed. by F.R. Althaus, H. Hilz, and S. Shall)
© Springer-Verlag Berlin Heidelberg 1985

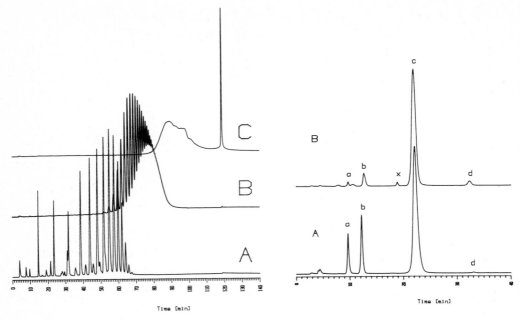

Fig. 1

Fig. 2

Fig. 1. HPLC resolution of short chains (*A*), medium chains (*B*), and long chains (*C*)

Fig. 2. Separation of products of phosphodiesterase digestion of short (*A*) and long (*B*) poly(ADP-ribose)

$$n = \frac{[AMP] + [PR\text{-}AMP] + [Ado(P)Rib(P)Rib(P)] + [end\ fraction^*]}{[AMP] - [Ado(P)Rib(P)Rib(P)]}.$$

Results

HPLC resolution of short chains is demonstrated in Fig. 1A. Each emerging peak was monitored by its UV spectrum (A_{220} to A_{340} not shown, see [5]). Twelve major peaks separated (ending at retention time of 63 min) and each fraction exhibited characteristic adenine absorbance. Each fraction was collected and analyzed further. Peak 1 is AMP and peaks 2 to 4 short oligomers (n 2–4). Peaks 5–10 were sufficient for chain length analyses, and results are given in Table 1. Increasing chain length coincides with an increase in apparent branching from 0.34% (n = 7.1) to 0.5% (n = 11.0). Separation and spectral identification of peak fractions were carried out for medium and long polymers as illustrated in Fig. 1B,C, respectively. Although partial resolution into individual peaks takes place and spectral identification of each component as an

---

* End fraction = probable attachment site to protein acceptors of as yet unknown structure

Table 1. HPLC analysis of phosphodiesterase digests of ADP-ribose oligomers

Sample	a	%AMP	%PR-AMP	%Ado(P)Rib(P)Rib(P)	Chain length
Short					Chains:
Peak 5	8.15	17.38	74.46	–	5.8
Peak 6	6.22	14.35	79.09	0.34	7.1
Peak 7	5.31	12.51	81.81	0.36	8.2
Peak 8	4.85	11.08	83.65	0.41	9.3
Peak 9	3.98	10.20	85.30	0.46	10.3
Peak 10	3.40	9.59	86.49	0.50	11.0
Medium:	1.26	6.51	91.10	1.14	18.6
Long:	1.79	5.29	89.80	3.02	44.0
Long I:	3.07	4.60	89.95	2.37	44.8
Long II:	1.57	5.21	90.1	3.10	47.4

[a] Protein attachment fragment

adenine nucleotide clearly defined the oligomers, base line separation, as obtained for short chains (Fig. 1A) did not occur. However, more than 18 individual peaks are discernable, but their individual chain lengths cannot be determined because of incomplete separation. The exact reason for this behavior is not known, but it can be assumed that the ion exchange separation is incomplete because with longer chain lengths the percentage charge differences between consecutive polymers diminishes. As seen from Table 1 the apparent average branching in the medium chain length fraction is 1.14%, thus branching increases proportionally with polymer size.

When the pH of the eluant was decreased from pH 4.5 to 2 (Fig. 1C), one broader area (Long I) and a sharp peak (Long II), most probably including a family of long chain polymers, were obtained. It is important that at 75° as employed in HPLC experiments thus far, recovery of all chains, including the long ones was nearly quantitative (85–95%), and there was no evidence of degradation at that elevated temperature. As given in Table 1, the average chain lengths of polymers eluting in long peaks I and II differ somewhat and apparent branching is increased from 2.37% to 3.1%. The most probable explanation for the behavior of long chains in this HPLC system is a simultaneous contribution of both interhelical forces [6, 7] and increased branching, producing a network of polymers that fails to be resolved on the basis of charge separation.

The evidence for apparent branching was examined by separation of the products of phosphodiesterase digestion (Fig. 2). Figure 2B illustrates resolution of a digest of long chain polymers. AMP(b), PR-AMP(c), and Ado(P)Rib(P)Rib(P)(d) [2, 3] can be discerned by characteristic retention times. Peaks (a) and (x) are presently unidentified, but peak (a) is suspected to be a terminal fragment, comprising a protein attachment site. The existence of oligomer termini has been reported [8], but their exact chemical structure is yet unknown. The phosphodiesterase digests of short chain oligomer (A) ($n_{average}$ = 6.96) exhibits a significant component of (a), and only traces of a branching product (d). The isolation and identification of component (a) is in progress and will be reported elsewhere. On the other hand, long chains contain much smaller amounts of terminal fragment (a), but discernable quantities of Ado(P)Rib(P)Rib(P), identified by its specific elution [2, 3, 9], which in the ion exchange column follows PR-AMP.

Conclusions

Programmed ion exchange HPLC is able to separate individual short oligo(ADP-ribose) chains, but with increasing length complex physical and probably structural contributions, leading to a presumed network-like structure prohibit complete resolution. By varying the temperature some improvement of subfractionation of longer chains is possible [5]. Branching [2,3] has been confirmed and some as yet unidentified degradation products suggest tertiary structures, requiring further study.

Acknowledgments. This research was supported by F-49620-81-C-0007 (AFOSR) and HL-27317 (NIH). E.K. is a recipient of the Research Career Award of the United States Public Health Service.

References

1. Hakam A, McLick J, Kun E (1984) Separation of poly(ADP-ribose) by high-performance liquid chromatography. J Chromatogr 296:369—377
2. Miwa M, Saikawa N, Yamaizumi Z, Nishimura S, Sugimura T (1979) Structure of poly(adenosine diphosphate ribose): Identification of 2'-[1''-ribosyl-2''-(or 3''-)(1' ''-ribosyl)]adenosine-5',5'',5' ''-tris(phosphate) as a branch linkage. Proc Natl Acad Sci USA 76:595—599
3. Kanai M, Miwa M, Kuchino Y, Sugimura T (1982) Presence of branched portion in poly(adenosine diphosphate ribose) in vivo. J Biol Chem 257:6217—6223
4. Kawaichi M, Ueda K, Hayaishi O (1981) Multiple autopoly(ADP-ribosyl)-ation of rat liver poly(ADP-ribose) synthetase. J Biol Chem 256:9483—9489
5. Hakam A, Kun E (1985) High performance liquid chromatography of in vitro synthesized poly(ADP-ribose) on ion exchange columns, separation of oligomers of varying chain length and estimation of apparent branching. J Chromatogr (in press)
6. Minaga T, Kun E (1983) Spectral analysis of the conformation of polyadenosine diphosphoribose. J Biol Chem 259:725—730
7. Minaga T, Kun E (1983) Probable helical conformation of poly(ADP-ribose). J Biol Chem 258:5726—5730
8. Tanaka M, Miwa M, Hayashi K, Kubota K, Matsushima T, Sugimura T (1977) Separation of oligo(adenosine diphosphate ribose) fraction with various chain lengths and terminal structures. Biochemistry 16:1485—1489
9. Juarez-Salinas H, Mendoza-Alvarez H, Levi V, Jacobson MK, Jacobson EL (1983) Simultaneous determination of linear and branched residues in poly(ADP-ribose). Anal Biochem 131:410—418

Probable Helical Conformation of Poly(ADP-Ribose)

TAKEYOSHI MINAGA[1] and ERNEST KUN[2]

Introduction

It is well established that helical conformations of nucleic acids are recognizable from a proportionality between the UV absorbance and CD spectrum [1]. For a helix the component of the absorption band polarized perpendicular to the helix axis gives rise to a unique CD curve. The λ max at 258 nm and the crossover point through the 0 line of the CD spectrum at the same wave-length fulfill the theoretical prediction of the helical conformation of poly(ADP-ribose) [2, 3]. Based on the following results, helical conformation of poly(ADP-ribose) was postulated.

Results and Discussion

Highly purified polydisperse poly(ADP-ribose) was prepared by boronic acid affinity chromatography, from which long, medium, and short chains of the polymer were prepared by molecular filtration (Sephadex G-50). The long chain poly(ADP-ribose) clearly showed temperature-dependent hypochromicity as shown in Fig. 1. The isosbestic point was observed near 281 nm. It is apparent that there is a very small temperature dependent A280 change, however, this is negligible and the main temperature-dependent absorbance change is around 260 nm. By this data, the A280/A260 ratio, which has been used in the past as one of the criteria of purity of poly(ADP-ribose), was found to be an unsuitable idex of the purity [4]. The A280/A260 ratio was also influenced by the ionic strength and chain length. Figure 2 showed parameter-dependent hypochromicity in long chain poly(ADP-ribose). At high NaCl concentration, a very clear transition was observed, which indicates that a distinct conformational change occurs. CD analysis of long chain poly(ADP-ribose) in 1 M salt concentration, proved that CD analysis is a more sensitive method than UV-spectrum analysis

1   Cutter Japan, Ltd. the Department of Research and Development, Kobe International Friendship Bldg., 9-1, Minatojima Nakamachi, 6-chome, Chuo-Ku, Kobe, 650, Japan
2   The Cardiovascular Research Institute and the Department of Pharmacology, the University of California, San Francisco, CA 94143, USA

ADP-Ribosylation of Proteins
(ed. by F.R. Althaus, H. Hilz, and S. Shall)
© Springer-Verlag Berlin Heidelberg 1985

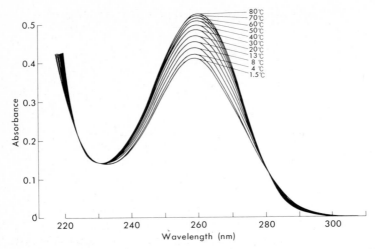

Fig. 1. The effect of increase in temperature on the absorption spectrum of long chain poly(ADP-ribose). An average chain length of 35 was used in the solution of 100 mM Tris-HCl (pH 7.0) containing 1 M NaCl

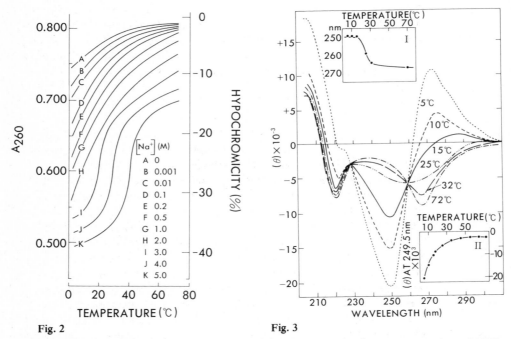

Fig. 2   Fig. 3

Fig. 2. Thermal denaturation of long chain poly(ADP-ribose) at various concentrations of NaCl. The 0% hypochromicity was defined as A260 value of 2'-(5''-phosphoribosyl)-5'-AMP that was the calculated basis of the specific activity of the nucleotide [9]. 50 μM of poly(ADP-ribose) was used for each of these assays

Fig. 3. The influence of increasing temperature on the CD spectrum of long chain poly(ADP-ribose) *Inset I:* variation of λ in nanometers of θ max (*ordinate*) at varying temperature. *Inset II:* the melting curve determined at 249.5 nm. 100 mM Tris-HCl (pH 7.0) containing 1 M NaCl was used as a solution

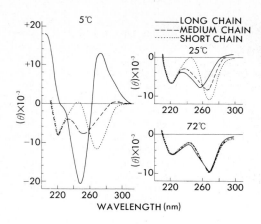

Fig. 4. The effect of temperature on CD spectra of poly(ADP-ribose) of long, medium, and short chain length. The same solution was used as in Fig. 3

to detect conformational changes as shown in Fig. 3. A distinct blue shift could be seen depending on the temperature, even at 1 M NaCl concentration (Fig. 3, inset I). However, the melting curve analysis of CD (Fig. 3, inset II) was similar to that of UV. The CD spectra of long, medium, and short chain poly(ADP-ribose) were determined at 4°, 25°, and 72°C and the results are summarized in Fig. 4. A temperature-dependent blue shift was observed with long and medium size poly(ADP-ribose) and the magnitude of this response was a function of chain length. With short chain poly(ADP-ribose), only the magnitude of θ exhibited temperature dependence. At 72°C, the difference between the CD spectra of all three species of poly(ADP-ribose) disappeared, indicating that conformation of heat-denatured long chain poly(ADP-ribose) was equivalent to that of short chains. The temperature decrease-dependent blue shift is generally interpreted to indicate breakdown of either hydrogen bonding of base pair-

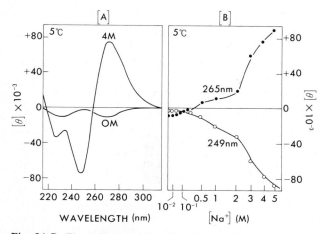

Fig. 5A,B. The influence of [Na$^+$] on θ determined at varying wave-lengths (A) and on the magnitude of θ determined at 249 nm and 265 nm (B) at 5°C. 150 μM of long chain poly(ADP-ribose) was used for each experiment

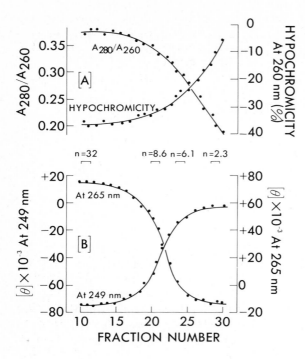

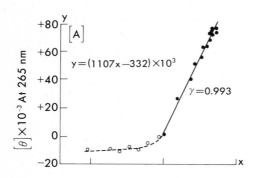

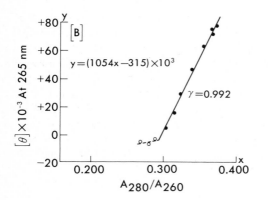

Fig. 6. Correlation between the chain length and spectral properties of poly(ADP-ribose) as determined at 5°C in 4 M NaCl. Molecular filtration of polydisperse poly(ADP-ribose) on Sephadex G-50 was carried out as described [10]. 0.75 ml aliquots of each of the tube's contents were added to 2.25 ml of 5 M NaCl solution (final concentration of NaCl was 4 M). The average chain length of poly(ADP-ribose) was determined in several fractions and is shown as "n"

Fig. 7A,B. Positive correlation between θ and A280/A260. A Paired values for each fraction shown in Fig. 6 were replotted. Results were obtained with oligomers of varying chain length. B The experimental values shown in Fig. 5 were replotted with the paired A280/A260. Results were obtained at various temperatures. γ = correlation coefficient; y = ax + b, expresses the least square linear regression curve

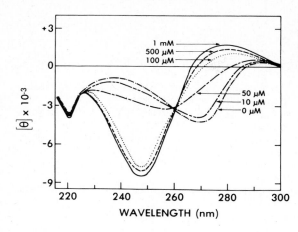

$[\theta] \times 10^{-3}$

WAVELENGTH (nm)

Fig. 8. The influence of Mg^{2+} on CD spectrum of long chain poly(ADP-ribose) at 5°C. The concentration of the polymer was 150 μM

ing or some other conformational transition of nucleic acid strands [5]. Ionic strength dependent-θ values and spectra are shown in Fig. 5. The phenomenological similarity of the CD spectra of long chain poly(ADP-ribose) to poly(A), poly(U), and poly(A+U), or DNA can be seen [6]. This again clearly indicates the helical conformation of poly-(ADP-ribose). Chain length dependent hypochromicity, A280/A260, and θ values at 249 nm and 265 nm are shown in Fig. 6, which shows very clear transition at an oligomer chain length of about 8 to 9. Positive correlations between A280/A260 (x-coordinate) and θ (y-coordinate) were obtained under varying condition (i.e., chain length, temperature, and ionic strength) and all equations describing regression curves are nearly identical. The θ values within an error of 5% can be calculated from either equations, as shown in Fig. 7, by substituting any numerical value of A280/A260 that is larger than 0.3, apparently indicating that below this value θ becomes independent of A280/A260. Therefore, in other words, the degree of the ordered form of poly-(ADP-ribose) is obviously reflected by the A280/A260 ratio, which should be considered as an excellent index. The influence of Mg^{2+} on the CD spectrum of long chain poly(ADP-ribose), dissolved in H_2O, is illustrated in Fig. 8. Addition of Mg^{2+} shifts the negative peak from 268 to 249 nm. Simultaneously, with this blue shift, there is a large increase in the magnitude of the negative peak. Saturation occurs at 1 mM. Complexation of Mg^{2+} by EDTA reverses the effect.

For the understanding of the secondary structure of poly(ADP-ribose), two important concepts have been reported by Sugimura's group. (1) According to ^{13}C-NMR data, it was suggested that poly(ADP-ribose) does not have as highly ordered conformation as poly(A) [7]; (2) how does the branched chain of poly(ADP-ribose) contribute to the conformation [8]? So far as the first point is concerned, the difference in our data interpretation was mainly due to the difference in experimental conditions, i.e., ionic strength and temperature. Our data under their experimental conditions do not conflict with them. So far as the second point is concerned, it is quite difficult to explain our data in the light of the existence of the branch. This is an interesting problem to examine further.

References

1. Tinoco I (1964) Circular dichroism and rotatory dispersion curves for helices. J Am Chem Soc 86:297–298
2. Minaga T, Kun E (1983) Spectral analysis of the conformation of polyadenosine diphospho-ribose – evidence indicating secondary structure. J Biol Chem 258:725–730
3. Minaga T, Kun E (1983) Probable helical conformation of poly(ADP-ribose) – the effect of cations on spectral properties. J Biol Chem 258:5726–5730
4. Kanai Y, Kawamitsu H, Tanaka M, Matsushima T, Miwa M (1980) A novel method for purification of poly(ADP-ribose). J Biochem (Tokyo) 88:917–920
5. Van Holde KE, Brahms J, Michaelson AM (1965) Base interactions of nucleotide polymers in aquous solution. J Mol Biol 12:726–730
6. Hashizume H, Imahori K (1967) Circular dichroism and conformation of natural and synthetic polynucleotides. J Biochem (Tokyo) 61:738–749
7. Miwa M, Saito H, Sakura H, Saikawa N, Watanabe F, Matsushima T, Sugimura T (1977) A ^{13}C-NMR study of poly (adenosine diphosphate ribose) and its monomer – evidence of α-(1''-2') ribofuranosyl ribofuradenosine residue. Nucleic Acids Res 4:3997–4005
8. Miwa M, Saikawa N, Yamaizumi Z, Nishimura S, Sugimura T (1979) Structure of poly(adenosine diphosphate ribose) = identification of 2'-[1''-ribosyl-2''-(or 3''-)C1' ''-ribosyl] adenosine-5'-5''-5' ''-tris(phosphate) as a branch linkage. Proc Natl Acad Sci USA 76:595–599
9. Minaga T, McLick J, Patlabiraman N, Kun E (1982) Steric inhibition of phenylboronate complex formation of 2'-(5''-phosphoribosyl)-5'-AMP. J Biol Chem 257:11942–11945
10. Minaga T, Romaschin AD, Kirsten E, Kun E (1979) The in vivo distribution of immunoreactive larger than tetrameric polyadenosine diphosphoribose in histone and nonhiston protein fraction of rat liver. J Biol Chem 254:9663–9668

Immobilized Poly(ADP-Ribose) Synthetase: Preparation and Properties

KUNIHIRO UEDA[1], JINGYUAN ZHANG[2], and OSAMU HAYAISHI[3]

Introduction

Poly(ADP-ribose) synthetase is a versatile nuclear enzyme that catalyzes all steps of poly(ADP-ribosyl)ation, i.e., initiation, elongation, and branching of poly(ADP-ribose) chains, on acceptor proteins [1–3]. The enzyme has been purified from various sources to apparent homogeneity and characterized by many investigators [2, 3]. The properties of the enzymes from different sources are almost identical, including an absolute dependency of the activity on DNA strand ends, and the capability of modifying itself (automodification) [2, 3]. We reported also that an NAD glycohydrolase activity is associated with this ADP-ribose transferring enzyme [4]. In order to investigate the mechanisms of this variety of enzyme actions, we recently immobilized the enzyme on a gel matrix [5], and analyzed its properties in comparison with free (nonimmobilized) enzyme.

Immobilization of Poly(ADP-Ribose) Synthetase

Immobilization was carried out using a homogeneous preparation of poly(ADP-ribose) synthetase obtained from calf thymus [6] and commercially available BrCN-activated Sepharose 4B. For successful immobilization of active enzyme, it was important to block a majority ($\sim$ two-thirds) of the active groups of the gel by pretreatment with dithiothreitol. The treated gel was mixed with the enzyme solution in the presence of 0.35 M KCl, 0.2 mM dithiothreitol, 20% glycerol, and 0.1 M K phosphate buffer (pH 8.0). Figure 1 shows the time course of immobilization of the enzyme activity. The decrease in the soluble enzyme activity reflects mainly immobilization of the enzyme, whereas the decrease in the immobilized enzyme activity is indicative of inactivation of the enzyme. In order to avoid the increase of inactive

1 Department of Medical Chemistry, Kyoto University, Faculty of Medicine, Sakyo-ku, Kyoto 606, Japan
2 Shanghai Institute of Biochemistry, Academia Sinica, Shanghai, China
3 Osaka Medical College, Takatsuki, Osaka 569, Japan

ADP-Ribosylation of Proteins
(ed. by F.R. Althaus, H. Hilz, and S. Shall)
© Springer-Verlag Berlin Heidelberg 1985

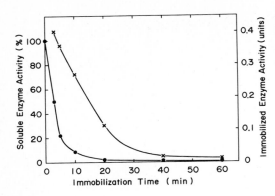

Fig. 1. Time course of immobilization of poly(ADP-ribose) synthetase. Immobilization was carried out as described in the text at 0–4°C. At intervals, aliquots were taken out and examined for enzyme activities remaining soluble (●—●) or bound to the gel (x—x)

enzymes in the immobilized enzyme population, we stopped the reaction at 3 min. Under these conditions, approximately one-half of the enzyme protein was bound to the gel; the other half was recovered in the solution together with the original (i.e., one-half of the total) enzyme activity. The gel-bound activity was stable for months in storage at −15°C in the presence of 2 M KCl, 1 mM dithiothreitol, 20% glycerol, and 50 mM Tris-HCl (pH 8.0).

Properties of Immobilized Poly(ADP-Ribose) Synthetase

Immobilized enzyme had a specific activity of about one-fourth to one-tenth of that of the original enzyme; the value fluctuated among preparations as well as with assay conditions used, such as the presence or absence of Mg^{++} and/or histone.

Table 1 shows the requirements for the immobilized enzyme activity. DNA was almost absolutely required for the activity, as also for free (nonimmobilized) enzyme activities [2, 3]. Histone H1, a potent activator of automodification of soluble enzyme [7], markedly inhibited the automodification of the immobilized enzyme in the presence of Mg^{++}, but slightly stimulated it in the absence of Mg^{++}. On the other hand, the histone, when included, served as a more efficient acceptor of poly(ADP-ribosyl)-ation than the gel-bound enzyme in the presence and absence of Mg^{++}. These results indicate that the histone served almost solely as an acceptor for poly(ADP-ribosyl)-ation by immobilized enzyme, which is in contrast to its actions as both an acceptor and an allosteric activator for free enzyme.

Another prominent difference between the immobilized enzyme and free enzyme was observed in the NAD glycohydrolase activity. The ratio of NAD hydrolysis to ADP-ribosylation is in a range of 0.3–0.5 for free enzyme under most conditions of poly(ADP-ribose) synthesis [4], whereas the ratio for the immobilized enzyme was 3.7 under certain conditions. The specific activity of releasing nicotinamide from NAD, that is a sum of ADP-ribosylation and NAD hydrolysis, was reduced to about one-half upon immobilization. Based on these figures, it was calculated that about two-thirds of the immobilized enzyme molecules lost the ADP-ribosylating activity and changed to NAD glycohydrolse. This change may be related to the observed

Table 1. Requirements for immobilized poly(ADP-ribose) synthetase activity[a]

| System | ADP-ribose incorporated | | |
| | Total | Gel-bound[b] | Soluble[c] |
	pmol	pmol	pmol
Complete	757	187	570
− Mg^{++}	697	171	526
− Histone Hl	465	437	28
− Mg^{++}, histone Hl	119	106	13
− Histone Hl, DNA	27	27	0

[a] After incubation for 8 min at 25°C, the reaction was terminated by the addition of 8 mM 3-aminobenzamide, and the mixture was filtered through a small polypropylene column to separate gel-bound and soluble products [5]
[b] Represents automodification of the enzyme
[c] Includes both histone-bound (where histone was included) and free poly(ADP-ribose) released from acceptors during the incubation; the latter amounted to ∿7% of total ADP-ribose incorporation

marked decrease in the acceptor sites of the enzyme, from >10 [8] to ∿0.5 site/ enzyme molecule, upon immobilization.

Intramolecular Mechanism of Automodification

In the absence of an exogenous acceptor, the immobilized synthetase catalyzed automodification, that is, poly(ADP-ribose) synthesis on gel-bound enzyme molecules (Table 1). The product polymer was a mixed population with a variety of chain lengths as revealed by polyacrylamide gel electrophoresis. The polymer had branching at a frequency of about once every 40 ADP-ribose residues. The average chain size was calculated as ∿100 ADP-ribose residues/polymer molecule in a 1 h incubation at 25°C.

Whether the automodification proceeds by an intramolecular mechanism (self-modification) or by an intermolecular mechanism (mutual modification) has not been investigated extensively. We approached this problem by examining the modification in the system containing either a free or immobilized enzyme that was active and one that had been inactivated by treatment with heat (60°C, 8 min) or N-ethylmaleimide (20 mM). Table 2 shows a typical result of analyses of products obtained by different combinations of active and inactive enzymes. Under neither conditions, was the modification of inactivated enzyme observed. Electrophoretic analyses of the products supported this view.

According to the generally accepted assumption [9] that enzyme molecules were immobilized separately from each other on the gel, these results, together with the fact that a considerable portion of the enzyme activity was retained through the immobilization procedure, suggested that poly(ADP-ribosyl)ation took place on the same

Table 2. No ADP-ribosylation of N-ethylmaleimide-treated poly(ADP-ribose) synthetase

		ADP-ribose incorporated		
Exp.	System	Total	Gel-bound	Soluble
		pmol	pmol	pmol
1.	Immobilized enzyme	464	434	30
	+ NEM-treated free enzyme[a]	458	434	24
2.	Free enzyme	696	–	696
	+ NEM-treated immobilized enzyme	673	13	660

[a] NEM, N-ethylmaleimide

molecule of enzyme, or automodification proceeded by an intramolecular mechanism, at least, under the conditions used. This conclusion appears to contrast with the finding of Kameshita et al. [10] and Holtlund et al. [11] that proteolytic fragments of the synthetase served as acceptors of poly(ADP-ribosyl)ation, that is suggestive of an intermolecular mechanism.

Concluding Remarks

In the present study, by making use of immobilized enzyme preparations, we presented evidence for the intrinsic NAD glycohydrolase activity associated with poly(ADP-ribose) synthetase and its manifestation by immobilization, and for the intramolecular mechanism of auto-poly(ADP-ribosyl)ation, including initiation, elongation, and branching of the polymer.

The immobilized enzyme is a very useful tool for studying many aspects of poly(ADP-ribosyl)ation reactions; it is physically stable, both in storage and reaction, and easily separable from soluble components, such as substrate, activator, inhibitor, exogenous acceptor, and soluble acceptor-bound or unbound products. By using this form of enzyme, isolation of, if any, intermediate(s) of the reaction, and thus analysis of the mechanism of enzyme action should be possible.

References

1. Ueda K, Kawaichi M, Oka J, Hayaishi O (1980) Biosynthesis and degradation of poly(ADP-ribosyl) histones. In: Smulson M, Sugimura T (eds) Novel ADP-ribosylations of regulatory enzymes and proteins. Elsevier/North-Holland, Amsterdam New York, pp 47–56
2. Ueda K, Kawaichi M, Hayaishi O (1982) Poly(ADP-ribose) synthetase. In: Hayaishi O, Ueda K (eds) ADP-ribosylation reactions: biology and medicine. Academic Press, New York London, pp 117–155
3. Ueda K, Hayaishi O (1985) ADP-ribosylation. Annu Rev Biochem 54:71–99

4. Kawaichi M, Ueda K, Hayaishi O (1981) NAD glycohydrolase activity of poly(ADP-ribose) synthetase. Seikagaku 53:877 (in Japanese)
5. Ueda K, Zhang J, Hayaishi O (1984) Poly(ADP-ribose) synthetase. Methods Enzymol 106: 500–504
6. Ito S, Shizuta Y, Hayaishi O (1979) Purification and characterization of poly(ADP-ribose) synthetase from calf thymus. J Biol Chem 254:3647–3651
7. Kawaichi M, Ueda K, Hayaishi O (1980) Initiation of poly(ADP-ribosyl) histone synthesis by poly(ADP-ribose) synthetase. J Biol Chem 255:816–819
8. Kawaichi M, Ueda K, Hayaishi O (1981) Multiple autopoly(ADP-ribosyl)ation of rat liver poly(ADP-ribose) synthetase. Mode of modification and properties of automodified synthetase. J Biol Chem 256:9483–9489
9. Chan WW-C (1977) Immobilized subunits. Methods Enzymol 44:491–503
10. Kameshita I, Matsuda Z, Taniguchi T, Shizuta Y (1984) Poly(ADP-ribose) synthetase. Separation and identification of three proteolytic fragments as the substrate-binding domain, the DNA-binding domain, and the automodification domain. J Biol Chem 259:4770–4776
11. Holtlund J, Lemtland R, Kristensen T (1983) Two proteolytic degradation products of calf-thymus poly(ADP-ribose) polymerase are efficient ADP-ribose acceptors. Implications for polymerase architecture and the automodification of the polymerase. Eur J Biochem 130: 309–314

The Domain Structure of Poly(ADP-Ribose) Synthetase and Its Role in Accurate Transcription Initiation

YUTAKA SHIZUTA[1], ISAMU KAMESHITA[1], MICHIKO AGEMORI[1], HIROSHI USHIRO[1], TAKETOSHI TANIGUCHI[1], MASAHIKO OTSUKI[2], KAZUHISA SEKIMIZU[2], and SHUNJI NATORI[2]

Introduction

Poly(ADP-ribose) synthetase, an enzyme localized in the nucleus of eukaryotic cells, catalyzes the polymerization of the ADP-ribose moiety of NAD to form poly(ADP-ribose) which is covalently bound to various nuclear proteins [1]. A unique feature of this enzyme is that it requires DNA for catalytic activity and that it is subjected to automodification during the reaction [2].

Recently, the enzyme has been purified to apparent homogeneity from various sources and extensively characterized [3]. Nevertheless, whether or not the difference of species, tissues, or organs causes differences of characteristic nature and function of the enzyme is not as yet fully clarified. In the present study, we attempted to elucidate the structural characteristics of the enzyme from three different sources and its function in accurate transcription initiation.

In this paper, we present evidence to show that the characteristic nature of the enzyme is common to various species and that the enzyme consists of three different functional domains, the first one of $M_r = 54,000$ possessing the site for the substrate binding, the second one of $M_r = 46,000$ retaining the site for DNA binding, and the third one of $M_r = 22,000$ containing the site for accepting poly(ADP-ribose). Furthermore, evidence is presented to show that the DNA-binding domain like the native enzyme preferentially suppresses random transcription initiation, resulting in the production of run-off RNA initiated from the correct initiation site on truncated DNA.

Procedures

Poly(ADP-ribose) synthetases were purified from calf thymus [4], mouse testis [5], and human placenta [6]. Proteolytic digestion was performed as described previously [7, 8]. 3-(Bromoacetyl)pyridine was prepared according to the method of Woenckhaus

---

1 Department of Medical Chemistry, Kochi Medical School, Kochi, 781-51, Japan
2 Faculty of Pharmaceutical Science, University of Tokyo, Tokyo 113, Japan

ADP-Ribosylation of Proteins
(ed. by F.R. Althaus, H. Hilz, and S. Shall)
© Springer-Verlag Berlin Heidelberg 1985

Table 1. Major characteristics of poly(ADP-ribose) synthetases

	Calf	Mouse	Human
M_r	120,000	120,000	120,000
$S_{20,w}$	5.0 S	–	4.6 S
$\bar{v}$	0.736	0.727	0.736
pI	9.8	–	10.0
Km for NAD	55 μM	47 μM	62 μM
Km for DNA	14 μg/ml	25 μg/ml	33 μg/ml

et al. [9]. Antisera to poly(ADP-ribose) synthetase were obtained by injecting BALB/C mice at multiple intradermal sites with 50–100 μg proteins emulsified with 50% Freund's complete adjuvant [10]. Booster injections were administered three times at 10-day intervals with 15–20 μg proteins. A HeLa cell lysate for RNA synthesis was prepared as described by Manley et al. [11]. The assay of accurate transcription was performed as described previously [12]. All other analytical methods were according to those previously described [7, 8].

Results and Discussion

Poly(ADP-ribose) synthetases were purified to apparent homogeneity from calf thymus, mouse testis, and human placenta. The major characteristics of these enzymes are presented in Table 1. In addition to molecular mass, sedimentation constant, isoelectric point, and partial specific volume, the apparent K_m for NAD and DNA as well as V_{max} of the reaction are all common to these three enzymes. Amino acid compositions of the enzymes are shown in Table 2. Here again, the numbers of each amino acid residue are very similar to each other, although some differences as denoted by star symbols are noted.

In order to investigate whether precursor proteins and/or endogenous degradation products of poly(ADP-ribose) synthetase in vivo exist, DNA-binding proteins were prepared from calf thymus, HeLa cells, mouse testis, and chicken liver and immuno-chemically stained with antisera against the calf enzyme after gel electrophoresis and transblotting (Fig. 1). In addition to the native enzyme of M_r = 120,000, proteins of M_r = 80,000 and M_r = 64,000 were detected in fresh calf thymus preparation. A faint band of M_r = 32,000 became dominant when the sample was prepared from aged calf thymus (data not shown). In contrast, degradation products of M_r = 32,000 in addition to the native enzyme of M_r = 120,000 were clearly detected in human, mouse, and chicken samples. These results indicate that the antigenic properties of poly(ADP-ribose) synthetases and their degradation products are all common to these animal species.

In the next experiments, we attempted to elucidate the domain structure of calf thymus enzyme by utilizing the technique of limited proteolysis. As shown in Fig. 2, we observed that mild digestion of the enzyme with papain led to the formation of

Y. Shizuta et al.

Table 2. Amino acid analysis of poly(ADP-ribose) synthetases

Amino acid	Calf	Mouse	Human
		mol/100 mol	
Lysine	12.10	10.70*	13.34
Histidine	2.23	2.18	2.06
Arginine	3.00*	3.80	3.43
Aspartic acid	10.30	9.10	9.94
Threonine	4.63	3.78*	4.48
Serine	7.57*	9.20*	8.46
Glutamic acid	13.10	12.30	10.96*
Proline	4.30	4.15	4.69
Glycine	9.07	10.50	7.43*
Alanine	7.20	7.83	6.94
Valine	6.57	6.30	6.22
Methionine	1.60	1.50	2.14*
Isoleucine	4.20	3.55*	4.61
Leucine	8.10	8.15	9.09
Tyrosine	2.97*	3.87	3.22
Phenylalanine	3.10	2.63	2.99

two different fragments of M_r = 74,000 and M_r = 46,000, whereas digestion with chymotrypsin resulted in the production of two fragments of M_r = 66,000 and M_r = 54,000. In order to elucidate the localization of the substrate binding site, the enzyme was modified with 3-(bromoacetyl)pyridine and then labeled with sodium boro[^{3}H]-hydride as shown in Fig. 3. When the labeled enzyme was subjected to SDS polyacrylamide gel electrophoresis after partial proteolytic digestion, the ratioactivity was

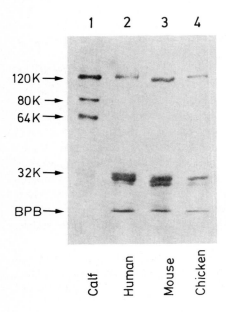

Fig. 1. Interspecies cross-reactivity of poly(ADP-ribose) synthetases as detected by immunoblotting. DNA binding proteins prepared from calf thymus (*lane 1*), HeLa cells (*lane 2*), mouse testis (*lane 3*), and chicken liver (*lane 4*) were immunostained with antisera against calf thymus poly(ADP-ribose) synthetase after SDS gel electrophoresis and transblotting

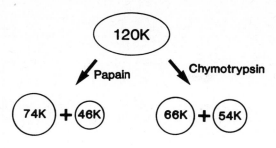

Fig. 2. Limited proteolysis of poly(ADP-ribose) synthetase

found in all of the chymotryptic fragments (M_r = 54,000 and 66,000) and the papain digests (M_r = 74,000 and 46,000) as shown in Fig. 4. Among these fragments, selective protection from the labeling was observed in the M_r = 54,000 and M_r = 74,000 fragments when the substrate, NAD, was included in the reaction mixture during alkylation. The covalent and nonspecific incorporation of tritium observed in the M_r = 66,000 and M_r = 46,000 fragments appears to be derived from acetylpyridine attached to some cysteinyl residues which are not required for the binding of the substrate, NAD. It is, therefore, concluded that the substrate binding site is localized on the M_r = 54,000 portion of the native enzyme.

In order to determine which fragments possess the DNA binding site and the automodification site, the enzyme was automodified with a low concentration of [*adenine-U-14C*]NAD and digested with chymotrypsin. The radioactivity was detected exclusively in the M_r = 66,000 fragment (Fig. 5). The M_r = 66,000 fragment thus labeled was further cleaved with papain into two fragments of M_r = 46,000 and M_r = 22,000. With these two fragments, the label was detected only in the M_r = 22,000 fragment, but not in the M_r = 46,000 fragment (Fig. 5).

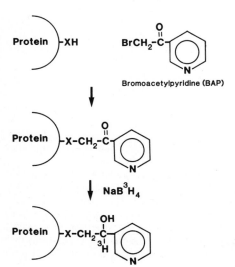

Fig. 3. Chemical modification of poly(ADP-ribose) synthetase with 3-(bromoacetyl)pyridine and NaB³H₄

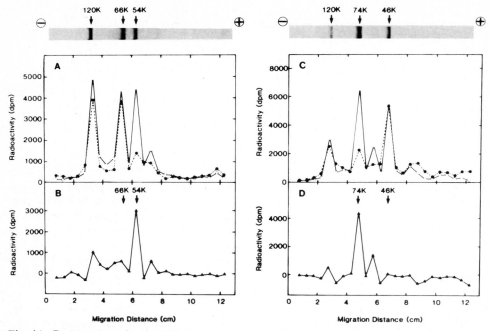

Fig. 4A–D. Digestion of poly(ADP-ribose) synthetase with chymotrypsin or papain after modification of 3-(bromoacetyl)pyridine in the presence and absence of NAD. The enzyme sample (25 μg), chemically modified in the presence (●) and absence (○) of NAD and labeled with NaB³H₄, was digested with 0.5 μg of α-chymotrypsin (*left*) or 0.5 μg of papain (*right*) and subjected to SDS gel electrophoresis. **A, C** Quantification of the radioactivity in SDS-gel; **B, D** the difference between the values denoted by ● and ○ in **A** and **C** was replotted

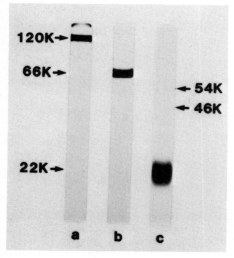

Fig. 5. Digestion of automodified poly(ADP-ribose) synthetase with α-chymotrypsin and with papain. The enzyme (40 μg) was automodified with 1.4 μM [*adenine*-U-¹⁴C]NAD (423 dpm/pmol) and then digested with α-chymotrypsin or further with papain. The radioactivity was detected by fluorography. *a* The automodified enzyme; *b* the enzyme after digestion with α-chymotrypsin; *c* the fragment obtained by digestion with papain after chymotryptic digestion

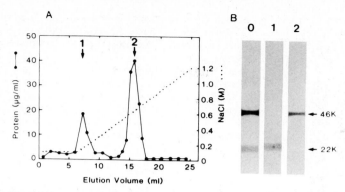

Fig. 6A,B. DNA-cellulose column chromatography of the enzyme fragment obtained by digestion of the M_r = 66,000 fragment with papain. **A** Elution profile of DNA-cellulose column chromatography; **B** SDS-polyacrylamide gel electrophoretic patterns showing the fragments obtained by digestion of the M_r = 66,000 fragment with papain (*0*), components of peak (*1*), and of peak (*2*), respectively

Using DNA-cellulose column chromatography, it was demonstrated that the M_r = 46,000 fragment binds to a DNA ligand with the same affinity as that of the native enzyme, while the M_r = 22,000 fragment and the M_r = 54,000 fragment have little affinity for the DNA ligand (Fig. 6). On the basis of these results, it is concluded that poly(ADP-ribose) synthetase consists of three separable domains and that the acceptor domain of M_r = 22,000 is located in between the two other domains for the binding of DNA and NAD. Figure 7 illustrates the above conclusion.

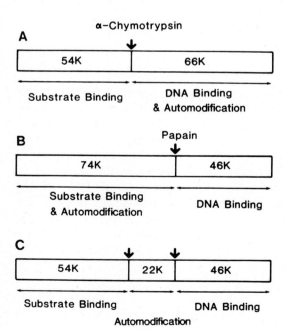

Fig. 7A–C. The domain structure of poly(ADP-ribose) synthetase. The chymotryptic and papain cleavage sites on the enzyme molecule are indicated by *arrows*

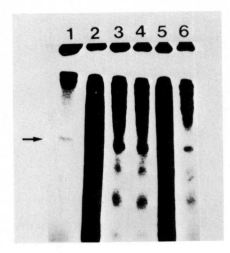

Fig. 8. Effect of poly(ADP-ribose) synthetase and its papain digested fragments on transcription in a HeLa cell lysate. Transcription was performed using 7.6 mg ml⁻¹ (*lane 1*) or 3.3 mg ml⁻¹ (*lanes 2 and 6*) of HeLa cell lysate with 0.8 μg of poly-(ADP-ribose) synthetase (*lane 3*), 0.8 μg DNA binding domain (*lane 4*), 0.8 μg autocatalytic domain (*lane 5*) and 0.8 μg each of both domains. The *arrow* indicates the position of the specific runoff transcript of 536 base RNA

In order to examine the role of the domain structure of poly(ADP-ribose) synthetase in RNA synthesis, the effect of papain digested fragments of the enzyme on accurate transcription initiation by RNA polymerase II was studied using a HeLa cell lysate and late promoter of adenovirus 2 as a template. Figure 8 shows that the DNA binding domain like the native enzyme, specifically suppresses random transcription initiation, resulting in the formation of the specific transcript of 536 base RNA, whereas the automodification and substrate binding domain, i.e., fragment of $M_r = 74,000$, has no effect.

Conclusions

Poly(ADP-ribose) synthetases have been purified to homogeneity from calf thymus, mouse testis, and human placenta. Amino acid compositions of these enzymes are very similar to each other and antisera against the calf thymus enzyme cross-react with mouse, chicken, and human enzymes. The native enzyme of $M_r = 120,000$ is cleaved by limited proteolytic digestion into three different domains, the first containing the site for DNA binding, the second containing the site for automodification, and the third containing the site for NAD binding. The DNA-binding domain ($M_r = 46,000$) thus cleaved, like the native enzyme, has the ability to preferentially suppress nick-induced random transcription initiation in a HeLa cell lysate, resulting in the production of runoff RNA initiated from the correct late promoter site on truncated DNA of adenovirus 2.

References

1. Hayaishi O, Ueda K (1977) Poly(ADP-ribose) and ADP-ribosylation of proteins. Annu Rev Biochem 46:95–116
2. Shizuta Y, Ito S, Nakata K, Hayaishi O (1980) Poly(ADP-ribose) synthetase from calf thymus. Methods Enzymol 66:159–165
3. Ueda K, Ogata N, Kawaichi M, Inada S, Hayaishi O (1982) ADP-ribosylation reactions. Curr Top Cell Regul 21:175–187
4. Ito S, Shizuta Y, Hayaishi O (1979) Purification and characterization of poly(ADP-ribose) synthetase from calf thymus. J Biol Chem 254:3647–3651
5. Agemori M, Kagamiyama H, Nishikimi M, Shizuta Y (1982) Purification and properties of poly(ADP-ribose) synthetase from mouse testicle. Arch Biochem Biophys 215:621–627
6. Ushiro H, Shizuta Y (1984) Purification and properties of human poly(ADP-ribose) synthetase. Seikagaku 56:674
7. Nishikimi M, Ogasawara K, Kameshita I, Taniguchi T, Shizuta Y (1982) Poly(ADP-ribose) synthetase. The DNA binding domain and the automodification domain. J Biol Chem 257: 6102–6105
8. Kameshita I, Matsuda Z, Taniguchi T, Shizuta Y (1984) Poly(ADP-ribose) synthetase. Separation and identification of three proteolytic fragments as the substrate-binding domain, the DNA-binding domain and the automodeification domain. J Biol Chem 259:4770–4776
9. Woenckhaus VC, Berghäuser J, Pfeiderer G (1969) Markierung essentieller Aminosäurereste der Lactat-Dehydrogenase aus Schweineherz mit [Carbonyl-^{14}C]3-[2-Brom-acetyl]-pyridin. Hoppe Seyler's Z Physiol Chem 350:473–483
10. Kameshita I, Yamamoto H, Fujimoto S, Shizuta Y (1984) Monoclonal antibody to poly(ADP-ribose) synthetase: its antigenic determinant and interspecies cross-reactivity. Seikagaku 56: 674
11. Manley JL, Fire A, Cano A, Sharp PH, Gefter ML (1980) DNA-dependent transcription of adenovirus genes in a soluble whole-cell extract. Proc Natl Acad Sci USA 77:3855–3859
12. Ohtsuki M, Sekimizu K, Agemori M, Shizuta Y, Natori S (1984) Effect of the DNA-binding domain of poly(ADP-ribose) synthetase on accurate transcription initiation in a HeLa cell lysate. FEBS Lett 168:275–277

DNA-Poly(ADP-Ribose) Polymerase Complex: Isolation of the DNA Wrapping the Enzyme Molecule

MARIE E. ITTEL[1], JENNY JONGSTRA-BILEN[2], CLAUDE NIEDERGANG[1], PAUL MANDEL[1], and ETIENNE DELAIN[3]

1. Introduction

Since the discovery of poly(ADP-ribose) it has been known that DNA is required for the synthesis of the polymer from NAD^+ [1, 2]. It was confirmed later with purified poly(ADP-ribose) polymerase that double-stranded DNA is necessary to express its activity in vitro [3–7]. Covalently-closed circular double-stranded DNA does not have any effect on the enzyme activity [7, 8]. It has been found by Ohgushi et al. [8] that the relative efficiency of enzyme activation by various DNA samples is closely related to their capacity to bind to the enzyme. These authors have suggested that poly(ADP-ribose) polymerase is activated by, and binds to, nicks or ends on the DNA. The highly purified calf thymus poly(ADP-ribose) polymerase preparation obtained in our laboratory [9, 10] contains a fraction of DNA, called sDNA, which has been shown to activate the enzyme more efficiently than total calf thymus DNA [6]. A DNA fraction called "active DNA" with high enzyme binding and activating capacity has also been isolated during the purification of bovine thymus poly(ADP-ribose) polymerase by Yoshihara and co-workers [5, 11]. The high affinity of this enzyme for the "active DNA" has been attributed to its small size and its high content of nicks [8].

Several lines of evidence indicate that poly(ADP-ribose) polymerase is autoADP-ribosylated during its catalytic reaction [12–14]. Moreover, the autoADP-ribosylation reaction of the purified enzyme has been followed by electron microscopy [15]. Auto-(ADP-ribosylated) enzyme molecules appear as dense structures stained by uranyl-acetate, increasing in size as a function of time. At prolonged reaction times, when the rate of the auto(ADP-ribosylation) reaction decreases, the detachment of the auto-ADP-ribosylated enzyme from sDNA occurs. These observations are in agreement with the biochemical investigations [14, 16], suggesting that automodification of the enzyme leads to a decrease of its affinity for DNA in parallel to a gradual decrease of the enzymatic activity [13, 16].

Considering the requirement of DNA in the poly(ADP-ribose) polymerase reaction we proceeded to a morphological investigation of the interaction of the purified

1 Centre de Neurochimie du CNRS, 5 rue Blaise Pascal, 67084 Strasbourg Cedex, France
2 Dept. of Neurobiology, Stanford University, School of Medicine, Stanford, CA 94305, USA
3 Laboratoire de Microscopie Cellulaire et Moléculaire, LA 147 du CNRS, Institut Gustave Roussy, 94805 Villejuif, France

ADP-Ribosylation of Proteins
(ed. by F.R. Althaus, H. Hilz, and S. Shall)
© Springer-Verlag Berlin Heidelberg 1985

enzyme with its activating copurified sDNA by electron microscopy [15]. We showed that the purified enzyme-DNA complex possesses a nucleosome-like structure with the DNA wrapped around the enzyme molecule. The wrapping of DNA around the histone core in nucleosomes has been successfully probed by nuclease digestion studies (for reviews, see [17–19]. This report describes our studies on the digestion of the DNA-poly(ADP-ribose) polymerase complex by micrococcal nuclease in an initial attempt to characterize the DNA fragment wrapping around the enzyme molecule.

2. Characterization of the Poly(ADP-Ribose) Polymerase-sDNA Complex by Electron Microscopy

Previous electron microscopic observations of shadowed preparations of the enzyme-sDNA complex have shown that poly(ADP-ribose) polymerase molecules appear as globular dark dots irregularly spaced on sDNA, with a small number of enzyme molecules localized at the extremities of the DNA fragments [15]. Moreover, the electron density of the enzyme deprived of its DNA is less than that of the DNA-bound enzyme and the diameter of poly(ADP-ribose) polymerase is statistically smaller in the DNA-deprived enzyme as compared to the DNA-bound enzyme. These results support the notion that in the poly(ADP-ribose) polymerase-sDNA complex, a segment of the DNA helix is wrapped around the enzyme molecule, giving rise to a nucleosome-like structure similar to that described for DNA gyrase–DNA complexes [20, 21].

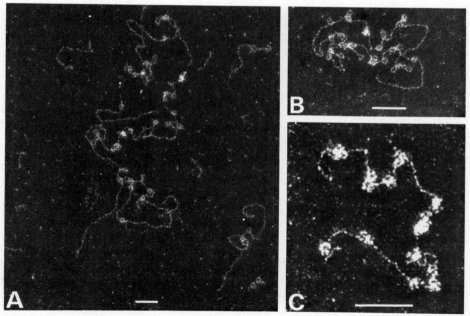

Fig. 1A–C. Dark field electron micrographs of positively stained DNA-bound enzyme at various magnifications showing the irregular DNA path around poly(ADP-ribose) polymerase molecules. Samples were spread as indicated in [15]. Bar represents 50 nm

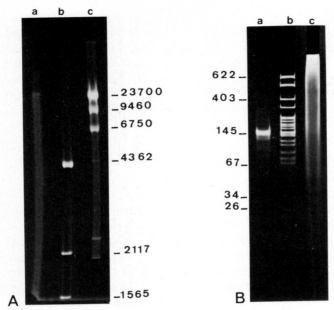

Fig. 2A,B. Analysis of sDNA by gel electrophoresis. **A** 1.4% agarose gel. Electrophoresis was per-formed for 16 h at 40 V in a horizontal slab gel apparatus [22] allowing determination of long sDNA fragments. *a* 2.5 μg sDNA isolated by hydroxylapatite chromatography [6]; *b* 160 ng pBR322 DNA digested with Eco RI and Rsa I; *c* 500 ng λ DNA digested with Hind III. The smaller DNA fragments are not visible since they have migrated off the gel. **B** 8% polyacrylamide Tris-borate-Mg^{+2} gel. Electrophoresis was performed at 110 V for 3 h as described in [23] allowing determina-tion of shorter sDNA fragments. *a* 600 ng purified nucleosome core particles DNA; *b* 1 μg pBR322 digested with Msp I; *c* 5 μg sDNA isolated by hydroxylapatite chromatography [6]. The lengths of some of the marker DNA fragments are indicated in base pairs

Figure 1 illustrates the morphological features of a poly(ADP-ribose) polymerase preparation observed by dark field electron microscopy. It appears that most of the uranyl acetate stain, which preferentially binds DNA, is located at the periphery of the enzyme-DNA complex indicating that the sDNA is wrapped around the enzyme molecule (Fig. 1A—C).

3. Analysis of sDNA

Examination of sDNA fragments by electron microscopy reveals a large heterogeneity in their size. Analysis on agarose (Fig. 2A) and polyacrylamide (Fig. 2B) gels stained with ethidium bromide of sDNA isolated by hydroxylapatite chromatography [6], confirm this observation. sDNA appears as a broad smear on both gels, except for a band corresponding to approx. 24,000 base pairs, which may be due to a lack of reso-lution in that part of the gel. The size of the sDNA fragments vary approx. between 24,000 and 26 base pairs (bp).

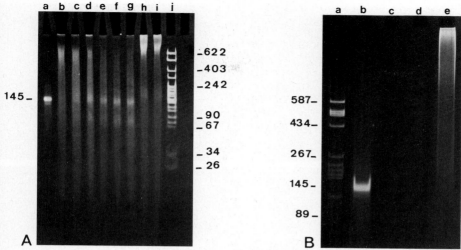

Fig. 3A,B. Micrococcal nuclease digestion of: A DNA-bound poly(ADP-ribose) polymerase; B sDNA, analyzed by polyacrylamide gels stained with ethidium bromide. A 8% polyacrylamide Tris-borate-Mg^{+2} gel [23]. Slot *a:* 100 ng purified nucleosome core particles DNA. Slots *b–i:* DNA bound poly(ADP-ribose) polymerase (60 μg ml^{-1}) was incubated without (slots *h–i*) or with (slots *b–g*) micrococcal nuclease (0.012 units ml^{-1}) at 30°C under conditions described in [20]. At different times of incubation: *b* 2 min, *c* 5 min, *d* 10 min, *e* 20 min, *f* 30 min, *g* 40 min; aliquots corresponding to 3 μg DNA-bound poly(ADP-ribose) polymerase were taken and DNA was purified by phenol extraction [24]; *j* marker DNA fragments generated after digestion of pBR322 DNA with MspI. B 8% polyacrylamide Tris-borate-EDTA gel [25]. Slot *a:* 150 ng pBR322 DNA digested with Hae III. Slot *b:* 500 ng purified nucleosome core particles DNA. Slots *c–e:* sDNA (12 μg ml^{-1}) digested with micrococcal nuclease under same conditions as described in **A**. At different times of incubation: *e* 0 min, *d* 5 min, *c* 10 min; aliquots corresponding to 2.4 μg sDNA were precipitated with ethanol before applying on polyacrylamide gel for 3 h at 120 V. The lengths of some of the marker DNA fragments are indicated in base pairs

4. Micrococcal Nuclease Digestion of the Poly(ADP-Ribose) Polymerase — sDNA Complex

4.1 Analysis by Gel Electrophoresis

The sDNA-poly(ADP-ribose) polymerase complex has been submitted to micrococcal nuclease digestions (for 2 to 40 min) and the phenol extracted DNA has been analyzed by an 8% polyacrylamide gel electrophoresis (Fig. 3). It appears that a fraction of the total sDNA is protected against micrococcal nuclease digestion in the enzyme-DNA complex (Fig. 3A), while protein deprived sDNA is digested totally under similar conditions (Fig. 3B). These results confirm the nucleosome-like structure of the sDNA-poly(ADP-ribose) polymerase complex observed by electron microscopy. As shown in Fig. 3A a DNA band of 160—140 bp appears within 5 min of digestion. Since the relative migration of this band is very similar to DNA purified from nucleosomal core particles which possess 145 bp, we shall refer to this protected fragment as the 145 bp fragment. A second broad band between 90—60 bp appears after 10 min of digestion.

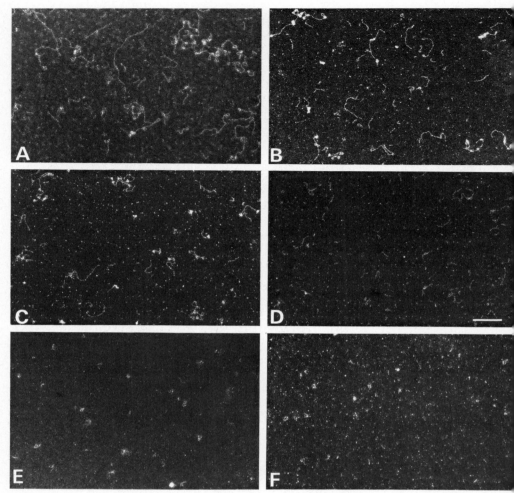

Fig. 4A–F. Visualization by dark field electron microscopy of the progressive micrococcal nuclease digestion of DNA-bound poly(ADP-ribose) polymerase [15]. DNA-bound poly(ADP-ribose) polymerase was incubated without **A** or with **B–F** micrococcal nuclease under the conditions described in Fig. 3. **B** 2 min; **C** 5 min; **D** 10 min; **E** 20 min, **A** and **F** 30 min. The bar represents 100 nm for all pictures

After longer digestion times a third faint band of about 300 bp becomes visible. In view of the fact that the 145 bp fragment appears more rapidly with increasing digestion time and that it is sharper than the lower DNA band (90–60 pb), it seems likely that the enzyme molecule protects DNA of 160–140 bp and that the lower DNA band (90–60 bp) derives from an additional cut in the 145 bp DNA. The faint fragment of about 300 bp appearing with prolonged incubation time may represent some aggregated structures.

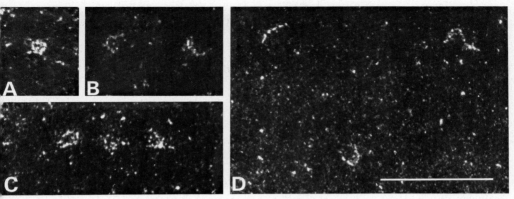

Fig. 5A–D. High magnification of DNA-bound poly(ADP-ribose) polymerase digested with micrococcal nuclease for 30 min. **A** shows an isolated complex with the same size and density as in undigested sample. No free DNA whiskers are detectable. **B** and **C** show some larger structures made of short bent DNA fragments with possible remnants of enzyme. In **D** the DNA is more extended and probably devoid of protein. The bar indicates 100 nm for all pictures

4.2 Electron Microscopic Observations

Micrococcal nuclease digestion of the poly(ADP-ribose) polymerase-sDNA complex realized in the same manner as for gel electrophoresis analysis, has also been followed by electron microscopic observations. Incubation without micrococcal nuclease does not change the aspect of the sDNA-poly(ADP-ribose) polymerase complex (Figs. 4A and 1) with its DNA heterogeneous in size and wrapping irregularly spaced enzyme molecules. As the incubation time proceeds (Fig. 4B–F) the DNA fragments become smaller; only very short segments remain after 20 min, most of them in a condensed form suggesting their binding to the enzyme molecule (Figs. 4F and 5). The length of the DNA fragments observed at the end of the digestion time course has been measured. Two populations of DNA have been determined: one of approximately the same number of base pairs (148 ± 21 bp) as the purified nucleosome core particle DNA (145 ± 5 bp) — which served as the reference — and another one of approx. 60 bp less (88 ± 10 bp). These results confirm the above findings on polyacrylamide gel analysis where two main bands were detected after micrococcal nuclease digestion performed under the same conditions.

5. Discussion

Present biochemical and electron microscopic studies provide information concerning the length of DNA fragments, protected against micrococcal nuclease digestions, of the poly(ADP-ribose) polymerase — sDNA complex. Two protected DNA fragments can be observed, approx. 145 bp and 90–60 bp, indicating the binding of the enzyme molecule to these DNA fragments. In view of these data, the following hypothesis can be postulated: (1) the enzyme molecule itself is surrounded by a DNA fragment of approx. 145 bp length, and the lower DNA band derives from an additional cut in the 145 bp DNA; (2) the shorter DNA fragment is indeed the unit of DNA protected by the enzyme molecule. (3) The enzyme–sDNA complex wrapped by the 145 bp DNA is split by a contaminating protease providing a fraction of the enzyme molecule which protects the shorter DNA fragment (90–60 bp). To test the third hypothesis, silver stained SDS–polyacrylamide gel electrophoresis has been performed on the enzyme–sDNA complex after 30 min digestion by micrococcal nuclease. Only one protein band at 120,000 daltons level, which corresponds to the molecular weight of the native poly(ADP-ribose) polymerase, has been detected with no bands at lower molecular weight levels (data not shown). These results suggest that proteolytic digestion does not occur. Electron microscopic observations and measurements of the DNA fragments obtained after micrococcal nuclease digestion of the enzyme-sDNA complex (Fig. 5) suggest that the enzyme molecule is stabilized by a DNA fragment of approx. 145 bp (Fig. 5A); when supplementary cuts appear in the surrounding DNA this structure is destroyed progressively (Fig. 5B–D) leading to shorter DNA fragments. These conclusions are also favored by the results obtained on polyacrylamide gel electrophoresis (Fig. 3) showing that the 145 bp DNA fragment appears more rapidly and sharper, while the lower, diffuse DNA band becomes visible with increasing digestion time.

Dark field micrographs of the undigested stained poly(ADP-ribose) polymerase-sDNA complex (Fig. 1) show round particles with the DNA wrapping a central core corresponding to the enzyme molecule. The direct observation of the enzyme–sDNA complexes is not sufficient to know the actual disposition of DNA surrounding the enzyme protein. Moreover electron microscopic observations of the undigested poly-(ADP-ribose) polymerase–sDNA complex (Fig. 1) show that very few enzyme molecules are located at the ends of the sDNA fragments as has been shown previously for autoADP-ribosylated enzyme molecules [15]. These results are in contradiction with those of Benjamin and Gill [7] who reported that double-stranded restriction fragments of pBR322 plasmid DNA are more efficient in activating poly(ADP-ribose) polymerase than circular pBR322 DNA in which nicks have been introduced by DNAse I. An explanation for this discrepancy may be that Benjamin and Gill [7] used a partially purified calf thymus poly(ADP-ribose) polymerase preparation. It is possible that the different activation levels of the enzyme by different types of breaks observed by these authors are the result of some preferential binding of a contaminant protein to internal DNA nicks, making less nicks available for poly(ADP-ribose) polymerase activation.

Preliminary results indicate that the DNA purified after micrococcal nuclease digestion of the enzyme–DNA complex activates more efficiently the DNA-free poly(ADP-ribose) polymerase than the total purified sDNA and much more than the purified nucleosomal core particles DNA (data not shown). Ohgushi et al. [8] and Benjamin and Gill [7] correlate the activation of poly(ADP-ribose) polymerase only to its binding to nicks or ends and not to a specific DNA sequence. Although our observations of poly(ADP-ribose) polymerase–sDNA complexes by electron microscopy and by polyacrylamide gel analysis do not exclude the possibility that the enzyme is activated preferentially by internal nicks on sDNA fragments, they raise the question whether some specific DNA sequences in sDNA fragments may be responsible for the strong activation of poly(ADP-ribose) polymerase by sDNA. The characterization of the isolated DNA fragment which wraps around the poly(ADP-ribose) polymerase molecule may elucidate the mechanism of activation of poly(ADP-ribose) polymerase by DNA.

Acknowledgments. The authors would like to thank F. Hog for excellent technical assistance and S. Ott for her fine typing of the manuscript.

References

1. Chambon P, Weill JD, Mandel P (1963) Nicotinamide mononucleotide activation of a new DNA-dependent polyadenylic acid synthesizing nuclear enzyme. Biochem Biophys Res Commun 11:39–43
2. Chambon P, Weill JD, Doly J, Strosser MT, Mandel P (1966) On the formation of a novel adenylic compound by enzymatic extracts of liver nuclei. Biochem Biophys Res Commun 25:638–643
3. Yoshihara K (1972) Complete dependency of poly(ADP-ribose) synthesis on DNA and its inhibition by actinomycine D. Biochem Biophys Res Commun 47:119–125
4. Okayama H, Edson CM, Fukushima M, Ueda K, Hayaishi O (1977) Purification and properties of poly(adenosine diphosphate ribose) synthetase. J Biol Chem 252:7000–7005
5. Yoshihara K, Hashida T, Tanaka Y, Ogushi H, Yoshihara H, Kamiya T (1978) Bovine thymus poly(adenosine diphosphate ribose) polymerase. J Biol Chem 253:6459–6466
6. Niedergang C, Okazaki H, Mandel P (1979) Properties of purified calf thymus poly(adenosine diphosphate ribose) polymerase. Eur J Biochem 102:43–57
7. Benjamin RC, Gill DM (1980) Poly(ADP-ribose) synthesis in vitro programmed by damaged DNA. J Biol Chem 255:10502–10508
8. Ohgushi H, Yoshihara K, Kamiya T (1980) Bovine thymus poly(adenosine diphosphate ribose) polymerase. J Biol Chem 255:6205–6211
9. Okazaki H, Niedergang C, Mandel P (1976) Purification and properties of calf thymus poly-adenosine diphosphate ribose polymerase. FEBS Lett 62:255–258
10. Mandel P, Okazaki H, Niedergang C (1977) Purification and properties of calf thymus poly-adenosine diphosphate ribose polymerase. FEBS Lett 84:331–336
11. Hashida T, Ohgushi H, Yoshihara K (1979) Highly effective DNA in stimulating poly(ADP-ribose) polymerase reaction. Biochem Biophys Res Commun 88:305–311
12. Okazaki H, Niedergang C, Mandel P (1980) Adenosine diphosphate ribosylation of histone H1 by purified calf thymus polyadenosine diphosphate ribose polymerase. Biochimie 62:147–157
13. Kawaichi M, Ueda K, Hayaishi O (1981) Multiple auto-poly(ADP-ribosylation) of rat liver poly(ADP-ribose) synthetase. J Biol Chem 256:9483–9489
14. Yoshihara K, Hashida T, Tanaka Y, Matsunami N, Yamagushi A, Kamiya T (1981) Mode of enzyme-bound poly(ADP-ribose) synthesis and histone modification by reconstituted poly-(ADP-ribose) polymerase-DNA-cellulose complex. J Biol Chem 256:3471–3478
15. De Murcia G, Jongstra-Bilen J, Ittel ME, Mandel P, Delain E (1983) Poly(ADP-ribose) polymerase automodification and interaction with DNA: electron microscopic visualization. EMBO J 2:543–548
16. Zahradka P, Ebisuzaki K (1982) A shuttle mechanism for DNA-protein interactions. Eur J Biochem 127:579–585
17. Kornberg RD (1977) Structure of chromatin. Annu Rev Biochem 46:931–954
18. Felsenfeld G (1978) Chromatin. Nature (London) 271:115–121
19. McGhee JD, Felsenfeld G (1980) Nucleosome structure. Annu Rev Biochem 49:1115–1156

20. Liu LF, Wang JC (1978) DNA-DNA gyrase complex: the wrapping of the DNA duplex outside the enzyme. Cell 15:979–984
21. Moore CL, Klevan L, Wang JC, Griffith JD (1983) Gyrase-DNA complexes visualized as looped structures by electron microscopy. J Biol Chem 258:4612–4617
22. Jongstra-Bilen J, Ittel ME, Niedergang C, Vosberg HP, Mandel P (1983) DNA topoisomerase I from calf thymus is inhibited in vitro by poly (ADP-ribosylation). Eur J Biochem 136:391–396
23. Klevan L, Wang JC (1980) Deoxyribonucleic acid gyrase-deoxyribonucleic acid complex containing 140 base pairs of deoxyribonucleic acid and an $\alpha 2\beta 2$ protein core. Biochemistry 19:5229–5234
24. Maniatis T, Fritsch EF, Sambrook J (1982) Molecular cloning, a laboratory manual. Cold Spring Harbor Laboratory, New York, p 458
25. Maniatis T, Jeffrey A, Van de Sande H (1975) Chain length determination of small double and single stranded DNA molecules by polyacrylamide gel electrophoresis. Biochemistry 14:3787–3794

The Role of Lysine Residues in the Catalytic Function and DNA Binding of Poly(ADP-Ribose) Polymerase as Determined by the Covalent Modification of the Enzyme Protein with Methyl Acetimidate

PAL I. BAUER and ERNEST KUN[1]

Introduction

Molecular mechanisms of protein–poly(ADP-ribosylation) require clarification at the level of the purified enzyme and in chromatin. The present paper is concerned with control reactions operative in the purified enzyme protein which depend on the coenzymic function of a copurified DNA species [1]. It is known that the polymerase enzyme has two catalytic functions: abortive or extra NAD-glycohydrolase and polymerase activities [2], and the enzyme can generate mono(ADP-ribose) and polymeric protein adducts. The correlation between these two distinct reactions, associated with the same enzyme protein, and the regulation of the chain length of polymers are not understood. We demonstrate that selective blocking of the lysine residues of the enzyme protein by methyl acetimidate [3] inactivates the polymerase reaction, but mono(ADP-ribose) protein adducts continue to be generated, simultaneously with a decrease in the binding of DNA to the enzyme, a binding which is essential for the polymerization process. The polymerase activity is more sensitive to covalent modification of the lysine residues than the extra NAD-glycohydrolase activity. The experimental evidence indicates that lysine end-groups participate in enzyme catalysis, but not in ADP-ribose binding to the protein. Lysine residues, which in globular proteins are generally located on molecular surfaces [4], play a critical polymer chain length regulatory function in poly(ADP-ribosylation) by way controling DNA binding to the enzyme.

Procedures

Poly(ADP-ribose) polymerase (95% homogeneous) and coenzymic DNA were isolated from calf thymus [1]. Its specific catalytic activity was 1066 nmol ADP-ribose incorporated per mg protein in 1 min. Details of isolation, DNA-cellulose binding as well as labeling of the enzyme with ^{125}I, a procedure that does not alter catalytic activity, are

---

1 Department of Pharmacology, Biochemistry and Biophysics and the Cardiovascular Research Institute, The University of California, San Francisco, CA 94143, USA

ADP-Ribosylation of Proteins
(ed. by F.R. Althaus, H. Hilz, and S. Shall)
© Springer-Verlag Berlin Heidelberg 1985

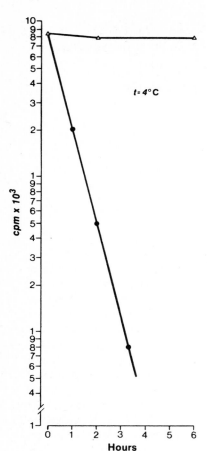

Fig. 1. Inactivation of the DNA dependent catalytic activity of poly(ADP-ribose) polymerase by methyl acetimidate

reported elsewhere [5]. NAD-glycohydrolase was determined by isolation of reaction products on PEI-cellulose with 0.1 M K-phosphate (pH = 3.4), or 0.9 M HAc, 0.3 M LiCl as developers followed by autoradiography and also direct radiochemical analysis. The lysine NH_2 groups were quantitatively determined by ^{125}I-labeled Bolton-Hunter reagent, and the extent of prior modification by methyl acetimidate was directly calculated (25 μg protein contained 24.5 pmol NH_3^+ groups). It is known that the polymerase protein contains 117 lysine residues and only one nonlysine end amino group [7]. Modification of lysine residues by methyl acetimidate was carried out at pH 9.2 and excess reagent removed by dialysis. The enzyme protein was covalently bound to Affigel 10 matrix in order to obtain matrix-bound enzyme (150 μg protein ml^{-1} resin). This bound enzyme was catalytically unstable (t1/2 = 1 week, -20°C). AMP, ADPR, PR-AMP, and poly(ADP-ribose) were quantitatively determined by high performance liquid chromatography [6]. Chain length analysis was carried out after phosphodiesterase digestion by HPLC [6] or by TLC separation of products.

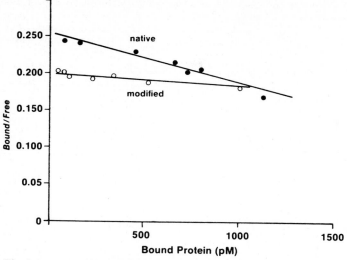

Fig. 2. Decrease in the DNA binding of the lysine modified enzyme compared to the native enzyme

Results

Figure 1 illustrates the inactivation of the DNA dependent catalytic activity of poly-(ADP-ribose) polymerase, which follows typical first-order kinetics ($t1/2 - 0.518$ h, $K_I = 1.34$ h^{-1}). Figure 2 shows the decrease in the binding of the lysine modified enzyme compared to the native enzyme (both labeled with ^{125}I) to DNA cellulose from $K_D = 15.7 \pm 1$ to K_D 47.7 ± 2 nM. Figure 3 illustrates that the modified enzyme can serve as polymer acceptor, when incubated with native polymerase, and the quantity of polymer protein adducts is proportional to the amount of lysine modified enzyme protein.

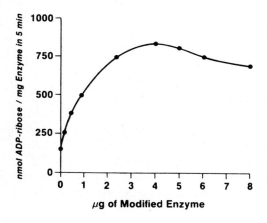

Fig. 3. Modified enzyme as polymer acceptor

Table 1. Enzymatic products formed by native and modified poly(ADP-ribose) polymerase and the effects of coenzymic DNA

	ADP-ribosylation (n mol mg^{-1} protein h^{-1})		Average chain length of oligomers
	mono(ADP-ribose)	poly(ADP-ribose)	
1. Poly(ADP-ribose) polymerase activity in the absence of DNA	41.6	9.96	6.1
2. Poly(ADP-ribose) polymerase activity in the presence of DNA	86.2	459.7	52.0
3. Methyl acetimidate modified poly(ADP-ribose) polymerase without DNA	26.3	0.89	2.9
4. Acceptor properties of modified enzyme protein incubated with DNA and matrix bound enzyme	37.6	2.2	3.9
5. Auto(ADP-ribosylation) of the matrix-bound poly(ADP-ribose) polymerase in the presence of DNA	6.9	13.8	14.2

The conversion of polymer to monomer formation is shown in Table 1. No. 1 indicates the mono- and poly(ADP-ribose) formation without added DNA and No. 2 the effect of DNA, both with the native enzyme. No. 3: The blocking of lysine end groups results in drastic decrease in polymer, but little change in monomer formation. In No. 4, matrix bound polymerase was incubated with lysine modified enzyme, and after the enzymatic reaction the modified free protein was reisolated and products analyzed. Similar to Nr. 3, monomer formation was close to the rate catalyzed by the native enzyme without DNA (No. 1), but oligomer formation was low. The matrix bound enzyme with added DNA synthesized relatively small quantities of polymers (No. 5), demonstrating that the physical state of the enzyme itself can modify the nature of products.

Conclusions

Lysine end groups of the polymerase protein are essential for the binding of coenzymic DNA. When these amino groups are modified by methyl acetimidate polymerization is inhibited, but the enzyme still can generate mono(ADP-ribose) adducts and possesses extra NAD-glycohydrolase activity. Modification of lysine end groups by, e.g., mono-alkylating drugs or by Schiff base formation with ADP-ribose [8,9], are likely to accomplish similar catalytic alterations of the enzyme as described here for methyl acetimidate. Physiological regulation of the polymer chain length in the nucleus may occur by enzymatic acetylation – de-acetylation of lysine end groups of the polymerase.

Acknowledgments. This work was supported by Grant F49620-81-C-0007 from the Office of Scientific Research of the United States Air Force and by Grant HL-27317 from the United States Public Health Service. P.I.B. is visiting scientist from the Semmelweiss University, 2nd Institute of Biochemistry, Budapest and E.K. is a recipient of the Research Career Award of the U.S. Public Health Service.

References

1. Yoshihara K, Hashida T, Tanaka Y, Ohgushi H, Yoshihara H, Kamiya T (1978) Bovine thymus poly(adenosine diphosphate ribose) polymerase. J Biol Chem 153:6459–6466
2. Kawaichi M, Ueda K, Hayaishi O (1981) Multiple autopoly(ADP-ribosyl)ation of rat liver poly(ADP-ribose) synthetase. J Biol Chem 256:9483–9489
3. Hunter MJ, Ludwig ML (1972) Amidination. Methods Enzymol 25:585–596
4. Stryer L (1968) Implication of X-ray crystallographic studies of protein structure. Annu Rev Biochem 37:25–50
5. Bauer PI, Hakam A, Kun E (to be published) The association of benzamide with DNA
6. Hakam A, McLick J, Kun E (1984) Separation of poly(ADP-ribose) by high performance liquid chromtography. J Chromatogr 296:369–377
7. Ito S, Shizuta Y, Hayaishi O (1978) Purification and characterization of poly(ADP-ribose) synthetase from calf thymus. J Biol Chem 258:3647–3651
8. Kun E, Chang ACY, Sharma MC, Herro AM, Nitecki D (1984) Covalent modification of proteins by metabolites of NAD⁺. Proc Natl Acad Sci USA 73:3131–3135
9. Hilz H, Koch R, Fanick W, Klapproth K, Adamietz P (1984) Nonenzymic ADP-ribosylation of specific mitochondrial polypeptides. Proc Natl Acad Sci USA 81:3929–3933

ADP-Ribosyltransferase from Hen Liver Nuclei: Purification, Properties, and Evidence for ADP-Ribosylation-Induced Suppression of Cyclic AMP-Dependent Phosphorylation

MAKOTO SHIMOYAMA, YOSHINORI TANIGAWA, TAKAHISA USHIROYAMA, MIKAKO TSUCHIYA, and RYOJI MATSUURA[1]

Introduction

ADP-ribosyltransferase catalyzes the transfer of the ADP-ribose moiety of NAD to acceptors, as arginine and other guanidino compounds and proteins, and forms mono-(ADP-ribose)-acceptor adducts. In eukaryotes, this enzyme was first detected in turkey erythrocyte by Moss and associates who went on to purify and characterize the enzyme [1].

In ontogenic studies of the ADP-ribosylation reactions, we observed that ADP-ribosyltransferase activity in the adult hen liver nuclei was much higher than in the embryo nuclei [2]. We purified this enzyme from hen liver nuclei [3] and noted the ADP-ribosylation-dependent suppression of protein phosphorylation [4, 5]. This article is an account of our recent studies on ADP-ribosyltransferase of hen liver nuclei and the novel properties of ADP-ribosylation-induced suppression of cAMP-dependent phosphorylation are described.

Purification and Some Properties of ADP-Ribosyltransferase from Hen Liver Nuclei

The hen liver nuclei contain both ADP-ribosyltransferase and poly(ADP-ribose) synthetase. To separate the ADP-ribosyltransferase from poly(ADP-ribose) synthetase, the 0.6 M potassium chloride extract from the nuclei was applied to a Sephadex G-200 column and eluted with the 0.1 M Tris-buffer, pH 8.0. Each fraction was incubated with 1 mM [adenine-^{3}H]NAD and 100 μg of whole histones, in a total volume of 0.2 ml containing 50 mM Tris-Cl$^-$ buffer, pH 9.0, and the radioactivity in the acid-insoluble fraction was determined. The result shows the two fractions containing the enzyme activities which catalyze the incorporation of the ADP-ribose moiety from NAD to the whole histones. From the hydroxyapatite column chromatographic analysis of the products formed by the respective fraction, we found that the former fraction contains poly(ADP-ribose) synthetase and the latter fraction contains poly-

---

1 Department of Biochemistry, Shimane Medical University, Izumo 693, Japan

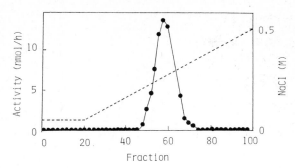

Fig. 1. Chromatography on a DNA-Sepharose 4B column. For enzyme assay each fraction was incubated with 1 mM [adenine-^{3}H]NAD and 100 μg of whole histones in a total volume of 0.2 ml (pH 9.0). Incubation was carried out at 25°C for 30 min. ●—● enzyme activity; --- NaCl concentrations. (Data from [3])

(ADP-ribose) synthetase and the latter fraction involves ADP-ribosyltransferase. To clarify that the radioactive compound formed by the latter fraction was indeed the mono(ADP-ribose) molecule, the acid-insoluble reaction product was treated with alkali at 37°C for 2 h. The radioactive material solubilized was adjusted to pH 7.0 and subjected to high performance liquid chromatography with reverse phase column. The eluate was monitored by UV and fractionated and radioactivity of the fraction was measured. The retention time of the radioactive product coincided with that of authentic mono(ADP-ribose). Furthermore, by treatment with snake venom phosphodiesterase this radioactive peak, tentatively considered to be ADP-ribose, migrated to the position corresponding to the 5'-AMP. These results indicate that hen liver nuclei contain ADP-ribosyltransferase. We purified this enzyme to a homogeneous state through salt extraction, gel filtration, hydroxyapatite, phenyl-Sepharose, Cm-cellulose, and DNA-Sepharose [3].

The last step of purification by DNA-Sepharose 4B column chromatography is shown in Fig. 1. The enzyme was eluted at approximately between 0.19 and 0.25 M sodium chloride, as a single peak. The finding that the enzyme protein binds to the DNA-Sepharose indicates that the ADP-ribosyltransferase may bind DNA in the nuclei.

A summary of a typical purification is shown in Table 1. An approximate 620-fold purification was achieved from the gel filtration of the initial extract of the nuclei with a recovery of 36.4%. The specific activity is three orders of magnitude less than that obtained with the purified turkey erythrocyte ADP-ribosyltransferase, as reported by Moss et al. [1].

We tested the purity of the enzyme obtained from the last purification step. When 5 μg of the enzyme protein were analyzed by SDS-polyacrylamide gel electrophoresis and stained with Coomassie blue, only one protein band was observed. The apparent molecular weight was estimated to be 27,500 ± 500. The optimum pH was observed at 9.0, when 100 μg of whole histones in a total reaction mixture of 0.2 ml were used as the acceptor protein. At each pH tested, the rate of nicotinamide release was somewhat higher than the rate of ADP-ribose incorporation into the acceptor proteins.

Next, we compared the acceptor specificities with regard to the release of nicotinamide and formation of ADP-ribose-protein adducts. As can be seen in Table 2, in addition to histones, protamine, casein, phosvitin, poly-L-arginine, agmatine, and arginine methyl ester also served as acceptors for the ADP-ribosylation. The concentrations of each acceptor were adjusted to give the maximum ADP-ribosylation. With

Table 1. Purification of ADP-ribosyltransferase from hen liver nuclei[a]

Fraction	Total protein	Total activity	Specific activity	Yield	Fold purification
	mg	μmol h^{-1}	μmol mg^{-1} h^{-1}	%	
Sephadex G-200 of nuclear KCl extract	18.9	0.66	0.03	100	1
Hydroxyapatite	2.5	0.49	0.20	74.2	6.7
Phenyl-Sepharose	0.9	0.43	0.46	66.5	15.2
Cm-cellulose	0.1	0.31	3.19	47.4	106.4
DNA-Sepharose	0.013	0.24	18.60	36.4	620.1

[a] The assays were carried out with [adenine-^{3}H]NAD as substrate in the presence of 100 μg of whole histones.
The purified nuclei (600 mg of DNA) prepared from 350 g (wet wt) hen liver nuclei were extracted with medium containing 0.6 M KCl, and the extract was applied to a Sephadex G-200 column to separate the ADP-ribosyltransferase from poly(ADP-ribose) synthetase. Assay conditions were the same as the legend to Fig. 1. (Data from [3])

Table 2. Comparison of acceptor specificities on the release of nicotinamide and formation of ADP-ribose-acceptor adducts[a]

Acceptors	ADP-ribose-acceptor adducts formed (b)	Nicotinamide released (a)	a/b
	nmol h^{-1}	nmol h^{-1}	
Whole histones (100 μg)	3.63	5.12	1.41
H1 (50 μg)	1.66	2.37	1.43
H2a (100 μg)	3.53	3.89	1.10
H2b (100 μg)	2.79	4.16	1.49
H3 (50 μg)	4.85	5.58	1.15
H4 (50 μg)	2.57	3.12	1.25
Protamine (20 μg)	2.28	3.76	1.65
Casein (400 μg)	2.08	3.67	1.76
Phosvitin (140 μg)	0.39	0.69	1.76
Poly-L-arginine (50 μg)	2.73	3.49	1.28
Agmatine sulfate (6.25 mM)	1.60	2.94	1.84
Arginine methyl ester (75 mM)	1.15	2.26	1.97

[a] The reaction mixture containing the purified enzyme (0.097 μg), [carbonyl-^{14}C]- or [adenine-^{3}H]NAD and the indicated amount of acceptors. Nicotinamide release was determined by thin layer chromatography. Other conditions were the same as given in the legend to Fig. 1. (Data from [3])

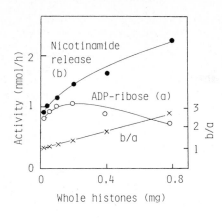

Fig. 2. Effect of increasing concentrations of whole histones on the release of nicotinamide and formation of ADP-ribose-histone adducts. Assay conditions were the same as described in the legend to Table 2 except the concentrations of whole histones. ●—● nicotinamide release; ○—○ ADP-ribose-histone adducts formed; x—x ratio of nicotinamide release to ADP-ribosylation. (Data from [3])

all the acceptors tested, the rate of nicotinamide release was in excess of the ADP-ribosylation and the ratio of nicotinamide release to the ADP-ribosylation varied with the acceptor. These results are basically similar to findings in the case of turkey erythrocyte ADP-ribosyltransferase reported by Moss et al. [1]. However, changes in the ratio of nicotinamide release to ADP-ribosylation seem to depend on the concentrations of the acceptor used. For example, as shown in Fig. 2, nicotinamide release increased when the concentrations of whole histones were increased. In the presence of acceptor proteins at concentrations of over 100 μg of whole histones in 0.2 ml reaction mixture, the rate of ADP-ribosylation of the acceptor proteins decreased. As a result, the ratio of nicotinamide release to the ADP-ribosylation increased. To obtain evidence that changes in the ratio of nicotinamide release to the ADP-ribosylation were not due to the decrease in the recovery of radioactive product with increasing concentrations of acceptor proteins, additional protein was included in the reaction mixture, before and after incubation in the presence of 100 μg of whole histones as the acceptor. Further addition of the histones after incubation did not alter the recovery of the reaction product. Similar results were also obtained with protamine. We confirmed that high concentrations of histones contained no activity of NADase, under the conditions used for ADP-ribosyltransferase assay. These results indicate that increase in the rate of nicotinamide release to the ADP-ribosylation was not due to changes in the recovery of ADP-ribose-histone adducts. Thus, the ratio of nicotinamide release to the ADP-ribose-acceptor protein conjugate formation increased with increasing concentrations of the acceptor proteins.

Determinations of Number and Site of the ADP-Ribose Molecule Linked to Histone H1

With prolonged incubation, we detected that 1 mol of ADP-ribose can be incorporated per mole of histone H1. [^{32}P]ADP-ribosylated histone H1 thus prepared was treated with N-bromosuccinimide and then subjected to gel filtration. The radioactivity was recovered in the fraction corresponding to the N2 fragment reported by Bustin and

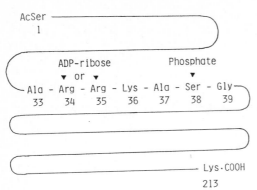

Fig. 3. Sites for ADP-ribosylation and phosphorylation of histone H1

Cole [6]. The N2 fragment is known to contain the cAMP-dependent phosphorylation site, serine residue 38 [7]. We confirmed this with cAMP dependently phosphorylated and subsequently N-bromosuccinimide-treated histone H1. The [^{32}P]ADP-ribosylated N2 fragment was further treated with three peptidases in the order of cathepsin D, aminopeptidase M, and carboxypeptidase B, and the product was analyzed by high performance liquid chromatography. The radioactive product was identified as ADP-ribose-arginine adduct [8].

An aliquot of radioactive ADP-ribosylated histone H1 was digested with cathepsin D and then TPCK-trypsin and the digested material was subjected to high performance liquid chromatography, and the amino acid composition of the peptide linking [^{32}P]-ADP-ribose was determined. The peptide around the ADP-ribosylated arginine is composed of lysine, proline, alanine, arginine, and glycine with a ratio of 1:1:2:2:1. Dansylation of this peptide showed the N-terminal amino acid to be lysine. We then searched for the corresponding peptide consisting of this amino acid composition and lysine as the N-terminal amino acid. This composition agreed with the amino acid sequence from lysine 29 to arginine 35 of the histone H1. We propose that among the two arginine residues 34 and 35, one arginine may be ADP-ribosylated, because 1 mol of ADP-ribose moiety was incorporated per 1 mol of histone H1 [8].

ADP-Ribosylation of Histone H1 Blocks Phosphorylation of the Histone

The data described above suggest that the ADP-ribosylated arginine may be sequenced near serine residue 38 which is phosphorylated by cAMP-dependent protein kinase (Fig. 3). In the experiment of protein kinase-catalyzed phosphorylation of synthetic peptide, Zetterqvist et al. [9] and Kemp et al. [10] proposed that two arginine residues are necessary at the N-terminal side of phosphate-accepting serine residue. Therefore, we speculated that the ADP-ribosylation may influence the cAMP-dependent protein kinase activity. Actually, we found that when whole histones from calf thymus were

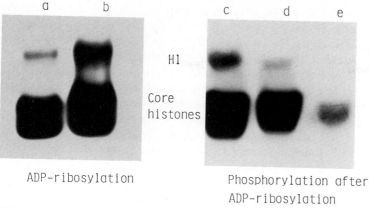

Fig. 4. Acid-urea polyacrylamide gel electrophoresis of ADP-ribosylated and/or phosphorylated whole histones. Whole histones (400 μg) from calf thymus were incubated with 1 mM [^{32}P]NAD and 1.6 μg purified ADP-ribosyltransferase at 25°C for 30 min and 60 min, respectively. The ADP-ribose-histone adducts were isolated, analyzed by acid-urea polyacrylamide gel electrophoresis and autoradiographed. The ADP-ribosylated histones prepared as above (except that the unlabeled NAD was used) were incubated with the purified catalytic subunit of cAMP-dependent protein kinase and [^{32}P]ATP at 30°C for 15 min. The phosphorylated samples were subjected to electrophoresis and autoradiography. *a, b* Histones incubated with [^{32}P]NAD for 30 and 60 min, respectively; *c–e* histones incubated with unlabeled NAD for 0, 30, and 60 min, respectively, and then incubated with [^{32}P]ATP. (Data from [5])

incubated with NAD and purified ADP-ribosyltransferase, the phosphorylation by cAMP-dependent protein kinase was suppressed. Figure 4 shows the results of autoradiography of the acid-urea polyacrylamide gel electrophoresis of the ADP-ribosylated and subsequently phosphorylated histones. The suppression of phosphorylation was detected with both ADP-ribosylated H1 and core histones with a prolonged incubation time for ADP-ribosylation (Fig. 4, lanes c–e). With [^{32}P]NAD, we confirmed that the ADP-ribosylation occurred in both H1 and core histones and was dependent on the incubation time (Fig. 4, lanes a and b). To investigate the nature of ADP-ribosylation-dependent suppression of phosphorylation, double reciprocal plots of protein kinase for H1 in the presence and absence of ADP-ribosylated histone H1 were determined, respectively, and we observed that ADP-ribosylated histone H1 did inhibit the phosphorylation of histone H1, in a competitive manner. Therefore, we presume that the ADP-ribosylation of arginine residue 34 or 35 in the histone H1 becomes a competitive inhibitor for the phosphorylation [8].

ADP-Ribosylation of Phosphorylase Kinase and Pyruvate Kinase and Kemptide (Synthetic Peptide) and Its Effect on Their Phosphorylation

We extended our research to see whether ADP-ribosylation occurs in certain enzymes, the activities of which are regulated by phosphorylation and dephosphorylation and whether the phosphorylation of the enzymes is influenced by ADP-ribosylation. An example is phosphorylase kinase. We purified this enzyme from rabbit skeletal muscle by the method of Cohen [11]. The phosphorylase kinase is composed of alpha, beta, gamma, and delta subunits. Interestingly, only alpha and beta subunits, which are phosphorylated by cAMP-dependent protein kinase, were ADP-ribosylated. As expected ADP-ribosylated phosphorylase kinase was less phosphorylated [12]. Another example is pyruvate kinase. We observed that this enzyme purified from hog liver is also ADP-ribosylated. Furthermore, this modification occurs at arginine residues of the synthetic heptapeptide [10], Leu-Arg-Arg-Ala-Ser-Leu-Gly, corresponding closely to the amino acid composition around the phosphate-accepting serine residue of the pyruvate kinase. A decrease in phosphorylation of this heptapeptide by cAMP-dependent protein kinase with increases in the ADP-ribosylation was also observed (Matsuura, R., Tanigawa, Y., and Shimoyama, M., unpublished observation).

Concluding Remarks

The physiological role of ADP-ribosyltransferase remains to be determined. Several important findings of alteration in acceptor proteins by ADP-ribosylation have been reported. For example, ADP-ribosylation inactivates the respective intrinsic GTPase of stimulatory guanyl nucleotide-binding protein of adenylate cyclase [13] and transducin [14]. With ADP-ribosylation, elongation factor 2 loses its binding activity to ribosomes [15]. As reported here, with ADP-ribosylation, histones, phosphorylase kinase, and pyruvate kinase lose their acceptor activities for cAMP-dependent phosphorylation. ADP-ribosylation of glutamate synthetase directly inhibits the enzyme activity [16]. All these findings are significant as they pave the way toward resolution of the role of the ADP-ribosylation of the several proteins, including certain enzymes. Studies on the effect of ADP-ribosylation of histones and certain enzymes have to be done to support the hypothesis that ADP-ribosylation produces a significant inhibition of cAMP-dependent phosphorylation, in vivo.

Acknowledgment. This work was supported in part by grants-in-aid for Scientific Research and Cancer Research from Ministry of Education, Science, and Culture, Japan.

References

1. Moss J, Stanley SJ, Watkins PA (1980) Isolation and properties of an NAD- and guanidino-dependent ADP-ribosyltransferase from turkey erythrocytes. J Biol Chem 255:5838–5840
2. Shimoyama M, Tanigawa Y, Kitamura A, Nomura H (1982) Mono(ADP-ribosyl)ation reaction in chicken liver nuclei. Abstr 12th Int Congr Biochem, p 180
3. Tanigawa Y, Tsuchiya M, Imai Y, Shimoyama M (1984) ADP-ribosyltransferase from hen liver nuclei: Purification and characterization. J Biol Chem 259:2022–2029

4. Tanigawa Y, Tsuchiya M, Imai Y, Shimoyama M (1984) Mono(ADP-ribosyl)ation of hen liver nuclear proteins suppresses phosphorylation. Biochem Biophys Res Commun 113:135−141
5. Tanigawa Y, Tsuchiya M, Imai Y, Shimoyama M (1983) ADP-ribosylation regulates the phosphorylation of histones by the catalytic subunit of cyclic AMP-dependent protein kinase. FEBS Lett 160:217−220
6. Bustin M, Cole RD (1969) Bisection of a lysine-rich histone by N-bromosuccinimide. J Biol Chem 244:5291−5294
7. Langan TA (1970) Localization of multiple phosphorylation sites in lysine-rich histone. J Cell Biol 47:115a
8. Ushiroyama T, Tanigawa Y, Tsuchiya M, Matsuura R, Ueki M, Sugimoto O, Shimoyama M (in press) Amino acid sequence of histone H1 at the ADP-ribose-accepting site and ADP-ribose-histone H1·adduct as an inhibitor of cyclic AMP-dependent phosphorylation. Eur J Biochem
9. Zetterqvist O, Ragnarsson U, Humble E, Berglund L, Engstron L (1976) The minimum substrate of cyclic AMP-stimulated protein kinase, as studied by synthetic peptides representing the phosphorylatable site of pyruvate kinase (type L) of rat liver. Biochem Biophys Res Commun 70:696−703
10. Kemp BE, Graves DJ, Benjamin E, Krebs EG (1977) Role of multiple basic residues in determining the substrate specificity of cyclic AMP-dependent protein kinase. J Biol Chem 252:4888−4894
11. Cohen P (1973) The subunit strcuture of rabbit-skeletal-muscle phosphorylase kinase, and the molecular bases of its activation reactions. Eur J Biochem 34:1−14
12. Tsuchiya M, Tanigawa Y, Ushiroyama T, Matsuura R, Shimoyama M (1985) ADP-ribosylation of phosphorylase kinase and block of phosphate incorporation into the enzyme. Eur J Biochem 147:33−40
13. Gilman AG (1984) G proteins and dual control of adenylate cyclase. Cell 36:577−579
14. Watkins PA, Moss J, Burns DL, Hewlett EL, Vaughan M (1984) Inhibition of bovine rod outer segment GTPase by *Bordetella pertussis*. Toxin 259:1378−1381
15. Bermek E (1976) Interactions of adenosine diphosphate-ribosylated elongation factor 2 with ribosomes. J Biol Chem 251:6544−6549
16. Moss J, Watkins PA, Stanley SJ, Purnell MR, Kidwell WR (1984) Inactivation of glutamine synthetase by an NAD:arginine ADP-ribosyltransferase. J Biol Chem 259:5100−5104

Poly(ADP-Ribos)ylation of Nuclear Enzymes

KOICHIRO YOSHIHARA[1], ASAKO ITAYA[1], YASUHARU TANAKA[1],
YASUHIRO OHASHI[1], KIMIHIKO ITO[1], HIROBUMI TERAOKA[2],
KINJI TSUKADA[2], AKIO MATSUKAGE[3], and TOMOYA KAMIYA[1]

1. Introduction

Poly(ADP-ribose) polymerase catalyzes a sequential transfer of an ADP-ribose portion of NAD^+ to various chromatin proteins [1] and to the polymerase itself (automodification [2]), forming a polymer of ADP-ribose, which is covalently bound to protein at one end [3]. Recent studies elucidated that two chromatin enzymes, Ca^{2+}, Mg^{2+}-dependent endonuclease [4, 5] and DNA topoisomerase [6, 7], were markedly inhibited as a result of poly(ADP-ribos)ylation of the enzyme proteins. RNA polymerase I [8] and DNA ligase II [9] also are suggested to be poly(ADP-ribos)ylated, although the latter enzyme seems to be activated after poly(ADP-ribos)ylation in vivo. Furthermore, bull seminal RNase [10, 29] and micrococcal nuclease [29] also have been shown to be the acceptors of ADP-ribose in the enzyme reaction in vitro. These results suggest a possibility that poly(ADP-ribose) polymerase randomly modifies many kinds of chromatin enzymes rather than it selecting a few kinds of specific enzymes as its targets. Thus, in order to study whether the modification reaction is specific only for the enzymes described above, we examined six nuclear enzymes, which are involved in metabolism or function of chromatin. After several unsuccessful trials using standard and modified conditions of the reconstituted ADP-ribosylating system, we found that all of these enzymes except DNA ligase I were markedly inhibited when the enzymes were incubated in an ADP-ribosylating reaction mixture containing a limited concentration of buffer (5 mM). Analysis of the mechanism of the inhibition of these enzymes (DNA pol α-primase, DNA polymerase α, DNA polymerase β, terminal deoxynucleotidyl transferase, and DNA ligase II) will be presented and a possible biological meaning of this event will be discussed.

1 Department of Biochemistry, Nara Medical University, Kashihara, Nara 634, Japan
2 Department of Pathological Biochemistry, Medical Research Institute, Tokyo Medical and Dental University, Kandasurugadai, Chiyoda-ku, Tokyo 101, Japan
3 Laboratory of Biochemistry, Aichi Cancer Center Research Institute, Chikusa-ku, Nagoya 464, Japan

ADP-Ribosylation of Proteins
(ed. by F.R. Althaus, H. Hilz, and S. Shall)
© Springer-Verlag Berlin Heidelberg 1985

2. Experimental Procedures

2.1 Reconstituted Poly(ADP-Ribos)ylating Enzyme System

Bovine thymus poly(ADP-ribose) polymerase was purified to near homogeneity (95%) as described previously [11]. The enzyme reaction was carried out principally as described in a previous report [5], except that the buffer concentration of the reaction mixture was decreased to 5 mM. The reaction mixture contained 5 mM Tris-HCl buffer, pH 8.0, 1 mM dithiothreitol, 10 mM MgCl$_2$, 10 μg of calf thymus DNA, 2 mM NAD$^+$, 5 μg of purified poly(ADP-ribose) polymerase, and an appropriate amount (1 to 15 μg protein) of various nuclear enzymes in a total volume of 0.2 ml. The mixture was incubated at 25°C for 40 min and the reaction was terminated by chilling the sample on ice or by the addition of a final concentration of 50 mM nicotinamide.

2.2 Enzyme Assays

DNA pol α with primase was assayed in a reaction mixture (0.2 ml) containing 50 mM Tris-HCl buffer, pH 7.4, 10 mM MgCl$_2$, 1 mM ATP, 2.5 μg of poly(dT), 20 μM [^{3}H]-dATP (35 cpm pmol^{-1}) and an appropriate amount of the enzyme. The reaction was carried out at 37°C for 1 h. DNA polymerase α and β activity was assayed with activated DNA as template-primer. Details of the assay will be described elsewhere [30]. The assay of DNA ligase [12, 13], Ca^{2+}, Mg^{2+}-dependent endonuclease [14], and terminal deoxynucleotidyl transferase [15] was carried out according to the respective reported method.

3. Results

As summarized in Table 1, various nuclear enzymes involved in the function of chromatin were strongly inhibited when they were incubated in a reconstituted poly(ADP-ribose) synthesizing enzyme system: purified poly(ADP-ribose) itself was slightly effective when it was added to the system in place of NAD$^+$, suggesting that poly-(ADP-ribos)ylation of the enzyme molecules caused the observed inhibition. Thus, the mechanism of the inhibition was studied for respective enzymes as described in the following sections.

3.1 Ca^{2+}, Mg^{2+}-Dependent Endonuclease

This enzyme has been shown to be poly(ADP-ribos)ylated and inhibited in a reconstituted poly(ADP-ribos)ylating system [4, 5]. Also, it has been demonstrated that the enzyme is markedly inhibited when rat liver cell nuclei or chromatin are incubated with NAD$^+$ [16]. As shown in Fig. 1, poly(ADP-ribos)ylated and inhibited enzyme was

Table 1. Inhibition of various nuclear enzymes by poly(ADP-ribos)ylation in vitro[a]

Enzymes	Enzyme activity (%)		
	Control I − NAD⁺	+ NAD⁺	Control II (+ Polymer)
Ca²⁺, Mg²⁺-dependent endonuclease	100	10	100
CT DNA pol α-primase	100	23	100
CT DNA polymerase α	100	50	100
CT DNA polymerase β	100	29	100
Mouse AH DNA polymerase β	100	47	——
Terminal deoxynucleotidyl transferase	100	11	90
DNA ligase II	100	45	90

[a] Poly(ADP-ribos)ylation was carried out as described in Sect. 2 except that the buffer concentration of the reaction mixture for ADP-ribosylation of Ca²⁺, Mg²⁺-dependent endonuclease was 25 mM. Enzyme activity of control samples incubated without NAD⁺ (control I) was set at 100%.
Control II samples contained 10 to 30 µg/0.2 ml of purified poly(ADP-ribose) in place of NAD⁺

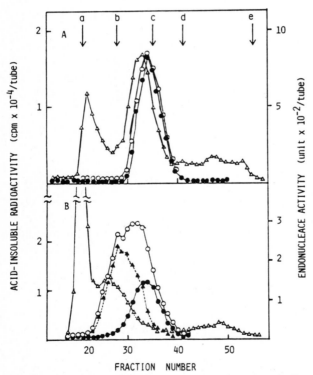

Fig. 1A,B. Sephadex G-100 column chromatography of oligo- and poly(ADP-ribos)ylated Ca²⁺, Mg²⁺-dependent endonuclease. **A** [³H]oligo(ADP-ribos)ylated enzyme, **B** its chased product with cold NAD⁺. Acid-insoluble radioactivity (△—△), endonuclease activity before (●—●) and after (○—○) alkali treatment, alkali-stimulated endonuclease (▲—▲). (Data from [5])

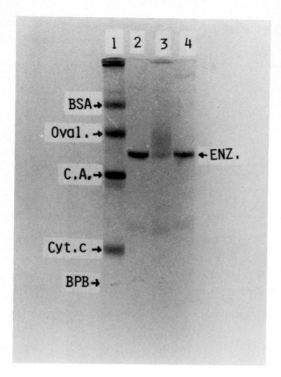

Fig. 2. SDS-polyacrylamide gel electrophoresis of poly(ADP-ribos)ylated Ca^{2+}, Mg^{2+}-dependent endonuclease. Lanes 1–4 correspond to marker proteins, unmodified control enzyme, modified enzyme, and alkali-treated modified enzyme, respectively. (Data from [5])

located in a much higher molecular weight region than unmodified enzyme on analysis by Sephadex G-100 column chromatography: the inhibited enzyme fraction was markedly activated by alkaline treatment, a procedure known to hydrolyze the carboxyl ester linkage between poly(ADP-ribose) and a protein [3]. This indicates that the covalent modification of the enzyme molecule with poly(ADP-ribose) is the basis of the observed inhibition of this enzyme. Analysis by SDS-polyacrylamide gel electrophoresis of the reaction product indicated that the modified endonuclease moved from the position of unmodified enzyme to a higher molecular region forming a broadly distributed band with low density: the broad poly(ADP-ribos)ylated endonuclease band was concentrated at the unmodified enzyme position after alkaline treatment (Fig. 2).

3.2 DNA Ligase II

It seems likely that in mammalian cells there are two forms of DNA ligase [17–19]. Recently, Creissen and Shall [27] reported that the activity of one of the two forms (DNA ligase II) was specifically activated in response to poly(ADP-ribos)ylation in vivo following DNA damage. They also have shown evidence suggesting that either DNA ligase II itself or a closely associated protein is ADP-ribosylated in vivo [9]. Most recently, Teraoka et al. [31] succeeded in obtaining a highly purified preparation of bovine thymus DNA ligase II composed of a single polypeptide with a mol. wt. of

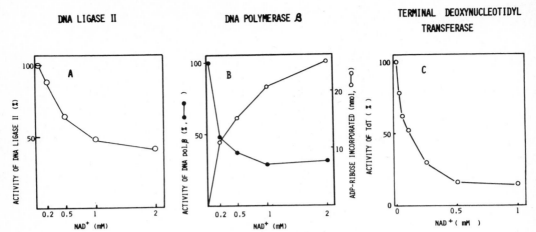

Fig. 3A–C. Inhibition of DNA ligase II, DNA polymerase β, and terminal deoxynucleotidyl transferase by poly(ADP-ribos)ylation. Highly purified DNA ligase II (5 μg, **A**), DNA polymerase β (5 μg, **B**), and terminal deoxynucleotidyl transferase (0.65 μg, **C**) from bovine thymus were incubated in a poly(ADP-ribos)ylating reaction mixture as described in Sect. 2 except that the concentration of NAD$^+$ was changed as indicated. After incubation, respective enzyme activity was assayed. The activity of a control sample incubated without NAD$^+$ in each experiment was set at 100%

70,000. Thus, we examined whether the enzyme could be activated by poly(ADP-ribos)ylation or not in vitro. As shown in Fig. 3A, however, DNA ligase II was not activated, but rather strongly inhibited with increasing concentration of NAD$^+$, when the enzyme was incubated in a reconstituted poly(ADP-ribose) synthesizing enzyme system. Analysis of the reaction products by SDS-polyacrylamide gel electrophoresis indicated that the density of the stained band of DNA ligase II with a mol. wt. of 70,000, which was calibrated with [^{3}H]AMP-DNA ligase complex [12], markedly reduced after poly(ADP-ribos)ylation (Fig. 4): the effect is probably due to a heterogeneous mobility of modified enzyme molecules carrying different sizes of poly(ADP-ribose) as also observed upon an analysis of poly(ADP-ribos)ylated Ca^{2+}, Mg^{2+}-dependent endonuclease (Fig. 2). The data also indicate that poly(ADP-ribose) polymerase itself (automodification [2]) and other contaminating proteins were poly(ADP-ribos)-ylated and disappeared from their original position.

We also tried to inhibit bovine thymus DNA ligase I [12] using the ADP-ribosylating system. In the preliminary trial, however, the enzyme activity was not affected by poly(ADP-ribos)ylation under the condition employed. Thus, DNA ligase I may not be the acceptor of the poly(ADP-ribose) polymerase reaction, although the final conclusion in this point should be reserved until further examination of this enzyme is completed.

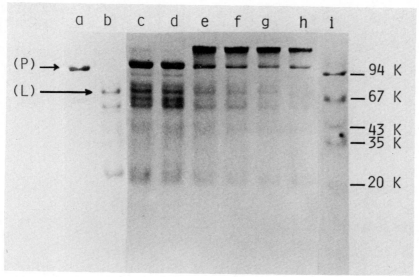

Fig. 4. SDS-polyacrylamide gel electrophoresis of poly(ADP-ribos)ylated DNA ligase II. Poly(ADP-ribos)ylation was carried out as described in Sect. 2 except that the NAD⁺ concentration was changed as indicated. *a* Purified poly(ADP-ribose) polymerase; *b* purified DNA ligase II; *c* a negative control sample incubated without NAD⁺; *d* another negative control sample incubated with 13 μg/0.2 ml of purified poly(ADP-ribose) in place of NAD⁺; *e–h* the samples incubated with 0.2, 0.5, 1, and 2 mM NAD⁺, respectively; *i* molecular weight markers. The position of poly(ADP-ribose) polymerase (*P*) and DNA ligase II (*L*) are indicated by *arrows*. The position of the latter enzyme was determined fluorographically by a parallel run of [³H]AMP-DNA ligase complex [12, 13]

3.3 DNA Pol α-Primase, DNA Polymerase α, DNA Polymerase β, and Terminal Deoxynucleotidyl Transferase

Yagura et al. [20] have reported that various eukaryotic cells contain at least two forms of DNA polymerase α. One possessed an ability to synthesize poly(dA) with poly(dT) as template and without primer in the presence of ATP and dATP [21]. They also showed that the unique property of this enzyme was due to the association of DNA primase activity with this enzyme [22]. As shown in Fig. 5, when bovine thymus DNA polymerase α fraction obtained from phosphocellulose column chromatography was further purified by DEAE-Sephadex A-50 column chromatography, the enzyme activity was separated into two major sharp peaks. Since one of the peaks was associated with the primase activity, we call this fraction DNA polymerase α with primase (DNA pol α-primase) and the other (fraction II in Fig. 5) DNA polymerase α in this study. When these two enzyme fractions were incubated in a reconstituted poly(ADP-ribos)ylating system, both DNA pol α-primase and DNA polymerase α were strongly inhibited (77 and 50% inhibition, respectively, Table 1). Thus, in order to clarify the mechanism of the inhibition, further study was carried out using mainly the DNA pol α-primase fraction. All the components of poly(ADP-ribos)ylation

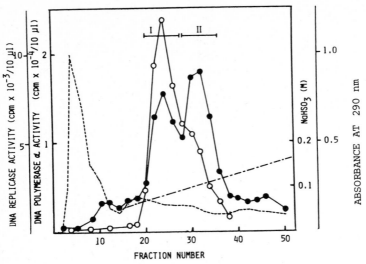

Fig. 5. DEAE-Sephadex A-50 column chromatography of DNA polymerase α. A crude preparation of DNA polymerase α fraction obtained from phosphocellulose column chromatography of bovine thymus extracts was further fractionated by DEAE Sephadex A-50 column chromatography. Details of the purification procedure will be described elsewhere [33]. *Dotted* and *broken lines* indicate absorbance at 290 nm and the concentration of NaHSO₃ in the elution buffer, respectively. *Open* and *closed circles* indicate DNA pol α-primase and DNA polymerase α activities, respectively

mixture, such as poly(ADP-ribose) polymerase, DNA, NAD⁺, and Mg²⁺, were essential for the observed inhibition of DNA pol α-primase, and inhibitors of poly(ADP-ribose) polymerase, nicotinamide, and 3-aminobenzamide, markedly reduced the extent of the inhibition [30]. Free poly(ADP-ribose) was ineffective when added to the ADP-ribosylating system and did not affect the assay of the two DNA polymerases. The inhibited DNA pol α-primase markedly restored its activity after treatment with an alkaline pH of 10 (approx. 3-fold activation was observed after the alkaline treatment for 6 h at 25°C) and the activity reached almost the same level as an unmodified control sample [30]. Since it has been shown that the restoration of the activity of poly(ADP-ribos)ylated and inhibited Ca²⁺, Mg²⁺-dependent endonuclease [5] and DNA topoisomerase [6] after alkaline treatment is due to the hydrolysis of the ester linkage between poly(ADP-ribose) and the enzymes, and that the polymer itself is rather stable at this pH [23], all of the above results obtained with DNA pol α-primase indicate that the inhibition was caused by direct poly(ADP-ribos)ylation of the enzyme, although its direct demonstration has not been carried out as yet. Details of the study on this enzyme will appear elsewhere [30].

The effect of poly(ADP-ribos)ylation on DNA polymerase β was also examined with partially purified enzyme from bovine thymus. In spite of our expectation that the enzyme may not be inhibited by poly(ADP-ribos)ylation, since the enzyme is considered to be main DNA polymerizing enzyme working in DNA repair [24] and poly(ADP-ribos)ylation in vivo stimulates the process [9], DNA polymerase β was also

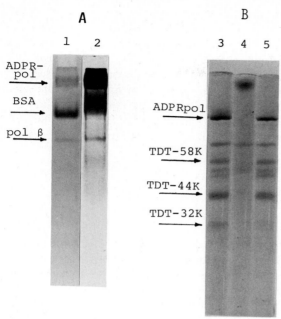

Fig. 6A,B. SDS-polyacrylamide gel electrophoresis of [³H]oligo(ADP-ribos)ylated DNA polymerase β and poly(ADP-ribos)ylated terminal deoxynucleotidyl transferase. [³H]oligo(ADP-ribos)ylated bovine DNA polymerase β (**A**) and poly(ADP-ribos)ylated terminal deoxynucleotidyl transferase (**B**) were analyzed. **A** Lanes 1 and 2 indicate stained protein band and the fluorography of the same gel. **B** Lanes 3–5 indicate protein staining of a control sample (lane 3) incubated without NAD⁺, a poly(ADP-ribos)ylated enzyme (lane 4), and an alkaline-treated sample of poly(ADP-ribos)ylated enzyme (lane 5)

strongly inhibited by poly(ADP-ribos)ylation in vitro (Fig. 3B). Details of the study on this enzyme will be described elsewhere [32]. The results are summarized as follows: (1) DNA, Mg²⁺, NAD⁺, and poly(ADP-ribose) polymerase were all essential for the inhibition; (2) purified poly(ADP-ribose) itself was ineffective; (3) inhibitors of poly-(ADP-ribose) polymerase essentially abolished the inhibition; and (4) the inhibited enzyme could be reactivated by an alkaline treatment (approx. 2.5-fold activation was observed). The inhibition of DNA polymerase β has been confirmatively observed also by using a homogeneously purified enzyme preparation [28] from rat ascites hepatoma cells (Yoshihara, K., and Matsukage, A., unpublished results). When highly purified bovine thymus DNA polymerase β was incubated in a ADP-ribosylating system under a limited concentration of [³H]NAD⁺ and the reaction product was analyzed by SDS-polyacrylamide gel electrophoresis and the fluorography of the gel, the protein band of DNA polymerase β with a mol. wt. of 44,000 was found to be associated with [³H]ADP-ribose (Fig. 6A).

The response of terminal deoxynucleotidyl transferase to poly(ADP-ribos)ylation in vitro was also quite similar to that of other enzymes already described. The enzyme activity showed time-dependent (data not shown) and NAD⁺-dependent (Fig. 3) decrease upon an incubation in the poly(ADP-ribos)ylating system. The components required for poly(ADP-ribos)ylation were essential for the suppression, the polymer itself was ineffective, and inhibitors of poly(ADP-ribose) polymerase abolished the inhibition [33]. Analysis of the reaction product by SDS-polyacrylamide gel electrophoresis indicated that the enzyme bands with mol. wts. of 58 K, 42–46 K, and 32 K [25] disappeared from their position after poly(ADP-ribos)ylation by the result of a very heterogeneous distribution of the modified enzyme on the gel and appeared again at their original position after the hydrolysis of the ester linkage between polymer and the enzyme by alkaline treatment (Fig. 6B). All of these results indicate that the enzyme was also inhibited as a result of poly(ADP-ribos)ylation of the enzyme molecule.

4. Discussion

The present results demonstrated that several nuclear enzymes involved in function or metabolism of chromatin could be inhibitied by poly(ADP-ribos)ylation in vitro. Although a direct demonstration for the binding of poly(ADP-ribose) to some of the enzyme molecules has not been performed as yet, various indirect evidences strongly support the hypothesis that all of these enzymes are inhibited as a result of the modification of the enzyme molecules.

The finding that six enzymes among seven so far tested could be inhibited by poly(ADP-ribos)ylation suggest that many other chromatin enzymes, if not all, also will be poly(ADP-ribos)ylated and inhibited if examined. Thus, the regulation of chromatin enzymes by poly(ADP-ribos)ylation does not seem to be a specific event limited to a very small number of enzyme species located on chromatin. Considering the reaction mechanism of poly(ADP-ribose) polymerase that the enzyme probably does not slide on chromatin, but is bound to a nick or a DNA end very tightly during the enzyme reaction [2, 26], the activated polymerase may randomly ADP-ribosylate various chromatin-bound enzymes, which were located, by chance, at or close to the damaged site on chromatin.

DNA polymerase β and DNA ligase II are supposed to be important enzymes working in DNA repair. In addition, Creissen and Shall [27] reported that DNA ligase II was activated in response to activation of poly(ADP-ribose) synthesis following DNA damage in cells presumably by poly(ADP-ribos)ylation of either the enzyme itself or a closely associated protein [9]. In spite of these findings, our results indicated that these two enzymes also were markedly inhibited by poly(ADP-ribos)ylation in vitro. At present, we cannot explain the discrepancy found between these two results. However, our results may be interpreted as follows: When poly(ADP-ribose) polymerase is activated once by binding to a damaged site on chromatin [26], the enzyme inhibits all or many chromatin enzymes located near the damaged site by poly(ADP-ribos)ylation to protect cells from abnormal metabolism of chromatin, the binding of these enzymes to DNA may be relaxed after the poly(ADP-ribos)ylation [2, 5, 6], and after the emergency process has been completed, an ordered process of DNA repair is carried out by repair enzymes including unmodified DNA ligase and DNA polymerase β. Thus, one of the roles of poly(ADP-ribose) polymerase in DNA repair may be to cause a kind of emergency halt of chromatin function at the damaged site to protect cells from an abnormal metabolism in chromatin, although the occurrence of poly(ADP-ribos)ylation of these enzymes in vivo remains to be proven.

The inhibition of some of the tested chromatin enzymes by poly(ADP-ribos)ylation could be demonstrated at a very low ionic strength in the ADP-ribosylating reaction mixture: increasing concentrations (25–50 mM) of salt or buffer in the mixture markedly suppressed the inhibition of DNA pol α-primase [30], in spite of the fact that the concentration of the buffer did not significantly affect the modification of Ca^{2+}, Mg^{2+}-dependent endonuclease [5]. Considering that DNA replicase and other tested enzymes have a relatively weak DNA-binding affinity compared to Ca^{2+}, Mg^{2+}-dependent endonuclease, higher ionic strength in the reaction mixture may interfere in the formation or the maintenance of a reaction complex composed of these enzymes, poly(ADP-ribose) polymerase, and DNA in the reconstituted poly(ADP-ribos)ylating reaction system.

Acknowledgments. The authors express their gratitude to the staff of Osaka City Meat Hygiene Laboratory and Dr. Isao Dota (Dept. of Bacteriology of the same laboratory) for donation of calf thymus, and to Miss Iyuko Matsuda for helping with the preparation of this manuscript. This work was supported in part by Grant in Aid 58580142 for scientific research and Grant in Aid 59010097 for Cancer Research from the Ministry of Education, Science and Culture of Japan.

References

1. Hayaishi O, Ueda K (1982) Poly- and mono(ADP-ribosyl)ation reactions: their significance in molecular biology. In: Hayaishi O, Ueda K (eds) ADP-ribosylation reactions. Academic Press, London New York, pp 3–16
2. Yoshihara K, Hashida T, Tanaka Y, Matsunami N, Yamaguchi A, Kamiya T (1981) Mode of enzyme-bound poly(ADP-ribose) synthesis and histone modification by reconstituted poly-(ADP-ribose) polymerase-DNA-cellulose complex. J Biol Chem 256:3471–3478
3. Nishizuka Y, Ueda K, Yoshihara K, Yamamura H, Takeda M, Hayaishi O (1969) Enzymic adenosine diphosphoribosylation of nuclear proteins. Cold Spring Harbor Symp Quant Biol 34:781–786
4. Yoshihara K, Tanigawa Y, Burzio L, Koide SS (1975) Evidence for adenosine diphosphate ribosylation of Ca^{2+}, Mg^{2+}-dependent endonuclease. Proc Natl Acad Sci USA 72:289–293
5. Tanaka Y, Yoshihara K, Itaya A, Kamiya T, Koide SS (1984) Mechanism of the inhibition of Ca^{2+}, Mg^{2+}-dependent endonuclease of bull seminal plasm induced by ADP-ribosylation. J Biol Chem 259:6579–6585
6. Ferro AM, Olivera BM (1984) Poly(ADP-ribosylation) of DNA topoisomerase I from calf thymus. J Biol Chem 259:547–554
7. Jongstra-Bilen J, Ittel ME, Niedergang C, Vosberg HP, Mandel P (1983) DNA topoisomerase I from calf thymus is inhibited in vitro by poly(ADP-ribosylation) Eur J Biochem 136:391–396
8. Müller WEG, Zahn RK (1976) Poly ADP-ribosylation of DNA-dependent RNA polymerase I from quail oviduct: dependence on progesterone stimulation. Mol Cell Biol 12:147–159
9. Shall S (1983) ADP-ribosylation, DNA repair, cell differentiation and cancer. In: Miwa et al. (eds) ADP-ribosylation, DNA repair, and cancer. Jpn Sci Soc Press, Tokyo, pp 3–25
10. Leone E, Farina B, Faraone-Mennella MR, Maura A (1980) ADP-ribosylation of ribonucleases. In: Holzer H (ed) Metabolic interconversion of enzymes 1980. Springer, Berlin Heidelberg New York, pp 294–302

11. Yoshihara K, Hashida T, Tanaka Y, Ohgushi H, Yoshihara H, Kamiya T (1978) Bovine thymus poly(adenosine diphosphate ribose) polymerase. J Biol Chem 253:6459–6466
12. Teraoka H, Tsukada K (1982) Eukaryotic DNA ligase purification and properties of the enzyme from bovine thymus and immunochemical studies of the enzyme from animal tissues. J Biol Chem 257:4758–4763
13. Teraoka H, Sawai M, Tsukada K (1983) DNA ligase from mouse Ehrlich ascites tumor cells. J Biochem 95:1529–1532
14. Hashida T, Tanaka Y, Matsunami N, Yoshihara K, Kamiya T, Tanigawa Y, Koide SS (1982) Purification and properties of bull seminal plasm Ca^{2+}, Mg^{2+}-dependent endonuclease. J Biol Chem 257:13114–13119
15. Yoshida S, Nakamura H (1981) Terminal deoxynucleotidyl transferase in gene engineering. Protein Nucleic Acid Enzymes 26:569–574
16. Yoshihara K, Tanigawa Y, Koide SS (1974) Inhibition of rat liver Ca^{2+}, Mg^{2+}-dependent endonuclease activity by nicotinamide adenine dinucleotide and poly(adenosine diphosphate ribose) synthetase. Biochem Biophys Res Commun 59:658–665
17. Söderhäll S, Lindahl T (1975) Mammalian DNA ligases, serological evidence for two separate enzymes. J Biol Chem 250:8438–8444
18. Teraoka H, Shomoyachi M, Tsukada K (1975) Two distinct polynucleotide ligases from rat liver. FEBS Lett 54:217–220
19. Teraoka H, Shimoyachi M, Tsukada K (1977) Purification and properties of deoxyribonucleic acid ligases from rat liver. J Biochem (Tokyo) 81:1253–1260
20. Yagura T, Kozu T, Seno T, Saneyoshi M, Hiruga S, Nagano H (1983) Novel form of DNA polymerase α associated with DNA primase activity of vertebrates. J Biol Chem 258:13070–13075
21. Yagura T, Kozu T, Seno T (1982) Mouse DNA polymerase accompanied by a novel RNA polymerase activity: purification and partial characterization. J Biochem (Tokyo) 91:607–618
22. Yagura T, Kozu T, Seno T (1982) Mouse DNA replicase: DNA polymerase associated with a novel RNA polymerase activity to synthesize initiator RNA of strict size. J Biol Chem 257:11121–11127
23. Sugimura T, Shimizu T (1968) Formation of poly(ADP-ribose) from NAD by nuclear enzyme preparations. Seikagaku 40:1–17
24. Waser J, Hübscher U, Kuenzle CC, Spadari S (1979) DNA polymerase β from brain neurons is repair enzyme. Eur J Biochem 97:361–368
25. Chang LMS, Plevani P, Bollum FJ (1981) Proteolytic degradation of calf thymus terminal depcynucleotidyl transferase. J Biol Chem 257:5700–5706
26. Ohgushi H, Yoshihara K, Kamiya T (1979) Bovine thymus poly(adenosine diphosphate ribose) polymerase: physical properties and binding to DNA. J Biol Chem 255:6205–6211
27. Creissen D, Shall S (1982) Regulation of DNA ligase activity by poly(ADP-ribose) Nature (London) 296:271–272
28. Ono K, Ohashi Y, Tanabe K, Matsukage A, Nishizawa M, Takahashi T (1979) Unique requirements for template-primers of DNA polymerase β from rat ascites hepatoma AH130. Nucleic Acids Res 7:715–726
29. Tanaka Y, Yoshihara K, Ohashi Y, Itaya A, Nakano T, Ito K, Kamiya T (1985) A method for determining oligo- and poly(ADP-ribosyl)ated enzymes and proteins in vitro. Anal Biochem 145:137–143
30. Yoshihara K, Itaya A, Tanaka Y, Ohashi Y, Ito K, Teraoka H, Tsukada K, Matsukage A, Kamiya T (1985) Inhibition of DNA polymerase α, DNA polymerase β, terminal deoxynucleotidyl transferase, and DNA ligase II by poly(ADP-ribosyl)ation reaction in vitro. Biochem Biophys Res Commun 128:61–67
31. Teraoka H, Sawai M, Tsukada K (in preparation)
32. Ohashi Y, Itaya A, Tanaka Y, Ito K, Yoshihara K, Matsukage Y, Kamiya T (in preparation)
33. Tanaka Y, Ito K, Yoshihara K (unpublished)

Kinetic Mechanism of Poly(ADP-Ribose) Polymerase

ROBERT C. BENJAMIN, PAUL F. COOK, and MYRON K. JACOBSON[1]

Abbreviations:

poly(dT)   poly(deoxythymidylic acid)
EDTA   ethylenediamine tetraacetic acid

1. Introduction

Poly(ADP-ribose) polymerase is a chromatin-associated enzyme which, in the presence of fragmented DNA, assembles branched homopolymers from the ADP-ribose moiety of NAD (reviewed in [1]). Fragmented DNA is an essential activator of the polymerase and is not modified by the reaction. This report presents data related to the kinetic mechanism of DNA activation of the poly(ADP-ribose) polymerase.

2. Materials and Methods

Poly(ADP-ribose) polymerase from calf thymus and plasmid pBR322 were prepared as previously described [2]. Nicotinamide, NADH, thymidine, and poly(dT) were obtained from the Sigma Chemical Company.

Reactions were carried out at 37°C in siliconized glass tubes containing 50 mM Tris-HCl, pH 8.0, 1 mM EDTA, 50 μl of enzyme preparation, and [^{32}P]-NAD and DNA as noted in a total volume of 200 μl. Incubations were terminated after 2 min with 20% TCA, collected on Whatman GF/A filters and acid-insoluble radioactivity determined by liquid scintillation counting.

1 Department of Biochemistry, Texas College of Osteopathic Medicine, North Texas State University, Denton, TX 76203, USA

ADP-Ribosylation of Proteins
(ed. by F.R. Althaus, H. Hilz, and S. Shall)
© Springer-Verlag Berlin Heidelberg 1985

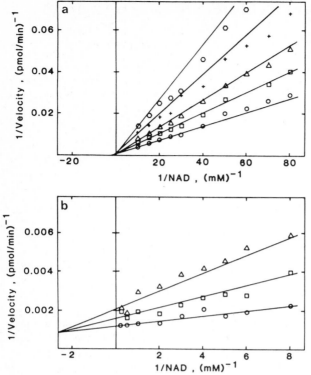

Fig. 1a,b. Initial velocity patterns of poly(ADP-ribose) polymerase as a function of NAD concentration in the presence of different concentrations of DNA. a NAD varied from 12.5 μM to 100 μM. Hae III digested pBR322 at 2.0 μg ml^{-1} (○); 0.8 μg ml^{-1} (□); 0.5 μg ml^{-1} (△); 0.3 μg ml^{-1} (+); and 0.2 μg ml^{-1} (o). b NAD varied from 100 M to 4 mM. Alkaline phosphatase treated Hae III digested pBR322 at 5.00 μg ml^{-1} (○); 0.50 μg ml^{-1} (□); and 0.25 μg ml^{-1} (△). K_{DNA} = 0.21 ± 0.02 μg ml^{-1}; K_{NAD} = 106 ± 10 μM; K_{iDNA} = 0.72 ± 0.09 μg ml^{-1}; K_{iNAD} = 369 ± 47 μM. K_{DNA} for Hae III digested pBR322 is 1 μg ml^{-1}

3. Results

3.1 Characterization of the Enzyme Preparation

The enzyme preparation is greater than 98% dependent on exogenous DNA for synthetic activity and contains no NAD. Under standard conditions at near saturating levels of NAD and DNA the preparation exhibits a linear time course for the conversion of radioactively-labeled NAD into acid-insoluble product over a period of at least 4–5 min. A plot of enzyme concentration vs initial velocity is linear over a more than tenfold range (5–60% enzyme preparation by volume) and passes through the origin. The preparation contains essentially no activities which degrade or modify DNA, NAD, or product under standard assay conditions. The K_{NAD} is approx. 100 μM, while the K_{DNA} for the Hae III digested pBR322 is about 1 μg ml^{-1} (1 × 10^{-9} M DNA fragments) and K_{DNA} for alkaline phosphatase treated Hae III fragments is 0.2 μg ml^{-1}.

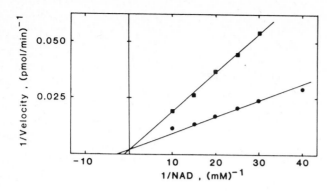

Fig. 2. Initial velocity pattern of poly(ADP-ribose) polymerase as a function of NAD concentration under control conditions (●) and in the presence of 28 μM nicotinamide (■). Hae III digested pBR322 was fixed at 1 μg ml^{-1}. The K_i for nicotinamide is approx. 15 μM

3.2 Initial Velocity Studies in the Absence of Added Products and Dead-End Inhibitors

Initial velocity studies where NAD was varied around its K_m at different fixed levels of DNA (Hae III digests of pBR322 or Hae III digests with terminal 5′PO$_4$ groups removed) around its K_m give a pattern which intersects to the left of the vertical axis (Fig. 1a,b). This is most apparent in Fig. 1b, where NAD concentrations of up to 40 times K_m were used. The data were fit to the theoretical equations for a sequential mechanism (v = VAB/$K_{ia}K_b$ + K_aB + K_bA + AB) and for an equilibrium ordered mechanism (v = VAB/$K_{ia}H_b$ + K_bA + AB) using the FORTRAN programs developed by Cleland [4]. V represents V_{max}, K_a and K_b represent K_m values for A and B, K_{ia} is the dissociation constant for A, and A and B represent substrate concentrations. Although the data obtained using relatively low NAD concentrations fit the equilibrium ordered equation nearly as well as the sequential equation, the data in Fig. 1b clearly show an intercept effect and provide a good fit only to the sequential model.

3.3 Initial Velocity Patterns in the Presence of Added Products and Dead-End Inhibitors

Initial velocity studies were conducted in the presence of a number of dead-end inhibitors of NAD and DNA binding. Nicotinamide, a product of the reaction which acts as a dead-end inhibitory analog of NAD, is competitive vs NAD (Fig. 2, cf. [3]) and noncompetitive vs DNA (not shown). NADH exhibits similar kinetic properties and is competitive vs NAD (not shown) and noncompetitive vs DNA (Fig. 3). Both monomeric and polymeric dead-end inhibitory analogs of DNA behaved similarly. Thymidine (Fig. 4) and poly(dT) (data not shown) are both competitive vs DNA. The polymeric form is seven times more effective as an inhibitor based on total thymine content (K_i values of 5 μM vs 35 μM). On a per molecule of inhibitor basis the difference is even more striking, the poly(dT) being more than 1000 times more effective than thymidine. Both poly(dT) (Fig. 5) and thymidine (data not shown, cf. [3]) also exhibited competitive kinetics vs NAD.

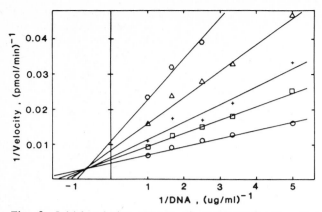

Fig. 3. Initial velocity pattern of poly(ADP-ribose) polymerase as a function of DNA under control conditions (○) and in the presence of 5 μM NADH (□); 10 μM NADH (+); 20 μM NADH (△); and 30 μM NADH (○). NAD was fixed at 200 μM. The K_i for NADH is approx. 10 μM

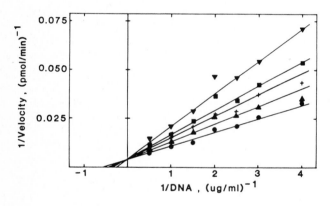

Fig. 4. Initial velocity pattern of poly(ADP-ribose) polymerase as a function of DNA concentration under control conditions (●) and in the presence of 20 μM (▲); 40 μM (+); 60 μM (■); and 100 μM (▼) thymidine. NAD was fixed at 200 μM; the K_i for thymidine was 35 μM

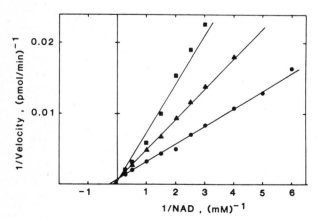

Fig. 5. Initial velocity pattern of poly(ADP-ribose) polymerase as a function of NAD concentration under control conditions (●), in the presence of 1 μg ml^{-1} poly-(dT) (▲) and in the presence of 2 μg ml^{-1} poly(dT) (■). Hae III digested pBR322 was fixed at 1 μg ml^{-1}

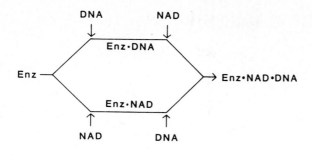

Fig. 6. Proposed kinetic mechanism for poly(ADP-ribose) polymerase

4. Summary

Initial velocity studies of the reaction catalyzed by poly(ADP-ribose) polymerase have been carried out under a variety of experimental conditions. An initial velocity pattern where NAD is varied at different fixed concentrations of DNA intersects to the left of the vertical axis. Nicotinamide, a product of the reaction, is competitive vs NAD and noncompetitive vs DNA. Initial velocity studies using dead-end inhibitors show that NAD analogs are competitive vs NAD and noncompetitive vs DNA, while DNA analogs are competitive vs both DNA and NAD. These data are most consistent with a random mechanism (Fig. 6). For as yet unknown reasons, DNA analogs do not appear to bind to enzyme:NAD.

Acknowledgments. This work was supported in part by NIH grant CA23994, the North Texas State University Faculty Research Fund and by the Texas Ladies Auxillary to the Veterans of Foreign Wars.

References

1. Pekala PH, Moss J (1983) Poly(ADP-ribosylation) of protein. Cur Top Cell Regul 22:1–49
2. Benjamin RC, Gill DM (1980) Poly(ADP-ribose) synthesis in vitro programmed by damaged DNA: A comparison of DNA molecules containing different types of strand breaks. J Biol Chem 255:10502–10508
3. Niedergang C, Okazaki H, Mandel P (1979) Properties of purified calf thymus poly(ADP-ribose) polymerase. Eur J Biochem 102:43–57
4. Cleland WW (1963) Statistical analysis of enzyme kinetic data. Methods Enzymol 63:103–138

Specificity of Poly(ADP-Ribose) Synthetase Inhibitors

MICHAEL R. PURNELL[1], WILLIAM R. KIDWELL[2], LINDSAY MINSHALL[3], and WILLIAM J.D. WHISH[3]

Introduction

Poly(ADP-ribose) synthetase inhibitors, and in particular 3-aminobenzamide, have been used extensively in recent years as probes to elucidate the function of poly(ADP-ribose) in the cell. Our initial report [1] on substituted benzamides as physiologically specific inhibitors was based on the observation that cells grew at an unchanged rate in the presence of 2 mM 3-aminobenzamide, a concentration 1,000 times the K_i value. In general, high concentrations are needed to elicit a cellular response compared to assays in vitro. Interpretation of results is complicated by the possibility of affecting a target other than poly(ADP-ribose) synthetase. Recently, processes other than ADP-ribosylation have been reported to be altered by benzamides. A major criticism of most work is that the studies entail the use of only one inhibitor, usually 3-amino-benzamide.

We have shown that poly(ADP-ribose) synthetase is inhibited by a series of benzamides (Table 1 [1]) and analysis of the structure of the inhibitors revealed that potency of inhibition did not correlate solely with any one property. On the assumption that another enzyme is unlikely to have the same spectrum of sensitivity as poly(ADP-ribose) synthetase to these compounds, then by using a series of inhibitors, it should be possible to answer definitively whether poly(ADP-ribose) synthesis regulates a cel-

Table 1. Inhibitors of poly(ADP-ribose) synthetase

Compound	Abbreviation	Symbol	$K_i(\mu M)$
3-Acetamidobenzamide	AAB	●	0.4
3-Aminobenzamide	AB	○	2.6
Benzamide	B	■	1.0
3-Hydroxybenzamide	HB	□	1.0
3-Methoxybenzamide	MB	▲	0.6
3-Nitrobenzamide	NB	△	9.8

1 Cancer Research Unit, Royal Victoria Infirmary, Newcastle-upon-Tyne NE1 4LP, Great Britain
2 Laboratory of Pathophysiology, NCI, NIH, USA
3 Biochemistry Group, University of Bath, Bath, Great Britain

ADP-Ribosylation of Proteins
(ed. by F.R. Althaus, H. Hilz, and S. Shall)
© Springer-Verlag Berlin Heidelberg 1985

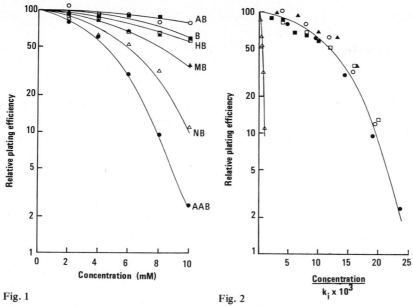

Fig. 1. Colony formation by CHO-Kl cells after 24 h exposure to benzamides

Fig. 2. Colony formation vs inhibitory potency towards poly(ADP-ribose)synthetase in vitro

lular process. For competitive inhibitors such as benzamides, the K_m term in the denominator of the Michaelis-Menten equation is modified by $(1 + I/K_i)$. Different compounds should therefore give the same biological response at the same I/K_i value. Using this criterion, we have measured a number of parameters to determine whether ADP-ribosylation is involved in their regulation.

The Effect of Benzamides on Colony Formation

In order to test for the cytotoxicity of benzamides, we chose an exposure time of 24 h. This was to minimise any cytostatic, but not cytotoxic lesions. Exponentially growing CHO-Kl cells, doubling time approx. 12–14 h, were added to six well dishes containing different concentrations of inhibitor for 24 h. Colonies were allowed to form during a subsequent 5 day incubation in normal growth medium (Ham's F12 + 10% FCS). Figure 1 shows that the inhibitors displayed a wide range of toxicity. When, however, data were expressed as relative plating efficiency vs poly(ADP-ribose) synthetase inhibition (Fig. 2), a remarkably good correlation was obtained for five of the six inhibitors; 3-nitrobenzamide is clearly different (see Discussion). It is apparent from Fig. 3 that these cells tolerate the presence of 10 mM 3AAB for 8 h with only minimal toxicity, suggesting either the lethal lesion can be repaired or it does not manifest itself except on prolonged exposure.

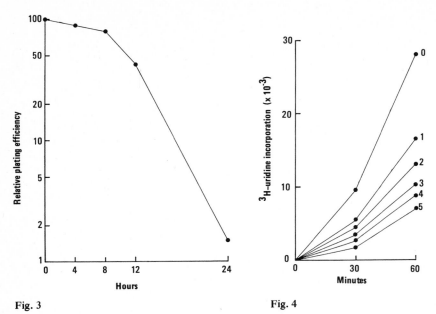

Fig. 3   Fig. 4

Fig. 3. Colony formation after treatment with 10 mM 3AAB for different times

Fig. 4. Inhibition of uridine incorporation by 0–5 mM 3AAB

The Effect of Benzamides on RNA Precursor Incorporation

In order to investigate the molecular basis for cytotoxicity, we examined the effect of benzamides on macromolecular syntheses. The most dramatic effect observed was on [^{3}H]-uridine incorporation (Fig. 4). The reaction time course was non-linear but extrapolation back to the origin gave a common intersection. For comparison of different benzamides, we measured incorporation for 30 min. Figure 5 shows the dose response curves for a number of compounds. 5-Methylnicotinamide also inhibited. It is interesting to note that all compounds gave biphasic curves. At present, we distinguish between two processes being affected or one process having a biphasic sensitivity. Uridine is not the only nucleoside whose incorporation is inhibited by benzamides. Adenosine incorporation (in the presence of hydroxyurea to inhibit DNA synthesis) is also inhibited by 3-acetoamidobenzamide and the two dose response curves are superimposable (Fig. 6); this suggests that the target is probably common to both pathways of incorporation, since it is unlikely that two different kinases, for example, are inhibited to the same extent by 3-acetamidobenzamide.

Analysis of the inhibition produced by 10 mM benzamides showed very poor correlation with a dose response curve for 3-acetamidobenzamide when the data were normalized for poly(ADP-ribose) synthetase inhibition (Fig. 7A). 5-Methylnicotinamide (▼), 3-nitrobenzamide and 3-methoxybenzamide deviated markedly from the AAB dose response curve. A much better correlation was obtained between inhibition

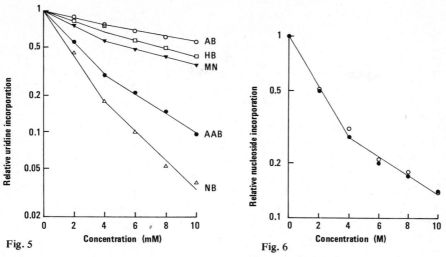

Fig. 5

Fig. 6

Fig. 5. Dose response curves for benzamides and 5-methylnicotinamide (MN)

Fig. 6. Inhibition of uridine (●) and adenosine (2 mM hydroxyurea) (○) incorporation by 3 AAB

and electron withdrawal by the substituent group (Hammet constant) as can be seen from Fig. 7B. Thus the effects observed are probably not poly(ADP-ribose)-mediated.

Figure 8 shows that one of the targets affected by benzamides is transport of nucleosides. At this preliminary stage, we cannot say whether transport is inhibited in a biphasic manner or another component of RNA metabolism is affected.

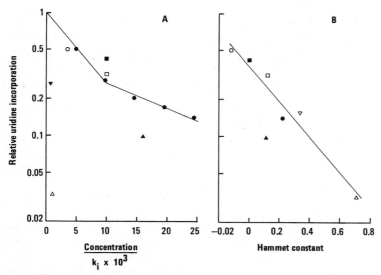

Fig. 7A,B. Inhibition of uridine incorporation by 10 mM benzamides vs (A) inhibition of poly-(ADP-ribose)synthetase or (B) Hammet constant

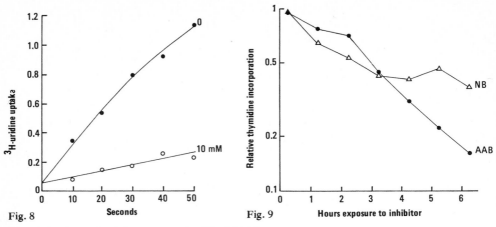

Fig. 8

Fig. 9

Fig. 8. Inhibition of uridine transport by 10 mM AAB

Fig. 9. Inhibition of thymidine incorporation by 10 mM AAB or 6 mM NB. Data expressed as incorporation in 30 min compared to controls

The Effect of Benzamides on Thymidine Incorporation

When [^{3}H]-thymidine incorporation was measured in the presence and absence of 10 mM 3AAB, no significant difference was observed. If, however, cells were preincubated in the presence of 10 mM 3AAB for increasing periods of time, a distinctly biphasic time course was observed (Fig. 9). At present we have no simple explanation for the initial slow decrease in incorporation. Blockage of cell cycle progress in G_1 at approx. 2 h before S-phase could account for the second portion of the curve. The observation that 3-nitrobenzamide at 6 mM (which gave an approx. equivalent inhibition of uridine incorporation) did not give this effect suggests such a block could be poly(ADP-ribose) synthetase mediated. To test whether this was indeed a G_1 block, we treated cells with 10 mM AAB for 8 h, a treatment which caused minimal toxicity (Fig. 3), and then released the block and monitored the recovery of [^{3}H]-thymidine

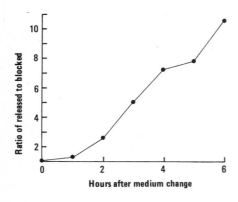

Fig. 10. Recovery of thymidine incorporation after treatment with 10 mM AAB for 8 h. Medium on treated cultures was replaced with fresh medium + AAB. Labelled thymidine was added for 30 min thereafter

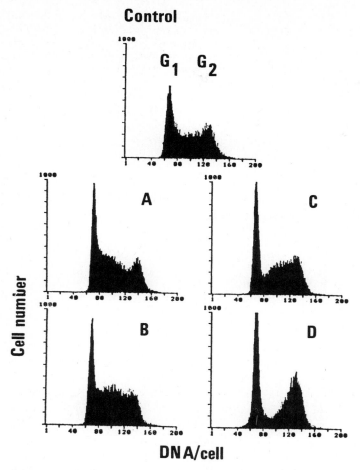

Fig. 11A–D. FMF analysis of cell cycle blockage by 10 mM NB: **A** 4 h treatment; **B** 8 h; or 10 mM AAB: **C** 4 h; **D** 8 h

incorporation. The data in Fig. 10, which showed an increase in incorporation 2 h after removal of inhibitor, are consistent with a G_1 blockade. A third method, viz. FMF analysis, also showed that entry of cells into S-phase was blocked by 10 mM AAB. In contrast, 10 mM 3NB caused an accumulation of cells in the S-phase (Fig. 11).

The Effect of 3-Aminobenzamide on DNA Methylation

Measurement of methylation of [^{14}C]-deoxycytidine incorporated into DNA in the presence of 5 mM 3AB showed no significant change (Fig. 12, Table 2). This is in contrast to the report of Morgan and Cleaver [2] who reported inhibition but did not present any experimental data.

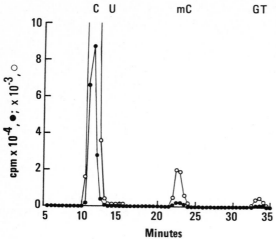

Fig. 12. HPLC resolution of Cyd and m5Cyd following isolation of ^{14}C-dCyd labelled cellular DNA and enzymic digestion to deoxynucleosides. Elution positions of markers are shown

Table 2. DNA methylation in CHO-K1 cells grown in the presence or absence of 5 mM 3AB for 48 h

Addition	m^{5}C/C
O	2.61
	2.63
	2.66
5 mM 3AB	2.63
	2.63

Discussion

It has clearly been shown that inhibition of poly(ADP-ribose) synthetase was responsible for the cytotoxicity exhibited by five of the six benzamides tested. 3-Nitrobenzamide was markedly more cytotoxic than would be expected on the basis of its K_i. Nitro-compounds have been shown to be cytotoxic and genotoxic. This is thought to arise by reduction to reactive intermediates which can then modify DNA [3]. When uridine incorporation was correlated with the Hammet constant, 3-nitrobenzamide fitted very well. The shorter exposure time in this assay minimises any metabolite effects.

It is of interest to note that such high levels of poly(ADP-ribose) synthetase inhibitors are required to exhibit cytotoxicity. One possible explanation is that in normal cycling cells, the poly(ADP-ribose) mediated step can take place over an extended period of time. For enhancement of cytotoxicity of alkylating agents which generally take place at lower inhibitor concentrations, the timing presumably is more critical.

Although probably not contributing to the cytotoxicity of the benzamides, inhibition of nucleoside transport is nevertheless important. A number of workers have used nucleoside incorporations in conjunction with inhibitors to define the role of ADP-ribosylation in the cell, e.g. UDS. Caution should, therefore, be exercised when evaluating such results, particularly as there may be synergy between agents which affect membrane function, e.g. bleomycin, and inhibitors of transport.

The most interesting new finding to emerge is the arrest of cells in G_1 and G_2 by high concentrations of 3-acetamidobenzamide, but not 3-nitrobenzamide. This opens the possibility of synchronizing cells at a poly(ADP-ribose)-mediated restriction point. Release of such a population would allow one to define which events are dependent on release. It should also make isolating cell-cycle specific acceptor proteins from cells considerably easier.

References

1. Purnell MR, Whish WJD (1980) Novel inhibitors of poly(ADP-ribose) synthetase. Biochem J 185:775–777
2. Morgan WF, Cleaver JE (1983) Effect of 3-aminobenzamide on the rate of ligation during repair of alkylated DNA. Cancer Res 43:3104–3107
3. Biaglow JE (1981) Cellular electron transfer and radical mechanisms for drug metabolism. Radiat Res 86:212–242

Inhibition of ADP-Ribosylation Reaction by 2′, 5′-Oligoadenylates

ENZO LEONE†[1], HISANORI SUZUKI[1], BENEDETTA FARINA[1],
ARMAN D. PIVAZIAN[2], and MARAT YA KARPEISKY[2]

1. Introduction

Many experimental data have been obtained regarding the physiological functions of poly(ADP-ribosyl)ation reactions. DNA replication, cell differentiation, transformation of cells, and DNA repair are among the processes in which poly(ADP-ribosyl)-ation is thought to be involved [1–3]. Inhibitors of ADP-ribosyltransferase (ADPRT) have been very useful in the attempt to ascribe a physiological role to these reactions in various cellular functions. 3-Aminobenzamide and 3-methoxybenzamide have been shown to be powerful competitive inhibitors of ADPRT [4]. Some experimental data obtained with the use of these inhibitors clearly indicate that poly(ADP-ribosyl)ation reactions are indeed crucial for cell differentiation [3] and the DNA strand break rejoining process [5]. On the other hand, the use of inhibitors often meets inconveniences in that their effects are not always limited to the desired metabolic pathway [6]. Tanaka et al. [7] have recently reported that diadenosine tetraphosphate, a ligand of a subunit of DNA polymerase α, could act as a very efficient inhibitor of ADPRT, but further data concerning its involvement in ADP-ribosylation reactions have not been reported so far. Polyamines, such as spermine and spermidine, are also known to have inhibitory effects on ADP-ribosylation of histone H1 [8], although their usefulness as a specific inhibitor of ADPRT remains doubtful. In the light of this circumstance, we have tried to examine the effect of various 2′,5′-oligoadenylates (2′,5′A) on the activity of ADPRT. 2′,5′A are synthesized by 2′,5′A polymerase in various tissues especially after viral infection or interferon treatment and are thought to be specific activators of cytoplasmic latent ribonuclease L which, once activated, would hydrolyze viral and cell ssRNA following completion of the antiviral state of the infected cells [9]. Recently, Suhadolnik et al. [10] suggested the possible relationship between interferon actions and ADP-ribosylation of proteins, without pointing out the possible mechanism by which interferon acts on ADP-ribosylation reactions. Hence, the present report is aimed, not only to seek new inhibitors of ADPRT, but also to elucidate the mechanism involved in the regulation of ADP-ribosylation of proteins by interferon.

1 Dipartimento di Chimica Organica e Biologica, Facoltà di Scienze, Università di Napoli, Via Mezzocannone 16, 80134 Napoli, Italy
2 Institute of Molecular Biology, the USSR Academy of Sciences, 32, Vavilov str., Moscow, V-334, USSR

ADP-Ribosylation of Proteins
(ed. by F.R. Althaus, H. Hilz, and S. Shall)
© Springer-Verlag Berlin Heidelberg 1985

2. Materials and Methods

[Adenine-^{14}C]NAD[+] was supplied by Radiochemical Centre, Amersham, England. pppA2'pA2'pA was synthesized enzymatically as described by Stark et al. [11]. All the other 2',5'-oligoadenylates were prepared by chemical synthesis [12]. ADPRT was extracted from bull testis as described by Farina et al. [13]. ADPRT activity was measured as described in [7]. 2',5'A-polymerase activity was assayed as in [11]. 2',5'A concentration was measured as in [14]. 10^7 cells, treated and untreated with mouse interferon (200 Units ml^{-1}) for 18 h, were homogenized in 20 mM HEPES buffer, pH 7.6, containing, in a total volume of 100 μl, 0.25 M sucrose, 0.02 M β-mercapto-ethanol, and 0.5% of the detergent Nonidet P-40 (Sigma); nuclei and cytoplasm were separated by differential centrifugation [12].

3. Results and Discussion

3.1 Effect of 2',5'-Oligoadenylates on ADPRT Activity

All of the tested oligoadenylates (except A2'pA) were shown to inhibit, to some extent, both histone-dependent and Mg^{2+}-dependent reactions catalyzed by ADPRT as shown in Table 1. Maximum inhibition was found with pppA2'pA2'pA which was effective at a micromolar concentration for both types of ADPRT reactions. Generally, a decrease of inhibition of ADPRT was observed with a decrease in the number, either of phosphate groups at 5'-terminus of the oligoadenylates or, of nucleoside units (Table 1). Dose response curves of pppA2'pA2'pA showed that the concentration of this compound, required to give 50% inhibition of initial activity, was around 10 μM and 50 μM for histone-dependent and Mg^{2+}-dependent reactions, respectively (data not shown). A2'pA2'pA was also shown to possess a significant inhibitory effect on the histone-dependent reaction (data not shown). Further kinetic studies were carried out to characterize the inhibitory action of these compounds. The inhibition of ADPRT

Table 1. Effect of different oligoadenylates on ADPRT[a]

2',5'-oligoadenylate	ADPRT activity, %	
	Histone-dependent	Mg^{2+}-dependent
—	100	100
A2'pA	100	94
pA2'pA	74	96
pppA2'pA	60	94
A2'pA2'pA	62	86
pA2'pA2'pA	70	83
pppA2'pA2'pA	15	17

[a] ADPRT assay as described [7]. Reaction mixtures contained 0.05 mUnits enzyme and 50 μM oligoadenylates

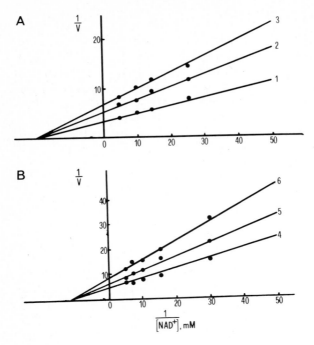

Fig. 1A,B. Inhibition of ADPRT by pppA2′pA2′pA. Of the two double reciprocal plots, **A** refers to experiments for histone-dependent reactions and **B** for Mg^{2+}-dependent reactions. Lines 1, 4 represent results without inhibitors; lines 2, 3, 5, and 6 correspond to 3×10^{-6} M, 5×10^{-6} M, 1×10^{-5} M and 2×10^{-5} M of inhibitor, respectively. ADPRT assays as described in [6], with 0.05 mUnits of enzyme

by these compounds was analyzed by the double reciprocal plot, $1/V$ against $1/S$ (Fig. 1). The constant of inhibition (K_i) with pppA2′pA2′pA was calculated to be 5 μM for the histone-dependent reaction and 20 μM for the Mg^{2+}-dependent reaction. pppA2′pA2′pA was a noncompetitive inhibitor of ADPRT for both reactions. The inhibition pattern of the histone-dependent reaction with A2′pA2′pA was shown to resemble the corresponding pattern of pppA2′pA2′pA, although the K_i was 50 μM (data not shown). This latter observation clearly demonstrates that pppA2′pA2′pA is the strongest noncompetitive inhibitor of ADPRT described so far and its natural occurrence suggests a possible physiological function.

3.2 ADPRT Activity in Mouse L Cells After Interferon Treatment

Endogenous ADPRT activity in mouse L cells was measured with or without interferon treatment (Table 2). A drop of ADPRT activity was seen in both nuclei and cytoplasmic fractions from interferon-treated cells. This decrease was observed both for endogenous enzyme activity as well as for "maximal" activity, i.e., the activity stimulated by DNAse I in nuclei or by the addition of DNA to cytoplasmic preparations.

Table 2. ADPRT levels (mUnits per incubation mixture) in fractions from L cells, before and after treatment with interferon[a]

Nuclei				Cytoplasm			
Nontreated		Treated		Nontreated		Treated	
Without DNAse	With DNAse	Without DNAse	With DNAse	Without DNA	With DNA	Without DNA	With DNA
476	881	238	357	57	119	11	90

[a] ADPRT was determined according to [7]. Incubation mixture, 125 μl final volume, contained nuclei or cytoplasm for 1×10^6 cells; where indicated 40 μg ml^{-1} DNAse I (DN-EP Sigma) or 10 μg ml^{-1} DNA were added, in order to obtain full activation of nuclear and cytoplasmic ADPRT, respectively (see Sect. 3.2)

3.3 Level of 2',5'-Oligoadenylates in Nuclei and Cytoplasm in Control and Interferon-Treated Mouse L Cells

The level of 2',5'A polymerase in nuclei and cytoplasm in L cells with or without interferon treatment was also measured. 2',5'A polymerase activity in nuclei was 10–30 times higher than in cytoplasm in both cases. The concentration of 2',5'A ($\sim$ 0.5 μM) in intact cells after interferon treatment was ten times higher than that found in control cells. This represents almost the same value as the one obtained by Knight et al. in mouse L cells [15]. Thus, although it is impossible to measure the precise concentration of 2',5'A in nuclei because of their permeability across nuclear membrane, it should be expected that the endogenous concentration of 2',5'A in nuclei is higher than the one in cytoplasm, reaching the same order of inhibition constant as calculated by us (5 μM with pppA2'pA2'pA for histone-dependent reaction).

In view of these data, it seems quite probable that the decrease of ADPRT activity found in L cells, especially in nuclei, can be explained as a consequence of enzyme inhibition by increased amounts of oligoadenylates following interferon treatment. The observation that glucocorticoid administration to lymphoblastoid cells increases the level of 2',5'A polymerase activity drastically [16], and that glucocorticoids enhance the expression of the prolactin gene in pituitary glands with concomitant decrease of ADP-ribosylation of nuclear nonhistone proteins HMG 14 and 17 [17], support our interpretation.

In conclusion, our results are indicative of a new pathway involving 2',5'-oligoadenylates in cell metabolism. In fact, such compounds can regulate ADP-ribosylation of proteins, which in turn may be crucial for the induction of the antiviral state of cells treated with interferon.

Acknowledgment. This paper is dedicated to Prof. Enzo Leone, who died in June 1984.

References

1. Mandel P, Okazaki H, Niedergang C (1982) Poly(adenosine diphosphate ribose). In: Waldo E. Cohn (ed) Progress in nucleic acid research and molecular biology, vol 27. Academic Press, London New York, p 1
2. Yamada M, Shimada T, Nakayasu M, Okada H, Sugimura T (1978) Induction of differentiation of mouse myeloid leukemia cells by poly(ADP-ribose). Biochem Biophys Res Commun 83:1325–1332
3. Shall S (1983) ADP-ribosylation, DNA repair, cell differentiation and cancer. In: Miwa M, Hayaishi O, Shall S, Smulson N, Sugimura T (eds) ADP-ribosylation, DNA repair and cancer. VNU Sci Press, Utrecht, p 3
4. Purnell MR, Whish WJD (1980) Novel inhibitors of poly(ADP-ribose)synthetase. Biochem J 185:775–777
5. Durkacz BW, Omidiji O, Gray DA, Shall S (1980) (ADP-ribose)$_n$ participates in DNA excision repair. Nature (London) 283:593–596
6. Cleaver JE, Bodell WJ, Borek C, Morgan WF, Schwartz JL (1983) Poly(ADP-ribose): spectators or participant in excision repair of DNA damage. In: Miwa M, Hayaishi O, Shall S, Smulson M, Sugimura T (eds) ADP-ribosylation, DNA repair and cancer. VNU Sci Press, Utrecht, p 195
7. Tanaka Y, Matsunami N, Yoshihara K (1981) Inhibition of ADP-ribosylation of histone by diadenosine 5′,5′′′-p^1, p^4-tetraphosphate. Biochem Biophys Res Commun 99:837–843
8. Faraone Mennella MR, Farina B, Leone E, Malanga M, Suzuki H (1984) ADP-ribosylation of polyamines and histones. In: Caldarera CM, Bachrach U (eds) Advances in polyamines in biomedical science. CLUEB, BO (Italy), p 69
9. Ball LA (1982) 2′-5′-oligoadenylate synthetase. In: Boyer PD (ed) The enzymes, vol XV. Academic Press, London New York, p 281
10. Suhadolnik RJ, Sawada Y, Gabriel J, Reichnlach NL, Henderson EE (1984) Accumulation of low molecular weight DNA and changes in chromatin structure in HeLa cells treated with human fibroblast interferon. J Biol Chem 259:4764–4769
11. Stark GR, Brown RE, Kerr IM (1981) Assay of (2′-5′)-oligoadenylic acid synthetase levels in cells and tissues: a convenient poly(I)·poly(C) paper-bound enzyme assay. In: Pestka S (ed) Methods in enzymology, vol 79. Academic Press, London New York, p 194
12. Pivazian AD, Suzuki H, Vartanian AA, Zhelkovsky AM, Farina B, Leone E, Karpeisky M Ya (1984) Regulation of poly(ADP-ribose)synthetase by 2′, 5′-oligoadenylates. Biochem Int 9: 143–152
13. Farina B, Faraone Mennella MR, Leone E (1979) Nucleic acids, histones and spermiogenesis: the poly(adenosine diphosphate ribose)polymerase system. In: Salvatore F, Marino G, Volpe P (eds) Macromolecules in the functioning cell. Plenum Press, New York, p 283
14. Nilsen TW, Wood DL, Baglioni C (1982) Presence of 2′,5′-oligo(A) and of enzymes that synthesize, bind, and degrade 2′,5′-oligo(A) in HeLa cell nuclei. J Biol Chem 257:1602–1605
15. Knight M, Cayaley PJ, Silverman RH, Wreschner DH, Gilbert CS, Brown RE, Kerr IM (1980) Radioimmune, radiobinding and HPLC analysis of 2′-5′A and related oligonucleotides from intact cells. Nature (London) 288:189–192
16. Krishnan I, Baglioni C (1980) Increased levels of (2′-5′)oligo(A)polymerase activity in human lymphoblastoid cells treated with glucocorticoids. Proc Natl Acad Sci USA 77:6506–6510
17. Tanuma S, Johnson LD, Johnson GS (1983) ADP-ribosylation of chromosomal proteins and mouse mammary tumor virus gene expression. J Biol Chem 258:15371–15375

Activity Gels of Poly(ADP-Ribose) Polymerase: Phylogenetic Studies and Variations in Human Blood Cells

ANNA I. SCOVASSI[1], ELISABETTA FRANCHI[1], PAOLA ISERNIA[2], ERCOLE BRUSAMOLINO[2], MIRIA STEFANINI[1], and UMBERTO BERTAZZONI[1]

Introduction

We have devised an activity gel procedure to recover the functional poly(ADP-ribose) polymerase catalytic activities after SDS-polyacrylamide gel electrophoresis [1]. Using this technique, we analyzed the catalytic polypeptides of this enzyme present in extracts obtained from a variety of organisms, belonging to a wide evolutionary scale. We also measured the enzyme in human lymphocytes before and after stimulation with mitogens and in human cells of different leukemia states.

Analysis of Poly(ADP-Ribose) Polymerase by the Activity Gel Procedure

The main steps of the activity gel assay are outlined in Table 1. The in situ detection of poly(ADP-ribose) polymerase activities includes gel electrophoresis in SDS, renaturation of proteins with appropriate buffers, incubation of the intact gel with [^{32}P]-NAD, removal of nonincorporated precursor by TCA, washing, and autoradiography. The method was developed by using extracts and partially purified fractions of the poly(ADP-ribose) polymerase. The catalytic peptides of the enzyme are identified as activity bands and their M_r determined by referring to protein markers [1].

Phylogenetic Studies

The phylogenetic survey has been essentially performed by determining the enzyme structure through the activity gel analysis. In Fig. 1 an autoradiogram of a typical activity gel loaded with different extracts is shown. A summary of the results obtained by using this procedure is presented in Table 2. A major activity band of 120,000 dal-

1 Istituto CNR di Genetica Biochimica ed Evoluzionistica, Via Abbiategrasso, 207, 27100 Pavia, Italy
2 Divisione di Ematologia, Policlinico S. Matteo, P.zzale Golgi, 27100 Pavia, Italy

ADP-Ribosylation of Proteins
(ed. by F.R. Althaus, H. Hilz, and S. Shall)
© Springer-Verlag Berlin Heidelberg 1985

Table 1. Activity gel procedure for poly(ADP-ribose) polymerase

---

Samples (extracts or purified fractions of the enzyme)
  ↓
Polyacrylamide-DNA gel
  ↓
Gel electrophoresis in SDS
  ↓
Washing out SDS from the gel
  ↓
Treatment of the gel with 6 *M* guanidine
  ↓
Renaturation of proteins within the gel
  ↓
Incubation of the gel with reaction mixture containing ^{32}P-NAD
  ↓
Washing out nonincorporated NAD with TCA
  ↓
Autoradiography

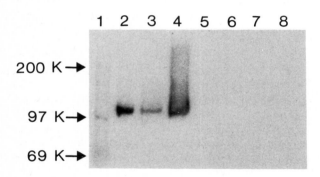

Fig. 1. Activity gel analysis of poly(ADP-ribose) polymerase activity in crude extracts from different organisms. Lanes: *1* M_r markers; *2* HeLa cells; *3* CHO cells; *4* chick embryo; *5* yeast; *6* E. coli; *7* rice cells; *8* carrot cells

Table 2. Phylogeny of poly(ADP-ribose) polymerase

Phylum	Class	Organism	Activity band
Vertebrates	Mammals	HeLa cells	+
	Birds	Chick embryo	+
	Reptiles	Viper gonads	+
	Amphibians	Frog oocytes	+
Arthropods	Insects	Med fly and *Drosophila* cells	+
	Crustaceans	*Artemia salina*	−
Mollusks	Gasteropods	Snail eggs and embryos	+
Echinoderms	Echinoids	Sea urchin embryos	−
Protozoans	Ciliates	*Chritidia fasciculata*	−
	Flagellates	*Trypanosoma brucei*	−
Thallophytes	Ascomycetes	Yeast	−
	Algae	*Chlamydomonas reinhardii*	−
Spermatophytes	Monocotyledones	Rice cells in culture	−
	Dicotyledones	Carrot cells in culture	−
Bacteriophytes	Bacteria	*Escherichia coli*	−

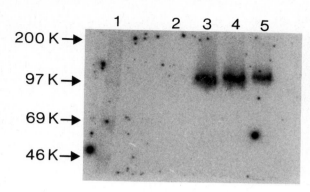

Fig. 2. Activity gel of poly(ADP-ribose) polymerase from human lymphocytes. Lanes: *1* M_r markers; *2* unstimulated human lymphocytes; *3–5* stimulated lymphocytes (after 4, 8, 12 days of stimulation with PHA, respectively)

tons was found in all vertebrate classes studied (mammals, birds, reptiles, amphibians) and in other animal organisms, such as insects and gasteropods. No evidence for the enzyme was so far obtained in crustaceans, echinoids, unicellular eukaryotes, and protozoans. The poly(ADP-ribose) polymerase was not detected in plant cultured cells and in bacteria.

Analysis of the Enzyme in Human Stimulated Lymphocytes

Human lymphocytes were stimulated for prolonged times with mitogens and analyzed for poly(ADP-ribose) polymerase using the activity gel technique. In Fig. 2 the stimulation by phytohemoagglutinin (PHA) is shown. No catalytic activity of the enzyme is evident before stimulation; after 4 days of treatment, a distinct activity band with $M_r = 116,000$ appears and remains constant up to 12 days from the stimulation.

Poly(ADP-Ribose) Polymerase in Human Leukemic Cells

To explore the possible utility of poly(ADP-ribose) polymerase as a marker enzyme in the diagnosis and prognosis of leukemia, we are determining, in concomitance with terminal deoxynucleotydil transferase and adenosine deaminase [2, 3], the levels of the enzyme in human leukemic cells of different origin. Preliminary observations indicate that the activity gel technique can be very usefully applied to this purpose, as outlined in Fig. 3. The results so far obtained are summarized in Table 3. In ANLL the percent of positive cases is about 40%; an interesting observation is that four of the five cases with a myelomonocytic morphology were positive for the enzyme. In the blastic transformation of CML, the enzyme is particularly elevated in about 30% of the cases, confirming the results reported by Ikai et al. [4].

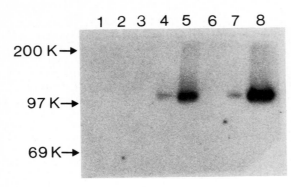

Fig. 3. Activity gel of poly(ADP-ribose) polymerase from human leukemic cells. Lanes: *1* M_r markers; *2* ANLL; *3* ANLL; *4* ALL; *5* BT-CML; *6* ANLL; *7* ANLL, myelomonocytic; *8* HeLa cells. (See also Table 3)

Table 3. Distribution of poly(ADP-ribose) polymerase in human leukemic cells

	Cases	Positive
Acute nonlymphoid leukemia (ANLL)	17	7
Acute lymphoid leukemia (ALL)	2	2
Blastic transformation of chronic myeloid leukemia (BT-CML)	9	3
ANLL after myelodysplastic syndromes (ANLL-MDS)	4	1
Myelodysplastic syndromes (MDS)	2	0
Chronic lymphocytic leukemia (CLL)	1	1

Discussion

We have recently described an activity gel technique for poly(ADP-ribose) polymerase which allows the in situ identification of catalytic polypeptides after SDS-PAGE and we have shown that the activity band of M_r = 120,000 is stimulated in response to treatment with a variety of mutagens [1].

We are now extending the utilization of the activity gel assay to the study of the possible presence and structure of the enzyme in many different organisms, from mammals to bacteria. In fact, a detailed description of enzyme activity in higher and lower eukaryotes is still lacking [5, 6]. From the phylogenetic survey which we are tracing, it appears that the structure of the enzyme is conserved in all vertebrate classes (including mammals, birds, reptiles and amphibians) showing the typical activity band at M_r = 120,000. The same pattern is obtained in some invertebrate animal cells (*Drosophila*, snail), but not in others (sea urchin, *Artemia*). In unicellular eukaryotes (yeast, *Chlamydomonas*), as well as in protozoans (*T. brucei* and *C. fasciculata*) and in cultured plant cells, the activity bands were not detectable. It seems that by descending the evolutionary scale, the enzyme has a tendency to become less distinct.

By using the activity gel procedure, we also analyzed the poly(ADP-ribose) polymerase in human stimulated lymphocytes and in leukemic cells. The enzyme is not detectable in these unstimulated lymphocytes; after treatment with mitogens, a notable activity band with M_r = 120,000 appears and persists up to 12 days from the stimulation. This suggests that the poly(ADP-ribose) polymerase activity is stimulated in coincidence with the increase in the DNA synthesis rate, but retains its full activity at very late times of stimulation, when the capacity to perform repair synthesis is elevated [7]. The requirement for the enzyme during prolonged stimulation of lymphocytes also suggests its involvement in a general mechanism of differentiation [8].

Initial observations on the distribution of the enzyme in human leukemias indicate that the determination of its activity could be usefully coupled to other marker enzymes to better characterize the leukemic phenotype.

Acknowledgments. This work was supported in part by contract BIO-428-81-I of the Radiation Protection Program of the Commission of the European Communities (contribution n° 2177).

References

1. Scovassi AI, Stefanini M, Bertazzoni U (1984) Catalytic activities of human poly(ADP-ribose)-polymerase from normal and mutagenized cells detected after sodium dodecyl sulfate-poly-acrylamide gel electrophoresis. J Biol Chem 259:10963–10967
2. Bertazzoni U, Brusamolino E, Isernia P, Scovassi AI, Torsello S, Lazzarino M, Bernasconi C (1982) Prognostic significance of terminal transferase and adenosine deaminase in acute and chronic myeloid leukemia. Blood 60:685–692
3. Brusamolino E, Isernia P, Lazzarino M, Scovassi AI, Bertazzoni U, Bernasconi C (1984) Clinical utility of terminal deoxynucleotydil transferase and adenosine deaminase determinations in adult leukemia with a lymphoid phenotype. J Clin Oncol 2:871–880
4. Ikai K, Ueda K, Fukushima M, Nakamura T, Hayaishi O (1980) Poly(ADP-ribose) synthesis, a marker of granulocyte differentiation. Proc Natl Acad Sci USA 77:3682–3685
5. Ueda K, Kawaichi M, Hayaishi O (1982) Poly(ADP-ribose)synthetase. In: Hayaishi O, Ueda K (eds) ADP-ribosylation reactions. Academic Press, London New York, pp 118–155
6. Mandel P, Okazaki H, Niedergang C (1982) Poly(adenosine diphosphate ribose). Prog Nucleic Acid Res Mol Biol 27:1–51
7. Bertazzoni U, Stefanini M, Pedrali Noy G, Giulotto E, Nuzzo F, Falaschi A, Spadari S (1976) Variations of DNA polymerases-α and -β during prolonged stimulation of human lymphocytes. Proc Natl Acad Sci USA 73:785–789
8. Johnstone AP, Williams GT (1982) Role of DNA breaks and ADP-ribosyl transferase activity in eukaryotic differentiation demonstrated in human lymphocytes. Nature (London) 300:368–370

The Relationship Between DNA Strand Breaks and ADP-Ribosylation

CHRISTOPHER J. SKIDMORE[1], JANET JONES[1], JANET M. OXBERRY[1],
ELIZABETH CHAUDUN[2], and MARIE-FRANCE COUNIS[2]

Abbreviations:

ADPRT	NAD^+:ADP-ribosyltransferase (EC 2.4.2.30)
AE-cellulose	amino-ethyl cellulose
nam	nicotinamide
PR-AMP	phosphoribosyl-5′-adenosine monophosphate
SVPDE	snake venom phosphodiesterase (EC 3.1.4.1)

Introduction

The nuclear NAD^+:ADP-ribosyltransferase of eukaryotes (EC 2.4.2.30–ADPRT) is a remarkable enzyme. It will catalyse at least two chemically distinct reactions – the transfer of ADP-ribose to protein and the transfer of ADP-ribose to the growing chain of (ADP-ribose)$_n$ [1]. It apparently effects its own short term regulation by auto-modification [2, 3]. In addition it interacts strongly with DNA and is activated by DNA strand breaks [4]. These properties would seem to indicate a certain complexity of structure in the enzyme and, indeed, recent proteolytic studies [5, 6] have indicated that it possesses at least three functional domains.

The interaction of the enzyme with DNA and with strand breaks in particular has been studied mainly with purified enzyme and defined DNAs. In these systems, there is a close correlation between DNA strand breaks and ADPRT activity. The increase in activity when the activating DNA is pretreated with endonucleases is proportional to the number of breaks induced [7] and one break is sufficient to activate one enzyme molecule [8]. Covalently closed circular DNA is incapable of activating ADPRT. Thus it is an attractive hypothesis that the active configuration of the enzyme is that associated with a strand break. Yet it is clear that not all breaks are equally effective as activating sites in vitro. In intact nuclei, the interaction of the enzyme with other chromatin components may be as important for its activity as that with DNA. The nature of the site at which ADPRT binds to chromatin is an open question.

---

1 Department of Physiology & Biochemistry, University of Reading, Whiteknights, Reading RG6 2AJ, Great Britain
2 INSERM U.118-CNRS ERA 842-Association Claude-Bernard, 29, rue Wilhem, 75016, Paris, France

ADP-Ribosylation of Proteins
(ed. by F.R. Althaus, H. Hilz, and S. Shall)
© Springer-Verlag Berlin Heidelberg 1985

Before precise questions can be asked at the molecular level about the important interactions of ADPRT with its substrates and with chromatin, further information is required concerning the response of the enzyme to stimuli under more nearly physiological conditions. To this end we have undertaken studies of the kinetics of ADPRT in permeabilised cells and have attempted to analyse the response of the enzyme to DNA strand breaks.

Procedures

Experiments with Cultured Cells: Cells were permeabilised and assayed by a method based on that described by Halldorsson et al. [17]. The reaction was stopped by making the solution 3 M in urea and 16 mM in nicotinamide. The reaction mixture was then placed on GF/C discs impregnated with 20% TCA and washed extensively in 5% TCA and then acetone.

For product analysis, TCA was added to the stopped reaction to a final 30% and the precipitate washed with 20% TCA and then with ethanol. It was resuspended in 9 M urea and digested in 0.1 M NaOH at 37° for 16 h, finally being neutralised with HCl. SVPDE digestion was performed as described by Stone et al. [18].

Experiments with Chick Lens Cells: Lenses were excised from Leghorn chicks and separated into epithelia and fibre cell masses as described elsewhere [19]. The cells were frozen at −80° and thawed prior to assay by Halldorsson's method [17].

Kinetics and Mechanism of ADP-Ribosyltransferase

The measurement of ADP-ribosyltransferase activity in situ, either in nuclei or in permeabilised cells is a procedure simple in execution but fraught with difficulties of interpretation. The incorporation of radioactivity from labelled NAD$^+$ into TCA insoluble products represents at best the net rate of ADP-ribosylation [ADPRT activity less poly(ADP-ribose) glycohydrolase and ADP-ribosylprotein lyase activities]. At zero time we may hope that the degradative enzymes, having no radiolabelled substrate, do not contribute to the rate.

ADPRT catalyses at least two reactions:

initiation
$$\text{Protein} + \text{NAD}^+ \rightarrow \text{Protein-(ADP-ribose)} + \text{nam}$$

elongation
$$\text{Protein-(ADP-ribose)}_n + \text{NAD}^+ \rightarrow \text{Protein-(ADP-ribose)}_{n+1} + \text{nam}$$
$$(n \geqslant 1)$$

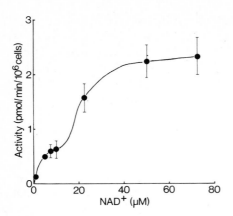

Fig. 1. The variation of ADPRT activity in permeabilised L1210 cells with NAD⁺ concentration

The second of these may be complicated if addition of ADP-ribose residues to create a branch point has a distinct mechanism.

We have attempted to dissect the incorporation of NAD$^+$ into these two reactions by snake venom phosphodiesterase (SVPDE) digestion. The rate of incorporation of label into 5′-AMP gives an estimate of the rate of initiation and the rate of incorporation into phosphoribosyl-AMP (PR-AMP) estimates the rate of elongation. AE-cellulose chromatography can be used to monitor the proportion of 5′-AMP deriving from mono- and poly(ADP-ribose) residues respectively. This provides an important check since, in permeabilised cells, mono(ADP-ribose) residues can be produced other than in the nucleus and also by non-enzymic reactions [9].

A complex reaction mechanism as proposed above will not a priori provide Michaelis-Menten kinetics. Analysis of the first reaction in particular is restricted by the fact that in nuclei the ratio of enzyme to acceptor protein cannot readily be manipulated.

Kinetic Analysis of ADPRT Activity in Permeabilised Cells

When the incorporation of [^{3}H]NAD$^+$ into TCA insoluble products was measured as a function of NAD$^+$ concentration in permeabilised L1210 cells, a pronouncedly biphasic plot was obtained (Fig. 1). This data was analysed using a Hofstee-Eadie transformation (v vs v/s) and was shown to divide quite clearly into two data sets, corresponding to NAD$^+$ concentrations above and below 10 μM. The kinetic constants calculated from this analysis are shown in Table 1. Below 10 μM the reaction has a K_m of 9.45 μM whereas above 10 μM the K_m is 29.9 μM.

If the fate of the radioactive label is analysed by SVPDE digestion then incorporation into 5′-AMP and into PR-AMP, considered separately, each behave in Michaelis-Menten fashion. The kinetic constants obtained for initiation (5′-AMP incorporation) are indistinguishable from those obtained from the low [NAD$^+$] data. The high [NAD$^+$] data and PR-AMP incorporation give estimates of K_m which are in close agreement, the V_{max} for PR-AMP is close to that for the total data less the initiation reaction (5′-AMP) (Table 1).

Table 1. Kinetic constants for ADP-ribosylation in permeabilised L1210 cells[a]

	K_m (μM)	V_{max} (pmol/min/10^6 cells)
Control		
measured by total counts		
[NAD$^+$] $\leqslant 10\ \mu M$	9.5 ± 2.5	1.1 ± 0.2
[NAD$^+$] $> 10\ \mu M$	29.9 ± 4.7	3.5 ± 0.1
after SVPDE digestion		
5'-AMP incorporation	13.9 ± 1.7	0.9 ± 0.2
PR-AMP incorporation	35.7 ± 1.7	2.2 ± 0.2
+ 25 μg/ml DNaseI		
measured by total counts		
[NAD$^+$] $\leqslant 10\ \mu M$	11.6 ± 2.1	3.9 ± 0.6
[NAD$^+$] $> 10\ \mu M$	12.2 ± 1.7	6.2 ± 0.5

[a] Data from Hofstee-Eadie analysis of kinetic data, the mean ± SE deriving from the appropriate linear regression

The kinetic constants for initiation and elongation reactions are thus distinct and initiation, with its lower K_m for NAD$^+$, is predominant at low NAD$^+$ concentrations.

When permeabilised L1210 cells are incubated with 25 μg ml^{-1} DNaseI, the kinetic distinction between low and high NAD$^+$ concentrations remains but the K_m for both data sets is in the region of 12 μM, statistically indistinguishable from that for initiation (Table 1). This can be interpreted as the initiation reaction becoming rate limiting under these conditions. DNA strand breaks have activated the enzyme to such an extent that the free acceptor proteins have become limiting. Further analysis to elucidate this point is at present in progress.

Effect of DNaseI on the Activity of ADPRT

The stimulatory effect of DNaseI on ADPRT activity was first noted by Miller [10]. Most often such treatment has been used to stimulate the enzyme and thus to reveal the 'maximal' activity. We were interested to examine the relationship between DNA breaks and ADPRT activity with the low levels of breaks seen under physiological conditions. Thus we titrated DNaseI concentrations downwards. For permeabilised M707 mouse erythroleukaemia cells we obtained the data shown in Fig. 2. A pronounced threshold was observed at DNaseI concentrations below 5 μg/assay. With larger amounts a dose-dependent increase in the activity of the enzyme was obtained, as expected. The same effect was obtained whether the cells were incubated for 10 min prior to assay with the DNaseI or whether digestion and assay took place over the same time course. Similar results were also observed in L1210 cells (not shown).

The obvious conclusion was that this was a phenomenon caused by insufficient permeabilisation of the cells resulting in the entry of the DNaseI molecules becoming

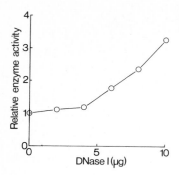

Fig. 2. The effect of low levels of DNaseI on the ADPRT activity of permeabilised M707 mouse erythroleukaemia cells

Table 2. The effect of Triton X-100 on the ADPRT activity of DNaseI-treated permeabilised mouse leukaemia cells

	Control[a]	With 0.05% Triton X-100[a]
M707 cells		
without DNaseI	1.00	3.14
with 5 μg DNaseI	1.20	2.02
with 10 μg DNaseI	3.26	3.05
L1210 cells		
without DNaseI	1.00	4.85
with 5 μg DNaseI	0.56	4.58
with 10 μg DNaseI	2.52	4.26

[a] Activities (expressed as pmol/min/10^6 cells) normalised with respect to the control incubation without DNaseI

the rate limiting step. Triton X-100 was therefore added to disrupt completely the cell membranes and the experiment was repeated. As shown in Table 2, this treatment increased the measured activities some fourfold but did not produce a stimulation by DNaseI of the enzyme over this range. Under these incubation conditions, single-strand breaks could indeed be observed in the DNA after alkaline sucrose density gradient sedimentation.

Thus there appears to be a threshold effect in the action of DNaseI on ADPRT in situ in permeabilised mouse leukaemic cells. There is a level of DNA breakage which does not stimulate the enzyme. A model in which two events, occurring at random, are necessary to activate the enzyme could account for such an effect. In this case a requirement for double-strand breaks would be an attractive hypothesis. Flush-ended double-strand breaks were most effective in activating the enzyme in Benjamin and Gill's experiments [7]. Alternatively, some aspect of chromatin structure could be involved. Digestion of nuclei with low levels of DNaseI preferentially digests 'active' DNA sequences and such areas could be depleted of the enzyme [11]. Berger and his coworkers have shown that in SV40 minichromosome preparations a similar threshold effect is seen in the presence of H1 but not in its absence [12].

Table 3. ADPRT activity in the developing chick lens

	Epithelia[a]	Fibres[a]
a) During the course of development		
6 day embryos	7.72 ± 2.47	7.41 ± 2.65
11 day embryos	6.58 ± 3.52	3.12 ± 2.20
15 day embryos	1.25 ± 0.49	4.08 ± 0.52
18 day embryos	1.35 ± 0.41	2.03 ± 0.95
b) After preincubation at 30° in assay buffer		
6 day embryos		
with 1 h incubation	26.4 ± 0.77	67.2 ± 10.2
with 2 h incubation	70.7[b]	152.7[b]
11 day embryos		
with 1 h incubation	35.4 ± 15.7	31.4 ± 11.7
with 2 h incubation	51.9[b]	23.3[b]

[a] Activities were measured in a crude nuclear preparation and are expressed as pmol/min/10^6 cells (mean ± SD)
[b] Single determination only

ADPRT and DNA Strand Breaks in Lens Cell Development

The foregoing observations used exogenous nucleases to induce DNA strand breaks in permeabilised cells. To investigate a more nearly physiological system we turned to the embryonic chick lens, in which DNA breaks arise endogenously during the course of development. Lens cells derive from a single epithelial layer and when post-mitotic, elongate into fibre cells and fill with specific proteins, the crystallins. During this process the nuclei become pycnotic, accumulating double-strand DNA breaks and eventually degenerating completely (for review see [13]). Pycnotic nuclei are first observed at 8 days of development [14] and double-strand breaks have been reported first on day 15 [15]. Lens cell nuclei contain an endonuclease, activated by Mg^{++} ions, which has greater activity in fibre cell nuclei than in epithelial nuclei [16].

Cells from the lens can be separated into epithelial and fibre fractions by mechanical methods, and a crude nuclear preparation obtained by freeze thawing. When ADPRT activity is measured at different stages of development, the results shown in Table 3a are obtained. The enzyme activity declines in both compartments through development. Only at 15 days is the activity in the fibre cells, which contain DNA breaks, significantly greater than that in the epithelial cells. Nevertheless, the endogenous nuclease is capable of causing considerable DNA breakage which in its turn will activate ADPRT. When nuclear preparations are preincubated in assay buffer for 1 or 2 h prior to assay up to 20-fold increases in ADPRT activity can be measured (Table 3b), accompanied by considerable degradation of the DNA (results not shown).

Again we have a case in which DNA strand breakage occurs, this time during the course of normal development, where no significant activation of ADPRT is observed. In both cases the steady state number of breaks is presumably low but in vitro in the presence of purified DNA, one break is sufficient to activate an ADPRT molecule [8].

This is clearly not the case in chromatin in situ and further investigations are proceeding to quantify and investigate the nature of the breaks involved.

Acknowledgments. This work was supported by the Cancer Research Campaign, by INSERM and Association Claude-Bernard. I also acknowledge the receipt of a travel grant from the Wellcome Foundation.

References

1. Ueda K, Kawaichi M, Oka J, Hayaishi O (1980) Biosynthesis and degradation of poly(ADP-ribosyl) histones. In: Smulson ME, Sugimura T (eds) Novel ADP-ribosylations of regulatory enzymes and proteins. Elsevier/North-Holland, Amsterdam New York, p 47
2. Zahradka P, Ebisuzaki K (1982) A shuttle mechanism for DNA-protein interactions: the regulation of poly(ADP-ribose) polymerase. Eur J Biochem 127:579–585
3. de Murcia G, Jongstra-Bilen J, Ittel M-E, Mandel P, Delain E (1983) Poly(ADP-ribose) polymerase auto-modification and interaction with DNA: electron microscopic visualization. EMBO J 2:543–548
4. Yoshihara K, Kamiya T (1982) Poly(ADP-ribose) synthetase-DNA interaction. In: Hayaishi O, Ueda K (eds) ADP-ribosylation reactions. Academic Press, London New York, p 157
5. Holtlund J, Jemtland R, Kristensen T (1983) Two proteolytic degradation products of calf-thymus poly(ADP-ribose) polymerase are efficient ADP-ribose acceptors: implications for polymerase architecture and the automodification of the polymerase. Eur J Biochem 130: 309–314
6. Kameshita I, Matsuda Z, Taniguchi T, Shizuta T (1984) Poly(ADP-ribose) synthetase: separation and identification of 3 proteolytic fragments as the substrate-binding domain, the DNA-binding domain and the automodification domain. J Biol Chem 259:4770–4776
7. Benjamin RC, Gill DM (1980) Poly(ADP-ribose) synthesis in vitro programmed by damaged DNA: a comparison of DNA molecules containing different types of strand breaks. J Biol Chem 255:10502–10508
8. Ohgushi H, Yoshihara K, Kamiya T (1980) Bovine thymus poly(adenosine diphosphate ribose) polymerase: physical properties and binding to DNA. J Biol Chem 255:6205–6211
9. Hilz H, Koch R, Fanick W, Klapproth K, Adamietz P (1984) Nonenzymic ADP-ribosylation of specific mitochondrial polypeptides. Proc Natl Acad Sci USA 81:3929–3933
10. Miller EG (1975) Stimulation of nuclear poly(adenosine diphosphate-ribose) polymerase activity from HeLa cells by endonucleases. Biochim Biophys Acta 395:191–200
11. Weintraub H, Groudine M (1976) Chromosomal subunits in active genes have an altered conformation. Science 395:191–200
12. Cohen JJ, Catino DM, Petzold SJ, Berger NA (1982) Activation of poly(adenosine diphosphate ribose) polymerase by SV40 minichromosomes: effects of deoxyribonucleic acid damage and histone H1. Biochemistry 21:4931–4940
13. Piatigorsky J (1981) Lens differentiation in vertebrates: a review of cellular and molecular features. Differentiation 19:134–153
14. Modak SP, Perdue SW (1970) Terminal lens cell differentiation: I. histological and micro-spectrophotometric analysis of nuclear degeneration. Exp Cell Res 59:43–56
15. Appleby DW, Modak SP (1977) DNA degradation in terminally differentiating lens fiber cells from chick embryos. Proc Natl Acad Sci USA 74:5579–5583
16. Muel A-S (1984) Structure et degradation de la chromatine lors de la differenciation terminale des cellules de cristallin d'embryon de poulet. Thesis, Univ Pierre & Marie Curie, Paris

17. Halldorsson H, Gray DA, Shall S (1978) Poly(ADP-ribose) polymerase activity in nucleotide permeable cells. FEBS Lett 85:349–352
18. Stone PR, Whish WJD, Shall S (1973) Poly(ADP-ribose) glycohydrolase in mouse fibroblast cells (LS cells). FEBS Lett 36:334–338
19. Counis M-F, Chaudun E, Courtois Y, Skidmore CJ (1985) Nuclear ADP-ribosylation in the chick lens during embryonic development. Biochem Biophys Res Commun 126:859–866

Mechanism of Ethanol Stimulation of Poly(ADP-Ribose) Synthetase

JAMES L. SIMS and ROBERT C. BENJAMIN[1]

Ethanol has recently been reported to potentiate the accumulation of poly(ADP-ribose) in SV40 transformed mouse 3T3 cells following hyperthermia or N-methyl-N′-nitro-N-nitroso-guanidine (MNNG) treatment [1, 2]. We became interested in understanding the mechanism by which ethanol potentiates polymer accumulation and how polymer synthesis might be affected by cellular redox metabolism in ethanol-treated cells and tissues. Figure 1 shows the effect of ethanol on the time course of poly(ADP-ribose) synthesis in nucleotide permeable cells in the presence and absence of very low doses of DNA damage. Polymer synthesis in undamaged control cells was essentially

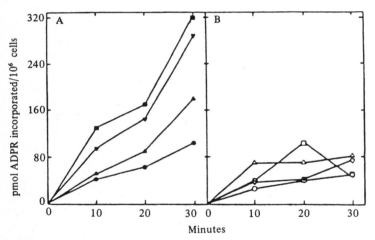

Fig. 1A,B. Effect of ethanol on poly(ADP-ribose) synthetase activity in nucleotide permeable cells. Cells were grown as previously described [3] and were made permeable to added [^{32}P]NAD (100 μM) by the procedure of Berger [4]. Assays were performed in the presence (*closed symbols, A*) and in the absence (*open symbols, B*) of 10 μg ml^{-1} DNase I and in the absence of ethanol (●, ○), and in the presence of 1% (■, □), 3% (▼,▽), and 5% ethanol (▲, △). Note that Triton X-100 was not used in these experiments as described elsewhere [4]. The results of one representative experiment are shown

---

1 Department of Biochemistry, Texas College of Osteopathic Medicine, North Texas State University, Camp Bowie at Montgomery, Ft. Worth, TX 76107, USA

ADP-Ribosylation of Proteins
(ed. by F.R. Althaus, H. Hilz, and S. Shall)
© Springer-Verlag Berlin Heidelberg 1985

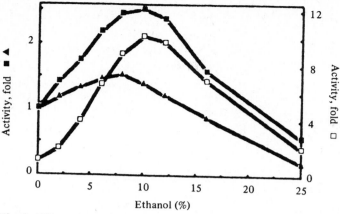

Fig. 2. Effect of ethanol on calf thymus poly(ADP-ribose) synthetase. Initial velocities were obtained by incubating partially purified enzyme with 100 μM [^{32}P]NAD in 0.2 ml reactions containing 50 mM Tris-HCl, pH 8.0, 5 mM EDTA, and 0.06 μg ml^{-1} Hae III fragments with 5′-hydroxyls (▲), 0.3 μg ml^{-1} Hae III fragments with 5′-phosphates (■), or 3 μg ml^{-1} Hpa II fragments (□). The results of one representative experiment are shown and the data for each DNA are expressed as activity relative to the activity in the absence of ethanol (165, 183, 137 pmol ADPR/2 min, respectively)

uneffected by ethanol up to 5%. This was in marked contrast to lightly damaged cells in which polymer synthesis was stimulated 3.5-fold by 1% ethanol. Higher ethanol concentrations also stimulated polymer synthesis, but were less effective than 1%. Control experiments (data not shown) indicated that this was not due to inactivation of DNAse I or altered polymer turnover under these conditions. In order to determine exactly how ethanol stimulates polymer synthesis, we conducted a series of kinetic experiments with partially purified enzyme and defined DNAs. We used restriction endonucleases to generate DNA fragments of pBR322 with flush ends and 5′-phosphates (Hae III fragments), flush ended-fragments with 5′-hydroxyls (alkaline phosphatase treated-Hae III fragments), and fragments with protruding 5′-phosphates (Hpa II fragments). Flush-ended fragments have been shown to be more effective stimulators of synthetase activity than fragments with protruding ends [5]. It has also been shown that flush-ended fragments with 5′-hydroxyls are more effective stimulators of the synthetase than the same fragments with 5′-phosphates [5]. As shown in Fig. 2, the ethanol dose response curve is biphasic. Ethanol was least effective in stimulating enzyme activity when Hae III fragments with 5′-hydroxyls were present compared to Hae III or to Hpa II fragments which have 5′-phosphates. When Hpa II fragments were present, enzyme activity was stimulated as much as 13-fold by 8% ethanol. Thus, ethanol greatly stimulates enzyme activity when relatively poor DNA activators are present and is less effective when the DNA is a very good activator of the enzyme. Concentrations of ethanol above 8% lead to progressive inhibition of enzyme activity for all DNAs tested. Figure 3 shows the results of a Lineweaver-Burk analysis of the initial velocity patterns from experiments carried out in the presence and absence of varying fixed concentrations of ethanol. Figure 3A shows that ethanol

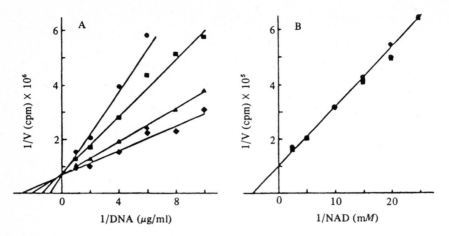

Fig. 3A,B. Effect of ethanol on K_m for NAD and the K_m for DNA. **A** The analysis of the initial velocity patterns for the effect of ethanol on K_{DNA} for assays performed at saturating NAD (2 mM). **B** The analysis of the initial velocity patterns for the effect of ethanol on K_{NAD} for assays performed at saturating DNA (4 μg ml^{-1} Hae III with 5′-OH). Assays were performed in the absence of ethanol (●), and in the presence of 1.5% (■), 3% (▲), and 6% ethanol (♦). The results of one representative experiment are shown

decreased the K_m for Hae III fragments with 5′-hydroxyls approximately threefold from 1 μg ml^{-1} in the absence of ethanol to 0.32 μg ml^{-1} in the presence of 6% ethanol. Figure 3B shows that ethanol did not affect the K_m for NAD (100 μM). Ethanol up to 8% did not effect V_{max}. Similarly, the relatively larger stimulation of enzyme activity by ethanol when Hae III and Hpa II fragments with 5′-phosphates are present is the result of even larger relative decreases in K_m for these DNAs (data not shown). In other experiments with ethanol concentrations above 8% (data not shown), the initial velocity patterns show that the decrease in enzyme activity is due to a large decrease in V_{max}, possibly due to denaturation of the enzyme. Although other effects cannot be ruled out in intact cells, these results strongly suggest that ethanol directly stimulates poly(ADP-ribose) polymerase by increasing its affinity for damaged DNA, its essential activator. It is not entirely clear why 1% ethanol is strongly stimulating in permeable cells while higher ethanol concentrations are needed for large effects with the defined DNAs. One possible explanation is that the breaks generated by DNase I are, in fact, rather poor stimulators of the enzyme and that low ethanol concentrations can greatly tighten the binding of enzyme to the low number of breaks produced in these experiments and produce the large stimulation of activity in a fashion similar to that observed for Hpa II fragments vs Hae III fragments with 5′-hydroxyls. Alternatively, low concentrations of ethanol may effect higher order structure of damaged chromatin in a manner that is strongly stimulating. It is possible that ethanol may also inhibit poly(ADP-ribose) synthetase in cell types that are active in the metabolism of ethanol due to a shift in the redox state and the production of NADH, a potent synthetase inhibitor [6]. The redox state of the transformed mouse 3T3 cell line was measured at various times following the addition of 1% ethanol to

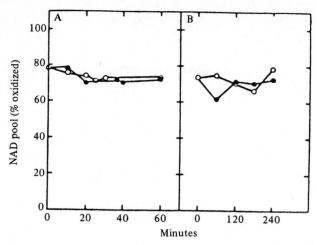

Fig. 4A,B. The effect of ethanol on the redox state of SV40 transformed 3T3 cells. Cells were seeded at $1 \times 10^5/35$ mm dish and 24 h later cells in log phase of growth were incubated in control medium (○) or medium containing 1% ethanol (●) for the indicated periods of time. Pyridine nucleotides were extracted and assayed as previously described [3, 7]. The results of two representative experiments are shown

the culture medium, a concentration which potentiates polymer accumulation in damaged whole cells [1, 2]. As shown in Fig. 4, 1% ethanol did not effect the redox state of these cells. Thus, in this cell line, ethanol appears to directly stimulate poly-(ADP-ribose) synthetase and does not lead to a significant alteration of cellular redox state. We, therefore, believe that the potentiation of polymer accumulation by ethanol in these cells reflects the direct effects of ethanol on the synthetase and is not complicated by the simultaneous production of a potent inhibitor, although direct effects on turnover in intact cells cannot be ruled out. NAD depleted cells have been reported to have increased sensitivities to various stresses [8, 9] and ethanol is known to sensitize cells to a variety of cellular stresses [10, 11]. Our results suggest that ethanol-treated cells may be much more susceptible to the cytotoxic effects of some stresses due, at least in part, to altered cellular NAD metabolism resulting from the stimulation of poly(ADP-ribose) synthetase. Although ethanol has been reported to have an effect on poly(ADP-ribose) synthetase activity [2, 12], this is, to our knowledge, the first report concerning the mechanism of stimulation and its possible significance to biological systems.

References

1. Juarez-Salinas H, Duran-Torres G, Jacobson MK (1983) Hyperthermia potentiates the accumulation of poly(ADP-ribose). Fed Proc 42:1337 (Abs)
2. Juarez-Salinas H, Duran-Torres G, Jacobson MK (1984) Alteration of poly(ADP-ribose) metabolism by hyperthermia. Biochem Biophys Res Commun 122:1381–1388

3. Jacobson EL, Lang RA, Jacobson MK (1979) Pyridine nucleotide synthesis in 3T3 cells. J Cell Physiol 88:417–425
4. Berger NA, Weber G, Kaichi AS (1978) Relation of poly(adenosine diphosphoribose) synthesis to DNA synthesis and cell growth. Biochim Biophys Acta 519:87–104
5. Benjamin RC, Gill DM (1980) Poly(ADP-ribose) synthesis in vitro programmed by damaged DNA. J Biol Chem 255:10502–10508
6. Gill DM (1972) Poly(adenosine diphosphate ribose) synthesis in soluble extracts of animal organs. J Biol Chem 247:5964–5971
7. Jacobson EL, Jacobson MK (1976) Pyridine nucleotide levels as a function of growth in normal and transformed 3T3 cells. Arch Biochem Biophys 175:627–634
8. Durkacz BW, Omidiji O, Gray DA, Shall S (1980) (ADP-ribose)$_n$ participates in DNA excision repair. Nature (London) 283:593–596
9. Jacobson EL, Smith JY, Mingmuang M, Meadows R, Sims JL, Jacobson MK (1984) Effect of nicotinamide analogues on recovery from DNA damage in C3H10T1/2. Cancer Res 44:2485–2492
10. Li GC, Shiu EC, Hahn GM (1980) Similarities in cellular inactivation by hyperthermia or by ethanol. Radiat Res 82:257–268
11. Mizuno S (1981) Ethanol-induced cell sensitization to bleomycin cytotoxicity and the inhibition of recovery from potentially lethal damage. Cancer Res 41:4111–4114
12. Kristensen T, Holtlund J (1978) Poly(ADP-ribose) polymerase from Ehrlich ascites tumor cells: properties of the purified enzyme. Eur J Biochem 88:495–501

Specific Proteolytic Processing of Poly(ADP-Ribose) Polymerase in Human Lymphocytes

NATHAN A. BERGER, CAROL S. SUROWY, and SHIRLEY J. PETZOLD[1]

Introduction

Four enzymatic systems have been described that directly and selectively affect cellular metabolism of poly(ADP-ribosyl)ated proteins. (1) The enzyme poly(ADP-ribose) polymerase cleaves nicotinamide from NAD and then forms a covalent linkage between the residual ADP-ribose moiety and specific protein acceptor molecules [1, 2]. The same enzyme can add successive ADP-ribose moieties to form homopolymers in excess of 100 residues with alternating phosphodiester and O-glycosidic linkages. (2) Poly(ADP-ribose) glycohydrolase degrades poly(ADP-ribose) at the O-glycosidic linkages in an exoglycosidic fashion [3, 4]. This appears to be the major cellular enzyme responsible for polymer degradation. (3) Phosphodiesterases degrade the polymer at the phosphodiester bonds. Some phosphodiesterases function as endonucleases and others as exonucleases [5, 6]. However, these enzymes do not appear to play an important role in cellular modulation of poly(ADP-ribose) levels. (4) Poly(ADP-ribose)-protein hydrolase specifically hydrolyzes the covalent linkage between proteins and ADP-ribose [7]. The physiological importance of this enzyme has recently been suggested by the identification of a lysosomal storage disease in which glutamyl ribose 5-phosphate accumulates in tissues, presumably due to a genetic deficiency of this enzyme [8]. Our studies in human lymphocytes reveal that a nucleotide and/or pyrophosphate stimulated proteolytic processing system constitutes a fifth enzymatic pathway for selective modulation of poly(ADP-ribosyl)ated proteins.

Results and Discussion

We have used permeabilized cells supplied with [^{32}P]NAD to label ADP-ribosylated proteins followed by gel electrophoresis and autoradiography to identify [^{32}P]ADP-ribosylated bands [9, 10]. Using this approach, we have shown that a number of proteins are labeled in resting human lymphocytes, including bands of mol. wt. 116,000,

---

1 Hematology/Oncology Division, Departments of Medicine and Biochemistry, Case Western Reserve University, School of Medicine and University Hospitals of Cleveland, Cleveland, OH 44106, USA

ADP-Ribosylation of Proteins
(ed. by F.R. Althaus, H. Hilz, and S. Shall)
© Springer-Verlag Berlin Heidelberg 1985

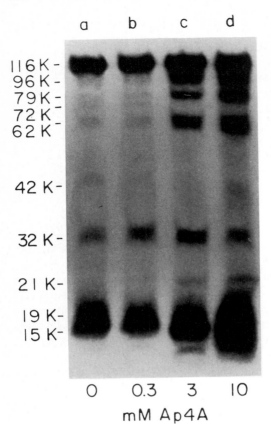

Fig. 1. Autoradiograph of the effect of Ap4A on acceptors of poly(ADP-ribose) in human lymphocytes. Freshly prepared human lymphocytes were treated with MNNG for 1 h, and then permeabilized and incubated for 60 min with 0.8 μM [^{32}P]NAD in the absence or presence of increasing concentrations of Ap4A. Molecular weights were determined by Coomassie blue staining of internal standards. Details of SDS-polyacrylamide gel electrophoresis and autoradiography in [10]

72,000, 42,000, 32,000 and several bands in the region of the core histones [9]. The poly(ADP-ribosyl)ated band at mol. wt. 116,000 is the enzyme poly(ADP-ribose) polymerase and the labeled band at mol. wt. 32,000 is histone H-1 [9]. Induction of DNA damage with MNNG increases the intensity of labeling of the mol. wt. 116,000, poly(ADP-ribose) polymerase, and also results in labeling of a new band at mol. wt. 62,000. The labeling in response to DNA damage is also increased on bands at 42,000, 32,000, and the core histones. Phytohemagglutin stimulation of lymphocytes results in the appearance of ADP-ribosylated bands at mol. wt. 96,000 and 37,000. These studies indicate that unique proteins are ADP-ribosylated in resting, proliferating, and DNA damaged cells [9, 10].

Since the conditions described above, i.e., intermitotic arrest, proliferation, and DNA damage are associated with dramatic changes in nucleotide pools, we conducted a study to determine the effect of nucleotides and their components on protein acceptors for poly(ADP-ribosyl)ation [11, 12]. Figure 1 shows the effect of increasing concentrations of Ap4A on poly(ADP-ribosyl)ated proteins in human lymphocytes. In the absence of Ap4A, the autoradiograph shows that the most prominent ADP-ribosylated proteins are poly(ADP-ribose) polymerase at mol. wt. 116,000, histone H-1 at 32,000, and the core histones between 15,000 and 21,000. With increasing concentrations of Ap4A, bands of increasing intensity appear at mol. wt. 96,000, 79,000, and 62,000. In pulse chase experiments, we added [^{32}P]NAD to first label the major bands then added an excess of cold NAD to prevent further labeling and then added Ap4A. These experiments showed that the bands at mol. wt. 96,000, 79,000, and 62,000 were all apparently derived from the band at 116,000. Examination of the Coomassie blue stained gel showed that the protein bands were essentially unchanged on incubation with Ap4A except for a decrease in a minor protein band at mol. wt. 116,000 as well as a decrease in a major protein band at mol. wt. 32,000.

The studies outlined above suggest that Ap4A stimulates the processing of poly(ADP-ribosyl)ated poly(ADP-ribose) polymerase into specific fragments of mol. wt. 96,000, 79,000, and 62,000. To confirm that the three ADP-ribosylated bands induced by incubation with Ap4A were all derivatives of the 116,000 mol. wt., poly(ADP-ribose) polymerase, we labeled purified lamb thymus poly(ADP-ribose) polymerase by incubating it with [^{32}P]NAD and then incubated the labeled enzyme with either lymphocytes, Ap4A, or with the combination. The effects were evaluated by gel electrophoresis and autoradiography. Ap4A alone had no effect on the size of the ADP-ribosylated enzyme, indicating that other cellular factors were required for processing. Incubation of the enzyme with permeabilized lymphocytes resulted in some processing and incubation with Ap4A plus lymphocytes resulted in marked processing of the ADP-ribosylated enzyme to labeled bands at mol. wt. 96,000, 79,000, and 62,000. Thus, the specific pattern of processing was reproduced by incubating [^{32}P] ADP-ribosylated-poly(ADP-ribose) polymerase with Ap4A-treated lymphocytes.

We found that the Ap4A-induced processing in lymphocytes could best be quantitated by using densitometry to measure the extent of the disappearance, on autoradiographs, of the ^{32}P-labeled, 116,000 mol. wt. ADP-ribosylated protein. Figure 2 shows the effect of a series of nucleotides and their components on this process. In addition to Ap4A, we found that Ap5A, Ap3A, ATP, dATP, p4A, α,β CH$_2$ATP, β, γ CH$_2$ATP, CTP, GTP, and UTP all induced disappearance of ADP-ribosylated-poly-(ADP-ribose) polymerase with relatively similar concentration effects. Thus, small effects were produced by 1 mM nucleotide, 30–50% disappearance was induced by 3 mM nucleotide, and 60–85% disappearance was induced by 10 mM nucleotide. In contrast to the agents which stimulate processing, compounds such as adenosine, cyclic AMP, 5′AMP, and ADPR were ineffective.

The common structural feature of all the compounds that stimulate processing is the presence of a pyrophosphate group. As shown in Fig. 2, when inorganic pyrophosphate was tested, it too was found to stimulate processing of the ADP-ribosylated-poly(ADP-ribose) polymerase. Intact pyrophosphate was required to activate the processing system since inorganic phosphate alone was ineffective. In addition, the nonphosphorylated compounds and nucleoside monophosphates were not able to stimulate processing. ADP-ribose was interesting in that it contained a pyrophosphate but was not very effective in stimulating processing. In this case, it is possible that the adenosine ribose on the one side and the ribose, on the other side, create steric hindrance which interferes with the ability of the pyrophosphate moiety to activate the processing system. NAD, which would have a similar steric hindrance problem could not be tested in this system because of its effect in diluting the labeled [^{32}P]NAD.

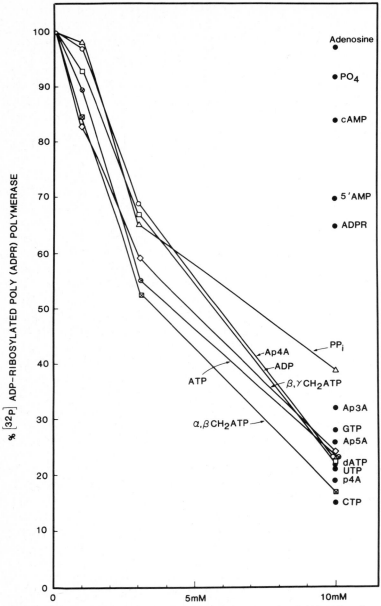

Fig. 2. Effect of various compounds on the processing of poly(ADP-ribose) polymerase. Lymphocytes were treated with MNNG, permeabilized, and incubated with [³²P]NAD for 1 h as in Fig. 1 along with the indicated concentration of each compound. The percentage of poly(ADP-ribose) polymerase remaining, relative to control was determined by densitometry of labeled bands on autoradiography of SDS polyacrylamide gels

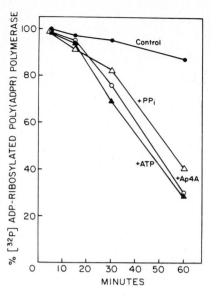

Fig. 3. Time course of effect of Ap4A, ATP, or pyro-phosphate on processing of poly(ADP-ribose) poly-merase. Lymphocytes were treated with MNNG, permeabilized, incubated with [^{32}P]NAD along with 10 mM Ap4A, ATP, or pyrophosphate for the indi-cated time periods. Products of each reaction were analyzed by SDS-polyacrylamide gel electrophoresis and autoradiography as discussed in Fig. 1 with quantitation performed as in Fig. 2. ● Control, △ PPi, ○ Ap4A, ▲ ATP

The pattern of fragments of poly(ADP-ribose) polymerase produced by processing was essentially the same when processing was stimulated by any of the effective com-pounds. Thus, human lymphocytes contain a system which is activated by nucleotides containing an accessible pyrophosphate moiety or by pyrophosphate itself to process poly(ADP-ribose) polymerase.

Figure 3 shows the time course of processing induced by Ap4A, ATP, or PPi. In all cases, the time courses were similar, indicating that extra time was not required for one of these nucleotides to be converted or degraded to another, thus, suggesting that each of these compounds directly activates the processing system.

To determine whether the processing of poly(ADP-ribose) polymerase is mediated by proteases, we evaluated the effects of different classes of protease inhibitors on the disappearance of the 116,000 mol. wt. ADP-ribosylated protein. Table 1 shows that inhibitors of trypsin-like enzymes, such as PMSF, DIFP, and soybean trypsin inhibitor all block the processing. Similarly, pepstatin A, which inhibits acid proteases, also blocks processing. In contrast, agents such as PHMB, N-ethylmaleimide, and iodo-acetamide which block sulfhydryl proteases and phosphoramidon, O-phenanthroline, and EDTA which block metaloproteases all failed to inhibit the processing. Thus, it appears that ApA4-stimulated processing of poly(ADP-ribose) polymerase involves specific proteases and that both a trypsin-like enzyme and an acid protease are involved. Examination of the bands that are labeled in the presence of these inhibitors demonstrates that no processing occurs when either trypsin inhibitors or acid protease inhibitors are present. Thus, activities of both these proteases are required for any processing of poly(ADP-ribose) polymerase to occur. The acid protease present in lymphocytes that is likely to be inhibited by pepstatin A is cathepsin D. We have demonstrated that human lymphocytes contain both trypsin-like and cathepsin D-like activities and will use these terms to refer to the two enzymes involved in the proteolytic processing of poly(ADP-ribosyl)ated-poly(ADP-ribose) polymerase.

Table 1. Protection against Ap4A induced decrease of poly(ADP-ribosyl)ated-poly(ADP-ribose) polymerase[a]

Agent	Protection (%)
Trypsin inhibitors	
PMSF	91
Diisopropylfluorophosphate	100
Soybean trypsin inhibitor	86
Acid protease inhibitors	
Pepstatin A	98
Sulfhydryl protease inhibitors	
PHMB	18
N-ethylmaleimide	19
Iodoacetamide	13
Metalo-protease inhibitors	
Phosphoramidon	8
O-phenanthroline	6
EDTA	8

[a] Proteolysis of poly(ADP-ribosyl)ated-poly(ADP-ribose) polymerase in human lymphocytes was induced with 10 mM Ap4A. All inhibitors were present at 5 mM except for soybean trypsin inhibitor which was at 50 μM

To determine whether the processing could be reproduced in vitro, we labeled purified lamb thymus poly(ADP-ribose) polymerase with [^{32}P]NAD and then treated it with trypsin or cathepsin D alone, in combination, and in the absence or presence of ATP. Trypsin cleaved ADP-ribosylated-poly(ADP-ribose) polymerase to a 62,000 mol. wt. labeled band, whereas cathepsin D had no apparent effect. No further cleavage was produced by combinations of these enzymes with or without added nucleotides. Thus, the in vivo system was partly reconstituted by incubating poly(ADP-ribose) polymerase with trypsin. However, other in vivo associations are required to observe the specific pattern of fragments produced in lymphocytes.

The identification of specific agents to inhibit proteolytic processing made it possible to determine whether native poly(ADP-ribose) polymerase could be processed in a similar fashion to the ADP-ribosylated enzyme and whether the resultant cleavage fragments retain enzymatic activity or acceptor capacity. Figure 4, lanes a and b show the patterns of ADP-ribosylated proteins that occur in cells incubated with [^{32}P]NAD in the absence and presence of ATP to induce processing. In lane c the cells were first incubated with ATP at 0°C for 30 min to induce processing, PMSF and pepstatin A were then added to inhibit further proteolysis and then [^{32}P]NAD was added. The cells were subsequently shifted to 37°C and incubated for a further 15 min before gel electrophoresis and autoradiography. Under these conditions, proteolysis of the polymerase was allowed to occur, but the polymerase activity was preserved by incubation at 0°C. Figure 4, lance c, shows that there was intense labeling of both the polymerase at mol. wt. 116,000 and the band at 62,000. There was also slight labeling of the bands at 96,000 and 79,000. Lane d is similar to lane c except that the preincubation with ATP was conducted at 37°C for 30 min. The results are qualitatively similar to lane c,

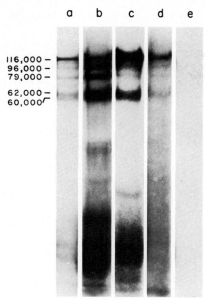

116,000 –
96,000 –
79,000 –

62,000 –
60,000 –

Fig. 4. Autoradiographs showing ADP ribosylation of fragmented poly(ADP-ribose) polymerase. Lymphocytes were treated with MNNG, permeabilized, and incubated with [³²P]NAD as in Fig. 1. *a* No ATP; *b* 10 mM ATP. *c–e* Cells were incubated with 10 mM ATP, for 30 min at 0°C (*c*) or 37°C (*d, e*). Then 5 mM PMSF and 5 mM pepstatin A were added to incubations followed by addition of [³²P]NAD and cells were incubated for a further 15 min at 37°C. Cells in *e* also received 10 mM 3-AB before the addition of [³²P]NAD. SDS gel electrophoresis and autoradiography was performed as in Fig. 1

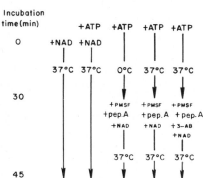

in that all bands were capable of being labeled after the proteolysis was complete. Lane e shows that addition of 3-aminobenzamide at the end of the proteolysis period prevented any labeling when cells were incubated with [³²P]NAD, thus confirming that the labeling in c and d was the result of ADP-ribosylation. These studies confirm that the 96,000, 79,000 and 62,000 mol. wt. cleavage fragments of poly(ADP-ribose) polymerase can serve as acceptors for poly(ADP-ribosyl)ation even after proteolytic processing. The intense labeling of the band at mol. wt. 62,000 suggests the possibility that this fragment also retains enzymatic capacity to synthesize poly(ADP-ribose). The possibility that cleavage fragments can retain enzyme activity is supported by the report of Nishikima et al., showing that a 74,000 mol. wt. fragment retained enzymatic activity after in vitro cleavage of purified poly(ADP-ribose) polymerase with papain [13]. It is interesting to note that in 1978, Tsopanakis et al. reported the purification, from pig thymus, of poly(ADP-ribose) polymerase with a mol. wt. of 63,500 [14].

Table 2. Proteolysis of histone H-1 and poly(ADP-ribose) polymerase by pyrophosphate, nucleo-tide-stimulated protease system

	Poly(ADP R) polymerase	Histone H-1
Proteolysis		
Native protein	+	+
ADP-ribosylated protein	+	+
T 1/2 of proteolysis	45 min	12 min
Activators		
Nucleoside diphosphate	+	+
Pyrophosphate	+	+
Nucleoside monophosphate	−	−
Nucleoside	−	−
Inhibitors		
Trypsin inhibitors	+	+
Acid protease inhibitors	+	−
Sulfhydryl protease inhibitors	−	−
Metalo-protease inhibitors	−	−

Poly(ADP-ribose) polymerase isolated from most other sources has been found to have a mol. wt. of 110,000 to 130,000 [15, 16]. Thus, our present findings suggest that the 63,500 mol. wt. enzyme from pig thymus may have resulted from a specific, in vivo proteolytic cleavage process.

We have demonstrated the presence of a proteolytic system in human lymphocytes which is activated by nucleotides containing an accessible pyrophosphate moiety or by pyrophosphate itself to proteolytically process poly(ADP-ribose) polymerase to specific fragments of mol. wt. 96,000, 79,000, and 62,000. This system is similar to our recent discovery of a nucleotide and pyrophosphate stimulated system which processes histone H-1 [17].

Table 2 summarizes our studies on the pyrophosphate and nucleotide stimulated proteolysis of poly(ADP-ribose) polymerase and histone H-1. The processing appears to be selective for these two chomosomal proteins since no other bands were noted to disappear from the Coomassie blue stained gel. This system is capable of proteolytically processing both the native and ADP-ribosylated forms of poly(ADP-ribose) polymerase and histone H-1. Histone H-1 processing (T 1/2 = 12 min) is significantly faster than that of poly(ADP-ribose) polymerase (T 1/2 = 45 min) suggesting that the attack on the histone may be direct, whereas some intermediate steps may be required for processing of poly(ADP-ribose) polymerase. The same set of activators stimulate processing of the enzyme and histone H-1. Thus, inorganic pyrophosphate and almost all nucleotides containing a pyrophosphate moiety are capable of stimulating proteolysis. ADPR and NAD where there is steric hindrance to the pyrophosphate moiety do not stimulate proteolysis. Nucleoside monophosphates and inorganic phosphates are also not effective at stimulating processing. Inhibitors of trypsin-like enzymes prevent proteolysis of both proteins. Pepstatin A, an inhibitor of acid proteases, prevents proteolysis of poly(ADP-ribose) polymerase, but has no effect on the processing of histone H-1. This observation suggests that an acid protease, such as cathespin D, may

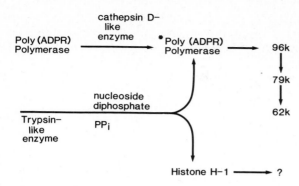

Fig. 5. Proposed pathway for pyrophosphate, nucleotide-stimulated proteolytic processing of poly(ADP-ribose) polymerase and histone H-1. *Poly(ADP-ribose) polymerase* indicates protein that has been acted on by cathepsin D-like enzyme

be required in the preliminary step for processing of poly(ADP-ribose) polymerase. This additional step may account for the slower time course of its processing compared to that of histone H-1.

Figure 5 presents a tentative sequence for the pyrophosphate-, nucleotide-stimulated proteolysis of histone H-1 and poly(ADP-ribose) polymerase that is consistent with all of our experimental results. The trypsin-like enzyme appears to be responsible for selectively processing histone H-1 and poly(ADP-ribose) polymerase. The cathepsin D-like enzyme is only required for processing of the poly(ADP-ribose) polymerase and this enzyme could either affect the trypsin-like enzyme or the poly(ADP-ribose) polymerase. Our preliminary studies indicate that the cathepsin D-like activity is only required early in the processing of poly(ADP-ribose) polymerase, whereas the trypsin-like activity is required throughout. These studies suggest that the cathepsin D-like activity is a prerequisite for further processing. If the cathespin D-like enzyme has a direct effect on poly(ADP-ribose) polymerase, then it must act very near to one of the termini of the enzyme, since no change in molecular weight is detected when the cathepsin D-like enzyme is allowed to act, while the trypsin-like enzyme is inhibited. The phyrophosphate, or nucleotide stimulation may affect the trypsin-like enzyme or may alter histone H-1 or poly(ADP-ribose) polymerase to render them susceptible to the trypsin-like enzyme. Our studies suggest that the trypsin-like enzyme cleaves the mol. wt. 116,000 poly(ADP-ribose) polymerase, in a sequential process to the species at 96,000, 79,000, and 62,000.

The role of this selective nucleotide and pyrophosphate stimulated proteolysis of histone H-1 and poly(ADP-ribose) polymerase in cellular processes remains to be determined. In this regard it is interesting to note that the 96,000 mol. wt. band seems to be selectively produced in mitogen-stimulated cells and the 62,000 band seems to be selectively produced in DNA damaged cells. Of particular relevance is the fact that this proteolysis is stimulated by physiological concentrations of ATP. In addition, it has recently been reported that stress, such as oxidation or heat shock, causes elevation of ATP levels into the concentration range where they could activate this processing system [18, 19], thus affecting two proteins important in modulating nuclear function.

Acknowledgments. These studies were supported by PHS Grants CA35983 and GM32647. Some of the studies were performed in facilities supported by the Sohio Foundation. A travel grant to C.S.S. from the Wellcome Trust, England, is acknowledged.

References

1. Hayaishi O, Ueda K (1977) Poly(ADP-ribose) and ADP ribosylation of proteins. Annu Rev Biochem 46:95−116
2. Hilz H, Stone P (1976) Poly(ADP-ribose) and ADP-ribosylation of proteins. Rev Physiol Biochem Pharmacol 76:1−57
3. Ueda K, Oka J, Narumiya S, Miyakawa N, Hayaishi O (1972) Poly ADP-ribose glycohydrolase from rat liver nuclei, a novel enzyme degrading the polymer. Biochem Biophys Res Commun 46:516−523
4. Miwa M, Tanuka M, Matsushima T, Sugimura T (1974) Purification and properties of a glycohydrolase from calf thymus splitting ribose-ribose linkages of poly(Adenosine Diphosphate Ribose). J Biol Chem 249:3475−3482
5. Matsubara H, Hasigawa S, Fujiwara S, Shima T, Sugimura T (1970) Studies on poly (Adenosine Diphosphate Ribose). J Biol Chem 245:3606−3611
6. Matsubara H, Hasigawa S, Fujiwara S, Shima T, Sugimura T, Futai M (1970) Studies on poly Adenosine Diphosphate Ribose). J Biol Chem 245:4317−4320
7. Okayama H, Honda M, Hayaishi O (1978) Novel enzyme from rat liver that cleaves an ADP-ribosyl histone linkage. Proc Natl. Acad Sci USA 75:2254−2257
8. Williams JC, Chambers JP, Liehr JG (1984) Glutamyl ribose 5-phosphate storage disease. A hereditary defect in the degradation of poly(ADP-ribosylated) proteins. J Biol Chem 259: 1037−1042
9. Surowy CS, Berger NA (1983) Unique acceptors for poly(ADP-ribose) in resting, proliferating and DNA-damaged human lymphocytes. Biochim Biophys Acta 740:8−18
10. Surowy CS, Berger NA (1983) Diadenosine $5',5'''$ p^1, p^4-tetraphosphate stimulates processing of ADP-ribosylated poly(ADP-ribose) polymerase. J Biol Chem 258:579−583
11. Tyrsted G, Munch-Petersen B (1977) Early effects of phytohemagglutinin on induction of DNA polymerase, thymidine kinase, deoxyribonucleoside triphosphate pools and DNA synthesis in human lymphocytes. Nucleic Acid Res 4:2713−2732
12. Berger NA, Berger SJ, Sikorski GW, Catino DM (1982) Amplification of pyridine nucleotide pools in mitogen-stimulated human lymphocytes. Exp Cell Res 137:79−88
13. Nishikimi M, Ogasawara K, Kemishita I, Taniguchi T, Shizuta Y (1982) Poly (ADP-ribose) synthetase. The DNA binding domain and the automodification domain. J Biol Chem 257: 6102−6105
14. Tsopanakis C, Leeson E, Tsopanakis A, Shall S (1978) Purification and properties of poly-(ADP-ribose) polymerase from pig-thymus nuclei. Eur J Biochem 90:337−345
15. Petzold SJ, Booth BA, Leimbach GA, Berger NA (1981) Purification and properties of poly-(ADP-ribose) polymerase from lamb thymus. Biochemistry 20:7075−7081
16. Carter SG, Berger NA (1982) Purification and characterization of human lymphoid poly-(adenosine diphosphate ribose) polymerase. Biochemistry 21:5475−5481
17. Surowy CS, Berger NA (1983) Nucleotide-stimulated proteolysis of histone H-1. Proc Natl Acad Sci USA 80:5510−5514
18. Lee PC, Bochner BR, Ames BN (1983) AppppA, heat-shock stress, and cell oxidation. Proc Natl Acad Sci USA 80:7496−7500
19. Bochner BR, Lee PC, Wilson SW, Cutler CW, Ames BN (1984) ApppA and related adenylated nucleotides are synthesized as a consequence of oxidation stress. Cell 37:225−232

Cytoplasmic Poly(ADP-Ribose) Synthetase in Rat Spermatogenic Cells

ILONA I. CONCHA, MARGARITA I. CONCHA, JAIME FIGUEROA,
and LUIS O. BURZIO[1]

Introduction

Poly(ADP-ribose) synthetase uses NAD as a substrate to catalyze the formation of a homopolymer of repeating ADP-ribose units [1–3]. The oligomers and polymers of ADP-ribose are synthesized in a covalent association with various proteins of which histones H_1 and H_{2B} [4–7] and the synthetase itself [8–10] are the best characterized.

The enzyme has been found ubiquitously in the nuclei of a great variety of eukaryotic cells, including human [11] and plant [12]. However, three previous studies have presented evidence that the enzyme is also localized in the cytoplasm. Roberts et al. [13] have found the enzyme associated with polysomes and ribosomes of HeLa cells. A poly(ADP-ribose) synthetase was also reported to occur in rat liver mitochondria [14]. More recently we found a significant level of enzymatic activity associated with testis mitochondria [15, 16]. The present paper is a follow-up of this study and it presents evidence that after the nuclear fraction, the most important level of poly(ADP-ribose) synthetase in spermatogenic cells is associated with the microsomal-ribosomal fraction.

Fractionation of Rat Testis Homogenate

Decapsulated testis from adult albino rats were homogenized in 10 vol of ice-cold medium containing 0.25 M sucrose (Sigma Grade I or ultra-pure grade from Bethesda Res. Labs.), 50 mM KCl, 5 mM $MgCl_2$, 50 mM Tris-HCl (pH 7.5), and 0.2 mM phenylmethylsulfonylfluoride. Homogenization was carried out with ten strokes in a motor-driven Potter-Elvehjem tissue grinder equipped with a Teflon pestle. The homogenate was centrifuged at 700 × g for 15 min at 4°C to obtain the crude nuclear fraction, and the resulting supernatant was centrifuged at 9,000 × g for 10 min at 4°C. The post mitochondrial supernatant was either centrifuged directly at 110,000 × g, or applied

1 Instituto de Bioquímica, Facultad de Ciencias, Universidad Austral de Chile, Valdivia, Chile

ADP-Ribosylation of Proteins
(ed. by F.R. Althaus, H. Hilz, and S. Shall)
© Springer-Verlag Berlin Heidelberg 1985

on top of a 2 M sucrose cushion (50 mM KCl, 5 mM MgCl$_2$, and 50 mM Tris-HCl, pH 7.5) and centrifuged at 110,000 × g for 2 h at 4°C [17]. The pellet obtained with the first procedure, or the compact layer at the 2 M sucrose interphase obtained with the second procedure, is referred to as the microsomal-ribosomal (M-R) fraction of the testis. The same fractionation procedure was utilized to prepare the M-R fraction from other rat tissues.

Poly(ADP-Ribose) Synthetase Assay

The assay was carried out in a final volume of 50 μl which contained 50 mM Hepes (pH 7.5), 10 mM MgCl$_2$, 0.1 mM [^{14}C]-NAD (105 cpm pmol^{-1}), and between 50 to 70 μg protein. The mixture was incubated at 25°C for 10 min and the radioactivity incorporated into the trichloroacetic acid-insoluble fraction was determined as described elsewhere [15].

The average chain length of the oligomers of ADP-ribose synthesized by the M-R fraction, was determined by a modification [4] of the procedure described by Shima et al. [18]. Protein was estimated by the procedure of Lowry et al. [19] using bovine serum albumin as a standard.

Electron Microscopy

The M-R fraction was fixed for 2 h in 2% glutaraldehyde buffered with 100 mM sodium phosphate at pH 7.5. After a buffer rinse, the samples were postfixed in 1% OsO$_4$ for 2 h, rinsed with phosphate buffer, dehydrated with ethanol and acetone, and then embedded in Epon-Araldite. Thin sections were stained with uranyl acetate and lead citrate. The specimens were examined in a Phillips EM-300 electron micro-scope.

Cellular Fractionation of Rat Testis and Distribution of Poly(ADP-Ribose) Synthetase

Table 1 shows the amount of enzyme units as well as the specific activity of poly-(ADP-ribose) synthetase in the different cell fractions of rat testis. These results show that a large proportion of enzyme is localized in the nuclear fraction and in the micro-somal-ribosomal fraction. However, the specific activity of the enzyme was larger in the latter fraction. In contrast, the specific activity of the enzyme in the microsomal fraction of rat liver, brain, and kidney was negligible (Table 2).

Electron microscopic studies of the microsomal-ribosomal fraction revealed the presence of abundant small vesicles and large aggregates of ribosomes (Fig. 1A). Care-

Table 1. Subcellular distribution of poly(ADP-ribose) synthetase in rat testis

Fraction	Total enzyme units[a]	Specific activity pmol ADP-ribose mg^{-1} protein
Homogenate	2,177	426
Nuclear pellet	2,652	1,174
9,000 × g pellet	47	270
110,000 × g pellet	580	1,443
110,000 × g supernatant	21	70

[a] One enzyme unit is the amount of enzyme which catalizes the incorporation of 100 pmol of [^{14}C]-ADP-ribose in 10 min at 25°C

Table 2. Poly(ADP-ribose) synthetase activity in the 110,000 × g pellet isolated from different rat tissues

110,000 × g pellet from:	Activity (pmol ADP-ribose mg^{-1} protein)
Liver	218
Brain	50
Kidney	43
Testis	1,990

ful analysis of this fraction showed total absence of nuclei or mitochondria. When the postmitochondrial supernatant was treated with 1% Triton X-100 and then centrifuged at 110,000 × g for 2 h, the pellet contained most of the activity. Moreover, the specific activity of the detergent-treated fraction was higher than the control (e.g., 1,400 vs 3,600 pmol mg^{-1} protein). Electron microscopic examination of this fraction revealed the presence of large aggregates of ribosomes and the total absence of the vesicles (Fig. 1B).

The traditional procedure to release the ribosome- and polysome-bound fraction is by treatment with sodium deoxycholate [17]. When this procedure was used, no poly(ADP-ribose) synthetase activity was found in the released ribosomes and poly-

Table 3. Effect of several compounds on the activity of poly(ADP-ribose) synthetases

Compound	Activity (pmol ADP-ribose mg^{-1} protein)
None	1,490
Nicotinamide (1 mM)	43
Thymidine (1 mM)	58
Theophylline (1 mM)	99
3-Aminobenzamide (1 mM)	35
Sodium Deoxycholate (9.6 mM)	132

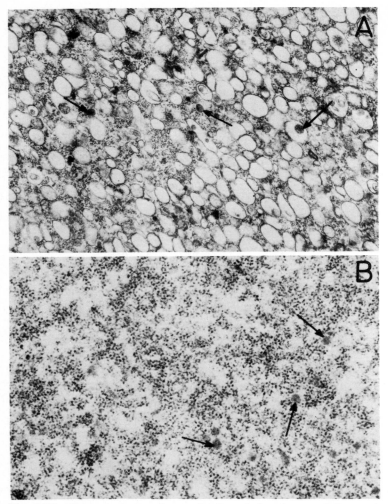

Fig. 1A,B. Electron microscopic examination of the microsomal-ribosomal pellet of rat testis. **A** Pellet of 110,000 × g (2 h) obtained from a control postmitochondrial supernatant. **B** The same pellet obtained after treatment of the postmitochondrial supernatant with 1% Triton X-100 (× 20,350)

somes or in the solubilized fraction. When this detergent was added to the assay mixtures of the microsomal-ribosomal fraction, a significant inhibition was observed (Table 3). To determine if this inhibitory effect was in fact a characteristic of the enzyme associated with the microsomal fraction of the testis, the detergent was added (final concentration 1%) to the assay of rat liver nuclei, and once again a similar level of inhibition was detected (data not shown).

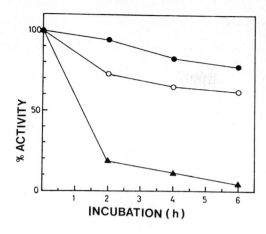

Fig. 2. Effect of DNAase I and RNAase on the activity of poly(ADP-ribose) synthetase. The microsomal-ribosomal fraction was pre-incubated at 25°C alone (●), with RNAase (○ 250 μg ml^{-1}) or with DNAase I (▲ 250 μg ml^{-1}). At the times indicated, aliquots were assayed for enzymatic activity

Properties of the Enzyme Activity

Some properties of the poly(ADP-ribose) synthetase activity associated with the M-R fraction are summarized in Table 3. The activity is strongly inhibited by nicotinamide, thymidine, theophylline, and 3-aminobenzamide. As discussed before, sodium deoxy-cholate is also inhibitory.

The apparent K_m for NAD of the enzyme associated with the M-R fractions was 105 μM, which is different from the K_m reported for the enzyme associated with the mitochondrial fraction and the nuclei [15, 16].

The average chain length of the oligomers synthesized by the enzyme associated with the microsomal fraction varied from 4.5 to 5.3 residues of ADP-ribose. This value is slightly less than the average chain length formed by rat testis nuclei (6.1 to 6.8 residues, data not shown).

DNA Requirements

One of the most intriguing, but well-established, properties of poly(ADP-ribose) synthetase is its requirement for DNA [20–24]. As reported by Benjamin and Gill [23] and Yoshihara's group ([21]; for review see [24]), the enzyme shows preference for DNA "ends". Thus, closed circular dsDNA (ColE$_1$ or pBR322) shows no activating potential on the enzyme. However, this ability increases proportionally with the number of nicks introduced by DNAase I treatment [21].

The activity of the enzyme associated with the microsomal-ribosomal fraction of the rat testis can be demonstrated by the absence of DNA in the assay. In the presence of DNA, the enzyme was stimulated three to five times. Treatment of the microsomal-ribosomal fraction with DNAase I at 25°C induced a considerable decrease in the enzymatic activity (Fig. 2). After a 2 h preincubation, less than 20% of the initial activity was detected. On the other hand, the samples preincubated in the presence

Table 4. Variation of poly(ADP-ribose) synthetase activity with the age of the rats

Age (days)[a]	Activity (pmol ADP-ribose mg^{-1} protein)
10	875
20	1,505
24	3,744
30	3,523
40	1,460
60	1,563

[a] At the age indicated, the microsomal-ribosomal fraction was prepared from two different pools of testis. Hence, the activities shown are the average of two determinations

or absence of RNAase, showed a small decrease in activity. These results suggest that preincubation with DNAase I digests DNA bound to the enzyme in the microsomal fraction and consequently, the activity of the enzyme is lost. However, it is also possible that some unknown proteolytic activity present in the DNAase I preparation, is responsible for the rapid decrease in activity. In order to dismiss this possibility, advantage was taken of the known property of DNAase I of being dependent on both Mg^{++} and Ca^{++} for maximal activity [25]. Accordingly, when the microsomal-ribosomal fraction of the testis was preincubated with DNAase I plus 0.25 mM EGTA, a stimulation of 24% of the enzymatic activity was detected, rather than a decrease in activity. Moreover, if the cellular fraction containing the enzyme was preincubated for 6 h with DNAase I and 0.25 mM EGTA and calf thymus DNA was added prior to the assay, about 1.5 times the activity of the control sample was found (data not shown). These experiments proved that the inactivation of the enzyme by preincubation with DNAase I was the result of digestion of the DNA bound to the enzyme.

Variation of the Activity with Age

Clermont and Perey [26] have shown that after birth, there is a good chronological relationship between the age of the rat and the type of spermatogenic cells present in the seminiferous tubules. This particular situation can be used to determine the type of spermatogenic cell in which poly(ADP-ribose) synthetase appears associated with the microsomal-ribosomal fraction. Accordingly, this cellular fraction was isolated from testis of rats at different time intervals after birth and the activity of the enzyme was measured. As shown in Table 4, the peak of activity was obtained with rats of 24- to 30-days-old. Thereafter, the specific activity of the enzyme reached the lower adult level.

These results indicate that the maximal activity of poly(ADP-ribose) synthetase associated with the microsomal-ribosomal fraction of the testis was attained in rats

24- to 30-days-old. At this age, the seminiferous tubules are quite rich in pachytene spermatocytes, which strongly suggests that it is in these cells that the enzyme reached the maximal level of activity in the microsomal-ribosomal fraction.

Discussion

Our results demonstrate that by using cell fractionation procedures, it is possible to measure quite a significant level of poly(ADP-ribose) synthetase in the microsomal-ribosomal fraction of the testis and not in other somatic tissues. However, a major criticism to this experimental approach, is that the activity detected might well be due to a contamination with the nuclear enzyme. One fact that supports this criticism is that the nuclei contain a considerable amount of enzyme. On the other hand, no matter which procedure one uses to homogenize the tissue, the possibility of damaging the nucleus and consequently releasing some amount of enzyme, is difficult to discard.

However, arguments suggesting that this activity is unlikely due to contamination with the nuclear enzyme were also presented. First, the apparent K_m for NAD was 105 μM for the microsomal activity, compared with a value of 210 μM for the nuclear enzyme [15, 16]. Second, the maximal specific activity of the enzyme associated with the microsomal-ribosomal fraction was found in testis of 24- to 30-days-old rats. If contamination with the nuclear enzyme is likely, one should expect that the activity in this fraction would parallel the activity in the nuclear fraction. On the contrary, developmental studies carried out by Momii and Koide [27] demonstrated that the activity of the enzyme in isolated nuclei increases with age and reaches a maximal value in the adult.

Independent of these arguments, the most direct and definitive proof that the enzyme is also localized in the cytoplasm of the germinal cells is to use immunocytochemical techniques. This approach has been carried out using an antibody raised against purified poly(ADP-ribose) synthetase of calf thymus. By using this antibody and the peroxidase-antiperioxidase technique [28] we have been able to demonstrate a positive reaction in the nuclei as well as in the cytoplasm of spermatocytes and young spermatids (J. Figueroa and L.O. Burzio, unpubl.). These experiments strongly suggest that in some spermatogenic cells poly(ADP-ribose) synthetase is also localized in the cytoplasm.

The enzyme associated with the microsomal-ribosomal fraction of the testis showed a considerable level of activity in the absence of added DNA. However, if calf-thymus DNA was added to the assay mixture, three to four times more stimulation of the activity was observed. Furthermore, the experiments carried out with DNAase I strongly suggest that the activity observed in this cytoplasmic fraction was due to the presence of DNA. Therefore, a pertinent question remains as to the nature of this cytoplasmic DNA responsible for the residual activity of the enzyme. To answer this, further work will be necessary.

Acknowledgments. The authors wish to express their gratitude to Dr. Kunihiro Ueda and Dr. Osamu Hayaishi for supplying the antibody against poly(ADP-ribose) synthetase, and to Dr. Sidney Shall for sending us 3-aminobenzamide. This work was supported by Grant RS-82-01 from the D.I., Universidad Austral de Chile and Grant 1071-83 from the Fondo Nacional de Desarrollo Cientifico y Tecnológico.

References

1. Hilz H, Stone PR (1976) Poly(ADP-ribose) and ADP-ribosylation of protein. In: Lynen F (ed) Physiol Biochem Pharmacol, vol 76. Springer, Berlin Heidelberg New York, pp 1–58
2. Hayaishi O, Ueda K (1977) Poly(ADP-ribose) and ADP-ribosylation of proteins. Annu Rev Biochem 46:95–116
3. Sugimura T, Miwa M (1982) Structure and properties of poly(ADP-ribose). In: Hayaishi O, Ueda K (eds) ADP-ribosylation reactions. Academic Press, London New York, pp 43–63
4. Riquelme PT, Burzio LO, Koide SS (1979) ADP-ribosylation of rat liver lysine-rich histone in vitro. J Biol Chem 254:3018–3028
5. Burzio LO, Riquelme PT, Koide SS (1979) ADP-ribosylation of rat liver nucleosomal core histones. J Biol Chem 254:3029–3037
6. Ogata N, Ueda K, Kagamiyama H, Hayaishi O (1980) ADP-ribosylation of histone H_1. J Biol Chem 255:7616–7620
7. Ogata N, Ueda K, Hayaishi O (1980) ADP-ribosylation of histone H_{2B}. J Biol Chem 255: 7610–7615
8. Yoshihara K, Hashida T, Yoshihara H, Tanaka Y, Ohgushi H (1977) Enzyme-bound early product of purified poly(ADP-ribose) polymerase. Biochem Biophys Res Commun 78:1281–1288
9. Kawaichi M, Ueda K, Hayaishi O (1981) Multiple autopoly(ADP-ribosylation) of rat liver poly(ADP-ribose) synthetase: Mode of modification and properties of automodified synthetase. J Biol Chem 256:9483–9489
10. Mandel P, Jongstra-Bilen J, Ittel ME, de Murcia G, Delain E, Niedergang C, Vosberg HP (1983) Some electron microscopy ascpects of poly(ADPR) polymerase-DNA interaction and of auto-poly(ADP-ribosyl)ation reaction. In: Miwa M, Hayaishi O, Shall S, Smulson M, Sugimura T (eds) ADP-ribosylation, DNA repair and cancer. Jpn Sci Soc Pross, Tokyo, pp 71–81
11. Burzio LO, Reich L, Koide SS (1975) Poly(adenosine diphosphoribose) synthase activity of isolated nuclei of normal and leukemic leukocytes. Proc Soc Exp Biol Med 149:933–938
12. Whitby AJ, Stone PR, Whish WJ (1979) Effect of polyamines and Mg^{2+} on poly(ADP-ribose) synthesis and ADP-ribosylation of histones in wheat. Biochem Biophys Res Commun 90: 1295–1304
13. Roberts JH, Stark P, Giri C, Smulson M (1975) Cytoplasmic poly(ADP-ribose) polymerase during the HeLa cells cycle. Arch Biochem Biphys 171:305–315
14. Kun E, Zimber PH, Chang AC, Puschendorf B, Grunicke H (1975) Macromolecular enzymatic product of NAD in liver mitochondria. Proc Natl Acad Sci USA 72:1436–1440
15. Burzio LO, Sáez L, Cornejo R (1981) Poly(ADP-ribose) synthetase activity in rat testis mitochondria. Biochem Biophys Res Commun 103:369–375
16. Burzio LO, Concha II, Figueroa J, Concha M (1983) Poly(ADP-ribose) synthetase associated to cytoplasmic structure of meiotic cells. In: Miwa M, Hayaishi O, Shall S, Smulson M, Sugimura T (eds) ADP-ribosylation, DNA repair and cancer. Jpn Sci Soc Press, Tokyo, pp 141–152
17. Noll H (1969) Polysomes: Analysis of structure and function. In: Campbell PN, Sargent JR (eds) Techniques in protein biosynthesis, vol II. Academic Press, London New York, pp 101–179
18. Shima T, Hasegawa S, Fujimura S, Matsubara H, Sugimura T (1969) Studies on poly(adenosine diphosphate-ribose). VII. Methods of separation and identification of 2'-(5''-phosphoribosyl)-5'-adenosine monophosphate. J Biol Chem 244:6632–6635

19. Lowry OH, Rosebrough NJ, Farr AL, Randall RJ (1951) Protein measurements with the Folin phenol reagent. J Biol Chem 193:265–275
20. Yamada M, Miwa M, Sugimura T (1971) Studies on poly(adenosine diphosphate-ribose). X. Properties of a partially purified poly(adenosine diphosphate-ribose) polymerase. Arch Biochem Biophys 146:579–586
21. Ohgushi H, Yoshihara K, Kamiya T (1980) Bovine thymus poly(adenosine diphosphate-ribose) polymerase. Physical properties and binding to DNA. J Biol Chem 255:6205–6211
22. Kristensen T, Holtlund J (1976) Purification of poly(ADP-ribose) polymerase from Ehrlich ascites tumor cells by chromatography on DNA-agarose. Eur J Biochem 70:441–446
23. Benjamin RC, Gill M (1980) A connection between poly(ADP-ribose) synthesis and DNA strand breakage. In: Smulson ME, Sugimura T (eds) Novel ADP-ribosylations of regulatory enzymes and proteins. Elsevier/North Holland, Amsterdam New York, pp 227–236
24. Ueda K, Kawaichi M, Hayaishi O (1982) Poly(ADP-ribose) synthetase. In: Hayaishi O, Ueda K (eds) ADP-ribosylation reactions. Academic Press, London New York, pp 117–155
25. Price PA (1975) The essential role of Ca^{2+} in the activity of bovine pancreatic deoxyribonuclease. J Biol Chem 250:1981–1986
26. Clermont Y, Perey B (1955) Quantitative study of the cell population of the seminiferous tubules in immature rats. Annu J Anat 100:241–268
27. Momii A, Koide SS (1980) RNA synthesis and poly(ADP-ribose) synthetase activity in developing mouse testis. Fed Proc 39:954
28. Sternberger LA, Hardy PH, Cuculis JJ, Meyer HG (1970) The unlabeled antibody-enzyme method of immunohistochemistry. Preparation and properties of antigenantibody complex (horseradish peroxidase-antihorseradish peroxidase) and its use in identification of spirochetes. J Histochem Cytochem 18:315–321

Poly(ADP-Ribose) Polymerase Associated with Cytoplasmic Free mRNP Particles: Effect of RNase A

HELENE THOMASSIN, LUC GILBERT, CLAUDE NIEDERGANG,
and PAUL MANDEL[1]

Introduction

A cytoplasmic poly(ADPR) polymerase activity was first detected by Roberts et al. in 1975 [1]. The enzyme associated with ribosomes and polysomes in HeLa cells requires both DNA and histones for activity. A mitochondrial enzyme has been described in rat liver and in testis (for review see [2, 3]). Finally, the existence of an active poly-(ADPR) polymerase associated with ribonucleoproteins in the microsomal-ribosomal fraction of testis has been reported. The enzyme shows a considerable activity without exogenous DNA and is insensitive to RNase A [3].

Recently, we have demonstrated the existence of a poly(ADPR) polymerase activity associated with cytoplasmic free messenger ribonucleoprotein particles (mRNP) isolated from mouse plasmacytoma cells [4]. The enzyme does not require DNA for activity and is able to produce an ADP-ribosylation of some of the mRNP proteins. We have extended our observations to Krebs II ascites-tumor cells and to rat liver. In the present report, we will discuss some properties of this enzyme, particularly the activation by RNase A.

Poly(ADPR) Polymerase Associated with Free mRNP; General Characteristics

Table 1 summarizes the poly(ADPR) polymerase activities measured in free mRNP isolated from mouse plasmacytoma, Krebs II ascites-tumor cells or rat liver. In contrast to the tumoral cells, the activity associated to rat liver free mRNP is very low; nevertheless, it is twelvefold higher than the activity associated with the microsomal-ribosomal fraction reported by Burzio et al. [3]. The addition of calf thymus DNA induces only an increase of about 20% in enzymatic activity.

Some interesting properties concerning the free mRNP enzyme can be demonstrated:

1 Centre de Neurochimie du CNRS, 5, rue Blaise Pascal, 67084 Strasbourg Cedex, France

ADP-Ribosylation of Proteins
(ed. by F.R. Althaus, H. Hilz, and S. Shall)
© Springer-Verlag Berlin Heidelberg 1985

Table 1. Poly(ADPR) polymerase activities in nuclei and free mRNP[a]

Tissue	nmol h^{-1} mg^{-1} protein Free mRNP		Nuclei
Plasmacytoma	36	$(42.3)^b$	15
Krebs II	23.4	$(27.6)^b$	5.7
Rat liver	2.2		1.6

[a] Postmitochondria supernatant was layered on a discontinuous D_2O-sucrose gradient and centrifuged 18 h at 200,000 g. Free mRNP particles are found in the 1.29 D_2O-sucrose density fraction. Poly(ADPR) polymerase activity was determined as in [4] at 30°C
[b] Reactions with additional calf thymus DNA (10 μg)

1. The poly(ADPR) polymerase activity associated with free mRNP is strongly inhibited by nicotinamide, thymidine, and 3-aminobenzamide with inhibition constants (Ki) of respectively, 55 μM, 139 μM, and 23 μM.
2. Pancreatic RNase A added in the incubation medium at a final concentration of 5 μg ml^{-1} increases the poly(ADPR) polymerase activity by a factor of two [4].
3. Poly(ADP-ribose) polymerase activity can be dissociated from the free mRNP particles by a high KCl concentration, as described for free mRNP associated kinases [5]. 0.6 M KCl is needed to entirely separate the poly(ADPR) polymerase activity from the mRNA, indicating that the enzyme is tightly associated with the particle. Moreover, the released enzyme becomes insensitive to RNase A (data not shown).
4. Incubation of mRNP particles with NAD, electrophoresis on lithium dodecyl sulfate polyacrylamide gels, and autoradiography reveal several ADP-ribosylated proteins with a predominant acceptor of mol. wt. 115,000 [4]. Using the protein blotting technique, we have observed that the antiserum directed against the nuclear calf thymus poly(ADPR) polymerase reacts with the 115,000 mol. wt. protein, associated with free mRNP particles.

These results strongly suggest some similarities between the nuclear and the mRNP poly(ADPR) polymerase, although several differences can be demonstrated.

Effect of RNase A on Free mRNP Associated Poly(ADPR) Polymerase

In order to characterize the association between poly(ADPR) polymerase and free mRNP we have investigated in detail the effects of DNase-free RNase A. Poly(ADPR) polymerase activity increases after addition of RNase A added to the incubation medium in a dose-dependent manner and reaches a plateau at 2 μg ml^{-1} of RNase A (Fig. 1). The enzymatic activity is not inhibited when the concentration of RNase is increased up to 500 μg ml^{-1}, thus suggesting no RNA dependence. The activation is time-dependent suggesting a relation with the degree of destruction of the mRNP architecture (data not shown). The apparent K_m for NAD^+ of the free mRNP poly-

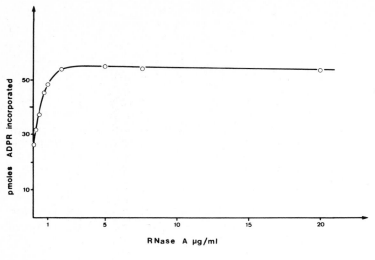

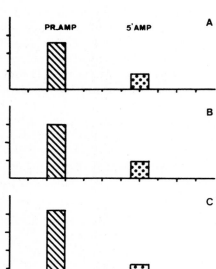

▲
Fig. 1. Effect of RNase A on free mRNP poly(ADPR) polymerase activity. The enzymatic assay was performed as described in [4] with the addition of variable amounts of RNase A

Fig. 2A–D. Effect of RNase A on the average chain length of poly(ADPR) synthetized by the mRNP poly(ADPR) polymerase. The reaction was carried out as described in [4] with addition of variable amounts of RNase A: A control; B 1.5 μg ml^{-1}; C 5 μg ml^{-1}; D 20 μg ml^{-1}. The reaction product was analyzed by thin layer chromatography on PEI cellulose after digestion by snake venom phosphodiesterase. Average chain length: A 3.9, B 4.2, C 4.5, D 4.7

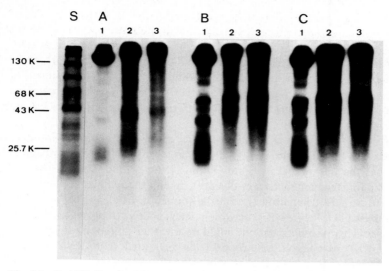

Fig. 3A–C. ADP-ribosylation of plasmacytoma free mRNP in presence of RNase A . After incubation of free mRNP with [^{32}P] NAD (1 μci) as described in [4] and variable amounts of RNase A (**A** control, **B** 0.5 μg ml^{-1}, **C** 2 μg ml^{-1}), the acid insoluble products were analyzed by 12% polyacrylamide gel containing LDS. *S:* Coomassie blue-stained proteins. *Slots 1 to 3:* autoradiograms at 1, 5 and 20 μM NAD

(ADPR) polymerase decreases from 200 μM to 150 μM in the presence of RNase A, demonstrating a greater accessibility of the enzyme and/or the ADP-ribose protein acceptors. This assumption has been confirmed by an increase of both the number and the average length of oligomeric chains of ADP-ribose when RNase A is added to the incubation medium (Fig. 2). In addition, in the presence of RNase A, lithium dodecyl sulfate gel electrophoresis and autoradiography show an increased ADP-ribosylation of the 115,000 mol. wt. protein acceptor and reveal new ADP-ribosylated proteins, particularly a protein of 50,000 mol. wt. (Fig. 3). This poly(ADP-ribose) acceptor is also detected by acid urea gel electrophoresis, but is absent in the proteins dissociated from the particle by 0.5 *M* KCl, suggesting it is basic in nature and is tightly associated with the mRNA (data not shown). Therefore, we can assume that the destruction of the mRNP conformation by the mRNA hydrolysis exposes additional protein sites for ADP-ribosylation, and autoADP-ribosylation sites of the enzyme.

Discussion

The results presented here demonstrate the association of an active poly(ADPR) polymerase with free mRNP particles, which ADP-ribosylates ribonucleoproteins tightly bound to the mRNP.

It is noteworthy that free mRNP particles are a temporary untranslatable form of mRNA in the cytoplasm and that proteins associated with the active polysomal mRNP or the repressed mRNP are different [6, 7]. There is now some evidence that ribonucleoprotein particles are dynamic structures and that protein exchanges occur between the cytoplasmic mRNA-associated proteins and free proteins. Involvement of mRNA-associated proteins in the regulation of protein synthesis has been considered [7−10] and post-translational modification of these proteins as a regulatory mechanism might be considered.

Protein kinases are known to be associated with free mRNP and phosphorylation of several mRNP proteins has been observed [5, 9]. It has recently been demonstrated in *Artemia salina* that the differential phosphorylation of the 38,000 mol. wt. poly(A) binding protein modifies the interaction of this protein with the poly(A) sequence and that this could act on polyadenylation and on the formation of the mRNP-ribosomal preinitiation complex [11]. Poly ADP-ribosylation which adds a strong negative charge to proteins compared to phosphorylation, could be involved in important structural changes in the mRNP particle. Ribonucleoproteins associated to hnRNA are also ADP-ribosylated [12]. It is conceivable that poly ADP-ribosylation could influence the arrangement and the exchange of proteins in the different ribonucleoprotein structures by modifying the protein-RNA and protein-protein interactions.

References

1. Roberts JH, Stark P, Giri CP, Smulson M (1975) Cytoplasmic poly(ADPR) polymerase during the HeLa cell cycle. Arch Biochem Biophys 171:305−315
2. Kun E, Kirsten E (1982) Mitochondrial ADP-ribosyltransferase system. In: Hayaishi O, Ueda K (eds) ADP-ribosylation reactions. Academic Press, London New York, p 193
3. Burzio LO, Concha II, Figueroa J, Concha M (1983) Poly(ADP-ribose) synthetase associated to cytoplasmic structures of meiotic cells. In: Miwa M, Hayaishi O, Shall S, Smulson M, Sugimura T (eds) ADP-ribosylation, DNA repair and cancer. Jpn Sci Soc Press, Tokyo/VNU Science Press, Utrecht, p 141
4. Elkaim R, Thomassin H, Niedergang C, Egly JM, Kempf J, Mandel P (1983) Adenosine diphosphate ribosyltransferase and protein acceptors associated with cytoplasmic free messenger ribonucleoprotein particles. Biochimie 65:653−659
5. Egly JM, Schmitt M, Elkaim R, Kempf J (1981) Protein kinases and their protein substrates in free messenger ribonucleoprotein particles and polysomes from mouse plasmacytoma cells. Eur J Biochem 118:379−387
6. Preobrazhensky AA, Spirin AS (1978) Informosomes and their protein components: the present state of knowledge. Prog Nucleic Acid Res Mol Biol 21:1−37
7. Vincent A, Goldenberg N, Standart N, Civelli O, Imaizumi-Scherrer T, Maundrell K, Scherrer K (1981) Potential role of mRNP proteins in cytoplasmic control of gene expression in duck erythroblasts. Mol Biol Rep 7:71−81
8. Schmid HP, Kohler K, Setyono B (1983) Interaction of cytoplasmic messenger-RNA with proteins: their possible function in the regulation of translation. Mol Biol Rep 9:87−90
9. Vincent A, Akhayat O, Goldenberg S, Scherrer K (1983) Differential repression of specific mRNA in erythroblast cytoplasm: a possible role for free mRNP proteins. EMBO J 2:1869−1876
10. Bag J (1984) Cytoplasmic mRNA-protein complexes of chicken muscle cells and their role in protein synthesis. Eur J Biochem 141:247−254

11. De Herdt E, Thoen C, Van Hove L, Roggen E, Piot E, Slegers H (1984) Identification and properties of the 38000-Mr poly(A)-binding protein of non-polysomal messenger ribonucleo-proteins of cryptobiotic gastrulae of Artemia salina. Eur J Biochem 139:155–162
12. Kostka G, Schweiger A (1982) ADP-ribosylation of proteins associated with heterogeneous nuclear RNA in rat liver nuclei. Biochim Biophys Acta 969:139–144

Effect of DNA Intercalators on Poly(ADP-Ribose) Glycohydrolase Activity

MANOOCHEHR TAVASSOLI and SYDNEY SHALL[1]

Abbreviations:

Ethacridine	2-thoxy-6,9 acridine diamine
m-AMSA	4′-(9-acridinylamino) methanesulphon-m-anisidide
Tilorone R10,556 DA	2,7-Bis(4-piperidinobutyryl)-9H-fluoren-9-one

Introduction

Poly(ADP-ribose) glycohydrolase is the major enzyme responsible for the degradation of poly(ADP-ribose) [1, 2]. This enzyme hydrolyses the ribosyl-ribose bond between the two ADP-ribose units and produces ADP-ribose. We have purified this enzyme to near homogeneity and characterised its properties [2].

It has been shown that DNA intercalators (ethidium bromide and proflavine) form complexes with single-stranded DNA and homopolynucleotides [3]. It was of interest, therefore, to determine the effect of intercalators on the degradation of the homopolymer (ADP-ribose)$_n$ by the poly(ADP-ribose) glycohydrolase.

Methods

Poly(ADP-ribose) Glycohydrolase Assay. Enzyme activity was determined by measuring the amount of ADP-ribose released by the enzyme, as previously described [2].

Phosphodiesterase Assay. The reaction mixture in a total volume of 300 μl, contained 15 μM [^{32}P] poly(ADP-ribose) (2×10^4 counts min^{-1}), 100 mM Tris-HCl buffer (pH 7.5) and 12 ng of snake venom phosphodiesterase. The reaction was carried out at 37° for 10 min and was terminated by placing the tubes in a boiling water bath for 2 min. After cooling in ice, the reaction products were completely digested with 1 unit

1 Cell and Molecular Biology Laboratory, University of Sussex, Brighton, East Sussex, BN1 9OG, Great Britain

ADP-Ribosylation of Proteins
(ed. by F.R. Althaus, H. Hilz, and S. Shall)
© Springer-Verlag Berlin Heidelberg 1985

Table 1. Inhibition of poly(ADP-ribose) glycohydrolase by intercalators in the absence or presence of DNA-histone complexes[a]

Intercalators	Concentration (μM)	Inhibition (%)[b]		Extra[c] inhibition
		Without DNA-histone	With DNA-histone	
Ethacridine	20	27	76	37
	50	97	85	46
Proflavine	20	0	74	35
	50	3	86	47
	100	66	ND	
Ellipticine	50	40	84	45
	100	64	ND	
Daunomycin	50	33	87	48
	100	79	ND	
Tilorone R10,556 DA	50	66	81	41
	100	100	ND	
m-AMSA	50	0	40	1
	100	15	ND	
Chloroquine	50	0	39	0
	100	2	ND	
Ethidium bromide	20	0	69	30
	50	2	87	48
	100	8	88	49

ND = Not done

[a] Histone (100 μg ml^{-1}) was added after mixing DNA (100 μg ml^{-1}) and poly(ADP-ribose) (12 ng ml^{-1}) followed by addition of the indicated concentrations of intercalators

[b] DNA-histone complexes (100 μg ml^{-1}) in the absence of intercalators, inhibit the enzyme activity by 39%

[c] The extra inhibition is calculated by subtracting the inhibition in the presence of DNA-histone alone (39%) from the inhibition by DNA-histone-dye mixtures

of alkaline phosphatase at 37° for 30 min. 60 μl of the reaction mixture (in 20 μl aliquots) was applied to a DEAE-cellulose paper and developed with 100 mM LiCl. The phosphodiesterase activity was estimated from the radioactivity in the inorganic phosphate which runs with solvent front.

Purification of Poly(ADP-Ribose) Glycohydrolase. Poly(ADP-ribose) glycohydrolase was purified 12,300-fold from pig thymus with an overall recovery of 8.5% [2].

Results

In the absence of DNA-histone complexes, ethacridine, proflavine, tilorone R10,556 DA, ellipticine and daunomycin inhibit the poly(ADP-ribose) glycohydrolase activity

Table 2. Effect of intercalators on degradation of poly(ADP-ribose) by snake venom phosphodiesterase

Intercalators	Concentration (μM)	% Remaining activity
None	–	100
Ethacridine	50	14
	100	0
Proflavine	100	108
	200	39
Tilorone R10,556 DA	100	92
	200	81
Ellipticine	200	93
Ethidium bromide	200	100

(Table 1). However, m-AMSA a potent antitumour agent, chloroquine, and ethidium bromide have essentially no effect on the enzyme activity.

From a double reciprocal plot we estimate K_i values of 7.3 ± 0.9 and 36 ± 9 μM for tilorone and proflavine, respectively.

The demonstration that proflavine and ethidium bromide bind to homopolynucleotides [3] suggests that the inhibition of poly(ADP-ribose) glycohydrolase by a number of intercalators might be due to the formation of a complex between these agents and the poly(ADP-ribose). If this is the case then these intercalators would also inhibit the degradation of poly(ADP-ribose) by phosphodiesterases. This is indeed so, and as shown in Table 2 ethacridine once again is the strongest inhibitor and at 50 μM reduces the phosphodiesterase activity by 86%. Tilorone, which strongly inhibits the glycohydrolase activity, has no significant effect on phosphodiesterase activity and proflavine is a poorer inhibitor.

The inhibition of poly(ADP-ribose) degradation by glycohydrolase and phosphodiesterase is not necessarily a reflection of binding of ethacridine and other inhibitors to the polymer. However, further evidence for the binding of ethacridine to poly(ADP-ribose), is the observation that in the presence of 100 μM ethacridine, ethanol-acetate fails to precipitate poly(ADP-ribose) from a 15 μM solution of poly(ADP-ribose), while this does occur in the absence of the inhibitor. None of the other intercalators tested displayed this property (data not shown). These observations suggest that inhibition of poly(ADP-ribose) glycohydrolase by ethacridine may be due to direct interaction with the substrate with consequent blocking of substrate binding.

Effect of DNA and DNA-Histone Complexes on the Inhibition of Poly(ADP-Ribose) Glycohydrolase by Intercalators. The inhibition of glycohydrolase activity by intercalators (proflavine, ellipticine, tilorone R10,556 DA) is relieved by the addition of 100 μg ml^{-1} calf thymus DNA (data not shown). This is probably due to the well-known binding of the intercalators to the added DNA. However, addition of DNA-histone complexes (100 μg ml^{-1} of each) slightly increases the inhibition of the glycohydrolase activity by those intercalators which are inhibitory (Table 1). These include ethacridine, tilorone R10,556 DA, ellipticine, daunomycin and proflavine.

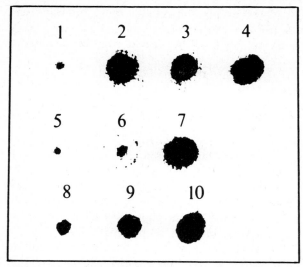

Fig. 1. Displacement of histone from DNA-histone complexes by intercalators: 100 μg of calf thymus DNA was added to 1 ml of 50 mM potassium phosphate buffer (pH 7.5) containing 100 μg of histone and incubated for 15 min at 37°C. To remove the free histone, the DNA-histone complexes were washed (by centrifugation) three times with 50 mM potassium phosphate buffer (pH 7.5), each time followed by 15 min incubation at 37°C. The washed complexes were suspended in 1 ml of 50 mM potassium buffer (pH 7.5) containing 200 mM sodium chloride and 100 μM of the indicated intercalator. After incubation at 37°C for 30 min, the insoluble complexes were precipitated by centrifugation (8000 g; 15 min, 4°C). 75 μl of the supernatant was spotted on Whatman filter paper in 15 μl aliquots. After drying, the filter was stained with 0.5% (w/v) Coomassie blue in 7% (v/v) acetic acid, 30% (v/v) ethanol and destained with 7% acetic acid. 30% ethanol. *1* none, *2* proflavine, *3* ethacridine, *4* tilorone R10,556 DA, *5* AMSA, *6* chloroquine, *7* ethidium bromide, *8–10* pure histone 0.5, 1.5 and 2 μg, respectively

Those intercalators which are not inhibitory do not become inhibitors in the presence of DNA-histone complexes, except for ethidium bromide. Although in the absence of DNA-histone complexes, ethidium bromide is not inhibitory even at 200 μM, in the presence of DNA-histone complexes, it inhibits the enzyme activity by 87% and 69% at 50 μM and 20 μM, respectively.

The increased inhibition of the glycohydrolase in the presence of DNA-histone could be due to the formation of dye-DNA-histone complexes or the release of inhibitory histone from DNA by the intercalators. To test this suggestion the ability of these intercalators to release histone from DNA-histone complexes was studied. The method is described in the legend to Fig. 1; the displaced histone was separated from the insoluble DNA-histone complex by centrifugation. As shown in Fig. 1, there is a correlation between the ability of the intercalators to release histone from the complexes and the increase in their inhibition of the glycohydrolase activity in the presence of DNA-histone. AMSA (spot 5) and chloroquine (spot 6), which do not inhibit the enzyme activity in the presence of DNA-histone complexes, did not release histone from these complexes. However, proflavine (spot 2), ethacridine (spot 3) and tilorone (spot 4) which are inhibitors of the glycohydrolase and also display increased inhibition of the enzyme activity in the presence of DNA-histone, did release histone from the complexes. Similarly, ethidium bromide (spot 7) which inhibits the glycohydrolase activity only if DNA-histone is present, was also able to release histone from the complexes. Spots 8, 9 and 10 are standards.

Discussion

The mechanism of inhibition by intercalators in the absence of DNA-histone complexes is not yet clear. However, the inhibition by ethacridine is probably due to its binding to the substrate, poly(ADP-ribose). This inference is drawn from the two observations that ethacridine inhibits the degradation of poly(ADP-ribose) not only by the glycohydrolase but also by the snake venom phosphodiesterase, and that ethacridine prevents the ethanol-acetate precipitation of poly(ADP-ribose). In the presence of DNA-histone complexes all inhibitory dyes produce a similar inhibition of the enzyme activity at equimolar concentrations. This suggests that the inhibition in these cases is due to the presence of the same amount of histones released by different intercalators.

The inhibition of purified poly(ADP-ribose) glycohydrolase by a number of intercalators raises the question as to the mechanism of action of these compounds in vivo. That is to say, how much of the effect of these intercalators on the cell is due to their inhibition of the glycohydrolase activity? Inhibition of ADP-ribosyl transferase retards DNA strand rejoining following DNA damage [4], inhibits cellular differentiation [5], but does not effect cell growth [4]. The consequences of inhibiting poly(ADP-ribose) glycohydrolase in intact cells is still to be elucidated.

Acknowledgments. This work was supported by the Cancer Research Campaign and the Medical Research Council.

References

1. Miyakawa N, Ueda K, Hayaishi O (1972) Association of poly(ADP-ribose) glycohydrolase with liver chromatin. Biochem Biophys Acta Res Commmun 49:239–245
2. Tavassoli M, Tavassoli MH, Shall S (1983) Isolation and purification of poly(ADP-ribose) glycohydrolase from pig thymus. Eur J Biochem 135:449–455
3. Gale EF, Cundliffe E, Reynolds PE, Richmond MH, Waring MJ (1981) The classical intercalating drugs. In: The molecular basis of antibiotic action, 2nd edn. Wiley, London New York, pp 274–306
4. Durkacz BW, Omidiji O, Gray DA, Shall S (1980) $(ADP-ribose)_n$ participates in excision repair. Nature (London) 283:593–596
5. Farzaneh F, Zalin R, Brill D, Shall S (1982) DNA strand breaks and ADP-ribosyl transferase activation during cell differentiation. Nature (London) 300:362–366

5'-ADP-3"-Deoxypentos-2"-Ulose. A Novel Product of ADP-Ribosyl Protein Lyase

KUNIHIRO UEDA[1], OSAMU HAYAISHI[2], JUN OKA[3], HAJIME KOMURA[4], and KOJI NAKANISHI[4]

Abbreviations:

FT-IR	Fourier Transformed Infrared Spectrogram
GC/MS	Gas Chromatography/Mass Spectrometry
HPLC	High Performance Liquid Chromatography

Introduction

Our previous studies [1, 2] revealed that the degradation of poly(ADP-ribosyl) proteins is carried out by consecutive actions of two enzymes, poly(ADP-ribose) glycohydrolase [3, 4] and ADP-ribosyl protein lyase (formerly termed ADP-ribosyl histone splitting enzyme) [5] (Fig. 1). The latter enzyme catalyzes removal of the last proximal ADP-ribosyl residue from acceptor protein. This report presents, after a brief review of ADP-ribosyl histones and the lyase, the identification of the enzymatic split product

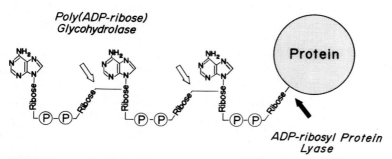

Fig. 1. Mode of degradation of poly(ADP-ribosyl) protein by poly(ADP-ribose) glycohydrolase and ADP-ribosyl protein lyase

1 Department of Medical Chemistry, Kyoto University Faculty of Medicine, Sakyo-ku, Kyoto 606, Japan
2 Osaka Medical College, Takatsuki, Osaka 569, Japan
3 The National Institute of Nutrition, Shinjuku-ku, Tokyo 162, Japan
4 Suntory Institute for Bioorganic Research, Shimamoto-cho, Mishima-gun, Osaka 618, Japan

ADP-Ribosylation of Proteins
(ed. by F.R. Althaus, H. Hilz, and S. Shall)
© Springer-Verlag Berlin Heidelberg 1985

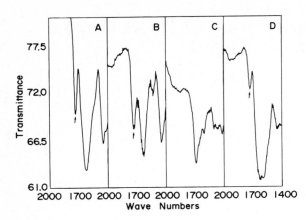

Fig. 2A–D. FT-IR spectra of **(A)** synthetic Pro-(α-ribosyl)Glu-Pro; **(B)** synthetic Pro-(β-ribosyl)Glu-Pro; **(C)** 5'-ADP; **(D)** a difference spectrum of Pro-(ADP-ribosyl)Glu-Pro-Ala-Lys of histone H2B and 5'-ADP

as a novel sugar derivative, and discusses its significance in poly(ADP-ribosyl) protein metabolism.

ADP-Ribosyl Carboxylate Esters of Histones

Among a number of nuclear proteins known to serve as acceptors of poly(ADP-ribosyl)ation [6], histones have been studied most extensively. ADP-ribosylation sites were identified in histones H1 and H2B; H1 had four sites, i.e., glutamic acid residues 2, 14, and 116, and a COOH-terminal lysine residue [7–10], while H2B had a single site, i.e., glutamic acid residue 2 [8, 10, 11]. In all these cases, the ADP-ribosyl protein bond appeared to be an ester (*O*-glycoside) of carboxylate, because the bond was, for the most part, very sensitive to both neutral $NH_2 OH$ and alkali [7–11]. A more explicit evidence for the ester bond was recently obtained by Fourier transform (FT)-IR analysis of ADP-ribosyl pentapeptide obtained from the NH_2-terminus of ADP-ribosylated H2B (Fig. 2). Computer-assisted subtraction of the FT-IR spectrum of ADP (Fig. 2C) from that of ADP-ribosyl pentapeptide visualized a prominent peak around the wave number of 1730 (Fig. 2D). A similar peak was observed with chemically synthesized α- (Fig. 2A) or β-ribosyl tripeptide (Fig. 2B), and was indicative of the presence of an ester bond. Similar treatment of FT-IR spectra of ADP-ribosyl pentapeptide at a smaller wave number region allowed us to tentatively assign its anomeric configuration as α (data not shown). It is noteworthy that there existed considerable heterogeneity in chemical stabilities of purified ADP-ribosyl (or phosphoribosyl or ribosyl) peptides [8, 9, 11, 12], which will be discussed later.

ADP-Ribosyl Protein Lyase

The enzyme was recently purified to apparent homogeneity from rat liver [12]. The purified preparationed exhibit a single protein band at the position of $M_r = 83,000$ upon

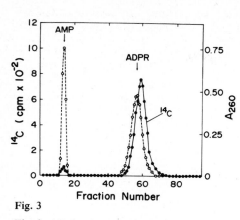

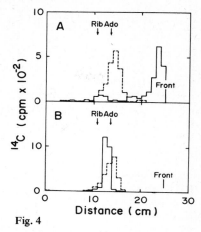

Fig. 3

Fig. 4

Fig. 3. AG 1 column chromatography of the reaction product of ADP-ribosyl protein lyase. [Ade-^{14}C]ADP-ribosyl histone H2B was treated with the enzyme, and 20% Cl$_3$CCOOH-soluble product was analyzed by chromatography on an AG 1-X2 (formate form) column; the column was eluted with a gradient of 0−6 M formic acid. ○ A$_{260}$; ● ^{14}C (cpm/0.9 ml). (Taken from Oka et al. [12])

Fig. 4A,B. Paper chromatography of compound X treated with (**A**) 10 mM phenylhydrazine (pH 4.5, 90 min, 25°C) or (**B**) 83 mM NaBH$_4$ (pH 6.5, 60 min, room temperature) (——) or no addition (− − −). Solvent system, 1-butanol/acetic acid/H$_2$O (52:13:35, organic layer). (Taken from Oka et al. [12])

SDS-polyacrylamide gel electrophoresis. The substrate specificity was rather loose with the protein portion; ADP-ribosyl histones H1 and H2B, peptide fragments of H2B, and nonhistone proteins (a mixture) served as substrates. In contrast, the specificity was very tight with the mono(ADP-ribosyl) portion and the carboxyl ester bond; poly- or oligo(ADP-ribosyl) histones were hardly split, and ADP-ribose·histone adducts formed chemically through Schiff base reduction [3] or ADP-ribosyl arginine bond formed by avian erythrocyte ADP-ribosyltransferase [14] did not serve as substrate.

5′-ADP-3″-Deoxypentos-2″-Ulose

The product of ADP-ribosyl histones split by the lyase has been shown to be close to, but not identical to, ADP-ribose [5]. Upon chromatography on an AG 1 column, the product did not co-elute with authentic ADP-ribose (Fig. 3). This discrepancy was in contrast to co-elution of nonenzymatically (alkali or NH$_2$OH) split products with authentic ADP-ribose [12].

This discrepancy proved to be due to an alteration in the terminal ribose moiety. When the enzymatic product was digested with snake venom phosphodiesterase and chromatographed on an AG 1 column, two products were recovered; one, derived from the AMP portion, co-eluted with authentic 5′-AMP, while the other, derived from the phosphoribose portion, eluted slightly after authentic ribose 5-phosphate

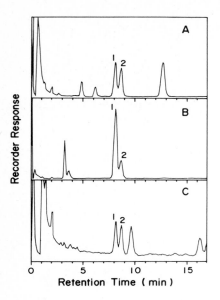

Fig. 5A–C. Gas chromatography of (A) 3-deoxy-pentitols-1,2-d_2-(Me$_3$Si)$_4$; (B) 3-deoxypentitols-1,4-d_2-(Me$_3$Si)$_4$; (C) reduced-X-d_2-(Me$_3$Si)$_4$. (Taken from Oka et al. [12])

(data not shown). Dephosphorylation of the latter product with alkaline phospho-monoesterase treatment gave a compound, designated as X, which did not comigrate with ribose or other known pentoses upon paper chromatography (cf. Fig. 4) [12]. Compound X was not produced by similar enzyme digestion of free ADP-[^{14}C]ribose treated or untreated with ADP-ribosyl protein lyase, suggesting that ADP-X was formed directly from histone-bound ADP-ribose during the splitting reaction, and not from once liberated ADP-ribose. Compound X retained original five carbon atoms of ribose, as indicated by identical yields of ADP-[^{14}C]X from ADP-[1''-^{14}C]ribosyl and ADP-[5'-^{14}C] ribosyl histone H2B's.

Compound X was reactive with phenylhydrazine to produce a distinct product under the conditions in which ribose was nonreactive (Fig. 4A). This hydrazone forma-tion suggested the presence of carbonyl group(s) other than the one participating in hemiacetal formation in compound X. Compound X also reacted with NaBH$_4$ under the conditions which reduced ribose to ribitol. The product (reduced-X) comigrated with 3-deoxyribitol (and 3-deoxyarabinitol) apart from various other pentitols in paper chromatography (Fig. 4B).

These results, together with the assumption that C-5 was engaged in phosphomono-ester formation in ADP-X, indicated that compound X is a derivative of 3-deoxyribose with additional keto function(s) at C-2 or C-4 or both. Analyses with gas chromato-graphy/mass spectrometry (GC/MS) after acetylation and with high-performance liquid chromatography (HPLC) after dansylhydrazone formation also supported this view [12].

Final identification of compound X was attained by GC/MS analysis in comparison with chemically synthesized 3-deoxypentos-2-ulose and 4-ulose after deuterium-reduc-tion and per-trimethylsilylation [12, 15]. Upon gas chromatography, all of these three compounds split into two peaks representing 3-deoxyribitol and 3-deoxy-D (or L)-

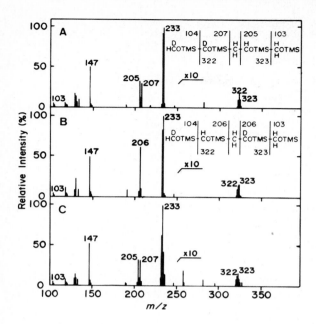

Fig. 6A–C. Mass spectra of Peak 1's of (A) Fig. 5A; (B) Fig. 5B; (C) Fig. 5C. *Inlets* show the chemcal structures with possible assignments for major fragment peaks. (Modified from Oka et al. [12])

arabinitol derivatives (Fig. 5). The mass spectra of each two peaks were essentially identical, and only those of peak 1's are shown in Fig. 6. Besides many peaks common to all three derivatives, diagnostic doublet peaks (m/z = 205 and 207) were given by the 3-deoxypentos-2-ulose derivative [3-deoxypentitol-1,2-d_2-(Me$_3$Si)$_4$] and the compound X derivative [reduced-X-d_2-(Me$_3$Si)$_4$] (Fig. 6A,C), whereas a singlet peak (m/z = 206) was given by the 3-deoxypentos-4-ulose derivative [3-deoxypentitol-1,4-d_2-(Me$_3$Si)$_4$] (Fig. 6B). These results clearly indicated that deuterium had been introduced to C-1 and C-2 of compound X, and thus that compound X was 3-deoxypentos-2-ulose or that ADP-X was 5'-ADP-3''-deoxypentos-2''-ulose (Fig. 7C). Taking into account possible tautomerism of dehydrated sugars, the original split product was inferred to be 5'-ADP-3''-deoxypent-2''-enofuranose (Fig. 7B). In view of the fact that ADP-ribose is bound to a carboxyl group of histone H2B through an ester bond (Fig. 7A), the production of ADP-3''-deoxypent-2''-ulose indicated that the splitting raction was not hydrolysis but elimination of the hydroxyl or carboxyl group at C-3 of the ribose.

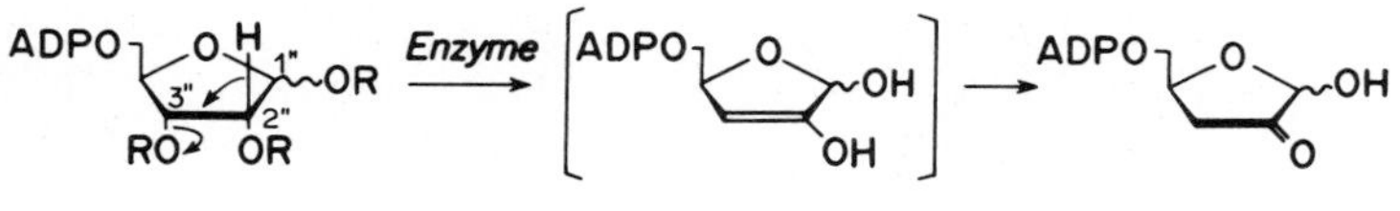

Fig. 7A–C. Possible reaction sequences involving (A) ADP-ribosylhistone carboxylate; (B) 5'-ADP-3''-deoxypent-2''-enofuranose; (C) 5'-ADP-3 ll-deoxy-D-*glycero*-pentos-2''-ulose. R = H or histone. (Modified from Komura et al. [15])

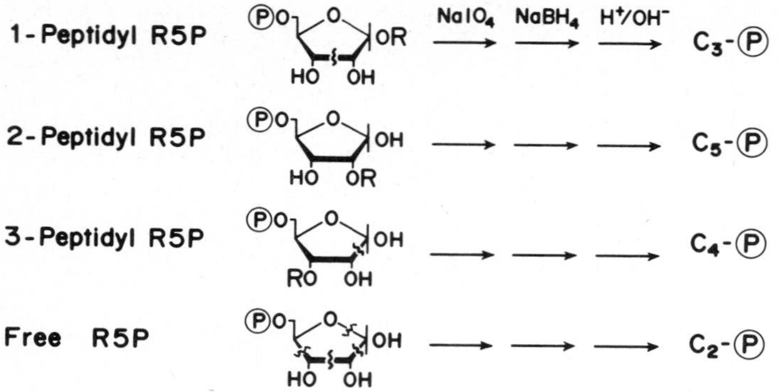

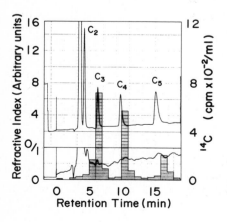

1-Peptidyl R5P ⟶ NaIO₄ NaBH₄ H⁺/OH⁻ ⟶ C_3-Ⓟ

2-Peptidyl R5P ⟶ ⟶ ⟶ C_5-Ⓟ

3-Peptidyl R5P ⟶ ⟶ ⟶ C_4-Ⓟ

Free R5P ⟶ ⟶ ⟶ C_2-Ⓟ

Fig. 8. Outlines of the method to identify isomers of phosphoribosyl pentapeptide. 1-, 2-, and 3-peptidyl ribose 5-phosphate are expected to give, after oxidation by $NaIO_4$ (pH 7, room temperature, 60 min), reduction by $NaBH_4$, and digestion with alkaline phosphomonoesterase, glycerol (C_3), ribitol (C_5), and erythritol (C_4), respectively. For reference, the expected production of ethylene glycol (C_2), which is mostly distilled out at the step of evaporation in vacuo, from free ribose 5-phosphate is also shown

Possible Isomers of ADP-Ribosyl Carboxylates

Referring to the structures of NAD and poly(ADP-ribose), the linkage between ADP-ribose and a carboxyl group of protein has so far been tacitly assumed to be through C-1 of the terminal ribose. The suggested mechanism of the lyase action, that is, elimination of the hydroxyl or carboxyl group from C-3 of the ribose, however, suggested a possibility of the linkage being through C-2 or C-3. We recently started to investigate these possibilities by analyzing $NaIO_4$-oxidized ADP-ribosyl carboxylates. The strategy of the experiment is outlined in Fig. 8. A preliminary result (Fig. 9) showed that a preparation of phosphoribosyl pentapeptide obtained from ADP-[U-^{14}C]ribosyl histone H2B comprized 1- and 3-peptidyl ribose 5-phosphate at a ratio of approximately 2:1 (corrected for the difference in the number of carbon atoms in

Fig. 9. HPLC patterns of a mixture of ethylene glycol (C_2), glycerol (C_3), erythritol (C_4), and ribitol (C_5) (*upper*), and $NaIO_4$/$NaBH_4$/phosphatase-treated 5-phospho[^{14}C]ribosyl pentapeptide of histone H2B (*lower*). —— Refractive index; ▨▨▨ [14]C (cpm ml⁻¹)

each fragment). Furthermore, the presence of a small amount of 2-peptidyl derivative was also indicated by the yield of nonoxidized ribose. Although precise quantitation of each of these isomers awaits further analyses, this result suggested, for the first time, possible migration of carboxylates on the ribose. This idea appears also to account for the observed heterogeneity of enzymic susceptibility or chemical stabilities of purified ADP-ribosyl (or phosphoribosyl) peptides [8, 9, 11, 12].

Concluding Remarks

Recently, Wielckens et al. [16] showed that a removal of primary ADP-ribosyl groups from acceptor proteins was the rate-limiting step in the overall turnover of poly(ADP-ribosyl) groups in cells treated with a DNA-damaging agent. Therefore, ADP-ribosyl protein lyase appears to play a critical role in poly(ADP-ribosyl)ation. The importance of this enzyme was further suggested by a recent report of Williams et al. [17]; they identified glutamyl ribose 5-phosphate, that is a core structure of ADP-ribosyl histone linkage, in a brain extract of a patient who succumbed at the age of 8 ys after a 6-yr course of progressive neurologic deterioration and renal failure, and diagnosed as a new type of lysosomal storage disease [18]. This is probably a congenital defect of ADP-ribosyl protein lyase [17], and his symptoms were suggestive of the adverse effect of the dysfunction of this enzyme in the human body.

References

1. Ueda K, Kawaichi M, Oka J, Hayaishi O (1980) Biosynthesis and degradation of poly(ADP-ribosyl) histones. In: Smulson M, Sugimura T, (eds) Novel ADP-ribosylations of regulatory enzymes and proteins. Elsevier/North Holland, Amsterdam New York, pp 47–56
2. Ueda K, Ogata N, Kawaichi M, Inada S, Hayaishi O (1982) ADP-ribosylation reactions. Cur Top Cell Regul 21:175–187
3. Ueda K, Oka J, Narumiya S, Miyakawa N, Hayaishi O (1972) Poly ADP-ribose glycohydrolase from rat liver nuclei, a novel enzyme degrading the polymer. Biochem Biophys Res Commun 46:516–523
4. Miwa M, Tanaka M, Matsushima T, Sugimura T (1974) Purification and properties of a glycohydrolase from calf thymus splitting ribose-ribose linkages of poly(adenosine diphosphate ribose). J Biol Chem 249:3475–3482
5. Okayama H, Honda M, Hayaishi O (1978) Novel enzyme from rat liver that cleaves an ADP-ribosyl histone linkage. Proc Natl Acad Sci USA 75:2254–2257
6. Ueda K, Hayaishi O (1985) ADP-ribosylation. Annu Rev Biochem 54:71–99
7. Riquelme PT, Burzio LO, Koide SS (1979) ADP ribosylation of rat liver lysine-rich histone in vitro. J Biol Chem 254:3018–3028
8. Ogata N, Ueda K, Hayaishi O (1980) ADP-ribosylation sites of histones H1 and H2B. In: Smulson M, Sugimura T (eds) Novel ADP-ribosylations of regulatory enzymes and proteins. Elsevier/North Holland, Amsterdam New York, pp 333–342
9. Ogata N, Ueda K, Kagamiyama H, Hayaishi O (1980) ADP-ribosylation of histone H1. Identification of glutamic acid residues 2, 14 and COOH-terminal lysine residue as modification sites. J Biol Chem 255:7616–7620

10. Burzio LO, Riquelme PT, Koide SS (1979) ADP ribosylation of rat liver nucleosomal core histones. J Biol Chem 254:3029–3037
11. Ogata K, Ueda K, Hayaishi O (1980) ADP-ribosylation of histone H2B. Identification of glutamic acid residue 2 as the modification site. J Biol Chem 255:7610–7615
12. Oka J, Ueda K, Hayaishi O, Komura H, Nakanishi K (1984) ADP-ribosyl protein lyase. Purification, properties, and identification of the product. J Biol Chem 259:986–995
13. Kun E, Chang ACY, Sharma ML, Ferro AM, Nitecki D (1976) Covalent modification of proteins by metabolites of NAD. Proc Natl Acad Sci USA 73:3131–3135
14. Moss J, Vaughan M (1978) Isolation of an avian erythrocyte protein possessing ADP-ribosyltransferase activity and capable of activating adenylate cyclase. Proc Natl Acad Sci USA 75:3621–3624
15. Komura H, Iwashita T, Naoki H, Nakanishi K, Oka J, Ueda K, Hayaishi O (1983) Structure and synthesis of 3-deoxy-D-*glycero*-pentos-2-ulose, an unusual sugar produced enzymatically from (ADP-ribosyl)histone H2B. J Am Chem Soc 105:5164–5165
16. Wielckens K, Schmidt A, George E, Bredehorst R, Hilz H (1982) DNA fragmentation and NAD depletion. Their relation to the turnover of endogenous mono(ADP-ribosyl) and poly(ADP-ribosyl) proteins. J Biol Chem 257:12872–12877
17. Williams JC, Chambers JP, Liehr JG (1984) Glutamyl ribose 5-phosphate storage disease. A hereditary defect in the degradation of poly(ADP-ribosylated) proteins. J Biol Chem 259:1037–1042
18. Williams JC, Butler IJ, Rosenberg HS, Verani R, Scott CI, Conley SB (1984) Progressive neurologic deterioration and renal failure due to storage of glutamyl ribose-5-phosphate. N Engl J Med 311:152–155

Glutamyl Ribose-5-Phosphate Storage Disease: Clinical Description and Characterization of the Stored Material

JULIAN C. WILLIAMS[1]

Introduction

Posttranslational processing of proteins involves conformational changes, formation and cleavage of peptide bonds, and chemical modification of amino acid residues. These events are assumed to have evolved for specific purposes, such as protection from degradation, binding to specific macromolecules, and regulation of physiological processes. Posttranslational covalent modification of amino acid moieties in proteins is a diverse process resulting in more than 130 derivatives of the 20 primary amino acids [1]. The formation of glycoproteins by the enzymatic addition of carbohydrate to amino acids and the subsequent elongation to form oligosaccharide chains has become a major area of investigation in recent years. In glycoconjugate synthesis, ADP-ribosylation is unique in that it involves an initial linkage of ADP-ribose via the ribosyl moiety and, in the case of poly(ADP-ribosylation), subsequent elongation with additional ADP-ribose units.

The carbohydrate-amino acid linkage regions of mammalian glycoconjugates are classified into N-linked and O-linked categories [2]. Covalent linkage of the amino nitrogen of asparagine to the reducing C-1 of N-acetylglucosamine occurs predominantly in plasma proteins. The O-linked group consists primarily of: (1) the xylose-serine linkage of proteoglycans; (2) galactose-hydroxylysine found in collagen; and (3) the N-acetylgalactosamine-serine linkage of epithelial tissue secretory proteins. Although much of the current attention on ADP-ribosylation is focused on its putative role in the regulation of gene expression and DNA repair, earlier work characterized the structure of this posttranslational modification and the nature of the ADP-ribose-protein linkages (for review see [3]). Goff demonstrated the linkage of ribose C-1 to the guanido nitrogen of arginine in the α-subunit of *E. coli* RNA polymerase [4]. Several bacterial toxins [5, 6] and an avian enzyme [7] perform a similar mono(ADP-ribosylation) of the arginine residues of nonnuclear proteins. Diphtheria toxin [8] and *Pseudomonas* exotoxin A [9] inactivate elongation factor 2 via modification of diphthamide at the imidazole N-1 via an α-glycosidic linkage of ADP-ribose [10–12].

---

1 Division of Metabolic Diseases, Department of Pediatrics, University of Texas Health Science Center, Houston, TX 77030, USA

ADP-Ribosylation of Proteins
(ed. by F.R. Althaus, H. Hilz, and S. Shall)
© Springer-Verlag Berlin Heidelberg 1985

ADP-ribosylation of nuclear proteins involves a small percentage of monomer adduct formation and predominantly poly(ADP-ribosylation), i.e., oligomers of ADP-ribose. The ADP-ribose nuclear protein conjugates are uniformly alkaline-labile [13], but vary in their sensitivity to neutral NH_2OH degradation. While poly(ADP-ribosylated) histones are sensitive to neutral NH_2OH, Hilz and collaborators have demonstrated that 10–30% of the modified nuclear proteins are NH_2OH resistant and contain monomer (or short oligomer) ADP-ribose protein conjugates [14–16]. Thus, in addition to the well characterized poly(ADP-ribose) histone linkage, there appears to be another, as yet unidentified, type of nuclear protein linkage. The sites of poly(ADP-ribosylation) of histones were initially identified as the glutamic acid residues at positions 2 and 116 in histone H1 [17, 18]. Detailed studies by Hayaishi's group extended the modified sites to also include glutamic acid residues 14 of histone H1 and 2 of histone H2B, as well as the COOH-terminal lysine of histone H1 [19, 20]. These latter experiments proved the linkage of the γ-carboxyl of glutamic acid and the α-carboxyl of lysine to the ribose moiety via an ester linkage. Periodate consumption studies suggested that the linkage was via the ribose C-1 carbon. The anomeric specificity (α vs β) of the carbohydrate-amino acid linkage has not been elucidated.

Characterization of the oligosaccharides and carbohydrate-amino acid conjugates found in the urine and tissues of patients with hereditary deficiencies of lysosomal enzymes has aided the structural elucidation of mammalian glycoconjugates [21]. Although several disorders due to deficiencies of glycosidases specific for various moieties of glycoprotein oligosaccharide chains have been identified, only one is known which results in the accumulation of a carbohydrate-amino acid linkage — aspartylglucosaminuria [22, 23]. This disease is due to a deficiency of 2-N-acetamido — (4'L-aspartyl)-2-deoxy-β-D-glucosylamine amidohydrolase (EC 3.5.1.26) and results in the accumulation of N-acetylglucosamine-asparagine derived from N-linked glycoproteins. Recently, we have described the accumulation of glutamyl ribose-5-phosphate in a patient with electron microscopic evidence of a lysosomal storage disease [24, 25]. The clinical description and characterization of the stored material are reviewed below and reader is referred to the original papers for details of the experimental methods. We have proposed that this disorder is a glycoproteinosis due to the failure to degrade poly(ADP-ribosylated) proteins by ADP-ribosyl protein lyase.

Clinical Studies

This male infant had normal physical and intellectual development until 2 yrs of age when abnormalities of speech were noted. Medical evaluation was delayed until the onset of seizures at 3 yrs. Diagnostic assessment at that time revealed microcephaly, mental retardation, and the presence of protein in the urine. Over the next 3 yrs he was followed for his kidney disease and signs of a progressive deterioration of neurological function. Our initial evaluation at that time revealed a retarded child with multiple signs of neurological dysfunction — microcephaly, decreased muscle tone and reflexes, optic atrophy with decreased visual acuity, a speech deficit, and a complex seizure disorder. In addition, he had growth failure and renal disease. Electron micro-

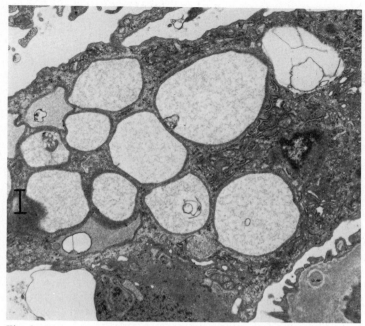

Fig. 1. Glomerular epithelial cell demonstrating enlarged lysosomes filled with granular material and occasional lamellar structures (*Bar* = 1 μm, uranyl acetate). (From [25] with permission)

Table 1. Lysosomal enzyme activities (nmol h^{-1} mg^{-1} protein) and sialic acid content. (From [25] with permission)

	Leukocytes		Fibroblasts	
	Patient	Controls (Mean ± 2 SD)	Patient	Controls (Mean ± 2 SD)
β-Hexosaminidase[a]	1342	913 ± 350	3439	2600 ± 690
α-Glucosidase[a]	35	30 ± 11	47	33 ± 17
β-Glucosidase[a]	2.4	3.9 ± 1.4	53	33 ± 13
α-Mannosidase[a]	246	156 ± 48	64	50 ± 19
α-Galactosidase[a]	40	22 ± 6	27	28 ± 6
β-Galactosidase[a]	155	83 ± 19	242	259 ± 63
α-Fucosidase[b]	47	52 ± 17	58	97 ± 22
β-Glucuronidase[b]	112	268 ± 47	52	120 ± 31
Arylsulfatase A[c]	86	99 ± 25	269	400 ± 150
Sialidase[a]	ND	ND	59	42 ± 203
Total sialic acid (nmol mg^{-1} protein)	ND	ND	10.2	8.3
Free sialic acid (nmol mg^{-1} protein)	ND	ND	2.1	3.0

[a] Determined with 4-methylumbelliferyl derivatives
[b] Determined with p-nitrophenyl derivatives
[c] Determined with p-nitrocatechol sulfate

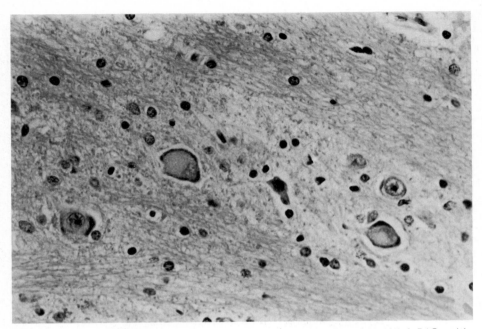

Fig. 2. Brain stem ganglion cells swollen with cytoplasmic storage material which is PAS-positive. (From [25] with permission)

scopy of a kidney biopsy indicated enlarged lysosomes containing granular material and rare lamellar structures (Fig. 1). Laboratory evaluation resulted in normal findings relevant to the known lysosomal storage diseases — negative mucopolysacchariduria and $^{35}SO_4$ accumulation in cultured skin fibroblasts, the absence of increased urinary oligosaccharides, and normal assays in leukocytes and fibroblasts for the lysosomal enzymes whose deficiency is associated with known hereditary defects (Table 1). Over the next 2 yrs, the patient's neurological function deteriorated to a vegetative state and his renal disease worsened, resulting in his death.

The patient's family history indicated that a maternal uncle had died of an identical disease. This suggested that this disorder was inherited as an X-linked recessive condition.

Postmortem studies revealed enlarged lysosomes in multiple tissues and brain stem ganglion cells distended by the cytoplasmic storage of PAS-positive material (Fig. 2). Thin-layer chromatographic analysis of brain gangliosides was normal. Gel filtration of an aqueous brain extract indicated the presence of a large peak of carbohydrate-containing material of small molecular weight (Fig. 3).

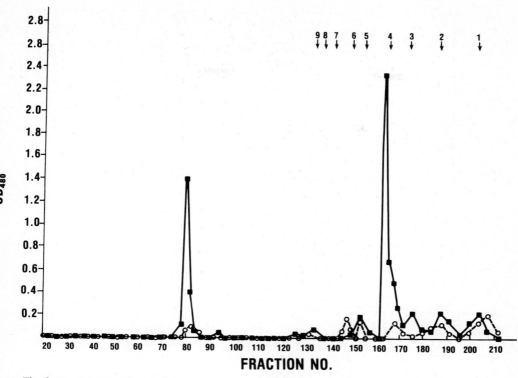

Fig. 3. Bio-Gel P-2 chromatography of an aqueous extract of brain assayed for carbohydrate with phenol-H_2SO_4. *Black boxes* (■) indicate patient vs control (○). *Arrows* indicate the elution volumes of glucose oligomers. (From [24] with permission)

Chemical Characterization of the Stored Material

The material isolated by gel filtration was further purified by batch chromatography on anion exchange resin (Dowex 1-Cl⁻) and gave a single spot on thin-layer chromatography [silica gel in butanol:acetic acid:H_2O (2:1:1)] when sprayed with carbohydrate and amino acid sensitive reagents. Further preliminary evidence as to its structure was gained from (1) binding to Dowex 50-H⁺; (2) absence of reaction with alkaline silver nitrate unless previously treated with periodic acid; and (3) a decreased electrophoretic mobility after incubation with alkaline phosphatase. These studies suggested the material to be an amino acid covalently linked to the reducing end of a sugar phosphate.

Amino acid analysis of a strong acid hydrolysate (2 N HCl at 100°C for 3 h) identified only glutamic acid and this was confirmed by gas chromatography-mass spectroscopy (Fig. 4). Mild acid hydrolysis (0.5 N HCl at 80°C for 1 h) followed by direct insertion mass spectroscopy gave a spectrum identical to authentic ribose-5-phosphate (Fig. 5). Borate electrophoresis of the alditol derived from the stored material and of its acid

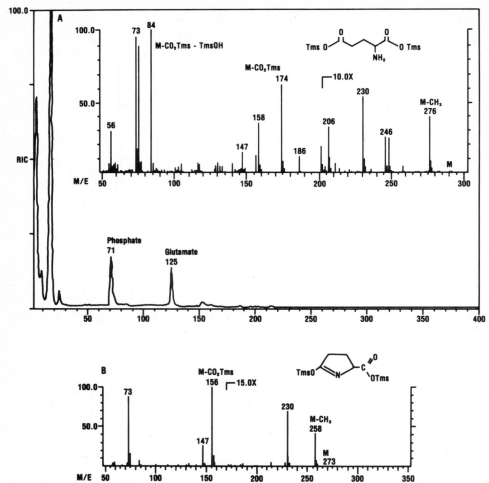

Fig. 4A,B. Gas chromatography – mass spectroscopy of the acid hydrolysate of the stored material. **A** The ion chromatogram of a 2 N HCl hydrolysate with the *inset* depicting the fragmentation pattern and structure of the glutamic acid derivative. **B** The spectrum and structure of pyroglutamic acid seen in a 0.5 N HCl hydrolysate. (From [24] with permission)

hydrolysis or alkaline phosphatase-treated product indicated it to be ribose-5-phosphate and ribose or arabinose, respectively (Fig. 6). The identity of the ribose moiety was confirmed by paper chromatography. Enzymatic degradation of the alditol phosphate was consistent with localization of the phosphate group to C-5 of ribose and quantitative analysis of the original material indicated glutamic acid:ribose:phosphate (1:1:1). The amount of glutamyl ribose-5-phosphate isolated from brain and kidney was 0.96 and 0.60 μmol g^{-1} wet wt., respectively.

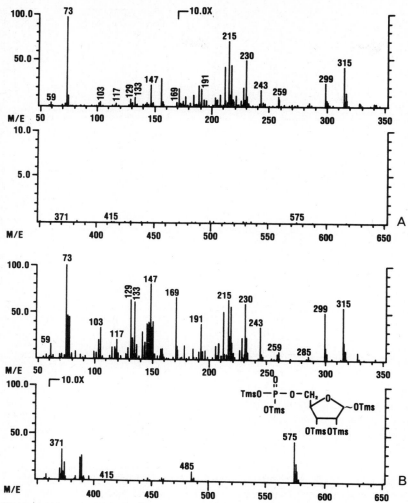

Fig. 5. **A** Direct probe mass spectrum of a 0.5 N HCl hydrolysate of the stored material. **B** An authentic sample of ribose-5-phosphate. (From [24] with permission)

Discussion

From the structural analysis, we have concluded that the stored material is ribose-5-phosphate linked via the C-1 position by an ester linkage to the γ-carboxyl of glutamic acid (Fig. 7). The possible sites of covalent bonding are the 1, 2, and 3 position of ribose. The assignment to C-1 was based on the failure of the material to react with silver reagent, which requires a free reducing end, and its reactivity to periodic acid, which requires adjacent free hydroxyl groups. The ninhydrin reactivity of the intact material indicates that the α-amino and α-carboxyl groups are free. The nature of the

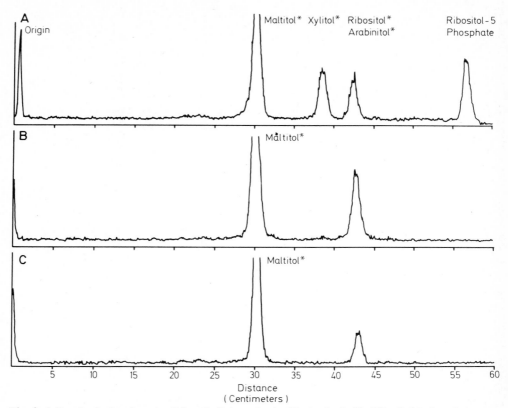

Fig. 6. A Borate electrophoresis of the alkaline product with standards (*). Ribositol-5-phosphate standard elutes at the same position. **B** The product of 2 N HCl hydrolysis of the alkaline NaB^3H_4-treated material. **C** The product of alkaline phosphatase treatment. (From [24] with permission)

ribose-glutamic acid linkage is speculated to be an ester linkage based on the instability of the stored material in 0.15 M NaHCO$_3$. The anomeric specificity of the linkage has not been determined. When completion of enzymatic studies of the patient's tissues allows use of the remaining samples for isolation of large amounts of the stored material, NMR analysis should yield a definitive structure.

Histological and electron microscopic studies indicated that presence of enlarged lysosomes and cytoplasmic storage of PAS-positive material. This would suggest the classification of this disorder as a lysosomal storage disease. Hers defined this disease category as the lysosomal storage of macromolecules due to a hereditary deficiency of a lysosomal enzyme [26]. This concept has been extended to included entities with lysosomal storage due to the hereditary deficiency of other proteins as in cystinosis (cystine storage due to the deficiency of a lysosomal transport protein) [27]. As indicated in Fig. 7, we have proposed that the enzyme deficiency in this patient is ADP-ribose protein hydrolase [28], which has recently been shown to be a lyase [29]. The available evidence indicates that ADP-ribosyl protein lyase is a cytosolic and/or

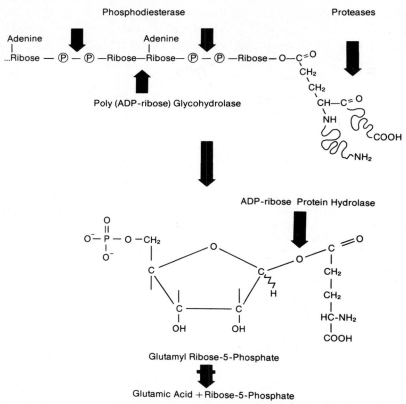

Fig. 7. Degradative pathway of poly(ADP-ribosylated) proteins and the proposed enzymatic deficiency resulting in accumulation of glutamyl ribose-5-phosphate. (From [24] with permission)

nuclear enzyme. A deficiency of this enzyme could result in lysosomal uptake of ADP-ribosylated histones or ribosyl-5-phosphate histones, depending on the rate of phosphodiesterase action, and subsequent proteolytic degradation. In the absence of any known lysosomal enzyme which cleaves the ribose-glutamic acid linkage, glutamyl ribose-5-phosphate would accumulate and be stabilized by the acidic lysosomal milieu. Thus, a lysosomal storage disease would result from a non-lysosomal enzyme deficiency. Enzymatic studies of the patient's cultured skin fibroblasts in collaboration with Hayaishi and Ueda will resolve this question.

The accumulation of ribose-5-phosphate linked only to glutamic acid suggests that cleavage of ADP-ribose-arginine and the NH_2OH-resistant ADP-ribose-nuclear protein linkages may be due to other, as yet undefined, enzymes. The occurrence of this disease also raises several other interesting lines of investigation. What is the pathway for transport of nuclear proteins to the lysosome for degradation? Will the introduction of label in the patient's cells create a nondegradable marker for the identification of low abundance or transiently poly(ADP-ribosylated) proteins? Can the initial rate of poly(ADP-ribosylation) and the timing of this process during physiological regulation be determined with such a marker? What is the turnover rate of the core proteins which are poly(ADP-ribosylated)? Hopefully, the use of point mutations to study ADP-ribosylation will be as fruitful as it has been in other areas of biochemistry.

Acknowledgments. The author is indebted to his collaborators [24, 25] for their many contributions and to the National Foundation – March of Dimes and the National Institutes of Health for grant support.

References

1. Wold F (1980) Post-translational covalent modification of proteins. In: Smulson ME, Sugimura T (eds) Novel ADP-ribosylations of regulatory enzymes and proteins. Elsevier/North Holland, Amsterdam New York, pp 325–332
2. Zinn AB, Plantner JJ, Carlson DM (1977) Nature of linkages between protein core and oligosaccharides. In: Horowitz MI, Pigman W (eds) The glycoconjugates, vol I. Academic Press, London New York, pp 69–85
3. Burzio LO (1982) ADP-ribosyl protein linkages. In: Hayaishi O, Ueda K (eds) ADP-ribosylation reactions. Biology and medicine. Academic Press, London New York, pp 103–116
4. Goff CG (1974) Chemical structures of a modification of the *Escherichia coli* ribonucleic acid polymerase α polypeptides induced by bacteriophage T_4 infection. J Biol Chem 249:6181–6190
5. Vaughn M, Moss J (1981) Mono(ADP-ribosyl)transferases and their effects on cellular metabolism. Curr Top Cell Regul 20:205–246
6. Moss J, Vaughn M (1977) Mechanism of action of choleragen. Evidence for ADP-ribosyltransferase activity with arginine as an acceptor. J Biol Chem 252:2455–2457
7. Moss J, Stanley SJ, Watkins PA (1980) Isolation and properties of an NAD- and guanidine-dependent ADP-ribosyltransferase activity from turkey erythrocytes. J Biol Chem 255:5838–5840
8. Pappenheimer AM (1977) Diphtheria toxin. Annu Rev Biochem 46:69–94
9. Iglewski BH, Kabat D (1975) NAD-dependent inhibition of protein synthesis by *Pseudomonas aeruginosa* toxin. Proc Natl Acad Sci USA 72:2284–2288
10. Van Ness BG, Howard JB, Bodley JW (1980) ADP-ribosylation of elongation factor 2 by diphtheria toxin. NMR spectra and proposed structures of ribosyl-diphthamide and its hydrolysis products. J Biol Chem 255:10710–10716
11. Van Ness BG, Howard JB, Bodley JW (1980) ADP-ribosylation of elongation factor 2 by diphtheria toxin. Isolation and properties of the novel ribosyl-amino acid and its hydrolysis products. J Biol Chem 255:10717–10720
12. Oppenheimer NJ, Bodley JW (1981) Diphtheria toxin. Site and configuration of ADP-ribosylation of diphthamide in elongation factor 2. J Biol Chem 256:8579–8581
13. Nishizuka Y, Ueda K, Honjo T, Hayaishi O (1968) Enzymic adenosine diphosphate ribosylation of histone and poly adenosine diphosphate ribose synthesis in rat liver nuclei. J Biol Chem 243:3765–3767
14. Adamietz P, Hilz H (1975) Covalent linkage of poly(adenosine diphosphate ribose) to nuclear proteins of rat liver by two types of bonds. Biochem Soc Trans 3:1118–1120
15. Hilz H, Stone R (1976) Poly(ADP-ribose) and ADP-ribosylation of proteins. Rev Physiol Biochem Pharmacol 76:1–58
16. Adamietz P, Bredehorst R, Hilz H (1978) ADP-ribosylated histone H1 from HeLa cultures. Fundamental differences to (ADP-ribose)$_n$ – histone H1 conjugates formed in vitro. Eur J Biochem 91:317–326
17. Riquelme PT, Burzio LO, Koide SS (1979) ADP-ribosylation of rat liver lysine-rich histone in vitro. J Biol Chem 254:3018–3028

18. Burzio LO, Riquelme PT, Koide SS (1979) ADP ribosylation of rat liver nucleosomal core histones. J Biol Chem 254:3029–3037
19. Ogata N, Ueda K, Hayaishi O (1980) ADP-ribosylation of histone H2B. Identification of glutamic acid residue 2 as the modification site. J Biol Chem 255:7610–7615
20. Ogata N, Ueda K, Kagamiyama H, Hayaishi O (1980) ADP-ribosylation of histone H1. Identification of glutamic acid residues 2, 14, and the COOH-terminal lysine residue as modification sites. J Biol Chem 255:7616–7620
21. Montreuil J (1980) Primary structure of glycoprotein glycans. Basis for the molecular biology of glycoproteins. Adv Carbohydr Chem Biochem 37:157–223
22. Pollitt RJ, Pretty KM (1974) The glycoasparagines in urine of a patient with aspartylglycosaminuria. J Biol Chem 254:1513–1515
23. Maury P (1979) Accumulation of two glycoasparagines in the liver in aspartylglycasaminuria. J Biol Chem 254:1515–1515
24. Williams JC, Chambers JP, Liehr JG (1984) Glutamyl ribose-5-phosphate storage disease. A hereditary defect in the degradation of poly(ADP-ribosylated) proteins. J Biol Chem 259:1037–1042
25. Williams JC, Butler IJ, Rosenberg HS, Verani R, Scott CI, Conley SB (1984) Progressive neurologic deterioration and renal failure due to storage of glutamyl ribose-5-phosphate. N Engl J Med 311:152–155
26. Hers HG (1965) Inborn lysosomal diseases. Gastroenterology 48:625–633
27. Jonas AJ, Smith ML, Allison WS, Laikind PK, Greene AA, Schneider JA (1983) Proton-translocating ATPase and lysosomal cystine transport. J Biol Chem 258:11727–11730
28. Okayama H, Honda M, Hayaishi O (1978) Novel enzyme from rat liver that cleaves and ADP-ribosyl histone linkage. Proc Natl Acad Sci USA 75:2254–2257
29. Oka J, Ueda K, Hayaishi O, Komura H, Nakanishi K (1984) ADP-ribosyl protein lyase. Purification, properties, and identification of the product. J Biol Chem 254:986–995

ADP-Ribosylation and Chromatin Structure

Effect of Poly(ADP-Ribosyl)ation on Native Polynucleosomes, H1-Depleted Polynucleosomes, Core Particles, and H1-DNA Complexes

ANN HULETSKY[1], GILBERT DE MURCIA[2], ALICE MAZEN[2],
PETER LEWIS[3], DAE G. CHUNG[3], DANIEL LAMARRE[1], REMI J. AUBIN[4],
and GUY G. POIRIER[1]

Abbreviations:

TCA (Cl₃AcOH)	Trichloroacetic acid
PAGE	Polyacrylamide gel electrophoresis
Poly(ADP-ribose)	Polymer of ADP-ribose
ADP-ribose	Adenosine (5′) diphospho (5)-β-D-ribose
EDTA	Ethylenediaminetetraacetate
TEACL	Triethanolamine chloride
PPi	Inorganic pyrophosphate
DMS	Dimethyl sulfate

Introduction

There is now evidence that poly(ADP-ribosyl)ation of nuclear proteins might be involved in DNA repair [1], DNA replication [2, 3] and cellular differentiation [4, 5]. A common function of nuclear protein poly(ADP-ribosyl)ation in these various events might be the alteration of chromatin structure [6, 7].

At high NAD concentrations, i.e. ⩾ 10 μM, we have shown previously that the main histone acceptor in polynucleosomes is histone H1 [8, 9]. However recently, we have found that core histones could be poly(ADP-ribosyl)ated by the intrinsic activity of the enzyme at low NAD concentrations, i.e. ⩽ 1 μM [10]. We have also suggested that, depending on the automodification state of the enzyme, the core histones and other core nucleosome proteins may become more accessible to poly(ADP-ribose) polymerase [10].

1 Chromatin Research Unit, Centre de Recherche sur les Mécanismes de Sécrétion, Département de Biologie, Faculté des Sciences, Université de Sherbrooke, Sherbrooke, Québec, Canada J1K 2R1
2 Laboratoire de Biophysique, Institut de Biologie Moléculaire et Cellulaire, 15 rue Descartes, 67084 Strasbourg Cédex, France
3 Department of Biochemistry, Medical Science Building, University of Toronto, Toronto, Ontario, Canada M5S 1A8
4 Health Sciences Division, Chalk River Nuclear Laboratories, Atomic Energy of Canada Limited, Chalk River, Ontario, Canada K0J 1J0

ADP-Ribosylation of Proteins
(ed. by F.R. Althaus, H. Hilz, and S. Shall)
© Springer-Verlag Berlin Heidelberg 1985

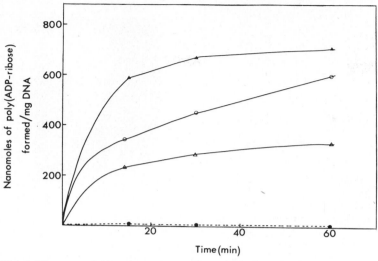

Fig. 1. Time course study of poly(ADP-ribosyl)ation of chicken erythrocyte core particles (O—O), calf thymus native chromatin (▲—▲), and H1-depleted chromatin (△—△) by purified calf thymus poly(ADP-ribose) polymerase at 30°C. Poly(ADP-ribose) polymerase (2 μg/OD unit), which was purified free of its DNA according to Zahradka and Ebisuzaki [15], was incubated with various chromatin preparations at 200 μM NAD. At various times, the reaction was stopped by the addition of 10% TCA (Cl₃AcOH)/2% PPi, and the activity was determined according to Aubin et al. [8, 9]. ●—● represents the endogenous activity of the poly(ADP-ribose) polymerase found on calf thymus native chromatin

In order to elucidate the nature of the poly(ADP-ribose) polymerase interaction with the various components of chromatin, biochemical and morphological studies on various levels of organization of histone-DNA complexes were carried out. The poly-(ADP-ribosyl)ation of histone H1-DNA complexes, core particles, chromatin depleted of histone H1, as well as native chromatin, were systematically studied using purified DNA-free calf thymus poly(ADP-ribose) polymerase in an in vitro reaction system.

Biochemical and Morphological Study of the Poly(ADP-Ribosyl)ation of Native Chromatin, Histone H1 Depleted Chromatin, and Core Particles

First, we have analyzed the stimulatory effect of various levels of chromatin structure on the enzymatic activity of purified calf thymus poly(ADP-ribose) polymerase free of its DNA (Fig. 1). We found that native chromatin stimulates the purified polymerase maximally. Chromatin depleted of non-histone proteins with 0.3 M NaCl has almost the same capacity to stimulate the polymerase as native chromatin (data not shown). Core particles were also found to stimulate the activity of the purified polymerase, but to a lesser degree. Finally, chromatin depleted of histone H1 was the least stimulatory, having one-third the effect that native chromatin has on the polymerase. No

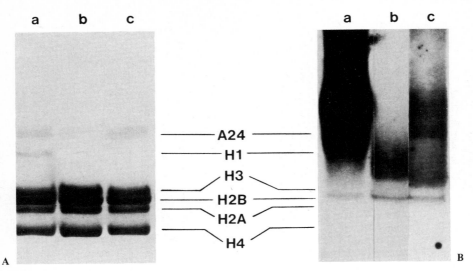

Fig. 2A,B. Acid-urea gel electrophoresis of [^{32}P] poly(ADP-ribosyl)ated histones isolated from native chromatin (*a*), core particles (*b*) and H1-depleted chromatin (*c*). These different fractions were incubated as described in the legend of Fig. 1 and precipitated with 20% TCA (Cl$_3$AcOH). Histones were extracted as described previously [8, 9] and separated by acid-urea gel electrophoresis according to Panyim and Chalkley [17]. **A** Stained gel, **B** autoradiogram of the gel

appreciable amount of endogenous enzymatic activity was detected in native chromatin.

We have tried to reconstitute the stimulatory activity onto H1-depleted chromatin with the extracted material by dialysis [11]. It was found that this reconstituted chromatin was not as stimulatory as native chromatin. This observation seems to indicate that histone H1 does not reconstitute properly even though the reconstituted chromatin appears morphologically the same as the native material as judged by electron microscopy. The larger stimulatory capacity of core particles can be explained by the presence of a high level of double-stranded breaks on the DNA which are known to optimally activate the polymerase as shown in the elegant work of Benjamin and Gill [12, 13]. Histone H1 has also been shown to stimulate the activity of poly(ADP-ribose) polymerase [14, 15] but it might necessitate to adopt a very strict conformation on chromatin to be stimulatory.

Furthermore, the acceptor proteins were analyzed by acid urea polyacrylamide gel electrophoresis (Fig. 2). Histones H2B, H1 and protein A24 were found to be poly-(ADP-ribosyl)ated in native chromatin while in histone H1-depleted chromatin and core particles it was found that histone H2B and protein A24 were poly(ADP-ribosyl)-ated. The presence of the hyper(ADP-ribosyl)ated forms of histone H2B on each of these various levels of chromatin structure have been confirmed by western blot analysis and by two-dimensional polyacrylamide gel electrophoresis (data not shown). The poly(ADP-ribosyl)ation of protein A24 have also been demonstrated by this last technique.

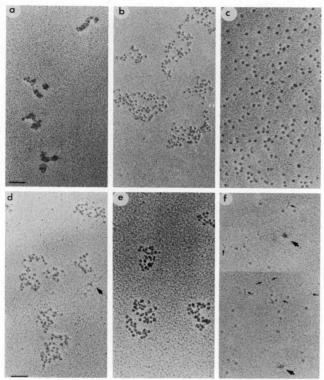

Fig. 3. Electron microscopic visualization of control and poly(ADP-ribosyl)ated native chromatin, H1-depleted chromatin and core particles. Following poly(ADP-ribosyl)ation at 200 μM NAD as described in the legend of Fig. 1, the samples were diluted to 0.01 OD unit at 260 nm and fixed for 1 h at 20°C in buffer containing 40 mM NaCl, 10 mM TEACL, 0.2 mM EDTA and 0.1% (v/v) glutaraldehyde and processed for electron microscopy according to Poirier et al. [6]. Control native chromatin, H1-depleted chromatin and core particles (**a–c**). Poly(ADP-ribosyl)ated native chromatin, H1-depleted chromatin and core particles (**d–f**). Notice the dissociation of the DNA from the nucleosome cores in the poly(ADP-ribosyl)ated core particles. Also notice the relaxation of chromatin structure in the native chromatin. *Big arrows* indicate the automodified enzyme and *small arrows* indicate the DNA (145 bp) free after poly(ADP-ribosyl)ation. The *bars* indicate 1000 Å

Concomitantly, structures resulting from the poly(ADP-ribosyl)ation of native chromatin, chromatin-H1 and core particles were examined by electron microscopy. We found, as described earlier for pancreatic chromatin [6, 7], that calf thymus chromatin adopts a more relaxed conformation upon poly(ADP-ribosyl)ation by purified calf thymus poly(ADP-ribose) polymerase free of its DNA (Fig. 3a,d). It was also found that this chromatin exhibited a lower sedimentation velocity as compared to control chromatin [6]. And recently, it has been shown that DNA polymerase α activity is more than twofold higher in the presence of pancreatic polynucleosomes ADP-ribosylated as compared to control polynucleosomes [16]. In striking contrast, no ultrastructural effect was observed when chromatin depleted of histone H1 was poly(ADP-ribosyl)ated (Fig. 3b,e).

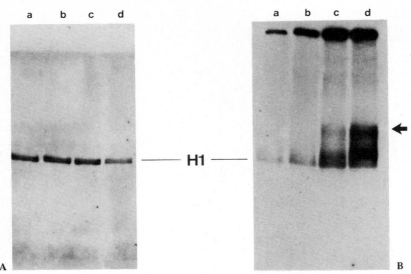

Fig. 4A,B. Acid-urea PAGE of [^{32}P]-poly(ADP-ribosyl)ated histone H1 isolated from histone H1-DNA complexes. Histone H1-DNA complexes (H1/DNA:1/1) were produced according to Hsiang and Cole [21] and were poly(ADP-ribosyl)ated by the purified calf thymus poly(ADP-ribose) polymerase (2 μg/OD unit) at 200 μM NAD and precipitated by 20% TCA (Cl$_3$AcOH). Histone H1 was extracted as described by Aubin et al. [8, 9] and was subjected to electrophoresis on 2.5 M urea − 0.9 N acetic acid vertical slab gels according to Panyim and Chalkley [17]. *a* 30 s pulse, *b−d* 5 min, 15 min, 30 min 200 μM NAD chase at 30°C, respectively. **A** Stained gel, **B** autoradiogram of the gel. The *arrow* indicates hyper(ADP-ribosyl)ated forms of histone H1. Histone H1 and DNA have been isolated from calf thymus tissue

Interestingly poly(ADP-ribosyl)ation of chicken erythrocyte core particles resulted in the dissociation of nucleosomal DNA from the histone octamer (Fig. 3c,f). The dissociation coincided with the generation of the hyper(ADP-ribosyl)ated forms of histone H2B (Fig. 2). This result is noteworthy because on exposing hepatoma cells to DMS, histone H2B becomes hyper(ADP-ribosyl)ated [18]. The dissociation of nucleosomal DNA from the histone octamer after poly(ADP-ribosyl)ation explains why nucleosomal DNA becomes more accessible to micrococcal nuclease [19]. The increased accessibility of nucleosomal core DNA caused by poly(ADP-ribosyl)ation could explain in part the increased accessibility of DNA repair patches to micrococcal nuclease observed during the early phase of DNA repair [20].

Poly(ADP-Ribosyl)ation of Histone H1-DNA Complexes

To further understand the relaxation of the chromatin structure caused by the endogenous and exogenous enzyme activity [6, 7], we have used a simple model which permits the study of the cooperative binding of histone H1 to DNA. Indeed, at a specific ratio of Histone H1 to DNA, it was found that toroidal structures are formed [21, 22]. As

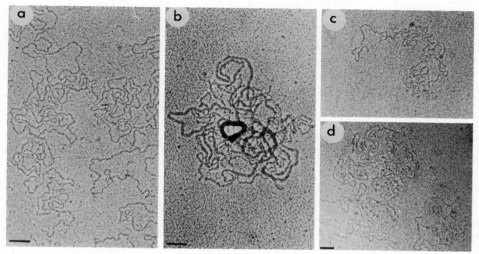

Fig. 5. Electron microscopic visualization of poly(ADP-ribosyl)ated histone H1-DNA complexes. Histone H1-DNA complexes (H1/DNA:1/1) were formed and poly(ADP-ribosyl)ated as described in the legend of Fig. 4. Samples were then treated for electron microscopy as described in the legend of Fig. 3. **a** DNA, **b** histone H1-DNA complexes (H1/DNA:1/1), **c, d** poly(ADP-ribosyl)ated histone H1-DNA complexes. Notice the dissociation of histone H1-DNA complexes after poly-(ADP-ribosyl)ation for 30 min at 200 μM NAD. The *bars* indicate 1000 Å

shown in Fig. 4, poly(ADP-ribose) polymerase is able to generate hyper(ADP-ribosyl) ated forms of histone H1 as described previously in pancreatic polynucleosomes [8, 9].

As shown in Fig. 5a, calf thymus DNA adopts a random conformation on the grid. However, as the ratio of histone H1-DNA is increased to 1, we can observe the formation of toroidal structures (Fig. 5b). These histone H1-DNA complexes were poly-(ADP-ribosyl)ated with DNA-free calf thymus poly(ADP-ribose) polymerase.

This causes a relaxation of these structures (Fig. 5c,d) and this correlates well with the formation of the hyper(ADP-ribosyl)ated forms of histone H1 (Fig. 4). It is tempting to suggest that the destabilization of these toroidal structures, resulting from the interaction between histone H1 and DNA, is similar to what has been observed in the relaxation of the solenoid conformation by poly(ADP-ribose) polymerase [6, 7]. Our results strongly suggest that poly(ADP-ribosyl)ation of the chromatin and its subsequent structural relaxation are probably tightly coupled to the interaction between histone H1 and DNA.

Interaction Between Poly(ADP-Ribose) and Core Particles

We have studied the interaction between poly(ADP-ribose) and purified chicken erythrocyte core particles which had been prepared according to Erard et al. [23] or Lutter [24]. When core particles and poly(ADP-ribose) were incubated and then separated by velocity sedimentation (Fig. 6), some interaction between polymer and core particles was observed.

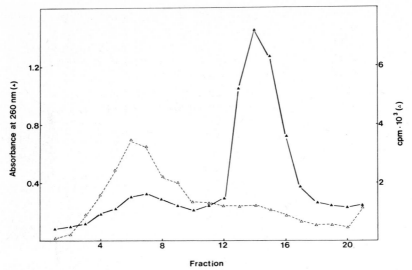

Fig. 6. Interaction between poly(ADP-ribose) and chicken erythrocyte core particles. [^{32}P]-poly-(ADP-ribose) was synthesized by purified poly(ADP-ribose) polymerase incubated with calf thymus polynucleosomes and the polymer was extracted as described by Miwa et al. [26]. Core carticles and the polymer were allowed to interact and then centrifuged on a 5−20% (w/v) iso-kinetic sucrose gradient containing 40 mM NaCl, 10 mM Tris, 0.2 mM EDTA at pH 7.4. Centrifugation was for 14 h at 40,000 rpm (4°C) in a SW41Ti Beckman rotor

This type of interaction between poly(ADP-ribose) and core particles is reminis-cent of the data obtained by Kunzler et al. [25] who demonstrated that polyglutamic acid can produce proper spacing of reconstituted chromatin although it was not shown to interact with the nucleosomes.

This interaction between polymer and core particles and the fact that the structure of core particles can be dissociated by poly(ADP-ribosyl)ation leads us to suggest that during DNA replication poly(ADP-ribose) can serve as a deposition mechanism for histone octamer.

Many lines of evidence indicate that poly(ADP-ribose) polymerase might be physi-cally associated with the replication fork. It has been found to copurify with DNA topoisomerase I [27] and recently also with DNA polymerase (G. de Murcia, D. Lamarre, J. Ménissier, unpublished results). Furthermore the DNA which copurifies with the enzyme contains replication fork-like DNA [28].

Relationship Between Histone H1 and Histone H2B
Poly(ADP-Ribosyl)ation on the Chromatin Structure

Even though histone H1 is mainly mono(ADP-ribosyl)ated during DNA repair [29], some poly(ADP-ribosyl)ation occurs as well. This pattern of modification suggests that during DNA repair the high rate of polymer turnover on histone H1 might cause a local relaxation of chromatin structure which could make histone H2B accessible to poly(ADP-ribose) polymerase. We have also shown recently that the degradation in vitro of the poly(ADP-ribose) on histone H1 by the purified bull testis glycohydrolase causes the recondensation of chromatin superstructure [30].

Indeed, we have found in chromatin depleted of histone H1 [which has a structure similar to chromatin relaxed by poly(ADP-ribosyl)ation] that (1) histone H2B is easily poly(ADP-ribosyl)ated (Fig. 2) and (2) during in vitro polymer turnover on poly(ADP-ribosyl)ated nucleosomes, the polymer on histone H2B is much more resistant to glycohydrolase than the polymer present on hyper(ADP-ribosyl)ated histone H1 [31].

In order to determine the effect of core histone poly(ADP-ribosyl)ation on chromatin structure, we have incubated histone H1 depleted chromatin with purified poly-(ADP-ribose) polymerase. This chromatin was then reconstituted with histone H1 at a ratio of histone H1 to DNA of 1. It was found that this chromatin did not recondense as the control chromatin did (A. Huletsky and G. de Murcia, unpublished observations). These results strongly suggest a close contact between histone H2B and histone H1 and that modification of the N-terminal part of histone H2B, which was found to be poly(ADP-ribosyl)ated [32, 33], affects the integrity of the chromatin structure. In this process histone H2B would have an important role to play together with histone H1.

Indeed, these results correlate well with other results we have obtained when polynucleosomes were poly(ADP-ribosyl)ated with their intrinsic activity [10]. In these experiments, we found that when the enzyme has a low level of modification, it modifies histones H2B, H2A and protein A24. However, when it was allowed to modify chromatin at high substrate concentrations, it appeared to have access to histone H1 only. Thus, it is tempting to suggest that a histone H1 containing chromatin domain, which has also some histone H2B modification, will not recondense.

Furthermore, some of our preliminary studies indicate that the turnover of poly-(ADP-ribosyl)ated proteins on polynucleosomes leads in time to an extensive modification of histone H2B. Thus, it is conceivable that during DNA repair in vivo, histone H2B modification results in destabilization of condensed chromatin and thereby permits increased accessibility to repair enzymes. Similarly, in active chromatin, where H1 levels are greatly reduced, the damaged DNA might be more accessible to repair enzymes because of histone H2B hyper(ADP-ribosyl)ation.

Acknowledgments. The authors would like to thank Dr. Jean Pouyet for helpful discussion, J. Dunand and D. Buhr for excellent technical assistance and C. Rancourt for typing the manuscript.

References

1. Berger NA, Sikorski GW (1981) Poly (adenosine diphosphoribose) synthesis in ultra-violet irradiated Xeroderma Pigmentosum cells reconstituted with micrococcus luteus UV endonuclease. Biochemistry 20:3610–3614
2. Kidwell WR, Mage MG (1976) Changes in poly (adenosine diphosphate ribose) polymerase in Synchronous HeLa cells. Biochemistry 15:1213–1217

3. Mandel P, Okazaki H, Niedergang C (1982) Poly (adenosine diphosphate ribose). Prog Nucleic Acid Res Mol Biol 27:1–51
4. Caplan AI, Rosenberg MJ (1975) Interrelationship between poly(ADP-ribose) synthesis, intracellular NAD levels and muscle or cartilage differentiation from mesodermal cells of embryonic chick limb. Proc Natl Acad Sci USA 72:1852–1857
5. Althaus FR, Lawrence SD, Hey Z, Sattler G, Tsukada Y, Pitot HC (1982) Effects of altered (ADP-ribose)$_n$ metabolism on expression of fetal functions by adult hepatocytes. Nature (London) 300:366–368
6. Poirier GG, de Murcia G, Jongstra-Bilen J, Niedergang C, Mandel P (1982) Poly (ADP-ribosyl)-ation of polynucleosomes causes relaxation of chromatin structure. Proc Natl Acad Sci USA 79:3423–3427
7. Aubin RJ, Fréchette A, de Murcia G, Mandel P, Lord A, Grondin G, Poirier GG (1983) Correlation between endogenous nucleosomal hyper (ADP-ribosyl)ation of histone H1 and the induction of chromatin relaxation. EMBO J 2:1685–1693
8. Aubin RJ, Dam VT, Miclette J, Brousseau Y, Poirier GG (1982) Chromosomal protein poly (ADP-ribosyl)ation in pancreatic nucleosomes. Can J Biochem 60:295–305
9. Aubin RJ, Dam VT, Miclette J, Brousseau Y, Huletsky A, Poirier GG (1982) Hyper (ADP-ribosyl)ation of histone H1. Can J Biochem 60:1085–1094
10. Huletsky A, Niedergang C, Fréchette A, Aubin R, Gaudreau A, Poirier GG (1985) Sequential (ADP-ribosyl)ation pattern of nucleosomal histones. Eur J Biochem 146:277–285
11. Thoma F, Koller TH, Klug A (1979) Involvement of histone H1 in the organization of the nucleosome and of the salt-dependent superstructure of chromatin. J Cell Biol 83: 403–427
12. Benjamin RC, Gill DM (1980) ADP-ribosylation in mammalian cell ghosts: dependence of poly (ADP-ribose) synthesis on strand breakage in DNA. J Biol Chem 255:10493–10501
13. Benjamin RC, Gill DM (1980) Poly (ADP-ribose) synthesis in vitro programmed by damaged DNA: a comparison of DNA molecules containing different types of strand breaks. J Biol Chem 255:10502–10508
14. Ferro AM, Olivera BM (1982) Poly (ADP-ribosyl)ation in vitro: Reaction parameters and enzyme mechanism. J Biol Chem 257:7808–7813
15. Zahradka P, Ebisuzaki K (1982) A shuttle mechanism for DNA-protein interactions. The regulation of poly (ADP-ribose) polymerase. Eur J Biochem 127:579–585
16. Niedergang C, Ittel ME, de Murcia G, Pouyet J, Mandel P (1985) Kinetics of nucleosomal histone H1 hyper (ADP-ribosyl)ation and polynucleosomes relaxation. This volume
17. Panyim S, Chalkley R (1969) High resolution acrylamide gel electrophoresis of histones. Arch Biochem Biophys 130:337–346
18. Adamietz P, Rudolph A (1984) ADP-ribosylation of nuclear proteins in vivo. Identification of histone H2B as a major acceptor for mono and poly (ADP-ribose) in dimethyl sulfate treated hepatoma AH 7974 cells. J Biol Chem 259:6841–6846
19. Ueda K, Ohashi Y, Hatakeyama K, Hayaishi O (1983) Inhibition of DNA ligase activity by histones and its reversal by poly (ADP-ribose). In: Miwa M et al. (eds) ADP-ribosylation, DNA repair and cancer. J Sci Soc Press, Tokyo/VNU Science Press, Utrecht, pp 175–182
20. Zolan ME, Smith CA, Calvin NM, Hanawalt PC (1982) Rearrangement of mammalian chromatin structure following excision repair. Nature (London) 299:462–464
21. Hsiang MW, Cole DR (1977) Structure of histone H1-DNA complex: Effect of histone H1 on DNA condensation. Proc Natl Acad Sci USA 74:4852–4856
22. von Mickwitz CU, Grade K, Lindigkeit R (1982) Ultrastructure and solubility of the non-histone proteins HMG 14 and HMG 17 in complexes with DNA and histone H1/DNA. Stud Biophys 92:111–118
23. Erard M, de Murcia G, Mazen A, Pouyet J, Champagne M, Daune M (1979) Ethidium bromide binding to core particles: comparison with native chromatin. Nucleic Acid Res 6:3232–3253
24. Lutter LC (1978) Kinetics analysis of DNase I cleavages in the nucleosome cores, evidence for a DNA superhelix. J Mol Biol 124:391–420
25. Kunzler P, Stein A (1983) Histone H5 can increase the internucleosome spacing in dinucleosomes to native like values. Biochemistry 22:1783–1789

26. Miwa M, Tanaka M, Matsushima T, Sugimura T (1974) Purification and properties of a glyco-hydrolase from calf thymus splitting ribose-ribose linkages of poly (adenosine diphosphate ribose). J Biol Chem 249:3475–3482

27. Jongstra-Bilen J, Ittel ME, Niedergang C, Vosberg HP, Mandel P (1983) DNA topoisomerase I from calf thymus is inhibited in vitro by poly (ADP-ribosyl)ation. Eur J Biochem 136:391–396

28. de Murcia G, Jongstra-Bilen J, Ittel ME, Mandel P, Delain E (1983) Poly (ADP-ribose) poly-merase automodification with DNA: Electron microscopic visualization. EMBO J 2:543–548

29. Kreimeyer A, Wielckens K, Adamietz P, Hilz H (1984) DNA repair associated ADP-ribosylation in vivo. J Biol Chem 259:890–896

30. de Murcia G, Huletsky A, Lamarre D, Gaudreau A, Pouyet J, Poirier GG (1985) Poly (ADP-ribose) glycohydrolase activity causes recondensation of relaxed poly (ADP-ribosyl)ated poly-nucleosomes. This volume

31. Gaudreau A, Ménard L, de Murcia G, Poirier GG (submitted) Differential turnover of poly (ADP-ribosyl)ated proteins on polynucleosomes by poly (ADP-ribose) glycohydrolase

32. Ogata N, Ueda K, Hayaishi O (1980) ADP-ribosylation of histone H2B: Identification of glutamic acid residue 2 as the modification site. J Biol Chem 255:7610–7615

33. Burzio LO, Riquelme PT, Koide SS (1979) ADP-ribosylation of rat liver nucleosome core histones. J Biol Chem 254:3029–3037

Poly(ADP-Ribose) Glycohydrolase Activity Causes Recondensation of Relaxed Poly(ADP-Ribosyl)ated Polynucleosomes

GILBERT DE MURCIA[1], ANN HULETSKY[2], DANIEL LAMARRE[2], ALAIN GAUDREAU[2], JEAN POUYET[1], and GUY POIRIER[2]

Introduction

Polyadenosine diphosphate ribose poly(ADP-ribose) polymerase catalyzes the incorporation of the ADP-ribose moiety of NAD into a homopolymer of repeating ADP-ribose units covalently bound to histones and other nuclear protein acceptors [1, 2]. This DNA-dependent enzyme, highly stimulated by nicks and DNA fragmentation [3, 4], is thought to be involved in several basic functions of the chromatin, especially in DNA repair [5–7].

We have recently shown that, in vitro, poly(ADP-ribosyl)ation of nucleosomes either by purified poly(ADP-ribose) polymerase [8] or by the endogenous chromatin bound enzyme [9] leads to relaxation of the chromatin superstructure through histone H1 modification. To test the reversibility of this phenomenon, we have used partially purified bull testis poly(ADP-ribose) glycohydrolase [10], which splits the ribosyl-ribose bond between two ADP-ribose units and produces ADP-ribose [11, 12]. Changes of the nucleosome structure were examined by electron microscopy and by ultracentrifugation, and the poly(ADP-ribosyl)ated histones were characterized.

Time Course of Synthesis and Degradation of Poly(ADP-Ribose)

Figure 1 illustrates a typical time course of incorporation of labeled NAD into poly-(ADP-ribose) using freshly prepared calf thymus nucleosomes (20–40 N) and poly(ADP-ribose) polymerase purified according to the method of Zahradka and Ebisuzaki [13]. After 60 min the ADP-ribosylation reaction was stopped with 10 mM nicotinamide, then poly(ADP-ribose) glycohydrolase was added. As indicated by the rapid decrease of acid insoluble radioactivity, the polymer produced either by the automodification reaction [14, 15] or by the modification of nucleosomal acceptors [16, 17] was hydrolyzed up to 95% within 60 min under our incubation conditions (Fig. 1).

1 IBMC du CNRS, Laboratoire de Biophysique, 15 rue Descartes, Strasbourg Cédex, France
2 Faculté des Sciences, Université de Sherbrooke, Sherbrooke, Québec, Canada J1K 2R1

ADP-Ribosylation of Proteins
(ed. by F.R. Althaus, H. Hilz, and S. Shall)
© Springer-Verlag Berlin Heidelberg 1985

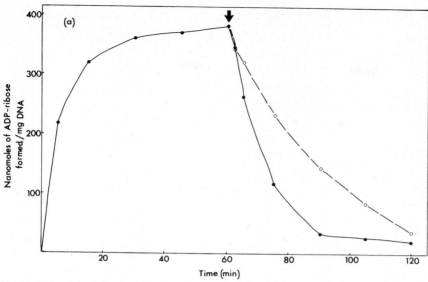

Fig. 1. Time course of poly(ADP-ribose) synthesis and degradation. Calf thymus nucleosomes were incubated at 25°C in the presence of purified poly(ADP-ribose) polymerase (2 μg/OD unit) [^{32}P] NAD (200 μM final). After 60 min (indicated by the *arrow*) the synthesis reaction was stopped with nicotinamide (10 mM final) partially purified glycohydrolase was added (*solid line:* 20 μg/OD unit, *broken line:* 10 μg/OD unit). Then the reaction mixture was incubated for another 60 min to follow the degradation of poly(ADP-ribose). At different time intervals samples were precipitated and transferred on Whatman GF/C filters

Electron Microscopy

In parallel experiments, cold NAD$^+$ (200 μM) was used in order to visualize the conformational changes of chromatin induced by the poly(ADP-ribosyl)ation reaction. After 0 min (Fig. 2a), 60 min (Fig. 2b), and 120 min of incubation (Fig. 2c), an aliquot (0.01 OD$_{260}$ unit) of the incubation medium was fixed for 1 h at 0°C in 5 mM triethanolamine buffer (pH 7.4) containing 10 mM nicotinamide, 0.2 mM EDTA, 40 mM NaCl, 0.1% glutaraldehyde. The samples were spread on positively charged carbon-coated grids and processed as described previously [18]. Following 60 min of incubation, poly(ADP-ribosyl)ated nucleosomes (Fig. 2b) exhibit a fully opened structure [8] as compared to the control (Fig. 2a). Furthermore, highly automodified poly(ADP-ribose) polymerase molecules are also visible (Fig. 2b, arrows). They are detached from DNA and surrounded by long chains of poly(ADP-ribose) as described previously [19]. When poly(ADP-ribosyl)ated nucleosomes were incubated with the poly(ADP-ribose) glycohydrolase preparation (Fig. 2c), the chromatin structure was found recondensed in a manner similar to the control. In this case (Fig. 2c), most of the poly(ADP-ribose) polymerase molecules were probably bound to DNA, since automodified enzyme molecules were no longer visible.

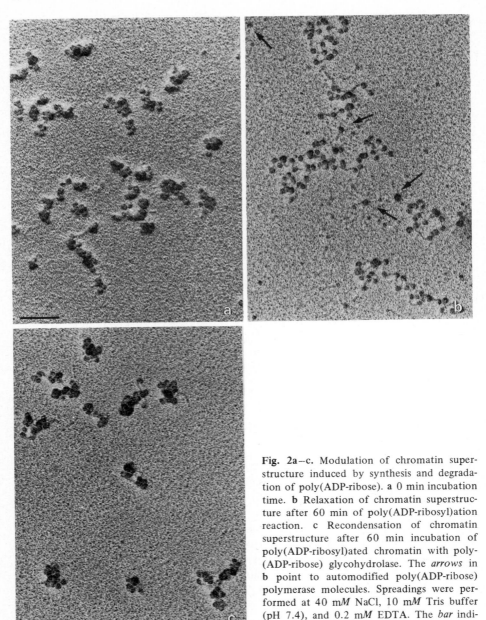

Fig. 2a–c. Modulation of chromatin superstructure induced by synthesis and degradation of poly(ADP-ribose). **a** 0 min incubation time. **b** Relaxation of chromatin superstructure after 60 min of poly(ADP-ribosyl)ation reaction. **c** Recondensation of chromatin superstructure after 60 min incubation of poly(ADP-ribosyl)ated chromatin with poly-(ADP-ribose) glycohydrolase. The *arrows* in **b** point to automodified poly(ADP-ribose) polymerase molecules. Spreadings were performed at 40 m*M* NaCl, 10 m*M* Tris buffer (pH 7.4), and 0.2 m*M* EDTA. The *bar* indicates 1000 Å

Table 1. Sedimentation coefficient of poly(ADP-ribosyl)ated chromatin samples

Reaction conditions	Time	$S_{20,w}^0$
Chromatin + polymerase + 200 μM NAD	0 min	61.5 ± 0.3 S
Stop poly(ADP-ribosyl)ated reaction with 10 mM nicotinamide	60 min	53.3 ± 0.5 S
Addition of poly(ADP-ribose) glycohydrolase at	60 min	
End of degradation reaction at	120 min	59.1 ± 1.2 S

Analytical Ultracentrifugation Studies

Table 1 shows the sedimentation coefficients of the three chromatin samples corresponding to Fig. 2a—c, respectively. One can see that the relaxation of the chromatin superstructure is correlated with a decrease of the S value as previously shown [8]. A complete description of this process is also given in this volume [20]. At the end of the polymer hydrolysis reaction (120 min) the sedimentation coefficient value is again very close to the control, thus demonstrating that the compact structure seen by electron microscopy also exists in solution.

Inhibition of Glycohydrolase Activity by ADP-Ribose

In order to demonstrate that the recondensation effect observed on poly(ADP-ribosyl)-ated nucleosomes was essentially due to the glycohydrolase activity, we decided to add the glycohydrolase preparation to the reaction mixture in combination with 10 mM ADP-ribose, which is a known inhibitor of this enzyme activity [11]. Figure 3 shows that 70% of the glycohydrolase activity was inhibited under these conditions. At the end of the reaction (120 min), an aliquot was fixed and visualized by electron microscopy (Fig. 3, insert). Similarly, an inhibition of chromatin structure recondensation was observed.

Acid-Urea Gel Electrophoresis of ADP-Ribosylated Histones

Calf thymus nucleosomes were incubated as indicated in the legend of Fig. 1. After 0, 60 and 120 min of incubation, histones were extracted, subjected to electrophoresis and autoradiographed as described previously [16]. After 60 min of ADP-ribosylation (Fig. 4, slot c), the hyperADP-ribosylated form of histone H1 corresponding to H1 "dimer" [12, 21] was predominant (see also [20]). After 120 min of incubation, (Fig. 4, slot d) most of the hyperADP-ribosylated forms of histone H1 were converted into H1-intermediates migrating faster than the H1 "dimer".

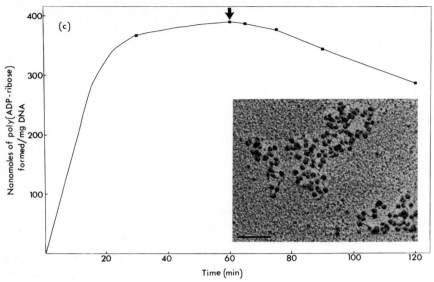

Fig. 3. Inhibition of glycohydrolase activity by ADP-ribose. Nucleosomes were incubated as described in Fig. 1. After 60 min (*arrows*), the reaction was stopped with 10 m*M* nicotinamide and 10 m*M* ADP-ribose was added. Then 20 μg/OD unit of glycohydrolase were added and the reaction mixture was incubated for 60 min more. At 120 min, an aliquot was taken and visualized by electron microscopy (*inset*)

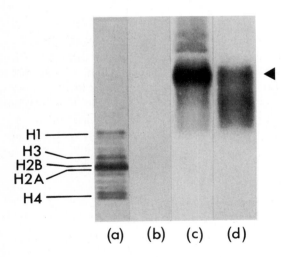

Fig. 4. Acid-urea polyacrylamide gel electrophoresis of [^{32}P]-(ADP-ribosyl)-ated histones. Nucleosomes were incubated as described in Fig. 1. At different reaction times, histones were extracted and electrophoresis was performed in a 1 *M* acetic acid, 6 *M* urea, 15% polyacrylamide gel. The gel was dried and autoradiographed. *a* Stained gel; *b–d* autoradiogram after 0, 60 and 120 min. The *arrow* points to the hyper(ADP-ribosyl)ated form of histone H1

Discussion

Both electron microscopic and sedimentation results demonstrate that, in vitro, chromatin superstructure can be successively relaxed and recondensed following poly(ADP-ribosyl)ation and glycohydrolysis, respectively. Since neither protease nor DNase activities could be detected during the incubation period, it appears that the conformational changes observed are exclusively due to the activities of the enzymes added to the reaction mixture. As already described in detail elsewhere [22], there is a positive correlation between the appearance of the hyper(ADP-ribosyl)ated form of histone H1 and a complete opening of the chromatin superstructure. conversely, when this highly modified form of histone H1 is digested by the glycohydrolase preparation, the chromatin superstructure can be reversibly refolded. However, as indicated both by the autoradiogram and the sedimentation coefficient, complete reversibility could not be obtained under our experimental conditions. This could be for two reasons: (1) removal of the last histone bound ADP-ribose monoresidue requires the presence of another degradation enzyme, namely ADP-ribosyl protein lyase [23]; (2) bidimensional polyacrylamide gel electrophoresis of the chromatin samples performed at 0, 60, and 120 min (data not shown) reveal that under our experimental conditions (200 μM NAD), histone H2B is slightly poly(ADP-ribosyl)ated (not visible in Fig.4) and that this modification is rather resistant to the glycohydrolase activity. A more detailed description of H2B modification is given in this volume [24]. From our results we conclude that histone H1 seems to play a major role in the relaxation-recondensation process of chromatin superstructure. However, we cannot completely exclude the participation of the modification of the N-terminal part of H2B in this phenomenon since it has been shown that core histone tails also contribute to the stability of the solenoidal structure [25]. Regarding the process of DNA repair and considering that the availability of DNA for the action of DNA polymerases and for repair enzymes could be dependent on the conformational state of chromatin organization, we propose that the relaxation effect could be one of the early events facilitating DNA accessibility. Subsequently, the native superstructure would be fully restored. To our knowledge, such a reversible change caused by covalent modification of histones has never been demonstrated in vitro for other postsynthetic structural modifications, i.e. acetylation and/or phosphorylation.

Acknowledgments. The authors wish to thank Dr. H. Ohlenbusch for a careful revision of the English text.

References

1. Hayaishi O, Ueda K (1977) Poly(ADP-ribose) and ADP-ribosylation of proteins. Annu Rev Biochem 46:95–116
2. Mandel P, Okazaki H, Niedergang C (1982) Poly(adenosine diphosphate ribose). Prog Nucleic Acid Res Mol Biol 27:1–51
3. Ohgushi H, Yoshihara K, Kamiya T (1980) Bovine thymus poly(adenosine diphosphate ribose) polymerase, physical properties and binding to DNA. J Biol Chem 255:6205–6211
4. Benjamin RC, Gill DM (1980) Poly(ADP-ribose) synthesis in vitro programmed by damaged DNA. J Biol Chem 255:10502–10508
5. Thi Man N, Shall S (1982) The alkylating agent, dimethylsulfate stimulates ADP-ribosylation of histone H1 and other proteins in permeabilised mouse lymphoma (L1210) cells. Eur J Biochem 126:83–88

6. Althaus FR, Lawrence SD, Sattler GL, Pitot H (1982) ADP-ribosyltransferase activity in cultured hepatocytes, interactions with DNA repair. J Biol Chem 257:5528–5535
7. Jacobson EL, Antol KM, Juarez-Salinas H, Jacobson MK (1983) Poly(ADP-ribose) metabolism in ultraviolet irradiated human fibroblasts. J Biol Chem 258:103–107
8. Poirier GG, de Murcia G, Jongstra-Bilen J, Niedergang C, Mandel P (1982) Poly(ADP-ribosyl)-ation of poly nucleosomes causes relaxation of chromatin structure. Proc Natl Acad Sci USA 79:3423–3427
9. Aubin RJ, Fréchette A, de Murcia G, Mandel P, Lord A, Grondin G, Poirier GG (1983) Correlation between endogenous nucleosomal hyper(ADP-ribosyl)ation of histone H1 and the induction of chromatin relaxation. EMBO J 2:1685–1693
10. Gaudreau A, de Murcia G, Poirier GG (to be published) Purification of Bull testis glycohydrolase
11. Miyakawa N, Ueda K, Hayaishi O (1972) Association of poly ADP-ribose glycohydrolase with rat liver chromatin. Biochem Biophys Res Commun 49:239–245
12. Stone PR, Lorimer III WS, Kidwell WR (1977) Properties of the complex between histone H1 and poly(ADP-ribose) synthesised in HeLa Cell nuclei. Eur J Biochem 81:9–18
13. Zahradka P, Ebisuzaki K (1984) Zinc, DNA and poly(ADP-ribose)-polymerase. Eur J Biochem 142:503–509
14. Kawaichi M, Ueda K, Hayaishi O (1981) Multiple auto poly(ADP-ribosyl)ation of rat liver poly(ADP-ribose) synthetase. J Biol Chem 256:9483–9489
15. Ferro AM, Olivera BM (1982) Poly(ADP-ribosylation) in vitro. J Biol Chem 257:7808–7813
16. Aubin RJ, Dam VT, Miclette J, Brousseau Y, Poirier GG (1982) Chromosomal protein poly-(ADP-ribosyl)ation in pancreatic nucleosomes. Cancer J Biochem 60:295–305
17. Huletsky A, Niedergang C, Frechette A, Aubin R, Gaudreau A, Poirier G (1985) Sequential ADP-ribosylation pattern of nucleosomal histones. Eur J Biochem 146:277–285
18. de Murcia G, Koller T (1981) The electron microscopic appearance of soluble rat liver chromatin mounted on different supports. Biol Cell 40:165–174
19. de Murcia G, Jongstra-Bilen J, Ittel ME, Mandel P, Delain E (1983) Poly(ADP-ribose) polymerase automodification and interaction with DNA: electron microscopic visualization. EMBO J 2:543–548
20. Niedergang C, Ittel ME, de Murcia G, Pouyet J, Mandel P (1985) Kinetics of nucleosomal histone H1 hyper ADP-ribosylation and polynucleosomes relaxation. This volume
21. Lorimer IIIWS, Stone PR, Kidwell WR (1977) Control of histone H1 dimerpoly(ADP-ribose) complex formation by poly(ADP-ribose) glycohydrolase. Exp Cell Res 106:261–266
22. Niedergang C, de Murcia G, Ittel ME, Pouyet J, Mandel P (1985) Time course of polynucleosomes relaxation and ADP-ribosylation, correlation between relaxation and histone H1 hyper ADP-ribosylation. Eur J Biochem 146:185–191
23. Oka J, Ueda K, Hayaishi O, Komura H, Nakanishi K (1984) ADP-ribosyl protein lyase purification, properties and identification of the product. J Biol Chem 259:986–993
24. Huletsky A, de Murcia G, Mazen A, Lewis P, Lamarre D, Aubin R, Poirier G (1985) Effect of poly(ADP-ribosyl)ation on native polynucleosomes, H1 depleted polynucleosomes, nucleosome core particles and H1-DNA complexes. This volume
25. Allan J, Harborne N, Rau DC, Gould H (1982) Participation of core histone "tails" in the stabilization of the chromatin solenoid. J Cell Biol 93:285–297

Kinetics of Nucleosomal Histone H1 Hyper(ADP-Ribosylation) and Polynucleosomes Relaxation

CLAUDE NIEDERGANG[1], MARIE-ELISABETH ITTEL[1], GILBERT DE MURCIA[2], JEAN POUYET[2], and PAUL MANDEL[1]

Introduction

Poly(ADP-ribose) polymerase, a chromatin-bound enzyme, catalyzes postsynthetic modifications of various nuclear proteins through the covalent attachment of ADP-ribose units at the expense of the cellular NAD pool [1−3]. Poly(ADP-ribosylation) appears to be involved in DNA excision repair, cellular proliferation, and differentiation [1−3] and has been shown to induce architectural changes in chromatin [4−6].

Postsynthetic modifications of histones, i.e., phosphorylation, acetylation, and poly(ADP-ribosylation), have been suggested to be involved in the modulation of the structure and the function of chromatin [7, 8]. A central role has been attributed to histone H1 in forming and stabilizing the nucleosomal structure and the higher order folding of the polynucleosomal chain into a solenoid conformation [8−10].

Previous studies have shown that at NAD concentrations above 10 μM, histone H1 is the major histone acceptor of ADP-ribose in pancreatic nucleosomes [11, 12]. Burzio et al. [13] have also observed that the ADP-ribosylation of histone H1 decreases its affinity for DNA. Furthermore, it was recently shown that under conditions of extensive ADP-ribosylation, using exogenous purified poly(ADP-ribose) polymerase and 1 mM NAD, the polynucleosomal architecture is fully relaxed [4]. Histone H1 has been found to be the main histone acceptor of ADP-ribose and to be modified as hyper(ADP-ribosylated) forms [4]. The mechanism of the relaxation process is not known, but it has been shown with the endogenous poly(ADP-ribose) polymerase to be dependent on the concentration of NAD and correlated with the hyper(ADP-ribosylation) of histone H1 [5].

In this paper we investigate further the correlation between histone H1 ADP-ribosylation and chromatin relaxation by performing a time course study. Polynucleosomes which have been poly(ADP-ribosylated) using the purified calf thymus enzyme, and histone H1 modifications as well as changes in the polynucleosomes morphology have been analyzed at different time intervals. The template capacity of the poly(ADP-ribosylated) nucleosomes has also been determined along the ADP-ribosylation time course.

1 Centre de Neurochimie de CNRS, 5, rue Blaise Pascal, 67084 Strasbourg Cédex, France
2 Laboratoire de Biophysique de l'Institut de Biologie Moléculaire et Cellulaire, 15, rue René Descartes, 67084 Strasbourg Cédex, France

ADP-Ribosylation of Proteins
(ed. by F.R. Althaus, H. Hilz, and S. Shall)
© Springer-Verlag Berlin Heidelberg 1985

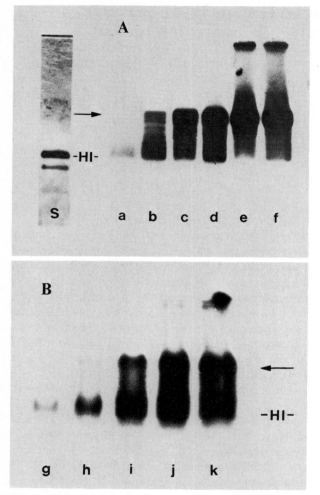

Fig. 1A,B. Time course of nucleosomal histone H1 ADP-ribosylation. Polynucleosomes were ADP-ribosylated with [32P]NAD as described in the text and histone H1 was selectively extracted [12], run on acid/urea/Triton X100 slab gels [15] and autoradiographed. **A** 200 μM NAD; **B** 20 μM NAD. The incubation times were 30 s (*a*), 1 min (*b, g*), 2 min 30 s (*c, h*), 5 min (*d, i*), 10 min (*e, j*), and 15 min (*f, k*). *S* Stained gel. *Arrow:* histone "H1 complex" position

Preparation and Poly(ADP-Ribosylation) of Pancreatic Polynucleosomes

Polynucleosomes were isolated from purified rat pancreatic nuclei digested by micrococcal nuclease to an acid solubility of 3 to 5% as previously described [4, 11]. They were poly(ADP-ribosylated by purified calf thymus DNA-bound poly(ADP-ribose) polymerase [14], using a ratio of 3 μg of the purified enzyme to 1 A_{260} unit of polynucleosomes and a NAD concentration of 200 μM under the same conditions as described before [4]. At appropriate time intervals, aliquots were taken and nicotinamide added to a final concentration of 20 mM. The nucleosomes were then treated as described below.

The concentration of 200 μM NAD was chosen in order to slow down the reaction rate. Indeed, polynucleosomes ADP-ribosylated with 1 mM NAD were almost fully relaxed after 3 min [4] and histone H1 appeared hyper(ADP-ribosylated) in less than 1 min. At a NAD concentration of 200 μM, incorporation of radioactivity into ADP-ribose was roughly linear for up to 15 min of reaction time.

Histone H1 Poly(ADP-Ribosylation)

The time course of ADP-ribosylation at 200 μM NAD showed a progressive shift of the autoradiographic bands relative to the position of stained bands suggesting the appearance of a hyper(ADP-ribosylated) form of histone H1 (Fig. 1A). A slightly ADP-ribosylated form of histone H1, corresponding to the position of the stained band, was observed after a very short reaction time. Following the appearance of increasingly modified intermediates, the hyper(ADP-ribosylated) form or "H1 complex" became predominant at reaction times longer than 10 min. This peculiar H1 complex seems to correspond to the "H1 dimer" described in vitro [16–18] and recently in vivo [19].

When the NAD concentration was decreased to 20 μM, the rate of ADP-ribosylation of histone H1 was considerably lower (Fig. 1B). The less modified form predominated for up to 5 min of incubation and, at the end of the time course study, the autoradiographic intensities of this form and of the H1 complex form were equivalent. In addition, the series of intermediates was still present.

Electron Microscopic Examination

The time course of ADP-ribosylation of polynucleosomes in the presence of 200 μM NAD is shown in Fig. 2A. Control nucleosomes adopt a native, condensed conformation and as the ADP-ribosylation reaction proceeds, individual nucleosomes connected by DNA filaments begin to be visible either at the extremities or within the polynucleosomal chains. This decondensation phenomenon is rapidly amplified leading to an almost fully relaxed conformation for 10 to 15 min. It is noteworthy that the relaxation appears complete at the same time as the H1 complex of poly(ADP-ribosylated) histone H1 becomes predominant.

When the ADP-ribosylation reaction was performed with 20 μM NAD, the relaxation process proceeded very slowly (Fig. 2B). At the end of the 15 min incubation period it had barely started.

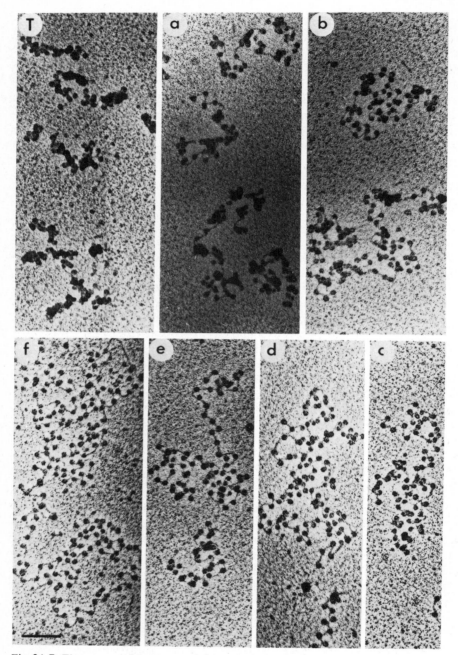

Fig. 2A,B. Time course of the structural changes of polynucleosomes induced by ADP-ribosylation. ADP-ribosylated polynucleosomes were fixed in buffer containing 20 mM NaCl and examined in bright field [20]. **A** 200 μM NAD; **B** 20 μM NAD. The incubation times are as indicated in Fig. 1; T corresponds to the zero time. *Bar* indicates 100 nm

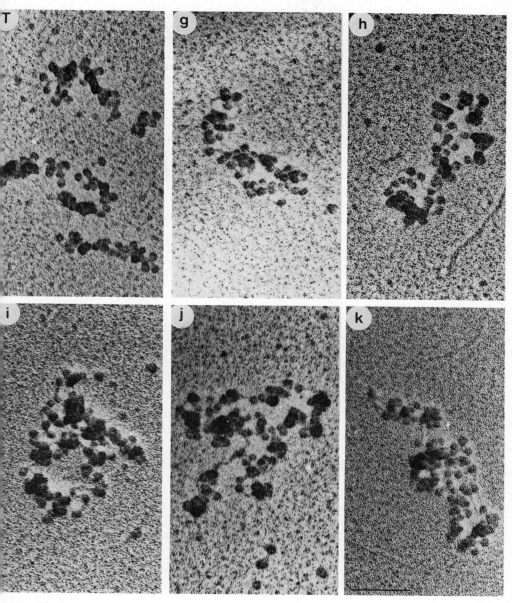

Fig. 2B

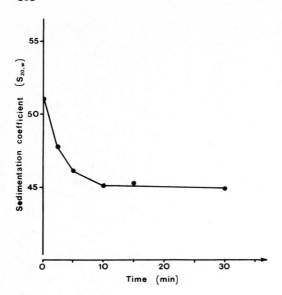

Fig. 3. Sedimentation velocity of ADP-ribosylated polynucleosomes. At various time intervals aliquots of ADP-ribosylated nucleosomes were taken, dialyzed, and the sedimentation coefficients were determined in 20 mM NaCl [21]

Sedimentation Velocity

Sedimentation coefficient determinations [4, 21] on polynucleosomes ADP-ribosylated in the presence of 200 μM NAD for different time intervals, gave values of $S_{20,w}$ varying from 51.5 ± 0.4 S for control nucleosomes to 44.9 ± 0.2 S for fully relaxed nucleosomes (Fig. 3). The time course was in good agreement with the relaxation process visualized by electron microscopy.

Template Capacity of ADP-Ribosylated Polynucleosomes

In order to investigate the modification of DNA accessibility in chromatin structures relaxed by ADP-ribosylation, the DNA polymerase α activity has been determined [22]. As shown in Table 1, this activity is more than twofold higher in polynucleosomes ADP-ribosylated for 25 min than in control polynucleosomes. Moreover, this increase in template capacity is correlated with the ADP-ribosylation induced relaxation time course as demonstrated by electron microscopy and sedimentation velocity analysis.

Discussion

It appears from our investigations that the polynucleosome relaxation process is closely correlated with the generation of ADP-ribosylated forms of histone H1. The kinetics as followed by histone H1 modification, electron microscopic visualization,

Table 1. DNA polymerase α activity as a function of the extent of polynucleosomal ADP-ribosylation

ADP-ribosylation time (min)[a]	DNA polymerase α activity [³H] dTTP incorporated (cpm)[b]	Increase (%)[c]
0	149	–
1	177	15.7
2.5	181	26.6
5	193	62.2
25	231	115.9

[a] Polynucleosomes have been ADP-ribosylated with 200 μM NAD as described in the text
[b] DNA polymerase α activity has been determined according to [22]
[c] Expressed in reference to control polynucleosomes incubated without NAD

and sedimentation coefficient determination share the same time characteristic. Moreover, the relaxation seems to be completed when essentially all modified histone H1 is in the H1 complex form. In view of these results and previous studies [4, 5], it can be suggested that the observed polynucleosomes relaxation is mediated by the poly-(ADP-ribosylation) of histone H1. Furthermore, it appears that ADP-ribosylation of H1 does not cause its dissociation from the nucleosomal chain [4, 5], but rather impairs its contribution to the maintenance of the chromatin structure.

How this modification of histone H1 affects its properties towards the chromatin structure may be explained on the basis of interactions between histone H1 and the nucleosomes. The native chromatin conformation results, and is maintained in part, by interactions of the N- and C-terminal portions of the H1 molecule with the nucleosomal DNA and/or with another H1 molecule [9, 10, 23, 24]. On the other hand, it is noteworthy that these regions of histone H1 are sites of ADP-ribosylation [25, 26]. Thus, the progressive modification of polarity of the basic regions of histones H1 will reduce their affinity for DNA and their cross-linking ability and, consequently, affect the higher conformation of chromatin. Therefore, it seems conceivable that the observed relaxation of the ADP-ribosylated polynucleosomes is modulated by the poly(ADP-ribosylation) of histone H1 favoring H1-H1 interactions through poly(ADP-ribose) chains at the expense of direct H1-H1 interactions on the nucleosomal DNA. The H1 hyper(ADP-ribosylated) form or H1 complex is not likely to represent H1 dimer [16–18] in relaxed nucleosomes. Most probably, considering the branched structure of poly(ADP-ribose) [1, 2], it may result from a steric hindrance leading to an arrangement favorable for polymers of a definite chain length, thus forming discrete ADP-ribosylated H1 complexes.

However, the ADP-ribosylation "in vivo" of other chromatin components, histones [27, 28], or nonhistone proteins [29], may also contribute to the relaxation process. Histone H2B has been shown to be ADP-ribosylated in isolated polynucleosomes at very low NAD concentrations [27] and also in hepatoma cells "in vivo" [28]. But, in contrast to H1, the interactions of H2B with DNA remain unchanged after ADP-ribosylation [13].

The stimulation of DNA polymerase α activity on ADP-ribosylated polynucleo-somes favors the assumption that the ADP-ribosylation-mediated relaxation of the chromatin may facilitate "in vivo" enzymatic events on the chromatin fiber. Indeed, an increased poly(ADP-ribose) polymerase activity has been associated with the DNA repair process [1–3]. However, "in vivo", the hyper(ADP-ribosylated) form of histone H1 has not yet been observed in DNA repair conditions [30], but only in synchronized cells at the S/G2 phase transition of the cell cycle [19]. This may be attributed to the very localized and transient modifications of chromatin structure associated with a high turnover of the poly(ADP-ribose) [31, 32]. It has also been shown that if only 10% of the total histone H1 is removed from the chromatin, regular higher order structures can no longer be observed [33].

The present study has been performed with NAD concentrations of 200 μM which is close to the physiological NAD concentration "in vivo" [34] in the absence of DNA damaging agents. However, in isolated polynucleosomes, the activity of glycohydrolase is negligible. Thus, the ADP-ribosylation reaction proceeds in an unregulated manner. Nevertheless, our results suggest that the poly(ADP-ribosylation) of histone H1 and the formation of a H1 complex is able to modulate the chromatin architecture. This may assign a role to the ADP-ribosylation reaction at the level of DNA rearrange-ments occurring during DNA replication, repair, or gene expression. It must be noted that histone hyperacetylation has been shown to have little effect on the higher order folding of chromatin [35]. The refolding of the nucleosomal chain, i.e., the condensa-tion of relaxed ADP-ribosylated polynucleosomes, by the glycohydrolase activity has recently been demonstrated [36].

Acknowledgments. We wish to thank Dr. G.G. Poirier for helpful discussions and F. Hog for his technical assistance. We also thank S. Ott for her excellent typing of the manuscript.

References

1. Mandel P, Okazaki H, Niedergang C (1982) Poly(adenosine diphosphate ribose). Prog Nucleic Acids Res Mol Biol 27:1–51
2. Ueda K, Ogata N, Kawaichi M, Inada S, Hayaishi O (1982) ADP-ribosylation reactions. Curr Top Cell Regul 21:175–187
3. Hilz H (1981) ADP-ribosylation – a multifunctional process. Hoppe Seyler's Z Physiol Chem 362:1415–1425
4. Poirier GG, De Murcia G, Jongstra-Bilen J, Niedergang C, Mandel P (1982) Poly(ADP-ribosyla-tion) of polynucleosomes causes relaxation of chromatin structure. Proc Natl Acad Sci USA 79: 3423–3427
5. Aubin R, Fréchette A, De Murcia G, Mandel P, Lord A, Grondin G, Poirier GG (1983) Cor-relation between endogenous nucleosomal hyper ADP-ribosylation of histone H1 and the induction of chromatin relaxation. EMBO J 2:1685–1693
6. Wong M, Malik N, Smulson M (1982) The participation of poly(ADP-ribosylated) histone H1 in oligonucleosomal condensation. Eur J Biochem 128:209–213
7. Bradbury EM, Matthews HR (1981) Histone variants, histone modifications and chromatin structure. In: Sarma R (eds) Proceedings of the 2nd SUNYA conversation in the discipline biomolecular stereodynamics, vol II. Adenine Press, New York, p 125

8. McGhee J, Felsenfeld G (1980) Nucleosome structure. Annu Rev Biochem 49:1115–1156

9. Thoma F, Koller T, Klug A (1979) Involvement of histone H1 in the organization of the nucleosome and of the salt-dependent super structures of chromatin. J Cell Biol 83:403–427

10. Hancock R, Boulikas T (1982) Functional organization in the nucleus. Int Rev Cytol 79:165–214

11. Aubin R, Dam V, Miclette J, Brousseau Y, Poirier GG (1982) Chromosomal protein poly(ADP-ribosylation) in pancreatic nucleosomes. Can J Biochem 60:295–305

12. Aubin R, Dam V, Miclette J, Brousseau Y, Huletsky A, Poirier GG (1982) Hyper ADP-ribosylation of histone H1. Can J Biochem 60:1085–1094

13. Burzio L, Puigdomenech P, Ruiz-Carillo A, Koide S (1981) Adenosine diphosphate ribosylation and regulation of nucleic acid synthesis. In: Niu M, Chuang H (eds) The role of RNA in development and reproduction. Proc 2nd Int Symp. Science Press, Beijing, China, p 114

14. Niedergang C, Okazaki H, Mandel P (1979) Properties of purified calf thymus poly(adenosine diphohsphate ribose) polymerase. Eur J Biochem 102:43–57

15. Bonner W, West M, Stedman J (1980) Two-dimensional gel analysis of histones in acid extracts of nuclei, cells and tissues. Eur J Biochem 109:17–23

16. Stone P, Lorimer W, Kidwell W (1977) Properties of the complex between histone H1 and poly(ADP-ribose) synthesised in HeLa cell nuclei. Eur J Biochem 81:9–18

17. Nolan N, Butt T, Wong M, Lambrianidou A, Smulson M (1980) Characterization of poly-(ADP-ribose)-H1 complex formation in purified polynucleosomes and chromatin. Eur J Biochem 113:15–25

18. Poirier GG, Niedergang C, Champagne M, Mazen A, Mandel P (1982) Adenosine diphosphate ribosylation of chicken erythrocyte histones H1, H5 and highmobility group proteins by purified calf thymus poly(adenosine diphosphate ribose) polymerase. Eur J Biochem 127:437–442

19. Wong M, Kanai Y, Miwa M, Bustin M, Smulson M (1983) Immunological evidence for the in vivo occurrence of a crosslinked complex of poly(ADP-ribosylated) histone H1. Proc Natl Acad Sci USA 80:205–209

20. De Murcia G, Koller T (1981) The electron microscopic appearance of soluble rat liver chromatin mounted on different supports. Biol Cell 40:165–174

21. De Murcia G, Mazen A, Erard M, Pouyet J, Champagne M (1980) Isolation and physical characterization of a stable core particle. Nucleic Acids Res 8:767–779

22. Knopf KW, Weissbach A (1977) Study of DNA synthesis in chromatin isolated from HeLa cells. Biochem 16:3190–3194

23. Thomas J, Khabaza A (1980) Cross-linking of histone H1 in chromatin. Eur J Biochem 112:501–511

24. Ring D, Cole D (1983) Close contacts between H1 histone molecules in nuclei. J Biol Chem 258:15361–15364

25. Riquelme P, Burzio L, Koide S (1979) ADP-ribosylation of rat liver lysine-rich histone in vitro. J. Biol Chem 254:3018–3028

26. Ogata N, Ueda K, Kagamiyama H, Hayaishi O (1980) ADP-ribosylation of histone H1. J Biol Chem 255:7616–7629

27. Huletsky A, Niedergang C, Frechette A, Aubin R, Gaudreau A, Poirier GG (1985) Sequential ADP-ribosylation pattern of nucleosomal histones. Eur J Biochem 146:277–285

28. Adamietz P, Rudolph A (1984) ADP-ribosylation of nuclear proteins in vivo. J Biol Chem 259:6841–6846

29. Jongstra-Bilen J, Ittel ME, Niedergang C, Vosberg HP, Mandel P (1983) DNA topoisomerase I from calf thymus is inhibited in vitro by poly(ADP-ribosylation). Eur J Biochem 136:391–396

30. Kreimeyer A, Wielckens K, Adamietz P, Hilz H (1984) DNA repair-associated ADP-ribosylation in vivo. J Biol Chem 259:890–896

31. Wielckens K, Schmidt A, George E, Bredehorst R, Hilz H (1982) DNA fragmentation and NAD depletion. J Biol Chem 257:12872–12877

32. Jacobson E, Antol K, Juarez Salinas H, Jacobson M (1983) Poly(ADP-ribose) metabolism in UV irradiated human fibroblasts. J Biol Chem 258:103–107

33. Thoma F, Koller T (1981) Unravelled nucleosomes, nucleosome beads and higher order structures of chromatin: influence of non histone components and histone H1. J Mol Biol 149: 709–733
34. Rechsteiner M, Hillyard D, Olivera B (1976) Magnitude and significance of NAD turnover in human cell line D 98/AH2. Nature (London) 259:695–696
35. McGhee J, Nickol J, Felsenfeld G, Rau D (1983) Histone hyperacetylation has little effect on the higher order folding of chromatin. Nucleic Acids Res 11:4065–4075
36. De Murcia G, Huletsky A, Lamarre D, Gaudreau A, Pouyet J, Poirier GG (1985) Poly(ADP-ribose) glycohydrolase activity causes recondensation of relaxed poly(ADP-ribosylated) polynucleosomes. This volume

DNA Strand Breaks and Poly(ADP-Ribosylated) Mediation of Transcriptionally Active Chromatin and Transforming Gene Stability

MARK E. SMULSON[1], CHRIS HOUGH[1], USHA KASID[1],
ANATOLY DRITSCHILO[1], and RON LUBET[2]

Introduction

This review will focus on two specific areas of recent interest in poly(ADP-ribosylation), both presumably involving lesions in DNA, rearrangement of DNA sequences in cells, and a possible involvement of poly(ADP-ribosylation) at sites of chromatin undergoing transcription.

The use of anti-poly(ADP-ribose)-Sepharose has been pivotal towards many of the experimental observations accomplished during the last year. Using this technique we were able to isolate sites adjacent to poly(ADP-ribosylation) from the bulk of chromatin [1]. It was shown that these nuclear domains contain significant amounts of *internal single strand breaks* compared to bulk chromatin, by several methods, including two-dimensional chromatography of isolated DNA [1] as well as the localization of in vivo incorporated Ara-C into chromatin sites adjacent to regions of poly(ADP-ribosylation) [4]. Additionally, a method to study poly(ADP-ribosylation) of H1 in vivo was developed using immunoblotting with anti-poly(ADP-ribose) [5] . It was shown that this histone becomes poly(ADP-ribosylated) to a larger complex, at the G2-S boundary of the cell cycle. The biosynthesis of the poly(ADP-ribose)-H1 complex was further elucidated by reconsitution of chromatin with partial peptide domains of histone H1 [6]. Finally, immunofractionation studies suggested that poly(ADP-ribosylation) may occur in the same regions of chromatin undergoing other nuclear protein modifications, such as phosphorylation [7] and acetylation [8, 9].

Since the acetylation modification has been associated with transcription, and based upon observations that DNA polymerase may require a transient single strand break in DNA as it transgresses the super-coiled DNA around the nucleosome during transcription, the studies outlined below were aimed at testing, whether actively transcribed gene sequences are adjacent or distal to sites of poly(ADP-ribosylation) in vivo.

---

1 Department of Biochemistry and Department of Radiation Medicine, Georgetown University, School of Medicine, Washington, D.C. 20007, USA
2 Microbiological Associates, Bethesda, MD, USA

ADP-Ribosylation of Proteins
(ed. by F.R. Althaus, H. Hilz, and S. Shall)
© Springer-Verlag Berlin Heidelberg 1985

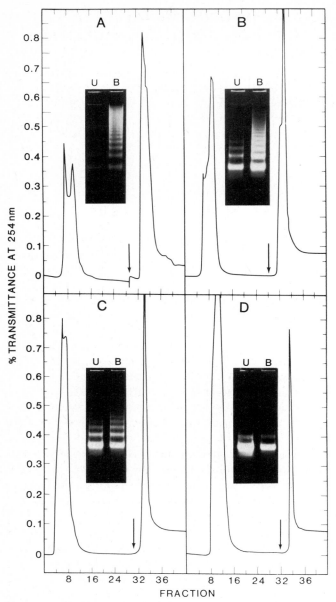

Fig. 1A–D. Immunofractionation of chromatin digested by micrococcal nuclease for various extents of time on an anti-poly(ADP-ribose) Sepharose column. HeLa nuclei from 10^8 cells were digested with micrococcal nuclease [50 units/10^8 nuclei at 37°C for 1 min (**A**), 5 min (**B**), 10 min (**C**), and 20 min (**D**)]. Soluble chromatin was subsequently prepared and immunofractionated on an anti-poly(ADP-ribose) Sepharose column. The columns were loaded and washed with phosphate buffered saline. Transmittance at 254 nm was continuously monitored. When no further UV-absorbing material was eluted, 6 M guanidinium hydrochloride was added to elute-bound material. The *insets* in each panel represent 1/50 of the DNA extracted from the pooled unbound (*U*) and bound (**B**) peak fractions subjected to electrophoresis on 1.5% agarose. The gel was stained with ethidium bromide. (Taken from [10])

Results and Discussions

In the data shown in Fig. 1, soluble chromatin was applied to a poly(ADP-ribose) antibody column in PBS, and washed extensively. In earlier studies, nucleosomes were labeled in vitro with NAD prior to immunofractionation. In the experiments described [1–3], no in vitro synthesis of exogenous poly(ADP-ribosylation) was performed and absorption to the antibody column relied upon the presence of in vivo synthesized poly(ADP-ribose) at the sites in chromatin where it presumably exists naturally. This laboratory has recently shown the natural presence of 15 units of poly(ADP-ribose) attached to histone H1 isolated from intact, unlabeled cells using the same antibody [5].

Continuous monitoring of UV transmittance at 254 nm showed a considerable fraction of UV absorbing material bound to the column in chromatin digested for a short time during nuclease digestion (Fig. 1A), and with increasing time of digestion, the total UV absorbance in the bound peak decreased. Thus, after 1 min of nuclease digestion, approximately 88% of the applied chromatin showed binding to anti-poly(ADP-ribose), whereas only 27% of chromatin digested for 20 min bound to the column. This latter level of binding corresponded to that obtained when chromatin was incubated with exogenous NAD prior to immunofractionation [1].

The insets in Fig. 1 show agarose gels of the DNA extracted from the chromatin which was unbound and bound to the anti-poly(ADP-ribose) Sepharose column. As expected, the general size of oligonucleosomes obtained using this method decreased in chain length with increased digestion. Thus, in the short digest (Fig. 1A) poly-nucleosomes exceeding 15 chain lengths were observed to be bound to the antibody column. At the longest period of nuclease digestion (Fig. 1D), nucleosomal chain lengths in the bound fraction were extremely short with the average being mainly mononucleosomal size. The decrease in binding of chromatin to the antibody column with increased nuclease digestion might thus be expected, since the probability that any chromatin fragment containing a poly(ADP-ribose) chain would increase with nucleosomal chain length. The data also confirm earlier observations that poly(ADP-ribosylated) regions of chromatin are nuclease sensitive. As nuclease digestion progressed, the ratio of limit digest product (i.e., mononucleosomes) in the bound vs unbound fraction was noted to decrease. In the late digest (Fig. 1D, inset) the ADP-ribosylated chromatin showed reduced amounts of mononucleosomes compared to the unmodified chromatin, while the reverse was true in the earlier digests.

Probing Immunofractionated Poly(ADP-Ribosylated) Chromatin for Active Genes

The probe for active genes was prepared by extracting total RNA from asynchronous, logarithmically growing HeLa cells, purifying the polyadenylated RNA by poly-U Sephadex chromatography, and reverse transcribing with AMV reverse transcriptase in the presence of [^{32}P]-labeled dCTP.

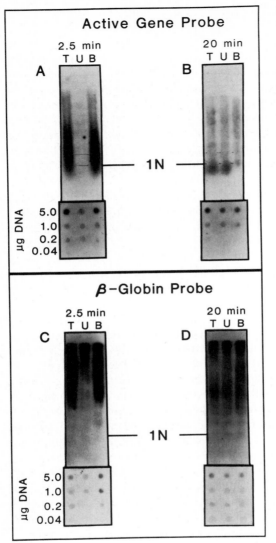

Fig. 2A–D. Dot-blot and Southern blot analyses of immunofractionated chromatin. The DNA extracted from immunofractionated chromatin was subjected to electrophoresis on 1.5% agarose and blotted to nitrocellulose by the method of Southern or spotted directly to nitrocellulose. The membranes were probed for active (**A, B**) and inactive gene (**C, D**) sequences. Unfractionated total (*T*), unbound (*U*), and bound (*B*) antibody column fractions are shown for DNA derived from chromatin digested for 2.5 (**A, C**) and 20 min (**B, D**) with micrococcal nuclease. (Taken from [10])

Both active and inactive DNA sequences may occur on the same chromatin chain. Thus, a fraction of chromatin binding to the antibody column would not be expected to show a total enrichment compared to unfractionated chromatin in DNA sequences which are associated with poly(ADP-ribosylated) proteins.

Different approaches were taken to obtain a representative view of the relative abundance of active and inactive genes in poly(ADP-ribosylated) and unmodified chromatin using dot-blot hybridization analysis of DNA derived from bound or unbound material [10]. In one approach, immunofractionated nucleosomes were obtained from either short or long digests with micrococal nuclease. No in vitro poly(ADP-ribosylation) was performed with these nucleosomal preparations prior to immunofractionation.

DNA was extracted from the unfractionated (total) and unbound or bound nucleosomes, and equal concentrations of the respective DNAs, as determined by both A_{260} and DAPI fluorescence assay were dot-blotted in serial dilution to nitrocellulose (Fig. 2, dot-blots). The nitrocellulose sheets were hybridized to the [^{32}P]-labeled cDNA to total active genes and subsequently rehybridized to the [^{32}P]-labeled beta globin probe. In most cases, there was a reasonable correlation between the hybridization signal and the amount of DNA applied to the nitrocellulose within a dilution series.

Active genes and β-globin gene occurred in both non(ADP-ribosylated) and (ADP-ribosylated) antibody column fractions. Both active and inactive gene sequences derived from chromatin digested for 20 min with micrococcal nuclease appeared to partition nearly equally between the unbound and bound antibody column fractions. This is especially evident by the dot-blot data shown in Fig. 2A, D.

Nucleosomal DNA Blots

In Southern transfers the same observations were apparent. Equal amounts of unfractionated, unbound and bound DNA from each digest time point were subjected to electrophoresis on 1.5% agarose, blotted onto nitrocellulose, and probed for active or inactive genes. Active genes, as detected by the cDNA probe, decreased in size and quantity with time as these genes were degraded. Also with time, the relative proportion of remaining actively transcribed DNA which bound as chromatin to the antibody column decreased. As observed by dot-blot hybridization, both active and inactive DNA sequences were enriched in the bound antibody column fraction in the early digestion. While the average size of DNA fragments hybridizing to the β-globin probe in this sample (Fig. 2C) was very large, the majority of actively (Fig. 2A) transcribed sequences were small, ranging from the size of a mononucleosome to that of about a trinucleosome.

These results are interpreted as showing that some, but not all, of actively transcribed chromatin contains associated poly(ADP-ribosylated) proteins. However since poly(ADP-ribosylated) proteins are associated with inactive genes, the function of this modification cannot be assigned solely to transcription. We view the function of poly-(ADP-ribosylation) to be more pleotropic in nature, all involving the signal generated by single strand breaks. In the study summarized below we have investigated another type of biological system, whereby DNA strain breaks must be illicited and where the modulation by poly(ADP-ribosylation) may be important.

The Relationship Between Poly(ADP-Ribosylation) and DNA Sequence Alteration as Measured by Transformation and Oncogene Analysis

Inhibition of poly(ADP-ribosylation) of nuclear proteins has been reported to increase the persistence of DNA strand breaks illicited by DNA damaging agents, and is marked

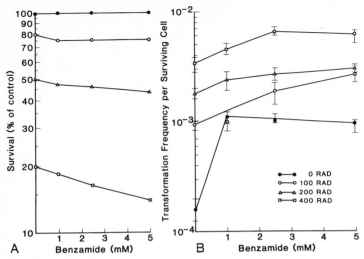

Fig. 3A,B. Dose-response curves for **(A)** cytotoxic effects and **(B)** transformation induced by various doses of X-rays and/or BA in BALB/3T3 clone A31-1-1 cell line. Exponentially growing cells were plated at 10^4 cells for the transformation assay and 100-300 cells for the assay of plating efficiency. 24 h later, the cells were treated with 0, 100, 200, or 400 rad of X-rays. *Immediately,* following irradiation, the cells were incubated in the absence or presence of 1.9, 2.5, or 5.0 mM BA in the culture medium for 48 h. (Taken from [11])

Table 1. Effects of 3AB on transformation of BALB/3T3 Cl A31-1-1 cells induced by MNNG[a]. (Taken from [11])

Treatment	Dose	Survival (% of control)	Transformation frequency[b] ($\times 10^{-4}$)
Acetone	2.0 μl ml^{-1}	100	0.19 (1/15)
3AB	3.0 mM	100	0.61 (3/14)
MNNG	0.7 μM	97	0.68 (3/13)
MNNG + 3AB	0.7 μM + 3.0 mM	60	3.80 (12/15)
MNNG	2.0 μM	86	2.44 (11/15)
MNNG + 3AB	2.0 μM + 3.0 mM	6	95.23 (30/15)

[a] Cells were seeded at 1×10^4 cells/dish, treated ± test chemical for 24 h, washed and cultured for 5 weeks with biweekly media changes. Cells were fixed (methanol), stained 10% aqueous Giemsa), and scored for type II and type III foci. Plating efficiency, (number of colonies counted/number of cells plated) $\times$ 100, of control cells was 35%

[b] Transformation frequency = total number of transformed foci/total number of surviving cells per condition. Numbers in parentheses are the absolute number of foci counted/total number of replicate dishes scored

Table 2. Characterization of the transformed cell line. (Taken from [11])

Code no.	Cell line	Origin[a] from BALB/3T3 clone A31-1-1 cell line after treatment with		Growth in soft Agar	Tumor induction in nude mice[b]
		X-ray (rad)	Benzamide (mM)		
1.	X3	400	–	+	+
2.	B1	–	1.5	+	+
3.	B2	–	2.5	+	+
4.	B3-1	–	5.0	+	+
5.	B3-2	–	5.0	+	+
6.	B3-3	–	5,0	+	–
7.	X3B1-1	400	1.0	+	N.D.
8.	X3B1-2	400	1.0	+	+
9.	X3B1-3	400	1.0	+	N.D.
10.	X3B2-1	400	2.5	+	N.D.
11.	X3B2-2	400	2.5	+	N.D.
12.	X3B3-1	400	5.0	+	+
13.	X3B3-2	400	5.0	+	N.D.
14.	X0B0	–	–	+	N.D.

[a] See Fig. 1 for details
[b] 10^6 cells were suspended in 0.1 ml of serum free medium and the cell suspension was injected subcutaneously into nu/nu (BALB/C) mice which were then monitored for tumor development (10–21 days). N.D. not determined

by an increase of sister chromatid exchanges. Accordingly, we wished to test at the molecular level, whether these cellular events cause rearrangements or alterations of specific sequences, such as oncogenes in DNA. We utilized transformation of BALB/3T3 cells as a selective system to obtain homogeneous samples of DNA after damaging cellular DNA by X-rays and/or inhibition of poly(ADP-ribosylation) (Fig. 3).

Inhibition of poly(ADP-ribosylation) by benzamide (BA) or 3-amino-benzamide (3AB) for a brief period (i.e., during which ADP-ribosylation following DNA damage is elevated in untreated cells) during MNNG-induced DNA damage to BALB/3T3 cells significantly (3-30-fold) enhanced transformation frequency (Table 1).

X-ray Induced Transformation in the Presence and Absence of Benzamide

Thirteen tranformed cell lines were established, after having been characterized for growth in soft agar and tumor induction in nude mice (Table 2). In order to determine the presence of a dominant transforming gene (s), high molecular weight DNAs were isolated from five representative transformed cell lines, exhibiting transformation due to BA or radiation and BA. Efficiency of transmission of transforming DNA sequences in the first cycle of transfection is shown in Table 3.

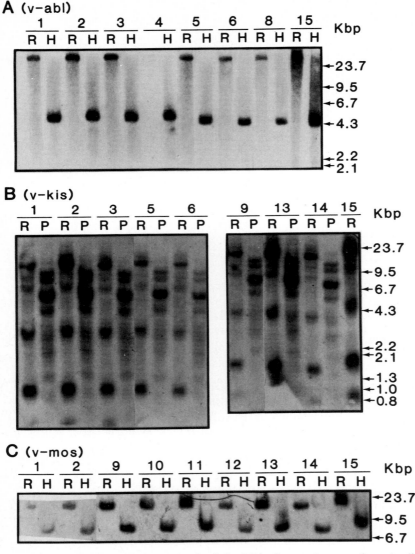

Fig. 4A–C. Southern blot analysis of cellular DNAs from various transformed cell lines digested with R, Eco R1; H, Hind III or P, Pvu II and probed with 0.7 Kb Bgl II fragment of v-abl; (A) 1.7 Kb Pvu II fragment of v-Ki-ras (B) or v-mos (C). Numbers 1–15 represent the Code Nos. of various cell lines. (Taken from [11])

Table 3. Transforming activity of DNAs from various transformed cell lines[a]. (Taken from [11])

Code no.[b]	Donor DNA[b]	Total foci/ total recipient cultures	Foci/ μg^{-1} DNA
2	B1	51/6	0.28
3	B2	55/10	0.18
4	B3-1	51/10	0.17
7	X3B1-1	64/10	0.23
12	X3B3-2	67/10	0.22
15	BALB/3T3	2/5	0.01
–	NIH 3T3	0/5	–
–	Mock (No DNA)	0/3	–

[a] Transfection assays were performed by the addition of 30 μg of high molecular weight DNA to 1.5×10^5 NIH/3T3 cells. Foci formation was scored 14–21 days later
[b] See legend to Fig. 2 for details of code numbers and cell lines

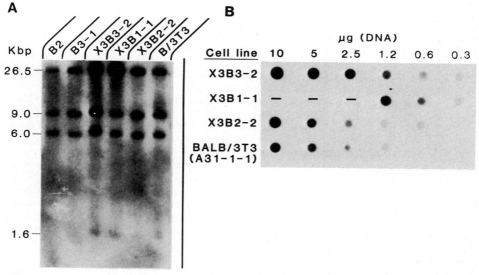

Fig. 5A,B. Analysis of c-Ha-ras sequences in BALB/3T3 cell lines transformed with BA or radiation and BA. **A** The DNA samples (10 μg) were digested with Eco RI, run on a 0.7% agarose gel, blotted, and hybridized to a 0.7 Kb SstI-PstI fragment of v-Ha-ras. **B** Estimation of c-Ha-ras amplification in radiation and BA transformed cell lines. (Taken from [11])

Oncogene Stability, Amplification, Expression, and Transfection

We tested whether any of the 13 transformed cell lines had acquired alterations in cellular oncogene structure or expression. No gross rearrangements of six representative cellular oncogenes, including the more frequently activated c-Ki-ras, was observed (Fig. 4).

In other experiments (Fig. 5A), the intensity of hybridization of v-Ha-ras to Eco RI restriction digests of DNAs of X3B1-1, X3B2-2, and X3B3-2 appeared to be greater than similar digests of the DNAs of the remaining transformed cells and of the control, a nontransformed cell line. Therefore, serial dilutions of the DNA from these cell lines were dot-blotted and hybridized to the ^{32}P-labeled v-Ha-ras probe (Fig. 5B). In X3B3-2 and X3B1-1, significant amplification (i.e., approx. four-fold) of the c-Ha-ras gene sequences over control DNA was observed.

Thus, the current study provides an approach to determine in the future, at the molecular level, whether inhibition of poly(ADP-ribosylation) or other mechanisms acts to reduce DNA repair alone, or in conjunction with damaging agents, can introduce alterations in genes sequences.

References

1. Malik N, Miwa M, Sugimura T, Thraves P, Smulson M (1983) Immuno-affinity fractionation of the poly ADP-ribosylated domains of chromatin. Proc Natl Acad Sci USA 80:2554–2558
2. Pieto-Soto A, Gourlie B, Miwa M, Piagiet, Sugimura T, Smulson M (1983) Polyoma virus minichromosomes: Poly (ADP-ribosylation) of associated chromatin proteins. J Virol 45:600–606
3. Smulson M (1984) Poly (ADP-ribosylation) of nucleosomal chromatin. Electrophoretic and immunological methods. Methods in Enzymology 106:933–943
4. Thraves PJ, Kasid U, Smulson ME (in press) Poly (ADP-rib) and DNA strand breaks. Cancer Research
5. Wong M, Kanai Y, Miwa M, Bustin M, Smulson M (1983) Immunological evidence for the in vivo occurrence of a crosslinked complex of poly (ADP-ribosylated) histone H1. Proc Natl. Acad Sci USA 80:205–209
6. Wong M, Allan J, Smulson M (1984) The mechanism of histone H1 crosslinking by poly ADP-ribosylation: Reconstitution with peptide domains. J Biol Chem 259:7963–7969
7. Wong M, Miwa M, Sugimura T, Smulson M (1983) Relationship between histone poly (adenosine diphosphate ribosylation) and histone H1 phosphorylation using anti-poly (adenosine diphosphate ribose) antibody. Biochemistry 22:2384–2389
8. Malik N, Smulson M (1984) A relationship between nuclear poly (adenosine diphosphate ribosylation) and acetylation post-translational modifications I. Nucleosome Studies Biochemistry 23:3721–3725
9. Wong M, Smulson M (1984) A relationship between nuclear poly (adenosine diphosphate ribosylation) and acetylation post-translational modifications II. Histone Studies Biochemistry 23:3726–3730
10. Hough C, Smulson ME (in press) The association of poly (ADP-ribosylated) nucleosomes with transcriptionally active and inactive regions of chromatin. Biochemistry
11. Kasid U, Stefanic D, Lubet R, Dritschilo A, Smulson M (submitted) Relationship between poly (ADP-ribosylation) and DNA sequence. Alterations as measured by transformation and oncogene analysis

Poly(ADP-Ribose) in Nucleoids

GORDANA BRKIĆ[1], ALEXANDER TOPALOGLOU[2], and HANS ALTMANN[2]

Introduction

Nucleoids are nuclei-like structures produced from cells lysed with nonionic detergents in the presence of high salt concentrations. By this treatment, histones and most of the nonhistone proteins are removed [3]. They contain naked, histone-free DNA, RNA, and a few proteins [1]. DNA is supercoiled and attached to a cage of residual proteins and RNA [5]. This relationship continues during replication and transcription and seems to be sequence specific [5]. Poly(ADP-ribose) is involved in the regulation of the activities of various enzymes, as well as structural proteins, and seems to be involved in the control of DNA and RNA synthesis [9, 11]. It was interesting for us to examine whether ADP-ribosylated proteins remained after generation of nucleoids. We found out that one of the residual protein fractions (possibly bound to DNA) was ADP-ribosylated.

Nucleoid Sedimentation

CHO-K_1 cells (American Type Culture Collection) were grown in F_{12}-Hepes medium (Flow Laboratory) with 10% fetal calf serum. While attached to the bottom of the culture flasks, cells were rendered permeable for NAD^+ by a cold hypoosmotic shock at 4°C. The permeabilization buffer consisted of: 0.01 M Tris, 0.25 M sucrose, 1 mM EDTA, 5 mM DTE, and 4 mM $MgCl_2$, pH 7.8. The incubation buffer for labeling of cells is made of 2.5 parts of permeabilization buffer and 1 part NAD^+ [adenine-2,8-H^3], 4.1 Ci mmol^{-1}, NEN) containing buffer with: 0.1 M Tris and 0.12 M $MgCl_2$, pH 7.8. Labeling of cells to obtain the stady state condition was done at 37°C for 30 min. Cells were harvested by means of a rubber policeman, washed, and centrifuged in a 15–30% neutral sucrose gradient with 1.95 M NaCl at pH 7.8, following lysis in: 0.5% Triton X-100, 2 mM Tris, 0.1 M EDTA, and 1 M NaCl [2]. The gradients were analyzed by flow photometry at 254 nm and fractionated into ten fractions. Radio-

1 Institute of Nuclear Science "Boris Kidric", Laboratory for Radiobiology, Belgrade, Yugoslavia
2 Institute for Biology, Research Center Seibersdorf, Austria

ADP-Ribosylation of Proteins
(ed. by F.R. Althaus, H. Hilz, and S. Shall)
© Springer-Verlag Berlin Heidelberg 1985

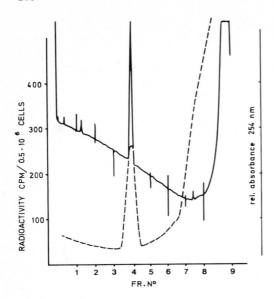

Fig. 1. Sedimentation of nucleoids from CHO-K$_1$ cells. —— absorbance at 254 nm; – – – radioactivity (CPM). Sucrose density gradients of histone-depleted cells labeled with [^{3}H]-NAD. The gradients range from 15 to 30% sucrose and contain 1.95 M NaCl at pH 7.8

activity was determined after cold 12% PCA precipitation (with 0.1% bovine serum albumin as a carrier) and 30 min hydrolyzation in 6% PCA at 90°C.

Electrophoretic Separation of Proteins

For preparative collection of nucleoids, the same procedure as described above was used. The pH of the gradient was decreased to 7 to reduce the potential loss of poly-(ADP-ribose) linked to acceptor proteins. After ultracentrifugation, nucleoids were collected and washed in 10 mM Tris–1 mM EDTA, pH 7, and digested with DNase I and RNase (bovine pancreas, type II-A, Sigma). In some samples, snake venom phosphodiesterase digestion of poly(ADP-ribose) was carried out in order to characterize the radioactivity. The volume was reduced by polyethyleneglycol and the concentration of proteins determined by a modification of the Lowry method [7]. Electrophoretic separation of proteins was performed in a 10% polyacrylamide gel system, either at pH 7 or at pH 2.4, in the presence of a SDS and phosphate-buffer [10], or LDS-citric acid buffer [6], respectively. Parallel gel lanes were sliced into 4 mm pieces and incubated at 37°C in 0.1% SDS overnight. Radioactivity was measured using a liquid scintillation counter.

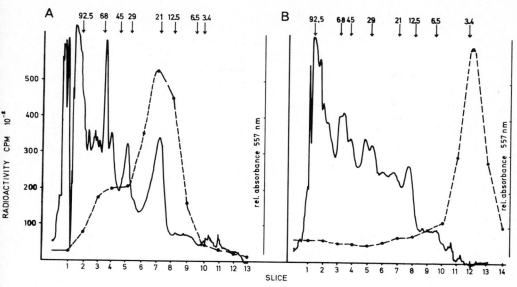

Fig. 2A,B. Electrophoretic separation of proteins associated with histone-depleted nucleoids from [³H]-NAD labeled CHO cells. —— absorbance at 570 nm; – – – radioactivity (CPM/slice). **A** LDS-10% gel electrophoresis at a low pH and temperature. **B** SDS-10% gel electrophoresis at neutral pH and room temperature. RNA and DNA in nucleoids are digested with DNase I and RNase. Marker proteins: phosphorylase B (92.5 kD), BSA (68 kD), egg albumin (45 kD), carboanhydrase (29 kD), trypsine inhibitor sojaseed protein (21 kD), cytochrome C (12.5 kD), trypsin inhibitor (lung) (6.5 kD), insulin B chain (3.4 kD)

Results

[³H]-NAD⁺-derived radioactivity was recovered from nucleoids. Following the analysis of ultracentrifugation runs, we found that a constant proportion of the total radioactivity (about 15% of total ADP-ribose synthetized in the cells) was comigrating with the nucleoid peak as detected by flow photometry at 254 nm. In the fractions preceding the peak, only background radioactivity could be measured. Likewise, in two fractions subsequent to the nucleoid peak practically no radioactivity was found. Only the nucleoid fractions contained significant radioactivity (Fig. 1). Following electrophoretic separation of the protein components of nucleoids we found three major fractions (Fig. 2). LDS polyacrylamide gel electrophoresis at low pH and 4°C allowed for the identification of an acceptor protein with an estimated mol. wt. of 12,000 (Fig. 2). SDS electrophoresis at neutral pH and room temperature released the radioactivity from acceptor proteins (Fig. 3). The poly(ADP-ribose) produced had a molecular weight which corresponds to the average chain length of 5–9 ADP-ribose units.

Partial phosphodiesterase digestion reduced both bound and nonbound poly(ADP-ribose) radioactivity to 50% (Fig. 3). A newly generated radioactive peak was comigrating with phosphodiesterase itself.

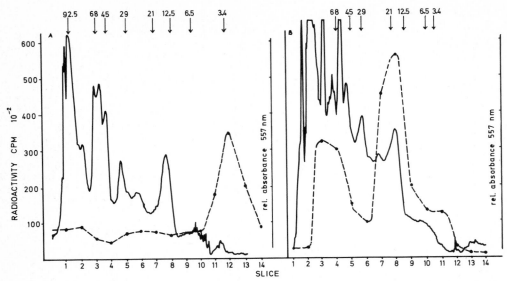

Fig. 3A,B. Electrophoretic separation of proteins associated with histone-depleted nucleoids; —— absorbance at 570 nm; – – – radioactivity (CPM/slice). **A** SDS-10% polyacrylamide gel electrophoresis at neutral pH and room temperature. Nucleoids were digested with DNase I, RNase, and snake venom phosphodiesterase. **B** LDS-10% polyacrylamide gel electrophoresis at a low pH and temperature

Discussion

In agreement with Cook and Brazell [3] and Berezney and Coffey [1] we found three major protein fractions in nucleoids. The residual protein in nucleoids cannot be removed by $2\,M$ NaCl. The ADP-ribosylated protein can be removed only after nuclease digestion. Therefore, we believe that the ADP-ribosylated protein was tightly bound to DNA. It may be that nuclei depleted of histones are contaminated by some unrelated material, but this material should not bind to DNA. In sucrose gradients, nuclei depleted of histones sediment as one peak (Fig. 1). Following electrophoresis, radioactivity was detected in a protein of roughly M_r 12,000 (Fig. 2). The unresolved question concerns the role of this modification. Many proteins, which are known to be ADP-ribosylated have probably been removed by our experimental treatment. However, the subunit of a larger nonhistone DNA protein complex, which is bound to DNA, may also be ADP-ribosylated. The linkage of poly(ADP-ribose) and acceptor proteins is unstable in a nonacid milieu [6]. Using two different kinds of electrophoretic systems, we confirmed those data (Fig. 2). With a partial snake venom phosphodiesterase digestion we additionally confirmed the polymer nature of our ADP-ribose (Fig. 3). Song and Adolph [12] have found poly(ADP-ribose) associated with nuclear scaffold proteins, but in their situation, acceptors exhibited a lower electrophoretic mobility compared to the protein charactized in this report.

We do not know whether ADP-ribosylation of nucleoids is involved in the regulation of transcription and replication or, as both processes are closely related to the nuclear cage [4, 8], we also do not exclude such a possibility.

Acknowledgments. This research work was performed in part in the frame of an IAEA fellowship. We thank Dr. R. Mukherjee for his help in the organization of the fellowship for G. Brkič.

References

1. Berezney R, Coffey DS (1975) Nuclear protein matrix: association with newly synthesized DNA. Science 189:291–293
2. Cook PR, Brazell IA (1975) Supercoils in human DNA. J Cell Sci 19:261–279
3. Cook PR, Brazell IA (1976) Detection and repair of single-strand breaks in nuclear DNA. Nature (London) 263:679–682
4. McCready SJ, Godwin J, Mason DW, Brazell IA, Cook PR (1980) DNA is replicated at the nuclear cage. J Cell Sci 46:365–386
5. McCready SJ, Jackson DA, Cook PR (1982) Attachment of intact superhelical DNA to the nuclear cage during replication and transcription. In: Natarajan AT, Obe G, Altmann H (eds) DNA Repair, chromosome alterations and chromatin structure. Elsevier Biomedicine Press, Amsterdam, pp 113–130
6. Dam VT, Faribault G, Poirier GG (1981) Separation of poly(ADP-ribosylated) nuclear proteins by polyacrylamide gel electrophoresis at acidic pH and low temperature. Anal Biochem 114:330–335
7. Hartree EF (1982) Determination of proteins: a modification of the Lowry method that gives a linear photometric response. Anal Biochem 48:422–427
8. Jackson DA, McCready SJ, Cook PR (1981) RNA is synthesized in nuclear cage. Nature (London) 292:522–555
9. Koide SS (1982) DNA replication and poly(ADP-ribosyl)ation. In: Hayashi O, Ueda K (eds) ADP-ribosylation reaction. Academic Press, London New York, pp 361–371
10. Murray FA, McGaugney RW, Yarus MD (1972) Blastokinin: Its size and shape and an indication of the existance of subunit. Fertil Steril 23:69–72
11. Slattery J, Dignam D, Matsui T, Roeder RG (1983) Purification and analysis of a factor which supresses nick-induced transcription by RNA polymerase II and its identity with poly(ADP-ribose)polymerase. J Biol Chem 258:5955–5959
12. Song MH, Adolph KW (1983) ADP-ribosylation of nonhistone proteins during HeLa cell cycle. Biochem Biophys Res Commun 115:938–945

ADP-Ribosylation of Nuclear Matrix Proteins. Association of Poly(ADP-Ribose) Synthetase with the Nuclear Matrix

JÓZEFA WĘSIERSKA-GĄDEK and GEORG SAUERMANN[1]

1. Introduction

The residual structures which remain after DNase digestion and high salt extraction of isolated nuclei, are termed nuclear matrices [1]. The nuclear matrix contains the structural elements of the peripheral lamina, the pore complexes, the nucleoli, and an internal fibrogranular network, and is considered to represent the main structural framework of the interphase nucleus. The DNA loops appear to be attached to the matrix at specific binding sites [2, 3]. Moreover, it was observed that basic biochemical reactions occur in tight association with this solid support. Thus, DNA replication is considered to proceed in association with the nuclear matrix [4, 5]. Furthermore, actively transcribed genes were found enriched in the nuclear matrix [6, 7]. It was observed that RNA synthesis and processing proceed in association with the matrix [8, 9] which may also play a role in certain transport processes [10]. Binding of steroid hormones [11] and of viral proteins [12] by the nuclear matrix and modification of matrix proteins by phosphorylation have been described [13, 14]. (For reviews, see [15, 16]).

The formation of mono-, oligo-, and of linear or branched poly(ADP-ribose) protein conjugates by poly(ADP-ribose) synthetase has been described by various investigators. This protein modification by ADP-ribosylation has been correlated with the biochemical and biological processes of DNA replication, DNA repair, chromatin condensation, regulation of gene expression, and cellular differentiation (for reviews, see [17–19]). While the modification of chromatin constituents, especially of the histones, has been extensively studied, information on the ADP-ribosylation of other subnuclear constituents is scarce [20].

The present experiments show that a number of ADP-ribosylated proteins in the range of 40,000–300,000 kD exist in nuclear matrices isolated from HeLa cells, and that the lamins A and B are probably modified. Furthermore, it was observed that a portion (approx. 1%) of the nuclear poly(ADP-ribose) synthetase is tightly associated with the isolated nuclear matrix.

---

1 Institut für Tumorbiologie-Krebsforschung der Universität Wien, Borschkegasse 8a, A–1090 Wien, Austria

ADP-Ribosylation of Proteins
(ed. by F.R. Althaus, H. Hilz, and S. Shall)
© Springer-Verlag Berlin Heidelberg 1985

2. Methods

HeLa cells were permeabilized by the method of Thi Man and Shall [21]. Nuclei were isolated by the method of Hodge et al. [22]. Using 1% Nonidet P-40 and 0.5% deoxycholate, nuclear matrices were prepared as described by Mariman et al. [8]. Briefly, the nuclei were subsequently digested with DNase I (500 μg ml^{-1}) and with RNase A (100 μg ml^{-1}) at 20°C for 15 min. After a sucrose step centrifugation, the nuclei were extracted with 0.4 M ammonium sulfate. All solutions contained 1 mM PMSF (phenylmethylsulfonyl fluoride).

The permeabilized cells were labeled by incubation with 0.1 mM [^{32}P]-NAD in 100 mM Tris. HCl (pH 8.0), 10 mM MgCl$_2$, 0.1 mM dithiothreitol (buffer A) at 25°C for 10 min.

To determine the poly(ADP-ribose) synthetase activity in the nuclear matrix, samples were incubated as above in 10 μM [^{32}P]-NAD in buffer A except that PMSF was omitted. For the assay of the enzyme activity in the subnuclear fractions, 25 μg DNA ml^{-1}, 25 μg histone H$_1$ ml^{-1}, and 60 μg DNase I ml^{-1} were additionally included.

Following acrylamide gel electrophoresis, the proteins were electrophoretically transferred to nitrocellulose sheets and the blots were exposed to X-ray films.

The ADP-ribose protein conjugates were purified on aminophenyl boronate columns as described by Adamietz et al. [23].

3. Results

3.1 Labeling of Matrix Components in Permeabilized Cells

HeLa cells were made permeable in order to allow the passage of the [^{32}P]-labeled ADP-ribose precursor NAD into the cell. After a short labeling period, the nuclei were isolated and the nuclear matrices were prepared. For complete removal of chromatin constituents and ribonucleoprotein complexes during the subsequent extraction with high salt buffer, nuclei were incubated with high concentrations of DNase I and RNase A.

Figure 1 shows the proteins contained in various subnuclear fractions. Histones were not detectable in the isolated nuclear matrix by Coomassie blue staining. But these known acceptor proteins for ADP-ribose were found in the ammonium sulfate extract. Autoradiography revealed that great amounts of the self-modified poly(ADP-ribose) synthetase — in accordance with the results of Table 1 — were released by DNase, RNase digestion of the isolated nuclei (not shown).

In order to disaggregate the ADP-ribose modified proteins, samples of the isolated matrix were dissolved in urea-SDS buffer and submitted to electrophoresis in a 7.5% polyacrylamide gel (Fig. 2). Densitometric scanning of autoradiograms revealed bands of modified proteins at 300, 220, 190, 160, 140, 116, 73, 69, 64, 60, 51, 46, 43, and 41 kD.

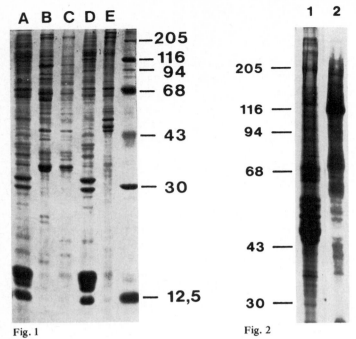

Fig. 1

Fig. 2

Fig. 1. Protein patterns of the subnuclear fractions. 12% polyacrylamide gel electrophoresis in 0.1% SDS. Coomassie blue staining. *A* Nuclei; *B* DNase-released; *C* RNase-released; *D* ammonium sulphate extract; *E* nuclear matrix

Fig. 2. [^{32}P]-ADP-ribosylated nuclear matrix proteins. Permeabilized cells were incubated in the presence of 0.1 mM [^{32}P]-NAD in buffer A for 10 min at 25°C. 7.5% polyacrylamide gel electrophoresis in 7 M urea, 0.1% SDS. *1* Coomassie blue staining; *2* autoradiography

The total labeled material of the nuclear matrix was further characterized by its sensitivity to snake venom phosphodiesterase and to neutral hydroxylamine. As characteristic for poly(ADP-ribose) protein conjugates, the labeled material was hydrolyzed by the enzyme. 69% of the labeled conjugates were found to be NH$_2$OH-sensitive, and 28% to be NH$_2$OH-resistant, indicating that both types of ADP-ribose protein linkages were present [24].

3.2 Isolation by Boronate Column Chromatography of in Vivo Existing ADP-Ribose Protein Conjugates

Since artifacts may be formed in the course of labeling experiments with permeabilized cells, the ADP-ribosylated proteins of the nuclear matrix were identified by another experimental approach. After labeling of cellular proteins by 24 h incubation with [^{35}S]-methionine, the matrix was isolated from the intact cells. In subsequent chromatography, the ADP-ribosylated proteins were selectively isolated due to the binding of the cis-diol grouping within the sugar component of the ADP-ribose conjugates onto the boronate column. While only 2% of the total matrix proteins were bound, control

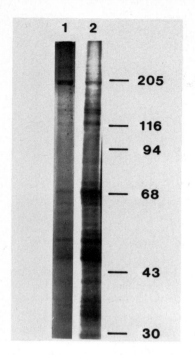

Fig. 3. Boronate-bound fraction of nuclear matrix proteins. Cells were labeled by 24 h incubation with [^{35}S]-methionine. PAGE as in Fig. 2. *1* Coomassie blue staining; *2* autoradiography

experiments revealed that 85% of the [^{32}P]-ADP-ribose labeled matrix proteins were bound by and recovered from the column.

The analysis of the boronate bound fraction by urea-SDS acrylamide gel electrophoresis is shown in Fig. 3. [^{35}S]-labeled proteins of the nuclear matrix with apparent mol. wts. of 300, 220, 200, 140, 116, 73, 69, 64, 60, 51, 46, 43, and 41 kD may be seen.

3.3 The Association of Poly(ADP-Ribose) Synthetase with the Nuclear Matrix

The main acceptor for ADP-ribose in the in vitro labeling experiments was a protein of mol. wt. of 115–125 kD (Fig. 2). It was assumed to be poly(ADP-ribose) synthetase, since this enzyme has been found to catalyze self-modification reactions [18]. This assumption was verified by incubating nuclear matrices isolated from unlabeled cells with labeled NAD. Figure 4 shows that the tracer was indeed incorporated into acid-insoluble material in a time-dependent reaction. The fact that a specific inhibitor of poly(ADP-ribose) synthetase, 3-aminobenzamide, almost completely inhibited the reaction, indicated that the reaction was catalyzed by this enzyme. This conclusion was also supported by electrophoretic analysis of the labeled products formed (not shown).

Table 1 shows the distribution of poly(ADP-ribose) synthetase activity in the different subnuclear fractions. The finding that the bulk of the poly(ADP-ribose) synthetase is released upon nuclease treatment of the nuclei, conforms with the previous observa-

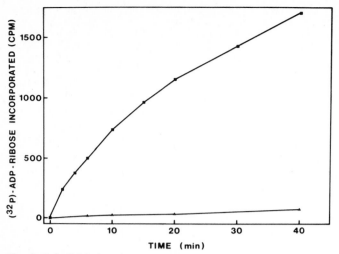

Fig. 4. Poly(ADP-ribose) synthetase activity in isolated nuclear matrices. Isolated nuclei matrices were incubated with 10 μM [^{32}P]-NAD in buffer A at 25°C for the times indicated. Assay of enzyme activity as in Methods. ■ Without further addition; ▲ in the presence of 10 mM 3-amino-benzamide

tions of Giri et al. [25] that the enzyme is localized in the internucleosomal linker region of chromatin. Only a small fraction of the total nuclear poly(ADP-ribose) synthetase was associated with the nuclear matrix (Table 1). Relative to protein content, the specific activity of the matrix was lower than that of the nucleus.

The question is raised as to the mechanism by which the enzyme is attached to the nuclear matrix. Even though the nuclei were digested with high DNase concentration, a residual amount of 0.06% of the original nuclear DNA was found associated with the matrix. Although the activity of the purified enzyme is known to be completely DNA dependent [18], it was observed that the synthetase reaction of the isolated matrix did not require addition of DNA.

Thus, the enzyme may be directly associated with the matrix, or may, alternatively, be indirectly bound by fixation to the residual DNA attached to the matrix. The data

Table 1. Poly(ADP-ribose) synthetase activity and DNA contained in subnuclear fraction[a]

Fraction	Nuclear values (%)			
	Protein	ADP-ribose synthetase activity	Radioactivity of nuclear DNA	
			Total	TCA insoluble
DNase, RNase-released	26	63	73	2
(NH$_4$)$_2$SO$_4$ released	47	28	20	11
Nuclear matrix	11	1.2	0.07	0.06

[a] Cellular DNA was labeled by incubation of cells with [H^3]-thymidine for 24 h

of Table 1 would point towards the interpretation that direct binding occurs because, in relation to the DNA content, the specific activity of the enzyme is 20 times greater in the nuclear matrix than in the nucleus.

4. Discussion

A comparison of the data resulting from the labeling experiments with permeabilized cells with the data obtained by analysis of the boronate-binding fraction suggests the existence of several ADP-ribosylated nuclear matrix proteins. Beginning with a weak, but distinct electrophoretic band at approximately 300 kD, there are modified proteins at 300, 220, 140, 116, 73, 69, 64, 60, 51, 46, 43, 41 kD.

Some bands in the range between 116 and 220 kD (i.e., at 190 and 160 kD) were observed only in the in vitro labeling experiments. The material is supposedly formed by automodification of poly(ADP-ribose) synthetase [18]. It is further remarkable that the intensity of the poly(ADP-ribose) synthetase band at 116 kD was relatively low in the boronate-bound fraction, but high in the autoradiogram after in vitro labeling.

The bands observed at 69 and 64 kD may be due to ADP-ribosylation of lamin A and lamin B, major constituents of the nuclear matrix, as was previously suggested by Song and Adolph [26]. They reported that nuclear scaffolds isolated by mild micrococcal nuclease digestion from in vitro labeled HeLa cell nuclei contained labeled proteins at 116 kD and in the 65 to 70 kD range.

At this point, reference to the modification of nuclear matrix proteins by phosphorylation may be of interest. Gerace and Blobel [13] observed the phosphorylation of lamins in the course of the reversible depolymerization of the lamina during cell division. Henry and Hodge [14] reported the occurrence in the matrix of HeLa cells of about ten proteins with mol. wts. between 200 and 19 kD which could be phosphorylated in vivo and in vitro.

The present results demonstrate that a portion of the nuclear poly(ADP-ribose) synthetase is found in tight association with the isolated nuclear matrix. The question is whether the enzyme, if present at this site in vivo, is required to maintain reactions proceeding in association with the nuclear matrix. These may, for example, be reactions involving DNA in the vicinity of the DNA attachment sites. In view of various indications that transcription of active genes occurs in association with the nuclear matrix [6—9], the results of Slattery et al. [27] are also of interest. The authors observed that poly(ADP-ribose) synthetase is identical with the factor TFIIC which, by inhibiting nick-induced transcription, eliminates random transcription by polymerase II.

References

1. Berezney R, Coffey DS (1977) Nuclear matrix. Isolation and characterizaton of a framework structure from rat liver nuclei. J Cell Biol 73:616–637
2. Vogelstein B, Pardoll DM, Coffey DS (1980) Supercoiled loops and eucaryotic DNA replication. Cell 22:79–85
3. Razin SV, Mantieva VL, Georgiev GP (1979) The similarity of DNA sequences remaining bound to scaffold upon nuclease treatment of interphase nuclei and metaphase chromosomes. Nucleic Acids Res 7:1713–1735
4. Pardoll DM, Vogelstein B, Coffey DS (1980) A fixed site of DNA replication in eucaryotic cells. Cell 19:527–536
5. Smith HC, Berezney R (1982) Nuclear matrix-bound deoxyribonucleic acid synthesis: an in vitro system. Biochemistry 21:6751–6761
6. Robinson SI, Small D, Idzerda R, McKnight GS, Vogelstein B (1983) The association of transcriptionally active genes with the nuclear matrix of the chicken oviduct. Nucleic Acids Res 11:5113–5130
7. Ciejek EM, Tsai M, O'Malley BW (1983) Actively transcribed genes are associated with the nuclear matrix. Nature (London) 306:607–609
8. Mariman ECM, Van Eekelen CAG, Reinders RJ, Berns AJM, Van Venrooij WJ (1982) Adenoviral heterogeneous nuclear RNA is associated with the host nuclear matrix during splicing. J Mol Biol 154:103–119
9. Ciejek EM, Nordstrom JL, Tsai M, O'Malley BW (1982) Ribonucleic acid precursors are associated with the chick oviduct nuclear matrix. Biochemistry 21:4945–4953
10. Baglia FA, Maul GG (1983) Nuclear ribonucleoprotein release and nucleoside triphosphatase activity are inhibited by antibodies directed against one nuclear matrix glycoprotein. Proc Natl Acad Sci USA 80:2285–2289
11. Barrack ER, Coffey DS (1980) The specific binding of estrogens and androgens to the nuclear matrix of sex hormone responsive tissues. J Biol Chem 255:7265–7275
12. Buckler-White AJ, Humphrey GW, Pigiet V (1980) Association of polyoma T antigen and DNA with the nuclear matrix from lytically infected 3T6 cells. Cell 22:37–46
13. Gerace L, Blobel G (1980) The nuclear envelope lamina is reversibly depolymerized during mitosis. Cell 19:277–287
14. Henry SM, Hodge LD (1983) Nuclear matrix: A cell-cycle-dependent site of increased intranuclear protein phosphorylation. Eur J Biochem 133:23–29
15. Barrack ER, Coffey DS (1982) Biological properties of the nuclear matrix: Steroid hormone binding. Recent Prog Horm Res 38:133–195
16. Berezney R, Basler J, Buchholtz LA, Smith HC, Siegel AJ (1982) Nuclear matrix organization and DNA replication. In: Maul GG (ed) Nuclear envelope and the nuclear matrix. Liss, New York, p 183
17. Hilz H (1981) ADP-ribosylation of proteins – A multifunctional process. Hoppe-Seyler's Z Physiol Chem 362:1415–1425
18. Ueda K, Kawaichi M, Hayaishi O (1982) Poly(ADP-ribose) synthetase. In: Hayaishi O, Ueda K (eds) ADP-ribosylation reactions. Academic Press, London New York, p 117
19. Mandel P, Okazaki H, Niedergang C (1982) Poly(adenosine diphosphate ribose). In: Cohn WE (ed) Progress in nucleic acid research and molecular biology, vol 27. Academic Press, London New York, p 1
20. Ueda K, Kawaichi M, Ogata N, Hayaishi O (1983) Poly(ADP-ribosyl)ation of nuclear proteins. In: Mizobuchi (ed) Nucleic acid research. Academic Press, London New York, p 143
21. Thi Man N, Shall S (1982) The alkylating agent, dimethyl sulphate, stimulates ADP-ribosylation of Histone H1 and other proteins in permeabilised mouse lymphoma (L1210) cells. Eur J Biochem 126:83–88
22. Hodge LD, Mancini P, Davis FM, Heywood P (1977) Nuclear matrix of HeLa S$_3$ cells. Polypeptide composition during adenovirus infection and in phases of the cell cycle. J Cell Biol 72:194–208

23. Adamietz P, Klapproth K, Hilz H (1979) Isolation and partial characterization of the ADP-ribosylated nuclear proteins from Ehrlich ascites tumor cells. Biochem Biophys Res Commun 91:1232–1238
24. Bredehorst R, Wielckens K, Gartemann A, Lengyel H, Klapproth K, Hilz H (1978) Two different types of bonds linking single ADP-ribose residues covalently to proteins. Eur J Biochem 92:129–135
25. Giri CP, West MHP, Ramirez ML, Smulson M (1978) Nuclear protein modification and chromatin substructure. Internucleosomal localization of poly(adenosine diphosphate-ribose) polymerase. Biochemistry 17:3501–3504
26. Song MH, Adolph KW (1983) ADP-ribosylation of nonhistone proteins during the HeLa cell cycle. Biochem Biophys Res Commun 115:938–945
27. Slattery E, Dignam JD, Matsui T, Roeder RG (1983) Purification and analysis of a factor which suppresses nick-induced transcription by RNA polymerase II and its identity with poly-(ADP-ribose) polymerase. J Biol Chem 258:5955–5959

Psoralen-Probing of Chromatin Organization in Intact Cells Undergoing DNA Excision Repair

GEORG MATHIS and FELIX R. ALTHAUS[1]

Introduction

A major problem with using isolated polynucleosomes to study chromatin organization in vitro is the uncertainty whether such preparations retain their native arrangement of chromosomal proteins and whether structural alterations in a selected chromatin preparation are representative of the situation in intact cells. The goal of the present study was to develop a technique which would allow us to determine the consequences of chromatin-associated ADP-ribosylation on chromatin organization of *intact* cells which undergo DNA excision repair.

Psoralen derivatives offer a number of distinct advantages as chemical probes for chromatin structure in intact cells [1–5, 9]: (1) Psoralens are readily taken up by intact cells; (2) a number of psoralens have been shown to intercalate specifically into free DNA. Upon activation by UV 365 nm, intercalated psoralens form stable monoadducts and some derivatives give rise to interstrand cross-links. These cross-links can be visualized by electron microscopy [1].

8-Methoxypsoralen (8-MOP) was selected for the present study because of its favorable physicochemical properties and because it forms a large proportion of divalent adducts with DNA [2]. These properties allow one to cross-link efficiently free DNA which makes these chromatin domains selectively amenable to isolation (vide infra).

Interaction of 8-Methoxypsoralen with Chromatin of Intact Hepatocytes Undergoing DNA Excision Repair

Because of the strict preference of psoralens for free DNA as opposed to core DNA [1, 3, 4] the amount of psoralens interacting with chromatin of intact cells under saturating conditions is a quantitative measure of the relative proportion of free DNA domains in a given experimental situation [5]. We have determined the psoralen acces-

1 University of Zürich, School of Veterinary Medicine, Institute of Pharmacology and Biochemistry, Winterthurerstraße 260, CH–8057 Zürich, Switzerland

ADP-Ribosylation of Proteins
(ed. by F.R. Althaus, H. Hilz, and S. Shall)
© Springer-Verlag Berlin Heidelberg 1985

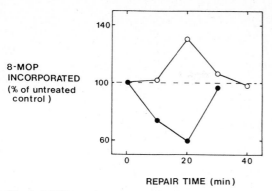

Fig. 1. 8-MOP reactivity of hepatocellular chromatin in early stages of DNA excision repair. Hepato-cytes were cultured [6] and exposed to UV light (254 nm, 45 J/m2) or N-acetoxy-2-acetylamino-fluorene (5×10^{-4} M, 10 min). At the times indicated in the *abscissa,* the cells were exposed to a saturating concentration of radioactively labeled 8-MOP (46 μM) and irradiated with 365 nm light in order to produce interstrand cross-links. The relative values on the *ordinate* are derived from the quantitative determination of 8-MOP radioactivity covalently associated with purified nuclear DNA of these cells. ○ UV irradiation; ● NAcAAF treatment

sibility of hepatocellular chromatin under conditions of carcinogen-induced DNA excision repair. Figure 1 demonstrates that UV irradiation (254 nm, 45 J/m2) of cul-tured hepatocytes [6] induces a rapid increase in the 8-MOP accessiblity of chromatin. Within 40 min after this UV treatment, the 8-MOP reactivity of hepatocellular DNA returned to control levels. By contrast, in hepatocytes treated with the chemical carcino-gen N-acetoxy-2-acetylaminofluorene (NAcAAF, 5×10^{-4} M, 10 min), the amount of 8-MOP reacting with hepatocellular DNA decreased initially and returned to control levels within a similar time frame. These alterations were dependent on the dose of carcinogen used (data not shown). These results suggest that changes in chromatin organization occur within minutes after infliction of DNA damage by a physical or chem-ical carcinogen. In addition, the pattern of changes observed with these two carcinogens is completely different. It should be noted that the distribution of DNA lesions relative to nucleosomal organization is also distinctly different for the two carcinogenic treat-ments. While pyrimidine dimers formed in response to $UV_{254\ nm}$-treatment are distrib-uted randomly in chromatin, the interaction of NAcAAF occurs primarily in linker DNA [7, 8]. Furthermore, the presence of these two types of lesions does not influence the 8-MOP accessibility of hepatocellular chromatin per se (data not shown). Thus, changes in chromatin organization as determined by 8-MOP accessibility appear to be a consequence of the DNA excision repair process initiated by these carcinogens.

Distribution of DNA Repair Patches in 8-MOP Accessible Domains of Hepatocellular Chromatin

In order to determine the proportion of newly synthesized DNA repair patches con-tained in 8-MOP accessible DNA as a function of repair time, it was necessary to isolate

Table 1. Effect of 8-MOP cross-linking on nuclease S1 digestion kinetics of hepatocellular DNA

Nuclease S1 digestion time (min)	Cell treatment (% DNA remaining)	
	None	8-MOP/UV 365
0	100	100
5	7	29
10	4	20
30	2	17
60	1	16

these chromatin domains on a preparative scale. Our approach is based on the observation by Ben-Hur et al. [9] who demonstrated differential renaturation kinetics of
psoralen cross-linked DNA sequences as compared to uncross-linked DNA. By virtue
of their faster renaturation kinetics following denaturation under alkaline conditions
or high temperatures, cross-linked DNA fragments become resistant to S1 nuclease
attack. We have modified this approach to purify 8-MOP cross-linked hepatocellular
DNA fragments in large quantities in order to be able to analyze their content of
newly synthesized repair patches. Table 1 summarizes the results of a typical experiment in which the nuclease S1 resistance of 8-MOP cross-linked hepatocellular DNA
was studied. The total proportion of 8-MOP cross-linked, nuclease S1-resistant DNA
("MOPS-DNA") was approximately 17%. This figure roughly corresponds to the
fraction of chromatin which is hypersensitive to micrococcal nuclease digestion (data
not shown).

We then determined the distribution of newly synthesized repair patches in MOPS-
DNA relative to random DNA as a function of repair time. In the experiment shown in
Table 2, 38% of repair incorporated radioactivity following a 20 min pulse-labeling
protocol in UV-irradiated hepatocytes was recovered with the MOPS-DNA fraction.
This value increased to 60% when the pulse-labeling was followed by a chase period of
another 100 min and it decreased to 52% when the chase period was extended to
160 min. These data suggest that the distribution of newly synthesized repair patches
relative to a selected chromatin fraction, i.e., MOPS-DNA, undergoes dramatic changes
as DNA excision repair proceeds from 20 to 120 min post-UV. These data can be used
to determine the kinetics of the rearrangement of repair patches in intact chromatin
and allow for the identification of modifying influences, such as those mediated by
chromatin-associated ADP-ribosylation [10].

Conclusions

Using 8-methoxypsoralen to probe chromatin organization of intact cells as they
recover from DNA damage, we have developed a model system to characterize modifying influences of chromatin-associated polyADP-ribosylation on the rearrangement of

Table 2. Distribution of newly synthesized repair patches in MOPS-DNA relative to random DNA of UV-irradiated hepatocytes

| Protocol | Distribution of repair radioactivity[a] | | |
	MOPS-DNA (cpm μg^{-1} DNA)	Random DNA (cpm μg^{-1} DNA)	% Contained in MOPS-DNA
20 min pulse	112	292	38
20 min pulse, 100 min chase	295	494	60
20 min pulse, 160 min chase	293	559	52

[a] Following UV irradiation (254 nm, 45 J/m2) of cultured hepatocytes [6] repair patches were radiolabeled with [methyl-^{3}H]thymidine for either 20 min (pulse period), 120 min (20 min pulse followed by a 100 min chase period with unlabeled thymidine), or 180 min (20 min pulse followed by 160 min chase). At the end of these treatments, repair incorporated radioactivity was quantified in MOPS-DNA and random DNA

repair patches [10]. This procedure allows to trap repair patches as they appear in free DNA and to follow their fate during excision repair. Because of the chemical stability of 8-MOP induced cross-links, these chromatin domains withstand fractionation procedures which are normally used to determine rearrangement of repair patches [11]. In addition, this procedure preserves chromatin domains which become destroyed during nuclease probing of native chromatin. Theoretically, results obtained on the rearrangement of repair patches using psoralen probing should be complementary to the data extracted from nuclease digestions experiments. Investigations regarding this prediction are currently under way in our laboratory.

Acknowledgments. We thank Drs. J.M. Sogo and T. Koller, ETH Zürich, Switzerland, for showing us some of their results prior to publication. We are grateful to Ralph Eichenberger for excellent technical assistance. This work was supported in part by Swiss National Foundation for Scientific Research Grant 3.375.082, the Sandoz Stiftung, and the Stiftung für wissenschaftliche Forschung der Universität Zürich.

References

1. Cech T, Pardue ML (1977) Cross-linking of DNA with trimethylpsoralen is a probe for chromatin structure. Cell 11:631–640
2. Isaacs ST, Shen CJ, Hearst JE, Rapoport (1977) Synthesis and characterization of new psoralen derivatives with superior photoreactivity with DNA and RNA. Biochemistry 16:1058–1064
3. Bohr V, Lerche A (1978) In vitro crosslinking of DNA by 8-methoxypsoralen visualized by electron microscopy. Biochim Biophys Acta 519:356–364
4. Hanson HV, Shen CJ, Hearst JE (1976) Cross-linking of DNA in situ as a probe for chromatin structure. Science 193:62–64
5. Sogo JM, Ness PJ, Widmer RM, Parish RW, Koller T (1984) Psoralen crosslinking of DNA as a probe for the structure of the active nucleolar chromatin. J Mol Biol 178:897–928

6. Althaus FR, Lawrence SD, Sattler GL, Pitot HC (1982) ADP-ribosyltransferase activity in cultured hepatocytes. J Biol Chem 257:5528—5535
7. Lang MC, de Murcia G, Mazen A, Fuchs RPP, Leng M, Daune M (1982) Non-random binding of N-acetoxy-N-2-acetylaminofluorene to chromatin subunits as visualized by immunoelectron microscopy. Chem Biol Interact 41:83—93
8. Kaneko M, Cerutti PA (1980) Excision of N-acetoxy-N-2-acetylaminofluorene induced DNA adducts from chromatin fractions of human fibroblasts. Cancer Res 40:4313—4318
9. Ben-Hur E, Prager A, Riklis E (1979) Measurement of DNA crosslinks by S1 nuclease: induction and repair in psoralen-plus-360 nm light treated E. coli. Photochem Photobiol 29:921—924
10. Althaus FR, Mathis G (1985) ADP-ribosylation, DNA repair and chromatin organization. This volume
11. Smerdon MJ, Lieberman MW (1980) Distribution within chromatin of deoxyribonucleic acid repair synthesis occurring at different times after ultraviolet radiation. Biochemistry 19:2992—3000

ADP-Ribosylation, DNA Repair, and Chromatin Organization

FELIX R. ALTHAUS and GEORG MATHIS[1]

Introduction

The formation of chromatin-associated poly(ADP-ribose) represents one of the earliest cellular reactions to DNA damage by physical and chemical carcinogens. Although the biological role of poly(ADP-ribose) formation is poorly understood, a large body of data is compatible with the idea that ADP-ribosylation acts to modify the biological expression of DNA damage during carcinogenesis [1]. As with other mechanisms whose functions become particularly obvious in the process of carcinogenesis, a biological role of polyADP-ribosylation reactions can also be demonstrated in normal cellular physiology and is generally related to the regulation of chromatin functions.

The lack of specific information regarding the critical target structures and mechanisms which are relevant in the biological action of poly(ADP-ribose) has prompted us to derive a working hypothesis from the following considerations:

1. Biological data suggest that ADP-ribosylation reactions are involved in all major chromatin functions, i.e., DNA replication, DNA repair, and transcription of genes [2]. Consequently, one can postulate that the target mechanism under ADP-ribosylation control is common to all these chromatin functions [20].
2. It is well established that poly(ADP-ribose) is always found covalently attached to chromatin proteins. It seems, therefore, reasonable to assume that this association with nuclear proteins is essential for the biological function of poly(ADP-ribose).
3. It is firmly established that the posttranslational modification of a protein with mono(ADP-ribose) is sufficient to regulate protein function [2]. By way of exclusion we can then postulate that the formation of large $(ADP\text{-ribose})_n$ polymers serves a different purpose, not related to the regulation of specific chromatin-associated enzyme proteins. In view of the considerable electronegative charges associated with long — and even branched $(ADP\text{-ribose})_n$ homopolymers, it seems conceivable to consider a role of these nucleic acid-like structures in modifying chromatin architecture. Alterations in chromatin organization have been described to accompany all major chromatin functions and if ADP-ribosylation has a role to

---

1 University of Zürich, School of Veterinary Medicine, Institute of Pharmacology and Biochemistry, Winterthurerstraße 260, CH–8057 Zürich, Switzerland

ADP-Ribosylation of Proteins
(ed. by F.R. Althaus, H. Hilz, and S. Shall)
© Springer-Verlag Berlin Heidelberg 1985

play in this process, this could account for the apparent involvement of this reaction in diverse chromatin functions.

4. In all systems examined so far, carcinogen-induced formation of poly(ADP-ribose) is observed immediately after formation of DNA strand breaks, i.e., if one considers the temporal sequence of reaction steps in DNA excision repair, ADP-ribosylation could have an impact on all repair steps concomitant with and/or subsequent to the formation of DNA strand breaks.

Based on these considerations we decided to address the following two questions: First, does ADP-ribosylation affect chromatin organization in intact cells while they undergo DNA excision repair? Secondly, what is the temporal relationship of such structural effects to the repair process?

(ADP-Ribose)$_n$-Dependent Alterations of Chromatin Organization During DNA Excision Repair

Most of the biochemical approaches to the investigation of chromatin organization rely on selective nuclease digestion of native chromatin. We have used micrococcal nuclease probing to determine the distribution of newly synthesized DNA repair patches relative to the nucleosomal organization of chromatin [3–6]. In addition, in order to overcome the principal shortcomings of this "destructive" procedure, we have modified and extended a technique which is normally used in connection with electron microscopic analysis, i.e., probing chromatin with psoralens [7]. We have developed a protocol which allows for the isolation of psoralen-accessible domains and subsequent biochemical analysis of the content of newly synthesized repair patches in these domains. The principle features of this new approach are discussed in the preceding chapter [8].

Table 1 demonstrates that newly synthesized repair patches are not uniformly distributed in monosomal and disomal DNA extracted from UV-treated hepatocytes in primary monolayer culture. These two DNA fractions were selected for examination because monosomal DNA consists exclusively of core DNA fragments, while two pieces of core DNA plus one linker DNA stretch is contained in a disomal DNA fragment. The nonuniform distribution of newly synthesized repair patches in these two DNA populations and an exchange of radioactivity between these two fractions following pulse-chase labeling of repair patches is, therefore, an indicator of rearrangements of nucleosomal organization which accompany DNA excision repair. However, these rearrangements do not necessarily occur between linker and core DNA, since micrococcal nuclease will recognize any piece of unprotected DNA. Therefore, in the interpretation of these results one has to consider the possibility that rearrangement reflects transient disruption of nucleosomal organization followed by reassembly of nucleosomes following complete repair of a particular chromatin region. We were impressed by the fact that this rearrangement is markedly affected by chromatin-associated ADP-ribosylation activity. Thus, when poly(ADP-ribose) formation was suppressed [9], the rearrangement process was markedly disturbed in UV-irradiated hepatocytes (Table 1).

Table 1. Consequences of inhibited ADP-ribosylation activity on the distribution of newly synthesized repair patches in oligosomal chromatin fractions

Treatment[a]	Total radioactivity (%) in	
	Monosomal DNA	Disomal DNA
20 min Repair		
Control	20	12
ABA	40	2
120 min Repair		
Control	9	13
ABA[b]	8	6

[a] Hepatocytes were cultured for 48 h in the presence or absence of ABA, UV-irradiated (254 nm, 45 J/m2) and newly synthesized repair patches were pulse labeled ([methyl-^{3}H] thymidine, 20 μCi/plate, 42 to 58 Ci/mmol). At the end of a 20 min labeling period ("20 min repair") or following a chase period (thymidine, 1 μM, "120 min repair"), the hepatocytes were harvested and nuclei were prepared [9]. Nuclei were treated for 3 min with micrococcal nuclease [10] and DNA was extracted. The digestion products were separated on 1.5% agarose gels. The percentage of radioactivity in monosomal and disomal DNA was then determined using a liquid scintillation counter. In separate experiments it was demonstrated that ABA did not affect the pattern of nucleosomal size classes obtained after micrococcal nuclease digestion of chromatin (F.R. Althaus, unpublished observation)
[b] ABA = 3-aminobenzamide, 8 mM

Table 2. Increase of repair radioactivity in MOPS-DNA and random DNA as a function of repair time[a]

Repair interval	Control		3-Aminobenzamide	
	MOPS-DNA (% increase)	Random-DNA (% increase)	MOPS-DNA (% increase)	Random-DNA (% increase)
20 min Pulse	111	291	64	110
100 min Chase	163	69	268	177
Additional chase of 60 min	− 1	13	3	13

[a] Hepatocytes were UV-irradiated (254 nm, 45 J/m2) and newly synthesized repair patches were labeled with [methyl-^{3}H]-thymidine in a pulse chase protocol [8]

Results complementary to those shown in Table 1 were obtained when the time course of the appearance of newly synthesized repair patches in isolated 8-methoxy-psoralen-accessible chromatin domains (MOPS-DNA [8]) was studied. Table 2 shows an example of such an experiment. When newly synthesized repair patches were pulse labeled followed by a chase period as repair proceeded for up to 120 min, an increasing proportion of the radioactivity was found contained in MOPS-DNA. The rate of this increase was significantly different in MOPS-DNA compared to random DNA. Inhibition of chromatin associated ADP-ribosylation greatly modified this process

(Table 2). Again, these data suggest that rearrangement of newly synthesized repair patches relative to the nucleosomal organization of chromatin is dependent on nuclear ADP-ribosylation activity.

Temporal Relationship of Nucleosomal Rearrangement to Individual Steps in DNA Excision Repair

We have examined the consequences of inhibition of chromatin-associated poly(ADP-ribosylation) on individual reaction steps of DNA repair elicited by a number of different carcinogens. In this analysis we could identify at least two ADP-ribose dependent reaction steps in the process of DNA repair, while others were found unresponsive to inhibition of nuclear ADP-ribosylation. Thus, for example, the rate of resynthesis of repair patches as well as the rate of ligation were increased in hepatocytes recovering from alkylation damage, pyrimidine dimers or bulky lesions caused by the ultimate carcinogen N-acetoxy-2-acetylaminofluorene (NAcAAF) ([9], F.R. Althaus, unpublished observations). Despite this apparent stimulation of intermediate steps of DNA repair, the removal of guanine adducts of NAcAAF (dG-8-AAF) was incomplete under conditions of inhibited ADP-ribosylation activity (F.R. Althaus and M.C. Poirier, unpublished observations).

These observtions are very intriguing and suggest that the suppression of chromatin-associated poly(ADP-ribosylation) disrupts the coordinated interaction of individual components in the repair process, which leaves a significant proportion of carcinogenic lesions unaffected by repair enzymes. It was, therefore, interesting to note that the manipulation of nuclear ADP-ribosylation activity affected the contribution of DNA polymerase α relative to DNA polymerase β in the resynthesis of repair patches (F.R. Althaus, unpublished observation). Considering our findings regarding the influence of ADP-ribosylation on alterations of chromatin organization during repair, one could argue that the local chromatin conformation determines the accessibility of damaged sites for individual repair enzymes. Although this is purely speculative at the moment, an influence of chromatin conformation on the activity of individual repair enzymes has already been demonstrated under in vitro conditions [11].

In view of these observations it was important to establish the exact temporal relationship of the effects of poly(ADP-ribosylation) on changes in chromatin organization on the one hand, and the actions on repair patch resynthesis and ligation, on the other hand. Table 3 shows repair activity at various time intervals following irradiation of hepatocytes with UV 254 nm (36 J/m2). Interestingly, the stimulating effects of ADP-ribosyltransferase inhibitors became manifest only 60 min after hepatocytes had been UV irradiated. In fact, during an initial repair period of about 20 min, less synthetic activity was observed in cells whose ADP-ribosylation capacity had been suppressed (G. Mathis and F.R. Althaus, unpublished observation). This early repair period coincided with distinct changes of chromatin organization (Fig. 1). Within 10 min after UV irradiation, the proportion of 8-methoxypsoralen accessible chromatin domains increased dramatically in hepatocytes treated with ADP-ribosyltransferase

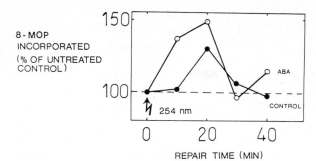

Fig. 1. Amount of 8-methoxy-psoralen (8-MOP) reacting with hepatocellular DNA at different times after UV irradiation (UV 254 nm, 45 J/m2) relative to an unirradiated control (= 100% value, *dashed line*). Experimental protocols were as described in [8]

Table 3. Activity of DNA repair synthesis at various time intervals following UV irradiation (254 nm, 36 J/m2) of cultured hepatocytes

Time interval following UV irradiation	[^{3}H]-Thymidine incorporation (%increase over unirradiated control)	
	UV	ABA + UV
0 min – 20 min	880	530
1 h – 4 h	550	890
20 h – 24 h	420	790

inhibitors and returned to control levels within 30 min post-UV. The time course of these changes was markedly different from control cells which were allowed uninhibited poly(ADP-ribose) biosynthesis in response to UV treatment (Fig. 1).

Conclusions

Alterations of chromatin organization represent the earliest poly(ADP-ribose) dependent reaction of hepatocytes in response to carcinogenic DNA damage measured so far. Further investigations demonstrated that the pattern of these changes differed with the carcinogen used to introduce DNA lesions [8]. Also, we could demonstrate that in the repair of lesions caused by the ultimate carcinogen N-acetoxy-2-acetyl-aminofluorene, none of these steps prior to repair patch resynthesis were affected by manipulations of ADP-ribosylation activity. Most importantly, the amount of DNA adduct formed was identical to those of untreated control cells (data not shown). Thus, considering the sequence of reaction steps involved in the repair of DNA lesions, ADP-ribosylation dependent changes of chromatin organization could have an impact on subsequent steps of the repair process. We, therefore, propose that the primary target mechanism of poly(ADP-ribose) action on DNA repair relates to structural changes of chromatin and all other influences on individual steps of repair, as also observed by other investigators, would represent secondary consequences of these structural changes. Such an explanation could also integrate a large body of biological data demonstrating an involvement of ADP-ribosylation in other chromatin functions, since these have also been shown to be accompanied by changes of chromatin organization [12]. Obviously, we cannot derive any conclusions from our data as far as the mechanism of such a poly(ADP-ribose) action is concerned. It is, however, important to keep in mind that the techniques applied in the present investigation allow for the analysis of chromatin in intact cells.

The in vitro model developed by Poirier et al. [13], however, could give a clue regarding possible mechanisms of action. These authors were able to demonstrate that in a reconstituted in vitro system the poly(ADP-ribosylation) of polynucleosomes induces reversible relaxation of chromatin structure. In this same system Niedergang et al. [11] demonstrated that DNA polymerase α activity is more than twofold higher in the presence of ADP-ribosylated pancreatic polynucleosomes compared to control polynucleosomes. This latter observation is also compatible with our results obtained in intact cells and lends support to the idea that the activity of enzymes involved in DNA rapair (and replication) is influenced by chromatin topology at the site of their action. Cleaver et al. [14] have recently claimed that the stimulation of repair synthesis by 3-aminobenzamide, a widely used competitive inhibitor of nuclear ADP-ribosyltransferase, is probably an artifact. Although this statement has been modified in a subsequent report [15], it should be clearly emphasized that enhancement of repair synthesis by 3-aminobenzamide or other ADP-ribosyltransferase inhibitors has been demonstrated in a variety of systems by a number of investigators [9, 15–19] and is clearly not an artifactual consequence of the use of hydroxyurea. Hydroxyurea was not used in any of the experiments shown in the present report. It is noteworthy that in our model system enhanced repair synthesis can already be observed at UV doses as low as 3 J/m2 (F.R. Althaus, unpublished observation). Therefore, UV excision repair in cultured hepatocytes is a very convenient model situation to determine the role of poly(ADP-ribose) in chromatin functions. Current investigations in our laboratory are now directed towards the identification of the target structures which are critically involved in the action of poly(ADP-ribose) on chromatin organization.

Acknowledgments. We are grateful to Ralph Eichenberger for excellent technical assistance. We thank Dr. J.E. Cleaver for sharing some of his results prior to publication. This work was supported in part by the Swiss National Foundation for Scientific Research Grant 3.375.082, the Sandoz Stiftung, and the Stiftung für wissenschaftliche Forschung der Unversität Zürich.

References

1. Miwa M, Hayaishi O, Shall S, Smulson M, Sugimura T (eds) (1983) ADP-ribosylation, DNA repair and cancer. Jpn Sci Soc Press, Tokyo/VNU Science Press, BV, Utrecht
2. Hayaishi O, Ueda K (eds) (1982) ADP-ribosylation reactions. Academic Press, London New York
3. Axel R (1978) Dissection of eukaryotic chromosome with deoxyribonucleases. Methods Cell Biol 18:41–54

4. Smerdon MJ, Lieberman MW (1980) Distribution within chromatin of deoxyribonucleic acid repair synthesis occurring at different times after ultraviolet radiation. Biochemistry 19:2992–3000

5. Bodell WJ, Cleaver JE (1981) Transient conformation changes in chromatin during excision repair of ultraviolet damage of DNA. Nucleic Acids Res 9:203–213

6. Zolan ME, Smith CA, Calvin NM, Hanawalt P (1982) Rearrangement of mammalian chromatin structure following excision repair. Nature (London) 299:462–464

7. Cech T, Pardue ML (1977) Cross-linking of DNA with trimethylpsoralen is a probe for chromatin structure. Cell 11:631–640

8. Mathis G, Althaus FR (1985) Probing chromatin organization with psoralens: Alterations during DNA excision repair. This volume

9. Althaus FR, Lawrence SD, Sattler GL, Pitot HC (1982) ADP-ribosyltransferase activity in cultured hepatocytes: Interactions with DNA repair. J Biol Chem 257:5528–5535

10. Butler PJG, Thomas JO (1980) Changes in chromatin folding in solution. J Mol Biol 140:505–529

11. Niedergang C, Ittel ME, de Murcia G, Pouyet J, Mandel P (1985) Kinetics of nucleosomal histone H1 hyper ADP-ribosylation and polynucleosomes relaxation. This volume

12. Reeves R (1984) Transcriptionally active chromatin. Biochim Biophys Acta 782;343–393

13. Poirier GG, de Murcia G, Jongstra-Bilen J, Niedergang C, Mandel P (1982) Poly(ADP-ribosyl)-ation of polynucleosomes causes relaxation of chromatin structure. Proc Natl Acad Sci USA 79:3423–3427

14. Cleaver JE, Bodell WJ, Morgan WF, Zelle B (1983) Differences in the regulation by poly(ADP-ribose) of repair of DNA damage from alkylating agents and ultraviolet light according to cell type. J Biol Chem 2581:9050–9068

15. Cleaver JE, Milam KM, Morgan WF (1985) Do inhibitor studies demonstrate a role for poly-(ADP-ribose) in DNA repair? Radiat Res 101:16–28

16. Althaus FR, Lawrence SD, Sattler GL, Pitot HC (1980) The effect of nicotinamide on unscheduled DNA synthesis in cultured hepatocytes. Biochem Biophys Res Commun 95:1063–1070

17. Bohr V, Klenow H (1981) 3-aminobenzamide stimulates unscheduled DNA synthesis and rejoining of strand breaks in human lymphocytes. Biochem Biophys Res Commun 102:1254–1261

18. Miwa M, Kanai M, Kondo T, Hoshino H, Ishihara K, Sugimura T (1981) Inhibitors of poly-(ADP-ribose) polymerase enhance unscheduled DNA synthesis in human peripheral lymphocytes. Biochem Biophys Res Commun 100:463–470

19. Sims JL, Sikorski GW, Datino DM, Berger S, Berger NA (1982) Poly (adenosine diphospho-ribose) polymerase inhibitors stimulate unscheduled deoxyribonucleic acid synthesis in normal human lymphocytes. Biochemistry 21:1813–1821

20. Althaus FR, Lawrence SD, He YZ, Sattler GL, Tsukada Y, Pitot HC (1982) Effects of altered (ADP-ribose)$_n$ metabolism on expression of fetal functions by adult hepatocytes. Nature (London) 300:366–388

ADP-Ribosylation, DNA Repair, and Xenobiotic Stress

Is There a Role for ADP-Ribosylation in DNA Repair?

WILLIAM F. MORGAN[1], MIRJANA C. DJORDJEVIC[1], KATHRYN M. MILAM[1],
JEFFREY L. SCHWARTZ[2], CARMIA BOREK[3], and JAMES E. CLEAVER[1]

1. Introduction

Synthesis of poly(ADP-ribose) results in depletion of cellular NAD^+ and transient ADP-ribosylation of DNA, chromatin, and other nuclear proteins. Synthesis of this homopolymer is stimulated by various kinds of DNA strand-breakage [1, 2], produced either directly, e.g., by X-rays [23], or indirectly through enzymatic excision of damaged bases [4, 9].

The function of poly(ADP-ribose) polymerase is largely defined in studies involving inhibitors of poly(ADP-ribose) synthetase, e.g., the benzamide analogues benzamide and 3-aminobenzamide (3AB), nicotinamide, thymidine, and various methylxanthines. From these inhibitor studies, poly(ADP-ribose) synthesis has been ascribed a role in the ligation stage of DNA repair [3, 4, 9–12, 14, 19, 23, 29]. The results from inhibitor studies, however, are difficult to incorporate into a single hypothesis defining the role of poly(ADP-ribose) polymerase in DNA repair. Furthermore, it has recently become apparent that many inhibitors have additional side effects, suggesting that some responses may be due to these side effects rather than to inhibiting poly(ADP-ribose) synthesis [15, 17].

In this section, we shall review evidence for a role of poly(ADP-ribose) in DNA repair.

2. Methodology

Chinese hamster ovary (CHO) cells were maintained as monolayers in McCoy's 5A medium supplemented with 10% fetal calf serum and antibiotics. Normal human lymphoid cells, WIL-2 were maintained as suspension cultures in RPMI medium containing fetal calf serum and antibiotics.

---

1 Laboratory of Radiobiology and Environmental Health, University of California, San Francisco, CA 94143, USA
2 Laboratory of Radiobiology, Harvard School of Public Health, Boston, MA 02115, USA
3 College of Physicians and Surgeons, Columbia University, New York, NY 10032, USA

ADP-Ribosylation of Proteins
(ed. by F.R. Althaus, H. Hilz, and S. Shall)
© Springer-Verlag Berlin Heidelberg 1985

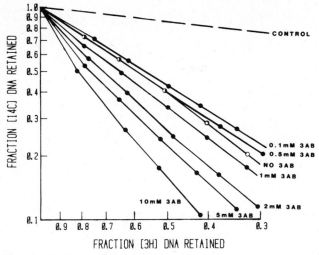

Fig. 1. Alkaline elution profiles of WIL-2 cells after 1 m*M* MMS (30 min) and 4 h growth with increasing concentrations of 3AB

DNA single-strand break frequency was measured by alkaline elution [16] and sister chromatid exchanges (SCEs) were assayed by the method described by Wolff [32]. Hypoxanthine-guanine phosphoribosyl transferase (HPRT; 6-thioguanine) and Na/K ATPase (ouabain) resistant mutants of CHO were determined by the methods described by Cleaver [7]. Repair replication after MMS treatment was measured in isopycnic gradients [8].

3. Results

3.1 DNA Single-Strand Breakage

When cells are exposed to an alkylating agent, single-strand breaks can be detected from both direct DNA strand-breakage and enzymatic action. WIL-2 cells were exposed to MMS (1 m*M*) for 1 h and then grown for 4 h with increasing concentrations of 3AB. At low concentrations of 3AB (below 0.5 m*M*), the frequency of MMS-induced strand-breaks was not changed; at 1 m*M* 3AB there was a small increase, but it was only at 2 m*M* 3AB and above that break frequency increased consistently (Fig. 1). Because DNA-protein cross-links can also be important parameters in alkaline elution, the effect of possible cross-linking induced by low concentrations of 3AB was also studied using proteinase K during cell lysis to negate DNA-protein cross-links. No consistent differences in elution profiles after MMS treatment and growth in 3AB was found with or without proteinase K treatments.

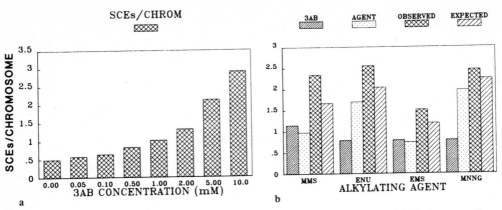

Fig. 2. a SCE frequency in CHO cells exposed to increasing concentrations of 3AB for two cell cycles. **b** SCE frequency in CHO cells exposed to different alkylating agents and 1 mM 3AB for two cell cycles. (From Morgan and Cleaver [18] and Schwartz et al. [28])

3.2 Sister Chromatid Exchange

Treatment for two complete cell cycles with increasing 3AB concentrations signifi-cantly elevated the SCE frequency in CHO cells (Fig. 2a). Exposure to the alkylating agents MMS (10^{-5} M), ethyl nitrosourea (ENU, 10^{-3} M), ethyl methanesulfonate (EMS, 10^{-4} M), or N-methyl-N'-nitro-N-nitrosoguanidine (MNNG, 10^{-7} M), in com-bination with 3AB resulted in synergistic increases in SCEs (Fig. 2b).

3.3 Repair Replication

Isopycnic gradients of bromodeoxyuridine plus [^{3}H]-thymidine-labeled DNA from cells exposed to MMS separated the DNA into heavy semiconservatively replicated DNA and normal density repair replicated DNA. The amount of repair in WIL-2 cells exposed to MMS was strongly dependent on both MMS dose and 3AB concentration (Fig. 3a,b). At low concentrations of 3AB there was a sharp increase in repair replica-tion which saturated at 2 mM (Fig. 3b).

The length of repair patches was determined from the increase in density of low molecular weight DNA in alkaline isopycnic gradients. The DNA was fragmented by staphylococcal nuclease digestion to obtain sizes of the order of 200 bases. Despite increases in repair replication observed by growth in 3AB, no change was detected in the length (30 to 80 bases) of the repair patches.

3.4 Rate of Ligation of Intracellular Repair Patches

The rate of ligation of MMS-induced repair patches was measured by first accumulat-ing a large proportion of incomplete patches using aphidicolin, a polymerase α inhibi-tor. These patches were then allowed to ligate in vivo by removal of aphidicolin.

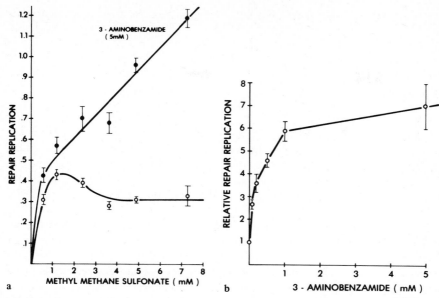

Fig. 3. a Repair replication in WIL-2 cells grown in [^{14}C]-thymidine, then exposed to increasing concentrations of MMS and grown for 4 h with (●) or without (○) 5 m*M* 3AB. **b** Repair replication in WIL-2 cells exposed to 4 m*M* MMS and grown for 4 h with various concentrations of 3AB. (From Cleaver [8])

Incomplete patches were detected by their sensitivity to rapid digestion by the 3′ to 5′ exonuclease activity of exonuclease III (exo III). Within 1 h of removal of aphidicolin, most repair patches were ligated and resistant to exo III. The presence of 3AB had no effect on the rate of loss of exo III sensitivity (Fig. 4), and therefore, did not delay

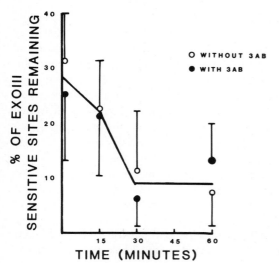

Fig. 4. The ligation of MMS (4 m*M*) induced repair patches in the presence (●) or absence (○) of 2 m*M* 3AB

the ligation of those patches dependent upon polymerase α for their resealing (up to 40%).

4. Discussion

Many different inhibitors of poly(ADP-ribose) synthetase have been used to define a role for that polymer in DNA repair. The consequences of inhibiting ADP-ribosylation is dependent upon the cell type studied, the kinds of DNA damage involved, and the particular biological endpoint in question. In spite of these differences, a general model for poly(ADP-ribose) in the regulation of ligation during the late stage of DNA repair has been proposed [10, 11, 14, 19, 29].

Initially, it appeared that this model could explain many of the observed phenomena. The inhibitors interacted with DNA ligase II [10], increasing the lifetime of individual breaks after DNA damage. Greater cytotoxicity would result [8–12, 14, 19, 23, 29], single-strand break frequencies would be increased, and delayed rejoining of individual breaks would permit repair polymerization to be more extensive.

A model based on regulation of ligation by stimulating ligase II also implies, however, that modulation of strand-break frequencies by 3AB (or any inhibitor) would occur to a similar extent in most cell types after a range of DNA damage. In our experiments, increased strand breakage only occurred at doses of 2 to 10 mM 3AB and was primarily found after exposure to alkylating agents [5, 9, 19], to a limited extent after X-ray exposure [20], and not at all after UV irradiation [5, 19].

This model has other problems. It cannot adequately account for the increase in SCE frequency induced by poly(ADP-ribose) inhibitors. These inhibitors increase SCE frequency in a variety of cell types [18, 21, 22, 24, 27, 28]. The stronger the inhibitor, the more SCEs induced [24]. Many DNA-damaging agents, mutagens and carcinogens, are known to increase SCE frequency, but inhibitors of poly(ADP-ribose) do not damage DNA per se [27] and are nonmutagenic [22, 26, 27]. It appears instead that these inhibitors increase SCE frequency by interacting with DNA that has incorporated bromodeoxyuridine, the thymidine base analogue necessary for the visualization of SCEs [21, 22], although increased SCEs have also been observed after autoradiographic analysis [21]. Morgan and Cleaver [19] first reported that combined treatment of an alkylating agent (MMS) and 3AB significantly increased SCE frequencies over that expected if the two events were additive. Similar synergistic increases in SCE rates have also been observed after combined treatment of 3AB and the alkylating agents MNNG, ENU, and EMS (Fig. 2b, [28]). When combined with X-rays [20] or UV light [19], on the other hand, 3AB has only an additive effect on SCE frequency.

At first sight, these results could be due to the persistence of DNA strand breaks induced by 3AB leading to increased SCEs. DNA strand breakage, however, is a very poor stimulus for SCE induction and there is a wealth of evidence indicating that DNA strand-breaks do not give rise to SCEs [20, 32].

Likewise, the observed effects of 3AB on mutation and transformation after exposure to a DNA-damaging agent are difficult to reconcile with a ligase effect. These two phenomena have been discussed more fully in an accompanying chapter (see Borek et al., this volume) and will not be considered further here.

Perhaps the most interesting area for probing the role of poly(ADP-ribose) polymerase in DNA repair is to investigate the role of this polymer in repair replication. Excision repair after alkylation damage involves recognition, incision, and removal of the damaged site (excision), resynthesis of new DNA (polymerization), and rejoining of the newly synthesized strand (ligation). Increased repair replication has been reported by many workers after treatment with different alkylating agents [3, 4, 8, 9, 12, 14, 29], but is most pronounced in lympoid cells [9] at low concentrations of 3AB (Fig. 3b, [8]). Increases in strand-break frequencies are detected at higher 3AB doses, i.e., greater than 1 mM (Fig. 1, [5, 14, 19, 20]). At lower doses of 3AB, strand-breaks are unaffected (Fig. 1).

Repair replication after MMS exposures in WIL-2 lymphoid cells was increased sevenfold by addition of 3AB (Fig. 4, [8]). This could be caused by an increased number of repaired patches or increased patch sizes. An increased number of patches at sites of original damage is unlikely because 3AB did not influence the excision of alkylated bases [9, 12]. Determination of the length of the repair patch indicated that 3AB had no effect on patch size [8]. This observation is inconsistent with the notion that 3AB causes longer patches to be synthesized because of inhibition of ligase II. Instead, there must be a greater number of patches which are not at sites of initial damage.

In an attempt to directly measure the role of poly(ADP-ribose) polymerase in ligation, incomplete repair patches induced by MMS were accumulated by aphidicolin, an inhibitor of polymerase α. Removal of aphidicolin allowed rapid resealing of the patch and the presence of 3AB at this time enabled direct determination of the effect of 3AB in the final stages of repair. 3AB did not delay the ligation of a large fraction of repair patches (up to 40%; Fig. 4). Although only a subset of the total repair patches can be assayed in this manner, there appears to be a single common ligation step for excision repair patches despite a variety of initial excision mechanisms. If we assume our results to be representative of the ligation process, this data demonstrates that ligation is not delayed by inhibiting poly(ADP-ribose) synthesis.

Our observations on the increase in repair replication caused by 3AB remain a mystery. The increase genuinely represents repair replication as defined by isopycnic gradient analysis and is not associated with increased excision of damaged sites nor with longer patches [8]. If ligation is not regulated by poly(ADP-ribose) polymerase, how are the observed effects on cytotoxicity, DNA strand breakage, SCE, mutation, transformation, and repair replication explained? It is unlikely that one single model will be sufficient. ADP-ribosylation has many varied effects within a cell. Attachment of the polymer inactivates Ca^{2+}, Mg^{2+} dependent endonucleases [33], retinal transducin [30], GTPases [6], diphtheria toxin [25], and topoisomerases [13]. Extensive ribosylation of nuclear proteins, histones, and the polymerase itself also occurs. Clearly, the integrity of DNA during excision repair has an important bearing on the consequences of DNA damage. Some effects may result from extensive ADP-ribosylation throughout the cell nucleus and cytoplasm unrelated to DNA repair. Furthermore, the inhibitors so widely used in defining a role for poly(ADP-ribose) are by no means specific in their inhibitory actions, but also influence other processes within the cell, e.g., incorporation of labeled methyl groups from the carbon-1 pool into DNA purines by de novo synthesis [5, 9, 15, 17]. Some effects, e.g., SCE and mutation, may result from a direct effect of the inhibitor on DNA precursor pool sizes and their subsequent de novo synthesis pathways to cause perturbations in DNA synthesis independent of ADP-ribosylation.

At present, the role of poly(ADP-ribose) in DNA repair is unclear, and our studies raise serious questions as to the validity of a current model for ADP-ribosylation in the regulation of ligation after DNA damage.

Acknowledgments. This work was supported by the U.S. Department of Energy (contract no. DE=ACO3-76-SF01012.

References

1. Benjamin RC, Gill DM (1980) ADP-ribosylation in mammalian cell ghosts. Dependence of poly(ADP-ribose) synthesis on strand breakage in DNA. J Biol Chem 255:10493−10501
2. Benjamin RC, Gill DM (1980) Poly(ADP-ribose) synthesis in vitro programmed by damaged DNA. A comparison of DNA molecules containing different types of strand breaks. J Biol Chem 255:10502−10508
3. Berger NA, Sikorski GW (1980) Nicotinamide stimulates repair of DNA damage in human lymphocytes. Biochem Biophys Res Commun 95:67−72
4. Berger NA, Sikorski GW, Petzold SJ, Kurohara KK (1979) Association of poly(adenosine diphosphoribose) synthesis with DNA damage and repair in normal human lymphocytes. J Clin Invest 63:1164−1171
5. Borek C, Morgan WF, Ong A, Cleaver JE (1984) Inhibition of malignant transformation in vitro by inhibitors of poly(ADP-ribose) synthesis. Proc Natl Acad Sci USA 81:243−247
6. Cassel D, Selinger Z (1977) Mechanism of adenylate cyclase activation by cholera toxia: inhibition of GTP hydrolysis at the regulatory site. Proc Natl Acad Sci USA 74:3307−3311
7. Cleaver JE (1977) Induction of thioguanine-- and ouabain-resistant mutants and single-strand breaks in the DNA of Chinese hamster ovary cells by ^{3}H-thymidine. Genetics 87:129−138
8. Cleaver JE (1985) Increased repair replication in human lymphoid cells by inhibition of polyadenosine diphosphoribose synthesis with no increase in patch sizes. Cancer Res 45:1163−1169
9. Cleaver JE, Bodell WJ, Morgan WF, Zelle B (1983) Differences in the regulation by poly(ADP-ribose) of repair of DNA damage from alkylating agents and ultraviolet light according to cell type. J Biol Chem 258:9059−9068
10. Creissen D, Shall S (1982) Regulation of DNA ligase activity by poly(ADP-ribose). Nature (London) 296:271−272
11. Durkacz BW, Omidiji O, Gray DA, Shall S (1980) (ADP-ribose)$_n$ participates in DNA excision repair. Nature (London) 283:593−596
12. Durrant LG, Margison GP, Boyle JM (1981) Effects of 5-methylnicotinamide on mouse L1210 cells exposed to N-methyl-N-nitrosourea: mutation induction, formation and removal of methylation products in DNA, and unscheduled DNA synthesis. Carcinogenesis 2:1013−1017
13. Ferro AM, Olivera BM (1984) Poly(ADP-ribosylation) of DNA topoisomerase I from calf thymus. J Biol Chem 259:547−554
14. James MR, Lehmann AR (1982) Role of poly(adenosine diphosphate ribose) in deoxyribonucleic acid repair in human fibroblasts. Biochemistry 21:4007−4013
15. Johnson GS (1981) Benzamide and its derivatives inhibit nicotinamide methylation as well as ADP-ribosylation. Biochem Int 2:611−617

16. Kohn KW, Ewig RAG, Erickson LC, Zwelling LA (1981) Measurement of strand breaks and cross-links by alkaline elution. In: Friedberg EC, Hanawalt PC (eds) DNA repair: A laboratory manual of research procedures, vol IB. Marcel Dekker, New York, pp 379–401

17. Milam K, Cleaver JE (1984) Inhibitors of poly(ADP-ribose) synthesis: effect on other metabolic processes. Science 233:589–591

18. Morgan WF, Cleaver JE (1982) 3-Aminobenzamide synergistically increases sister-chromatid exchanges in cells exposed to methyl methanesulfonate but not to ultraviolet light. Mutat Res 104:361–366

19. Morgan WF, Cleaver JE (1983) Effects of 3-aminobenzamide on the rate of ligation during repair of alkylated DNA in human fibroblasts. Cancer Res 43:3104–3107

20. Morgan WF, Djordjevic MC, Jostes RF, Pantelias GE (in press) Delayed repair of DNA single strand breaks does not increase cytogenetic damage. Int J Rad Biol

21. Morgan WF, Wolff S (1984) Induction of sister chromatid exchange by 3-aminobenzamide is independent of bromodeoxyuridine. Cytogenet Cell Genet 38:34–38

22. Natarajan AT, Csukas I, van Zeeland AA (1981) Contribution of incorporated 5-bromodeoxyuridine in DNA to the frequencies of sister-chromatid exchanges induced by inhibitors of poly(ADP-ribose)-polymerase. Mutat Res 84:125–132

23. Nduka N, Skidmore CJ, Shall S (1980) The enhancement of cytotoxicity of N-methyl-N-nitrosourea and of γ-radiation by inhibitors of poly(ADP-ribose) polymerase. Eur J Biochem 105:525–530

24. Oikawa A, Tohda H, Kanai M, Miwa M, Sugimura T (1980) Inhibitors of poly(adenosine diphosphate ribose) polymerase induce sister chromatid exchanges. Biochem Biophys Res Commun 97:1311–1316

25. Pappenheimer AMJ (1977) Diphtheria toxin. Biochemistry 46:69–94

26. Schwartz JL, Morgan WF, Brown-Lindquist P, Afzal V, Weichselbaum RR, Wolff S (1985) Comutagenic effects of 3-aminobenzamide in Chinese hamster ovary cells. Cancer Res 45:1556–1559

27. Schwartz JL, Morgan WF, Kapp LN, Wolff S (1983) Effects of 3-aminobenzamide on DNA synthesis and cell cycle progression in Chinese hamster ovary cells. Exp Cell Res 143:377–382

28. Schwartz JL, Morgan WF, Weichselbaum RR (in press) Different efficiencies of interaction between 3-aminobenzamide and various monofunctional alkylating agents in the induction of sister chromatid exchanges. Carcinogenesis

29. Sims JL, Sikorski GW, Catino DM, Berger SJ, Berger NA (1982) Poly(adenosine diphosphoribose) polymerase inhibitors stimulate unscheduled deoxyribonucleic acid synthesis in normal human lymphocytes. Biochemistry 31:1813–1821

30. Van Dop C, Tsuboka M, Bourne HR, Ramachandran J (1984) Amino acid sequence of retinal transducin at the site ADP-ribosylated by cholera toxin. J Biol Chem 259:696–698

31. Wolff S (1978) Chromosomal effects of mutagenic carcinogens and the nature of lesions leading to sister chromatid exchange. In: Evans HJ, Lloyd DC (eds) Mutagen induced chromosome damage in man. Edinburgh Univ Press, Edinburgh, pp 208–215

32. Wolff S (1981) Measurement of sister chromatid exchange in mammalian cells. In: Friedberg EC, Hanawalt PC (eds) DNA repair: A laboratory manual of research procedures, vol IB. Marcel Dekker, New York, pp 575–586

33. Yoshihara K, Tanigawa Y, Burzio L, Koide SS (1975) Evidence for adenosine diphosphate ribosylation of Ca^{2+}, Mg^{2+}-dependent endonuclease. Proc Natl Acad Sci USA 72:289–293

Poly(ADP-Ribosyl)ation, DNA Repair, and Chromatin Structure

KLAUS WIELCKENS[1], TANJA PLESS[2], and GITY SCHAEFER[1]

Abbreviations:

aphi	aphidicolin
BA	benzamide
DMS	dimethyl sulfate
ddt	dideoxythumidine
HU	hydroxyurea

Introduction

While it is now generally accepted that poly(ADP-ribosyl)ation is involved in DNA repair, its exact role remains unclear. Some years ago it was first postulated that poly-(ADP-ribosyl)ation is necessary for full activity of the DNA ligase step, the final step of DNA excision repair [1–4]. This was concluded from data showing a higher steady state concentration of DNA strand breaks in alkylated cells when inhibitors of poly-(ADP-ribose) synthesis were present. Additionally, no effect on other steps of DNA repair could be demonstrated, providing further evidence for this hypothesis [5, 6]. A direct modification of the DNA ligase II in vitro has also been suggested [7]. This is in contrast to in vivo experiments which have failed to demonstrate a modification of the DNA ligase by poly(ADP-ribose) [8, 9].

Therefore, it has to be taken into consideration that the higher steady state concentration of DNA strand breaks in alkylated/aminobenzamide treated vs alkylated cells do not necessarily result from a slow-down of the repair rate, but are the consequence of at least two other mechanisms:

1. A higher number of incisions either by repair endonucleases or by endonucleases not directly involved in DNA repair (Ca^{2+}, Mg^{2+}-dependent endonuclease).
2. Inhibition of DNA repair only at a limited number of sites which are only reparable when the poly(ADP-ribosyl)ation system is active.

---

1 Institut für Klinische Chemie, Medizinische Universitätsklinik, Martinistr. 52, 2000 Hamburg 20, FRG
2 Institut für Physiologische Chemie, Universität Hamburg, Martinistr. 52, 2000 Hamburg 20, FRG

ADP-Ribosylation of Proteins
(ed. by F.R. Althaus, H. Hilz, and S. Shall)
© Springer-Verlag Berlin Heidelberg 1985

In this chapter we describe data favoring the idea that the altered DNA repair kinetics in alkylated cells treated with the poly(ADP-ribose) synthesis inhibitor benzamide are the consequence of an inhibition of DNA repair at a limited number of sites in combination with an increase in the number of incisions. Both alterations could be explained by an involvement of poly(ADP-ribosyl)ation in the maintenance of chromatin structure in order to allow correct repair of the chromatin following alkylation. Preliminary studies analyzing the chromatin structure in cells cultivated in the presence of benzamide have indeed pointed to a participation of poly(ADP-ribosyl)ation in the organization of the chromatin, possibly in the chromatin/nuclear matrix interaction.

Poly(ADP-Ribosyl)ation: An Early Event During DNA Repair

In isolated nuclei or permeabilized cells, fragmentation of the DNA leads to a multifold stimulation of poly(ADP-ribose) formation [10–12]. Therefore, it was concluded that in intact cells the incision reaction of DNA excision repair represents the stimulating signal for poly(ADP-ribose) formation. Despite the "early" synthesis of poly(ADP-ribose), the action of this polymer was postulated to be in the last step, the DNA rejoining. Consequently, the question arose as to whether the protein-bound poly-(ADP-ribose) chain persists until the rejoining is completed, strongly connecting poly-(ADP-ribosyl)ation and DNA rejoining.

In order to obtain more information about the involvement of poly(ADP-ribosyl)-ation during DNA repair, studies have been performed using the DNA synthesis inhibitor aphidicolin. A blocker of the alpha-DNA-polymerase, aphidicolin at least partially inhibits DNA repair just after the incision step (Wielckens, unpublished). When HeLa cells were incubated in the presence or absence of 20 μg ml^{-1} aphidicolin under conditions where nearly all of the available NAD was consumed by inclusion of 200 μM DMS, no effect of aphidicolin on the poly(ADP-ribose) accumulation could be demonstrated (Fig. 1). Therefore, it appears that in the intact cell, as in vitro, the DNA incision represents the stimulating signal for poly(ADP-ribosyl)ation. On the other hand, the absence of a change of the polymer degradation argues against a direct connection between poly(ADP-ribose) degradation and the DNA ligase reaction.

Interestingly, there was no recovery of the NAD level in aphidicolin/DMS-treated cells compared to cells incubated with the alkylating agent alone. This finding substantiated the notion that the incision represents the stimulus for poly(ADP-ribose) formation and pointed to the crucial role of the processing of the repair site for signal deactivation. Further supporting evidence was provided by analysis of the poly(ADP-ribose) polymerase activity in digitonin-permeabilized cells: in aphidicolin/DMS-treated cells the elevated enzyme activity persisted, whereas in cells incubated with the alkylating agent alone the poly(ADP-ribose) polymerase returned to baseline levels within 4 h (Wielckens, unpublished). While these results do not rule out the possibility that poly(ADP-ribose) is necessary for the DNA rejoining step, they favor the idea of an indirect involvement of poly(ADP-ribosyl)ation in DNA repair, possibly by allowing correct repair of the chromatin by effects on chromatin structure.

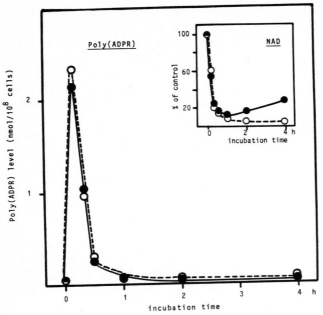

Fig. 1. Poly(ADP-ribose) and NAD in DMS-treated HeLa cells in the presence or absence of aphi-dicolin. HeLa cells in suspension culture were preincubated with 20 μg ml^{-1} aphidicolin for 15 min. Then 200 μM DMS were added and aliquots were removed at the appropriate times. Poly(ADP-ribose) and NAD were determined in the cells according to [13]. *Inset:* NAD concentration. ●——● DMS alone; ○— —○ DMS + aphidicolin

Poly(ADP-Ribosyl)ation and DNA Repair

It was first shown in S. Shall's laboratory that the nicotinamide analog aminobenzamide leads to an increase of the steady state concentration of DNA strand breaks during DNA repair [1, 2]. This phenomenon has been explained by a decreased rejoining rate.

If a general decrease in the rejoining rate does in fact occur, then the slow-down of the repair rate must result in an alteration of the slope of the repair curve (the number of strand breaks during DNA repair). Therefore, we have performed a careful analysis of the DNA repair kinetics by measuring the actual number of strand breaks. HeLa cells were treated with a moderate DMS dose (100 μM which reduced the colony-forming ability by a factor of two, and the degree of DNA fragmentation was determined with the aid of the alkali unwinding technique [14]. This treatment led to pronounced DNA damage with a maximum number of strand breaks occurring after 1 h; most of these were repaired after 12 to 14 h (Fig. 2). As postulated, benzamide increased the degree of DNA fragmentation after 1 h, but the number of strand breaks remained constantly higher during the entire repair time, "additional" breaks persisting even when the DNA repair was complete. Consequently, the slope of the repair curve was unaffected by the presence of benzamide. The fact that the overall repair rate was not decreased by inhibitors of poly(ADP-ribosyl)ation was also suggested by experiments

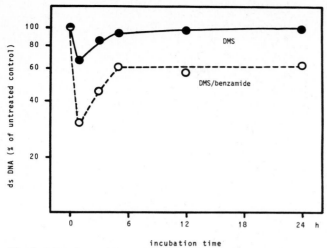

Fig. 2. DNA fragmentation in DMS-treated HeLa cells in the presence or absence of benzamide. HeLa cells were labeled overnight with [^{14}C]-thymidine, then medium was changed and the cells were treated with 25 μM DMS ± 5 mM benzamide. At the times indicated, aliquots were removed and the DNA fragmentation was determined with the aid of the alkali unwinding procedure

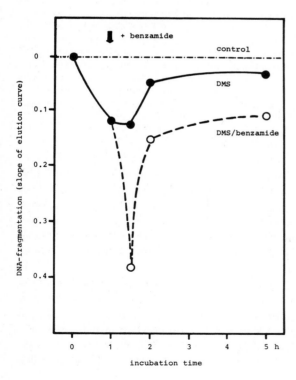

Fig. 3. DNA single-strand breaks in HeLa cells following treatment with 50 μM DMS ± benzamide. HeLa cells were labeled overnight with [^{14}C]-thymidine, the medium was changed and the cells were incubated with 50 μM DMS. Benzamide was added to the culture 1 h after DMS. Aliquots for the determination of DNA fragmentation by the alkaline elution technique [15] were removed at the times indicated

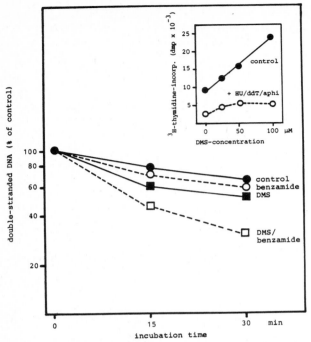

Fig. 4. Effect of benzamide on the DMS-induced DNA fragmentation in devil's mixture (= hydroxy-urea/dideoxythymidine/aphidicolin) pretreated cells. HeLa cells which were labeled overnight with [^{14}C]-thymidine were pretreated with 10 mM hydroxyurea and 0.1 mM dideoxythymidine for 15 min. Aphidicolin (20 μg ml^{-1}) was added and the cells preincubated for an additional 5 min. Benzamide (5 mM) and 2 μM DMS were then added. Aliquots were removed at the times indicated and the degree of DNA fragmentation was measured by the alkali unwinding technique. *Inset:* Unscheduled DNA synthesis in devil's mixture treated cells. Cells were pretreated with devil's mixture as described above and [^{3}H]-thymidine was added together with increasing amounts of DMS. After 4 h incubation, the acid-insoluble radioactivity was determined. Control cells received 10 mM hydroxyurea in place of devil's mixture

where benzamide was added 1 h after the alkylating agent (Fig. 3). The addition of benzamide led to a rapid burst of strand breaks which disappeared almost as fast as they appeared. These observations are in contradiction to the hypothesis that benzamide slows the rejoining velocity; more likely, poly(ADP-ribosyl)ation is only necessary for the repair of a limited number of sites which cannot be repaired in the presence of benzamide and consequently, persist ("persisting breaks"). That most of these breaks persist even after removal of the benzamide 12 h later has been also demonstrated (Wielckens, in preparation).

The appearance of "additional" breaks in cells which repaired under conditions where poly(ADP-ribose) synthesis was blocked raised the question of whether these breaks were only due to the inhibition of DNA repair at a limited number of sites or were additionally the consequence of a higher number of incisions. A series of experiments were, therefore, performed to allow differentiation of the two possibilities. HeLa cells were preincubated with a mixture of 10 mM hydroxyurea, 100 μM dideoxythymidine, and 20 μg ml^{-1} aphidicolin ("devil's mixture"), blocking most of the DNA repair synthesis (Fig. 4, inset) and consequently, the normal passage through the repair process. Therefore, benzamide would only increase the number of strand breaks in DMS-treated cells which were preincubated with devil's mixture if additional incisions occurred. Benzamide clearly increased the number of strand breaks in DMS-treated cells under this condition (Fig. 4), although the augmentation of DNA breakage was not as pronounced as in the absence of devil's mixture. The additional breaks were believed to be the consequence of an increased number of incisions together with a block of DNA repair at a limited number of sites. The nature of these persisting breaks is still unclear: they represent neither double-strand breaks nor DNA-protein cross-links (Wielckens, unpublished). Similarly, nothing is known about the mechanism of formation of the "additional" incisions. Preliminary experiments point to the occurrence of these breaks at nondamaged rather than damaged sites (Wielckens, unpublished results). In light of a possible higher sensitivity of the chromatin in benzamide-treated cells, more sites may be attacked by nonspecific endonucleases. Alternatively, the activity of the DNA topoisomerase I may be increased, leading to increased incidence of strand breaks. It remains to be determined what proportion of the persisting and additional breaks are identical.

Effect of Benzamide on the Cell Cycle of DMS-Treated Cells

B. Durkacs first demonstrated that inhibitors of poly(ADP-ribose) synthesis potentiate the toxicity of alkylating agents, thereby increasing the probability of lethal events in the alkylated cell [1]. We have observed the occurrence of persisting DNA strand breaks which appear in increasing numbers with increasing DMS doses, not only in DMS/benzamide-treated cells, but also in cells incubated with DMS alone. In the latter, the strand breaks occurred with a much lower probability pointing to the significance of the persisting breaks for the DMS-induced cytotoxicity (Wielckens, in preparation).

To better understand the role of poly(ADP-ribosyl)ation during DNA repair, we studied the consequences of a poly(ADP-ribose) synthesis block and, therefore, the consequences of the presence of persisting breaks on the normal passage of cells through the cell cycle. For this purpose HeLa cells were treated with DMS in the presence or absence of benzamide and the cell cycle distribution was analyzed 24 h later. At a lower DMS dose (25 μM, 90% colony-forming ability), no effect of DMS alone on the cell cycle distribution was found, but when benzamide was included, an accumulation in G2 occurred (Fig. 5). Application of a moderate DMS dose (100 μM, 50% colony-forming ability) resulted in G2-accumulation in the cells treated with the alkylating agent alone, in contrast to an apparent S-phase block produced by the DMS/benzamide combination. Further experiments analyzing the cell cycle distribution over a period of 6 days revealed that the cells were not actually blocked in the S-phase, but rather passed through the cell cycle much slower (T. Pless, unpublished data).

These data together with the results described in the preceding sections led us to speculate that poly(ADP-ribosyl)ation could be involved in the maintenance or

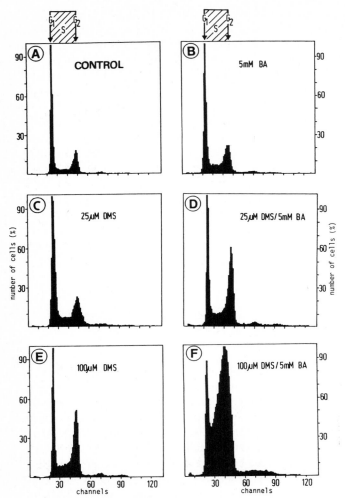

Fig. 5A–F. Cell cycle distribution of DMS ± benzamide-treated cells. HeLa cells were treated in Petri dishes with the indicated amounts of DMS ± 5 mM benzamide. 24 h later, the cells were scraped off with a rubber policeman and fixed with ethanol. After labeling the DNA with bisbenzimide the cell cycle distribution was determined by FACS

restoration of the normal chromatin structure necessary for efficient DNA replication and chromatin condensation before mitosis. Therefore, the inhibiton of poly-(ADP-ribose) formation could conceivably alter the chromatin condensation. Considering that the chromatin condensation is more sensitive than the DNA replication to structural alterations, this would explain the accumulation of cells in the G2 or the S-phase, respectively.

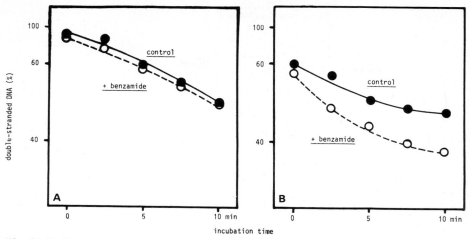

Fig. 6A,B. Susceptibility of HeLa cell chromatin from benzamide-treated cells to nuclease digestion. HeLa cells were labeled overnight with [^{14}C]-thymidine prior to 24 h incubation ± 5 mM benzamide. Nuclei were prepared and incubated with 0.1 U ml^{-1} micrococcal nuclease (A) or 5 U ml^{-1} DNAse I (B). At the times indicated aliquots were removed and the degree of DNA fragmentation was measured by the alkali unwinding technique

Poly(ADP-Ribosyl)ation and Chromatin Structure

In order to study the role of poly(ADP-ribosyl)ation in the maintenance of the normal chromatin structure, we omitted DMS and determined the effect of benzamide alone. This was necessary since the high number of strand breaks brought about by the DMS/benzamide combination would interfere with the structural analysis of the chromatin. However, if poly(ADP-ribosyl)ation is indeed a process involved in chromatin structure maintainance, restoration or alteration, benzamide alone should also influence the chromatin structure.

The first series of experiments were designed to study the accessibility of the chromatin in benzamide-treated vs control cells by measuring the rate of DNA fragmentation induced by nucleases, thereby providing information about changes in the chromatin package. HeLa cells were incubated for 24 h in the presence or absence of benzamide prior to preparation of nuclei according to the procedure of Hewish and Burgoyne [16]; on the basis of the DNA fragmentation these nuclei exhibit the same chromatin integrity as intact cells. The nuclei were then incubated either with 0.1 U ml^{-1} micrococcal nuclease or 5 U ml^{-1} DNase I. At different times aliquots were removed for the determination of the DNA fragmentation with the aid of the alkali unwinding technique. As seen in Fig. 6, no differences in the micrococcal nuclease digestion velocity were found in the nuclei of cells preincubated ± benzamide. By contrast, the digestability of the DNA by DNase I in nuclei of the benzamide-treated cells was increased, suggesting a better accessibility of the chromatin. Since DNase I preferentially splits active DNA sequences [17], we suspected that poly(ADP-ribosyl)ation could be involved in maintaining the structure of the active part of the chromatin.

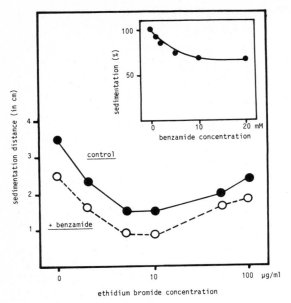

Fig. 7. Effect of ethidium bromide on the sedimentation of nucleoids from cells incubated ± benzamide. HeLa cells in suspension culture were incubated with 5 mM benzamide. 4 h after benzamide addition cells were sedimented, suspended in saline, and applied to isokinetic sucrose gradients (containing bis-benzimide and increasing amounts of ethidium bromide) overlaid with a lysis buffer containing Triton X-100. 15 min later the gradients were centrifuged for 30 min at 16,000 rpm. The sedimentation distance was determined by visual inspection under UV light. *Inset:* Effect of benzamide on the sedimentation of nucleoids from benzamide-treated cells. HeLa cells were incubated for 4 h with increasing amounts of benzamide and th the nucleoid sedimentation rate was determined in bis-benzimide containing sucrose gradients as described above

The active chromatin is proposed to be associated with a network of fibers called the nuclear matrix and this DNA/matrix complex is presumably the site of DNA replication [18] and transcription [19]. Therefore, we asked whether or not poly(ADP-ribosyl)ation is involved in the organization of the DNA/nuclear matrix complex. If this were the case, inhibitors of poly(ADP-ribosyl)ation should alter the structure of the DNA/matrix complex. This was studied with the nucleoid technique described by Cook and Brazell [20] which measures the sedimentation rate of the nuclei-like (nucleoids) DNA/nuclear matrix complexes in sucrose gradients after lysis of the cell by high salt/detergent treatment on top of the gradient. When HeLa cells were treated for 4 to 5 h with increasing amounts of benzamide it became evident that the sedimentation of the nucleoids from benzamide-treated cells were diminished in a dose-dependent fashion reaching a maximum between 3 and 4 mM (Fig. 7, inset). This lowering of the sedimentation rate could either be due to a higher degree of DNA fragmentation or to a change in the supercoiling of the histone-stripped DNA. This can be differentiated by titration of the sedimentation rate with the DNA intercalator ethidium bromide. This intercalation decreases the natural negative supercoiling of the DNA resulting in a full relaxation of the DNA at the appropriate ethidium bromide concentration. Further increases of the ethidium bromide concentration then induce positive supercoiling. If a decrease in the sedimentation rate is due to a higher degree of DNA fragmentation, the ethidium bromide titration should result in the same minimal sedimentation rate with a decrease in the maximal sedimentation with high doses. By contrast, if the decreased sedimentation rate is due to a change in the supercoiling, the concentration of ethidium bromide causing minimal sedimentation should be reached earlier. When the sedimentation of nucleoids from cells incubated in the presence or absence of benzamide was titrated with increasing amounts of ethidium bromide it became obvious that the effect of benzamide cannot be explained by either model, since the ratio between the sedimentation of nucleoids from benzamide-treated vs control cells remained constant over the entire range of ethidium bromide concentrations (Fig. 7). Apparently, the altered sedimentation of nucleoids from benzamide-treated cells is explainable by a more complex mechanism which changes the structure of the DNA/nuclear matrix structure, for example, a change of the length of the DNA loops attached to the nuclear matrix. Both types of experiments, the nuclease digestion as well as the nucleoid sedimentation, argue for an involvement of poly(ADP-ribosyl)ation in maintainance of the chromatin structure, possibly in the organization of the DNA/nuclear matrix complex.

Conclusions

As pointed out earlier, the data we have described favor the idea that poly(ADP-ribosyl)ation has a structural function in the maintenance or restoration of the chromatin integrity. Due to the rapid turnover of the polymer [21–23], the high consumption of NAD under conditions where the DNA is damaged argues for a modulation of an energy-rich structure rather than an activity controlling protein modification.

Considering that the active part of the chromatin, in contrast to the majority, is under torsional tension [24], the role of poly(ADP-ribosyl)ation could be inhibition of uncoiling induced by the incision or restoration of torsional tension. On the other hand, it is also conceivable that poly(ADP-ribosyl)ation enhances the accessibility of the damaged chromatin to repair-enzyme complexes. Regardless, poly(ADP-ribosyl)ation would not be directly involved in one of the repair steps. Therefore, we conclude that poly(ADP-ribosyl)ation is a general chromatin function serving to maintain, restore, or alter the normal chromatin structure. This would explain the involvement of poly(ADP-ribosyl)ation in differentiation and adaptive processes [22, 25, 26].

References

1. Durkacz BW, Omidiji O, Gray DA, Shall S (1980) (ADP-ribose)$_n$ participates in DNA excision repair. Nature (London) 283:593–596
2. Durkacz B, Irwin J, Shall S (1981) The effect of inhibition of (ADP-ribose)$_n$ biosynthesis on DNA repair assayed by the nucleoid technique. Eur J Biochem 121:65–69
3. James MR, Lehmann AR (1982) Role of poly(adenosine diphosphate ribose) in deoxyribonucleic acid repair in human fibroblasts. Biochemistry 21:4007–4013
4. Ohashi Y, Ueda K, Kawaichi M, Hayaishi O (1983) Activation of DNA ligase by poly(ADP-ribose) in chromatin. Proc Natl Acad Sci USA 80:3604–3607
5. Durrant LG, Margison GP, Boyle JM (1981) Effects of 5-methylnicotinamide on mouse L1210 cells exposed to N-methyl-N-nitrosourea: mutation induction, formation and removal of methylation products in DNA, and unscheduled DNA synthesis. Carcinogenesis 2:1013–1017
6. Durkacz BW, Irwin J, Shall S (1981) Inhibition of (ADP-ribose)$_n$ biosynthesis retards DNA repair but does not inhibit DNA repair synthesis. Biochem Biophys Res Commun 101:1433–1441
7. Creissen D, Shall S (1982) Regulation of DNA ligase activity by poly(ADP-ribose). Nature (London) 296:271–272
8. Kreimeyer A, Wielckens K, Adamietz P, Hilz H (1984) DNA repair-associated ADP-ribosylation in vivo. J Biol Chem 259:890–896
9. Adamietz P, Rudolph A (1984) ADP-ribosylation of nuclear proteins in vivo. J Biol Chem 259: 6841–6846
10. Whish WJD, Davies MI, Shall S (1975) Stimulation of poly(ADP-ribose) polymerase activity by the anti-tumour anitbiotic, streptozotocin. Biochem Biophys Res Commun 65:722–730
11. Miller EG (1975) Stimulation of nuclear poly(adenosine diphosphate ribose) polymerase activity from HeLa cells by endonucleases. Biochim Biophys Acta 395:191–200
12. Berger NA, Sikorski GW, Petzold SJ, Kurohara KK (1979) Association of poly(adenosine diphosphoribose) synthesis with DNA damage and repair in normal human lymphocytes. J Clin Invest 63:1164–1171
13. Wielckens K, Bredehorst R, Adamietz P, Hilz H (1981) Protein-bound polymeric and monomeric ADP-ribose residues in hepatic tissues. Eur J Biochem 117:69–74
14. Ahnstrom G, Erixon K (1981) Measurement of strand breaks by alkaline denaturation and hydroxyapatite chromatographie. In: Friedberg EC, Hanawalt PC (eds) DNA repair. Marcel Dekker, New York, pp 403–418
15. Kohn KW, Ewig RAG, Erickson LC, Zwelling LA (1981) Measurement of strand breaks and crosslinks by alkaline elution. In: Friedberg EC, Hanawalt PC (eds) DNA repair. Marcel Dekker, New York, pp 379–401
16. Hewish DR, Burgoyne LA (1973) The calcium dependent endonuclease activity of isolated nuclear preparations. Relationship between its occurence and the occurence of other classes of enzymes found in nuclear preparations. Biochem Biophys Res Commun 52:475–781
17. Weintraub H, Groudine M (1976) Chromosomal subunits in active genes have an altered conformation. Science 193:848–856
18. Pardoll DM, Vogelstein B, Coffey DS (1980) A fixed site of DNA replication in eucaryotic cells. Cell 19:527–536
19. Jackson DA, McCready SJ, Cook PR (1981) RNA is synthesized at the nuclear cage. Nature (London) 292:552–555
20. Cook PR, Brazell IA (1975) Supercoils in human DNA. J Cell Sci 19:161–279
21. Wielckens K, Schmidt A, George E, Bredehorst R, Hilz H (1982) DNA fragmentation and NAD depletion. J Biol Chem 257:12872–12877
22. Wielckens K, George E, Pless T, Hilz H (1983) Stimulation of poly(ADP-ribosyl)ation during Ehrlich ascites tumor cells "starvation" and suppression of concomitant DNA fragmentation by benzamide. J Biol Chem 258:4098–4104
23. Jacobson EL, Antol KM, Juarez-Salinas H, Jacobson MK (1983) Poly(ADP-ribose) metabolism in ultraviolett irradiated human fibroblasts. J Biol Chem 258:103–107

24. Sinden RR, Carlson JO, Pettijohn DE (1980) Torsional tension in the DNA double helix measured with trimethylpsoralen in living E. coli cells: analogous measurements in insect and human cells. Cell 21:773–783

25. Nagle WA, Moss AJ (1983) Inhibitors of poly(ADP-ribose) synthetase enhance the cytotoxicity of 42°C and 45°C hyperthermia in cultured Chinese hamster cells. Int J Radiat Biol 44:475–481

26. Caplan AI, Rosenberg MJ (1975) Interrelationship between poly(ADP-rib) synthesis, intracellular NAD levels, and muscle or cartilage differentiation from mesodermal cells of embryonic chick limb. Proc Natl Acad Sci USA 72:1852–1857

Isolation and Identification of Mono- and Poly(ADP- Ribosyl) Proteins Formed in Intact Cells in Association with DNA Repair

PETER ADAMIETZ[1]

Introduction

The biological role of poly(ADPR) has been related to the regulation of several processes including DNA replication, DNA repair, transcription as well as cellular differentiation. We have been engaged in the investigation of the various nuclear acceptor proteins that are modified by mono- and poly(ADPR) in association with chromatin functions. Recently we found that histone H2B serves as one of the major acceptors to be ADP-ribosylated in living rat hepatoma cells in response to a brief treatment with dimethyl sulfate (DMS) [1]. Under these conditions the cellular level of protein-bound poly(ADPR) was raised 20-fold, while that of the substrate NAD dropped to about half of its original level. At the same time a considerable increase in the degree of DNA damage could be detected using the nucleoid sedimentation technique [2].

Here I present evidence that the major nonhistone proteins being ADP-ribosylated in addition to H2B consist of a series of species with relatively high apparent molecular masses (170, 110, 100, 86, 66 and 45 kD) that all show a strong affinity to DNA. The most abundant conjugate, the 110 kD peptide, could be identified as the automodified poly(ADPR) synthase, another as topoisomerase I.

The Endogenous Mono- and Poly(ADPR) Nonhistone Proteins of Yoshida Hepatoma Cells can be Isolated by Aminophenyl Boronic Acid Chromatography

In order to assess the nuclear nonhistone proteins being ADP-ribosylated in association with DNA repair we have isolated the endogenous conjugates formed in DMS-treated cells by applying aminophenyl boronic acid chromatography to 0.2 M sulfuric acid precipitates of isolated nuclei [1]. With hepatoma cells, this procedure provided roughly 80% of the newly-formed conjugates in a highly purified form. The isolated fraction of histone-depleted (ADP-ribosyl)$_n$ proteins comprised about 1% of total nuclear

---

1 Institut für Physiologische Chemie, Universität Hamburg, Martinistr. 52, 2000 Hamburg 20, FRG

ADP-Ribosylation of Proteins
(ed. by F.R. Althaus, H. Hilz, and S. Shall)
© Springer-Verlag Berlin Heidelberg 1985

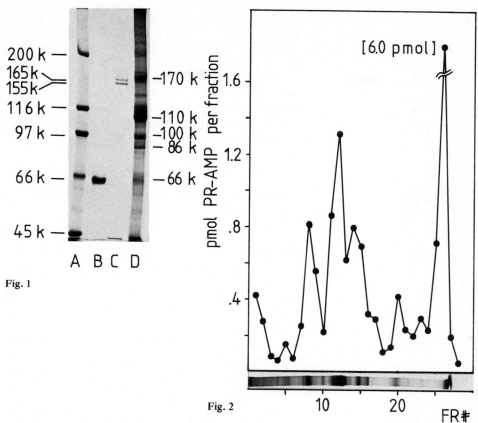

Fig. 1

Fig. 2

Fig. 1. Electrophoretic analysis of $(ADPR)_n$ nonhistone proteins from DMS-treated hepatoma cells: 10 ml of Yoshida hepatoma AH 7974 cells ($3-5 \times 10^8$ cells) grown for 7 days in the peritoneum of male Wistar rats (200 g) were withdrawn and incubated with 40 μl DMS dissolved in dimethyl sulfoxide (DMSO) for 20 min at 37°C. Nuclei were prepared according to [1] and freed from histones by precipitation with cold 0.2 M sulfuric acid. After dissolving the insoluble material in 6 M guanidine hydrochloride, 0.05 M sodium phosphate, pH 6.0, 5 mM DTT, the bulk of the DNA was removed by ultracentrifugation (15 h at 210,000 g and at 4°C). The $(ADPR)_n$ proteins were isolated utilizing aminophenyl boronic acid agarose chromatography as described recently [1]. *Lane D:* Electrophoretic analysis was performed with 10 μg of the conjugates using a 7% polyacrylamide gel according to Laemmli [8]. The gels were stained with Coomassie brilliant blue. *Lanes A−C:* Marker proteins used for estimation of apparent molecular weights: Myosin (200 kD), $\beta\beta'$-subunits of RNA-polymerase from *E. coli* (155, 165 kD), β-galactosidase from *E. coli* (116 kD), phosphorylase b (97 kD), bovine serum albumin (66 kD), ovalbumin (45 kD)

Fig. 2. Covalent attachment of poly(ADPR) to the nonhistone proteins being retained by boronate agarose chromatography. The fraction of nonhistone proteins being retained by the boronate resin was separated by sodium dodecyl sulfate electrophoresis (cf. Fig. 1). The stained gel was sliced and poly(ADPR) determined in individual fractions utilizing a specific radioimmunoassay [1]

protein and contained 0.4 nmol ADPR/mg cellular DNA in a polymeric form and 2 nmol mono(ADPR) mg^{-1} cellular DNA. When subjected to sodium dodecyl sulfate electrophoresis it became apparent that this fraction comprised several different proteins of apparent molecular masses ranging from 45 to 170 kD (Fig. 1). All of them carried covalently attached poly(ADPR) as can be seen from the distribution of the polymer on the gel (Fig. 2). None of the proteins seen on the gel in Fig.1 could be isolated by boronate chromatography when the DMS-induced synthesis of poly(ADPR) was suppressed by the additional presence of an effective inhibitor of the poly(ADPR) synthase, such as 10 mM benzamide (not shown).

Poly(ADPR) Synthase is the Major Poly(ADPR) Nonhistone Acceptor in the DMS-Treated Yoshida Hepatoma Cell

The apparent size of the major conjugate band (Fig. 1) suggested that it might represent poly(ADPR) synthase itself. Unequivocal proof of this was offered by the demonstration of its intrinsic enzymic activity. This was possible only after the completely denatured enzyme had been partially renatured, following essentially the protocol of Hager and Burgess [3]. As shown in Fig. 3A renaturation of the ADP-ribosylated poly-(ADPR) synthase took place gradually reaching a maximum after about 2 h at 22°C. The later decline in enzymic activity is certainly due to adverse effects which also influence the activity of the native enzyme when diluted to a similar extent (Fig. 3A). Finally, when the renaturation procedure was applied to individual fractions obtained by preparative gel electrophoresis of the purified poly(ADPR) nonhistone proteins it became evident that the bulk of the poly(ADPR) synthase was associated with the 110 kD peptide (fraction 2 in Fig. 3B). In addition, a significant amount of enzyme activity was found also in fraction 5 containing a faster moving band suggesting that it may represent a proteolytic product of the 110 kD enzyme.

Topoisomerase I is ADP-Ribosylated in the Living Hepatoma Cell in Response to DNA Damage

The second enzyme that could be identified among the endogenous acceptors for poly(ADPR) was topoisomerase I. Figure 4 shows the relaxation of supercoiled form I DNA (of phage phi x 174) catalyzed by a renatured sample of the isolated poly(ADPR) proteins. Although this assay was not performed with material extracted from the gel but with the whole conjugate fraction, it is very likely that the topoisomerase I is represented by the 100 kD band as this value agrees very well with the molecular size of eukaryotic type I topoisomerases.

The experiment shown in Fig. 4 also suggests that DNA endonucleases were not among the poly(ADPR) proteins since neither degradation nor nicking of the circular DNA occurred. The possible presence of topoisomerase II as argued from a 170 kD

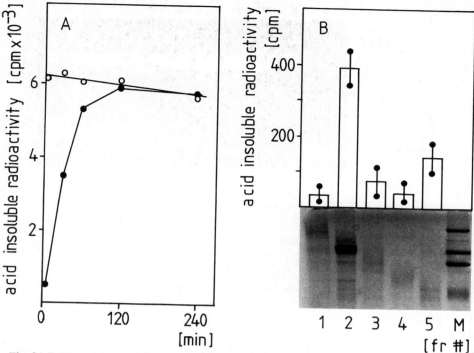

Fig. 3A,B. Identification of automodified poly(ADPR) synthase after renaturation. **A** 0.2–0.5 mg of poly(ADPR) nonhistones dissolved in 100 μl of 6 M guanidine hydrochloride, 50 mM Tris/HCl, pH 7.9, 1 mM DTT were renatured using the method of Hagar and Burgess [3]. It includes dilution by a 50-fold excess of renaturation buffer (50 mM Tris/HCl, pH 7.9, 0.15 M NaCl, 0.1 mg ml⁻¹ bovine serum albumin, 0.1 mM EDTA, 20% glycerol) and subsequent incubation at 22°C. The reaction mixture used for determination of poly(ADPR) synthase activity contained in a total volume of 3.0 ml: 100 mM Tris/HCl, pH 8.0, 10 mM MgCl₂, 1 mM DTT, 60 μg of calf thymus DNA (Sigma type I), 60 μg of lysine-rich histones, 100 μM [³H]NAD (3.9 Ci mol⁻¹) and 1.5 ml of renatured poly(ADPR) nonhistone proteins. Incubation was for 30 min at 25°C. At indicated times 250 μl aliquots were withdrawn, added to 50 μl of bovine serum albumin (10 mg ml⁻¹) and precipitated by 10% trichloroacetic acid (*closed circles*). Kinetic analysis of the native synthase as purified from hepatoma cells by DNA agarose chromatography is also shown (*open circles*). For comparison, these samples were diluted with renaturation buffer. **B** 200 μg of poly(ADPR) non-histone proteins were separated on a preparative gel and the individual components extracted after slicing. From five consecutive fractions the detergent was removed by acetone precipitation and the poly(ADPR) synthase activity determined as described above. In addition, 10 μl aliquots of the isolated fractions were analyzed on an analytical gel. *M* = marker proteins (*from top to bottom*): Myosin, β-galactosidase, phosphorylase b, bovine serum albumin

band in the conjugate preparation could not be substantiated. Adding ATP and spermidine to the reaction mixture, which should either enhance the rate of relaxation or stimulate DNA catenation gave no response (Fig. 4) [5]. These conclusions are, however, preliminary since a negative result may also be a consequence of inadequate renaturation.

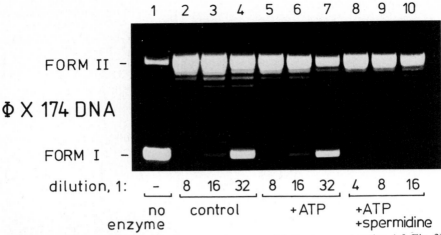

Fig. 4. Identification of ADP-ribosylated topoisomerase I following renaturation (cf. Fig. 3). The reaction mixture for the topoisomerase I assay contained in a total volume of 15 μl 50 mM Tris/ HCl, pH 7.5, 50 mM KCl, 10 mM MgCl$_2$, 0.5 mM DTT, 0.1 mM EDTA, 0.03 mg ml^{-1} BSA, 0.5 μg of form I phi $\times$ 174 DNA (BRL), and 10 μl of renatured and diluted poly(ADPR) nonhistone proteins. ATP and spermidine were included where indicated. The incubation was carried out for 30 min at 37°C and was terminated by the addition of 2 μl of proteinase K (Merck) (2 mg ml^{-1} of 1% sodium dodecyl sulfate). Electrophoresis was performed for 15 h at 50 V using an 1.4% agarose gel and 1 μg ml^{-1} ethidium bromide for staining. *Lane 1:* control without enzyme; *lanes 2–4:* 8-, 16-, and 32-fold diluted samples; *lanes 5–7:* 8-, 16-, and 32-fold diluted samples incubated in the presence of 1 mM ATP; *lanes 8–10:* 4-, 8-, and 16-fold diluted samples incubated in the presence of 1 mM ATP and 5 mM spermidine

ADP-Ribosylation of Nuclear Proteins in Association with DNA Repair Shows Species-Specific Differences

The alkylation-induced ADP-ribosylation of nuclear proteins as described above does not seem to be unique for the Yoshida hepatoma cell line AH7974. The endogenous (ADPR)$_n$ proteins isolated from DMS-treated normal hepatocytes showed principally the same molecular weights as observed in the tumor cell, suggesting that they represent functionally equivalent proteins (Fig. 5). On the other hand, the two cell types differed significantly with respect to the relative amounts of the newly-formed conjugates. Hence, the most abundant acceptor of the tumor cell, poly(ADPR) synthase, was found to be ADP-ribosylated less effectively in normal hepatocytes as compared to peptides with 90 to 100 kD (Fig. 5B).

What is the Role of Poly(ADPR) Synthesis in DNA Repair?

Since it is difficult to draw conclusions about the possible mechanism of poly(ADPR) function in association with DNA repair before having identified all acceptor proteins,

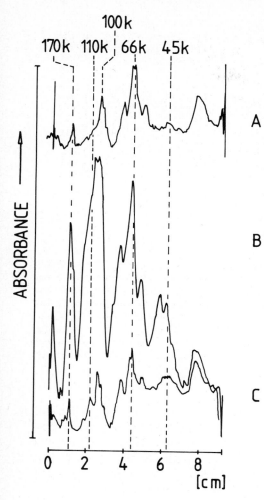

Fig. 5A–C. Effect of treatment with dimethyl sulfate on the formation of poly(ADPR) non-histones in cultured rat hepatocytes. Hepatocytes were obtained from thyroidectomized male Wistar rats (180 g) and seeded on fibronectin-coated petri dishes (5×10^6 cells/ 100 mm dish). They were kept at 37°C applying 10 ml per dish of hormone and serum free medium [9] except for the first 3 h when dexamethasone (10^{-7} M) and insulin (10^{-8} M) were present. After 44 h in culture 40 µl of 0.1 M DMS (dissolved in DMSO) were added and the cells incubated for another 20 min (**B**). Control dishes were treated with DMSO alone (**A**) or with DMS plus 10 mM benzamide (**C**). Poly(ADPR) non-histones were isolated as described in the legend of Fig. 1. Electrophoretic analysis was carried out in the presence of sodium dodecyl sulfate and urea using Tris/phosphate buffer at pH 6.0 [1]. Samples applied to the 7% gel corresponded to the content of 1×10^7 cells. Scans were performed from silver-stained gels (Bio-Rad kit) using a Hoefer densitometer (Hoefer Scientific, San Francisco)

it is probably relevant that all the isolated conjugates showed a pronounced affinity to DNA. Following renaturation they were tightly retained on a DNA-agarose column and could only be eluted by 0.4–0.6 M NaCl containing buffers (not shown).

Thus, any model designed to explain the biological role of poly(ADPR) has to consider that both structural constituents like histone H2B as well as DNA-binding enzymes like topoisomerase I and poly(ADPR) synthase are modified by poly(ADPR). Assuming that poly(ADPR) synthase remains attached to the chromatin when activated by a DNA break, I favor the idea that poly(ADP-ribosyl)ation of the isolated acceptors is primarily a function of their accessibility to the synthase. The acceptor proteins may merely function as matrix to permit the accumulation of relatively large amounts of poly(ADPR) at distinct sites of the chromatin adjacent to the stimulating event. Thus, poly(ADPR), probably in the form of a three-dimensional network, may represent a specific tool to introduce changes into the chromatin structure.

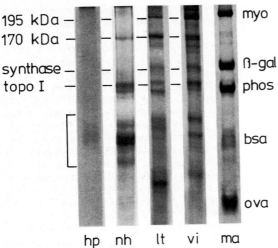

Fig. 6. Comparative analyses of endogenous $(ADPR)_n$ nonhistone proteins from different hepatic tissues. The endogenous poly(ADPR) nonhistones were isolated from different types of untreated liver cells. Electrophoretic analyses were performed on 7% polyacrylamide gels using the Tris/ phosphate buffer system [1] at pH 6.0 in the presence of 0.1% sodium dodecyl sulfate and 6 M urea. Gels were stained with silver stain (Bio-Rad, München). Samples applied to the gel correspond to the yield obtained from 10^7 cells. *hp* Yoshida hepatoma AH 7974; *nh* normal hepatocytes after 44 h in primary culture; *lt* normal liver tissue cells from whole animals; *ma* marker proteins from *top to bottom:* Myosin (200 kD), β-galactosidase (116 kD), phosphorylase b (92 kD), bovine serum albumin (66 kD)

As a consequence, DNA-binding enzymes may be prevented from interaction with the DNA as recently assumed for topoisomerase I [4] from in vitro data. The interaction between DNA and the nuclear matrix may also be affected. The ADP-ribosylation of nuclear lamina proteins [6] and of actin [7] have been suggested from in vitro studies and seem also likely to occur in vivo. $(ADPR)_n$ protein conjugates with corresponding apparent molecular masses (66 kD, 45 kD) are present among the isolated species of DMS-treated hepatocytes (Fig. 5) and hepatoma cells (Fig. 1). Whether the 170 kD and 195 kD proteins also represent components of the nuclear matrix requires further investigation.

Is Poly(ADPR) Synthesis also Involved in Other Chromatin Functions?

The finding that nearly all of the conjugates except modified histone H2B and poly-(ADPR) synthase itself are also present in untreated normal hepatocytes (Figs. 5 and 6), supports the view that these proteins may be involved in other chromatin functions in addition to DNA repair. It is tempting to speculate that the higher degree of ADP-ribosylation of most of these proteins in the normal hepatocyte (Fig. 6 nh and lt) as compared to the hepatoma cell (Fig. 6 hp) may be related to the terminally differentiated state of the normal liver cell.

References

1. Adamietz P, Rudolph A (1984) ADP-ribosylation of nuclear proteins in vivo. J Biol Chem 259: 6841–6846
2. Cook PR, Brazell IA, Jost E (1976) Characterization of nuclear structures containing superhelical DNA. J Cell Sci 22:303–324
3. Hager DA, Burgess RR (1980) Elution of proteins from sodium dodecyl sulfate-polyacrylamide gels, removal of sodium dodecyl sulfate, and renaturation of enzymic activity. Anal Biochem 109:76–86
4. Ferro AM, Higgins NP, Olivera BM (1983) Poly(ADP-ribosylation) of a DNA topoisomerase. J Biol Chem 258:6000–6003
5. Colwill RW, Sheinin R (1983) ts A1S9 locus in mouse L cells may encode a novobiocin binding protein that is required for DNA topoisomerase II activity. Proc Natl Acad Sci USA 80:4644–4648
6. Song MH, Adolph KW (1983) ADP-ribosylation of nonhistone proteins during the HeLa cell cycle. Biochem Biophys Res Commun 115:938–945
7. Kun E, Romaschin AD, Blasdell RJ, Jackowski G (1981) ADP-ribosylation of nonhistone chromatin proteins in vivo and of actin in vitro and effects of normal and abnormal growth conditions and organ-specific hormonal influences. In: Holzer H (ed) Proceedings of the international Titisee conference on metabolic interconversion of regulatory enzymes. Springer, Berlin Heidelberg New York, pp 280–293
8. Laemmli UK (1970) Cleavage of structural proteins during the assembly of the head of bacteriophage T4. Nature (London) 227:680–685
9. Süßmut W, Höppner W, Seitz H (in press) Permissive action of thyroid hormones in the cAMP-mediated induction of P-enolpyruvate carboxy kinase in hepatocytes in culture. Eur J Biochem

Mono(ADP-Ribosyl) Histone H1 in Alkylated Hepatoma Cells: Unusual Acceptor Site

ANDREAS KREIMEYER and HELMUTH HILZ[1]

1. Introduction

Treatment of cells with DNA-damaging agents like UV irradiation or alkylating chemicals is known to greatly stimulate mono- and poly(ADPR) formation [1, 2] and turnover [3, 4]. Although there are many attempts to elucidate the function of poly(ADP-ribosyl)ation in DNA repair, the role of cellular mono(ADP-ribosyl)ation in the course of DNA damage has received relatively little attention. Recently Kreimeyer et al. [5] were able to demonstrate that histones were extensively mono(ADP-ribosyl)ated in intact DMS-treated hepatoma AH7974 cells. 20–30% of DMS-induced mono(ADPR) was found to be associated with histone H1 and nearly 60% with core histones. Adamietz et al. [6] identified histone H2B as the main acceptor of mono(ADPR) under these conditions. Surprisingly, mono(ADPR)-H1 as well as mono(ADPR)-H2B conjugates seemed to be resistant towards neutral hydroxylamine, indicating a linkage between ADPR and histone that was different from the ester glycosidic bond observed with poly(ADPR) in liver nuclei [5].

This contribution presents data to further characterize the linkage involved. It also analyzes the location of the ADPR group(s) on the histone H1 molecule.

2. Results

2.1 Effect of DMS on Endogenous Mono(ADP-Ribosyl)ation of Proteins

When hepatoma AH7974 cells were incubated in the presence of 400 μM DMS for 30 min at 37°C, total mono(ADPR) residues were augmented by a factor of 1.5. As shown previously [5, 6], nearly all of the additionally synthesized mono(ADPR) groups were associated with histones. There were, however, not only quantitative changes. While in untreated hepatoma cells nearly 60% of total mono(ADPR)-protein conjugates and about 20% of histone H1-associated mono(ADPR) was sensitive towards 0.5 M neutral NH_2OH, treatment of the cells with dimethyl sulfate induced a

1 Institut für Physiologische Chemie der Universität Hamburg, Martinistr. 52, 2000 Hamburg 20, FRG

ADP-Ribosylation of Proteins
(ed. by F.R. Althaus, H. Hilz, and S. Shall)
© Springer-Verlag Berlin Heidelberg 1985

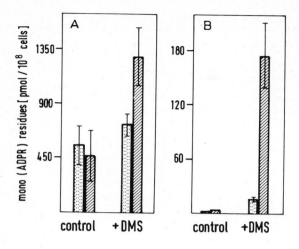

Fig. 1 A,B. NH_2OH sensitive and resistant mono(ADPR) residues in control and DMS-treated cells. **A** Total acid-insoluble fraction; **B** histone H1. *Dotted column:* 0.5 M NH_2OH sensitive; *striped column:* 0.5 M NH_2OH resistant. Mean values ± SD from three determinations

profound change (Fig. 1): total hydroxylamine-sensitive conjugates were reduced to 30–40%, and more than 90% of the H1 conjugates had become resistant towards NH_2OH.

In alkylated cells considerable amounts of free ADPR must be formed in consequence of the highly stimulated turnover of $(ADPR)_n$ proteins [3, 4]. Since free ADPR can form acid-insoluble protein conjugates [7, 8], mono(ADP-ribosyl) histone H1 could have been arisen by such a reaction. We, therefore, analyzed endogenous conjugates with respect to chemical stability and compared it with other conjugates of known structure as well as with "nonenzymic" ADPR-H1 adducts. In addition, location of the ADPR groups on the histone H1 molecule was investigated.

2.2 Chemical Stabilities of Mono(ADPR)-H1-Conjugates

Recently, several mono(ADPR)-protein conjugates differing with respect to the acceptor amino acid have been described. Hilz et al. [8] applied a set of chemical treatments, which allowed the differentiation of these types of conjugates. Using these methods, we investigated the linkage between histone H1 and mono(ADPR) groups. The results are shown in Table 1.

Mono(ADPR)-H1 conjugates as obtained from DMS-treated cells were completely resistant against 0.2 M NH_2OH, except for a small fraction that was released within 10 min (4–5%). Even in the presence of 3 M NH_2OH the conjugates were highly stable: after 3 h only 20% of ADPR residues were released. The half-life under these conditions was calculated to be 19 h. These conjugates also resisted treatment with picrylsulfonate at pH 8.2, as well as a 10 min exposure to 0.1 M NaOH at room temperature (8% release of ADPR). When "nonenzymic" histone H1 conjugates were synthesized by reaction of (^{3}H)ADPR with H1 and subjected to the same stability tests, a completely different pattern of ADPR release was found (Table 1).

Extension of these studies to the stability pattern of other "enzymic" and "nonenzymic" ADPR conjugates (unpublished experiments) revealed that ADP-ribosyl

Table 1. Chemical stabilities of histone H1 mono(ADPR) conjugates isolated from DMS-treated cells or synthesized nonenzymically by reaction of free [^{3}H]ADPR and H1. (For details see [16])

Treatment			Half-life (h)	
Chemical	Conc.	pH	Endogenous conjugates	"Nonenzymic" conjugates
NH$_4$Cl	0.2 M	7.0	180	3
NH$_2$OH	0.2 M	7.0	>180	10
NH$_4$Cl	3.0 M	7.0	61	4
NH$_2$OH	3.0 M	7.0	19	3
Hepes	0.1 M	8.2	>180	0.8
Picrylsulf.	0.1 M	8.2	>180	0.4

histone H1 (as well as ADPR-H2B) as formed in hepatoma cells in response to alkylating agents is neither the result of a nonenzymic reaction with free ADPR, nor is it linked through a carboxyl group or an arginine residue.

On the basis of these data, it is postulated that formation of mono(ADPR) histones in alkylated hepatoma cells occurs by the action of a specific (nuclear) ADP-ribosyl transferase, or by a chemical rearrangement of formerly more energy-rich bonds.

2.3 Location of Mono(ADPR) Residues in the H1 Molecule

The tertiary structure of histone H1 shows a globular "head" (residues 42–120) with a N-terminal "nose" (residues 1–42) and a C-terminal "tail" (residues 120–213) [9]. Specific functions have been assigned to these structural fragments [10, 11]. Whereas the globular part is believed to correctly align the H1 molecule to chromatin, the N-terminal and especially the C-terminal extensions were found to interact strongly with DNA. Poly(ADPR) residues formed in vitro on histone H1 were found to be linked to the N-terminal Glu 2 and Glu 14 and to C-terminal Lys 213 [12, 13], although in Ehrlich ascites tumor cells most of the (ADPR)$_n$ residues were present at the C-terminus [14].

To determine the intramolecular location of mono(ADPR) residues in histone H1, the conjugates extracted from alkylated hepatoma cells were cleaved with N-bromosuccinimide [15]. Two fragments were generated: an N-terminus (residues 1–72) and a C-terminus (residues 73–213). These fragments were separated by gel electrophoresis and the distribution of mono(ADPR) residues was determined with the aid of a radio-immunoassay [5]. The conjugate was also treated with thrombin, which leads to a scission at amino acids 121/122 (globular region [9]). The results of these experiments show that the H1 molecule as obtained from intact cells is ADP-ribosylated exclusively at the C-terminal, nonglobular tail (residues 122–213) which functions to anchor the H1 molecule to DNA [11].

3. Discussion

Treatment of hepatoma AH7974 cells with dimethylsulfate results in a drastic stimulation of mono(ADP-ribosyl)ation that surpassed that of total polymeric ADPR residues by a factor of 3. Assuming a mean chain length of 10, 30 times more acceptor sites will be occupied by monomeric ADPR residues than by poly(ADPR) chains. These acceptor sites are concentrated; 60% in histones H1 and H2B [5, 6]. The present report shows that the linkage of H1-associated mono(ADPR) is neither of the ester type nor does it involve an arginine residue. Furthermore, a comparison with nonezymically formed ADPR-H1 conjugates exclude a nonezymic generation of mono(ADPR)-H1 in DMS-treated cells.

The exclusive modification of the C-terminus of the H1 molecules by ADPR residues suggests that the mono(ADP-ribosyl)ation alters the interaction of the linker protein with DNA and, therefore, changes chromatin structure leading presumably to the relaxation of limited chromatin regions. These interpretations are further supported by preliminary results concerning H2B-mono(ADPR) conjugates which exhibited the same chemical stability as the H1 conjugates. In contrast to H1, H2B-associated mono(ADPR) residues were found exclusively in the N-terminal fragment. This interesting observation apparently relates to the fact that the N-terminus of H2B is equivalent to the C-terminus of H1 with respect to the interaction with DNA [11]. Mono(ADP-ribosyl)ation of the two major acceptors in response to alkylation, therefore, may affect primarily the interaction of histones H1 and H2B with DNA.

References

1. Juarez-Salinas H, Sims HL, Jacobson MK (1979) Poly(ADP-ribose) levels in carcinogen-treated cells. Nature (London) 282:740
2. Wielckens K, George E, Pless T, Hilz H (1983) Stimulation of poly(ADP-ribosyl)ation during Ehrlich ascites tumor cell "starvation" and suppression of concomitant DNA fragmentation by benzamide. J Biol Chem 285:4098–4104
3. Wielckens K, Schmidt A, George E, Bredehorst R, Hilz H (1982) DNA fragmentation and NAD depletion. J Biol Chem 257:12872–12877
4. Jacobson EL, Antol KM, Juarez-Salinas H, Jacobson MK (1983) Poly(ADP-ribose) metabolism in ultraviolet irradiated human fibroblasts. J Biol Chem 258:103–107
5. Kreimeyer A, Wielckens K, Adamietz P, Hilz H (1984) DNA repair-associated ADP-ribosylation in vivo. J Biol Chem 259:890–896
6. Adamietz P, Rudolph A (1984) ADP-ribosylation of nuclear proteins in vivo. J Biol Chem 259:6841–6846
7. Kun E, Chang ACY, Sharma ML, Ferro AM, Nitecki D (1976) Covalent modification of proteins by metabolites of NAD⁺. Proc Natl Acad Sci USA 73:3131–3135
8. Hilz H, Koch R, Fanick W, Klapproth K, Adamietz P (1984) Nonenzymic ADP-ribosylation of specific mitochondrial polypeptides. Proc Natl Acad Sci USA 3929–3933
9. Hartman PG, Chapman GE, Moss T, Bradbury EM (1977) Studies on the role and mode of operation of the very-lysine-rich histone H1 in eukaryote chromatin. Eur J Biochem 77:45–51

10. Allan J, Hartman PG, Crane-Robinson C, Aviles FX (1980) The structure of histone H1 and its location in chromatin. Nature (London) 288:675
11. Böhm L, Crane-Robinson C (1984) Proteases as structural probes for chromatin: the domain structure of histones. Biosci Rep 4:365–386
12. Riquelme PT, Burzio LO, Koide SS (1979) ADP ribosylation of rat liver lysine-rich histone in vitro. J Biol Chem 254:3018–3928
13. Ogata N, Ueda K, Kagamiyama H, Hayaishi O (1980) ADP-ribosylation of histone H1. J Biol Chem 255:7616–7620
14. Braeuer HC, Adamietz P, Nellessen U, Hilz H (1981) ADP-ribosylated histone H1. Eur J Biochem 114:63–68
15. Sherod D, Johnson G, Chalkley R (1974) Studies on the heterogeneity of lysine-rich histones in dividing cells. J Biol Chem 249:3923–3931
16. Kreimeyer A, Adamietz P, Hilz H (1985) Alkylation-induced mono(ADP-ribosyl) histones H1 and H2B. Biol Chem Hoppe-Seyl 366:537–554

ADP-Ribosylation Reactions in Biological Responses to DNA Damage

ELAINE L. JACOBSON, JANICE Y. SMITH, VIYADA NUNBHAKDI,
and DEBRA G. SMITH[1]

Abbreviations:

MBA   3-methoxybenzamide
MNNG N-methyl-N'-nitro-N-nitrosoguanidine
PLDR potentially lethal damage repair

Introduction

The synthesis and degradation of poly(ADP-ribose) is a rapid response to DNA damage [1, 2]. The focus of our recent studies has been to investigate the role of this metabolism in cellular responses to DNA damage, including cell death, mutagenesis, and malignant transformation. All of these biological responses to DNA damage are interrelated with DNA repair mechanisms, however, the relationships are poorly understood. We describe here recent studies in the in vitro transformation model, the C3H10T1/2 cell line. This cell line has been utilized extensively for quantitative studies of chemical carcinogenesis, mutational induction of ouabain resistance, and DNA repair. We have used inhibitors of ADP-ribosylation reactions to examine the possible involvement of poly-(ADP-ribose) metabolism in responses of C3H10T1/2 cells to N-methyl-N'-nitro-N-nitrosoguanidine (MNNG) treatment.

ADP-Ribosylation Reactions and Cytotoxic Responses to DNA Damage

Previous work from our laboratory [3] has shown that 3-methoxybenzamide (MBA) at 1 mM effectively inhibits poly(ADP-ribose) synthesis in intact C3H10T1/2 cells. We have investigated the effect of MBA on cell survival in exponentially dividing and quiescent cells following MNNG treatment. Table 1 shows the effect on survival of incubating both dividing and quiescent cells with MBA for 48 h after DNA damage. The data show that the inhibitor was strongly cocytotoxic in dividing, but not in

---

1   Department of Biology, Texas Woman's University, Denton, TX 76204, USA

ADP-Ribosylation of Proteins
(ed. by F.R. Althaus, H. Hilz, and S. Shall)
© Springer-Verlag Berlin Heidelberg 1985

Table 1. Cell survival in dividing and quiescent C3H10T1/2 cells

Treatment	Dividing cells			Quiescent cells		
	Colonies[a]	Plating eff. %	RCFA[b] %	Colonies	Plating eff. %	RCFA %
Control	295	81	100	223	56	100
MNNG[c]	75	76	27	92	51	44
MBA	334	92	113	194	42	115
MNNG and MBA[d]	0	67	0	86	49	43

[a] Numbers shown are colonies per five dishes
[b] Denotes relative colony forming ability (no. of colonies/plating eff. $\times$ 300 $\div$ no. of colonies in control/plating eff. $\times$ 300) $\times$ 100%
[c] Cells were treated with 34 μM MNNG for dividing cells and 68 μM MNNG for quiescent cells
[d] MBA was present for 48 h following treatment at which time cells were subcultured for colony-forming assays

quiescent cells. A number of studies have shown that the presence of ADP-ribosylation inhibitors following DNA damage results in an increased cytotoxicity [3–6] and an increased number of DNA strand breaks [6–8]. These strand breaks have been attributed to an inhibition of DNA repair which result in the cocytotoxic effect of ADP-ribosylation inhibitors. These observations have led to the hypothesis that ADP-ribosylation is required for recovery from DNA damage because it regulates the ligation step of DNA excision repair [9]. The presence of MBA in C3H10T1/2 cells caused the appearance of an additional number of DNA strand breaks during the entire 48 h period of DNA repair following MNNG in both dividing and quiescent cells (Fig. 1). Yet, during this 48 h period following MNNG treatment, quiescent C3H10T1/2 cells effected potentially lethal damage repair (PLDR) (Fig. 2). The process of PLDR in quiescent cells has been attributed exclusively to DNA excision repair, since post-replication repair is not occurring under these conditions. Thus, the inhibition of ADP-ribosylation has no detectable direct effect on DNA excision repair as measured by PLDR in quiescent cells.

ADP-Ribosylation Requirements in Cycling Cells

Poly(ADP-ribose) synthesis is an early event following MNNG treatment [3]. Yet our earlier studies showed that dividing C3H10T1/2 cells were sensitive to inhibitors of ADP-ribosylation for up to 36 h following MNNG treatment [3]. Investigations of macromolecular syntheses during this time frame revealed that a selective inhibition in DNA synthesis occurred beginning approx. 16 h after MNNG treatment in the presence of MBA (Fig. 3), while no alterations in RNA or protein synthesis were detected. Therefore, we employed DNA flow cytofluorimetric and autoradiographic analyses of synchronized cultures over a 72 h time period following DNA damage to determine the effect of ADP-ribosylation inhibitors on cell cycle progressions following DNA

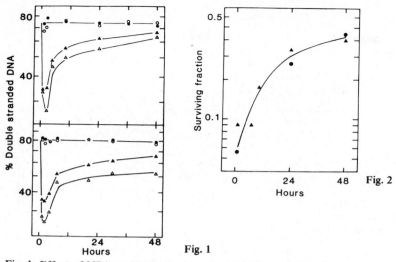

Fig. 2

Fig. 1

Fig. 1. Effect of MBA on DNA strand breaks in dividing and quiescent cells. Cells were prelabeled for 24 h with 0.05 μCi ml^{-1} [^{14}C]thymidine, treated with ($\bullet$) control medium, ($\circ$) 1 mM MBA, ($\blacktriangle$) 17 μM MNNG, or ($\triangle$) 17 μM MNNG + 1 mM MBA for 20 min and held in control medium (*closed symbols*) or medium containing 1 mM MBA (*open symbols*). At the indicated time, % double-stranded DNA was measured by the method of Anhstrom and Erixon [10]

Fig. 2. Time course of PLDR in MNNG-treated cells. Quiescent cells were treated with 68 μM MNNG for 20 min and cell survival was determined by colony-forming ability at the indicated times, relative to untreated control cells. ($\blacktriangle$) Experiment 1, ($\bullet$) experiment 2

damage. C3H10T1/2 cells were synchronized by holding the cells at confluence for 2 days and then reseeding them in isoleucine-free medium. Cultures were held in iso-leucine-free medium for 36 h and then released by adding complete medium. The cultures were treated 4 h later with MNNG ± MBA. The data in Fig. 4 show that following MNNG treatment MBA increased the length of S-phase from approx. 6.5 to 10 h and caused an accumulation of cells in G$_2$. Whether or not the cells were arrested in

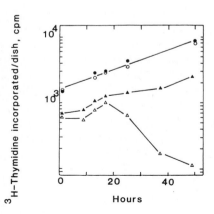

Fig. 3. Effect of MBA on DNA synthesis following MNNG treatment. Exponentially dividing cells were treated as in Fig. 1 and held in control medium (*closed symbols*) or medium containing 1 mM MBA (*open symbols*). At the indicated times cultures were pulse labeled for 2 h with [^{3}H]thymidine. Cpm/dish in ($\bullet$) control, ($\circ$) MBA, ($\blacktriangle$) MNNG, and ($\triangle$) MNNG + MBA-treated cultures

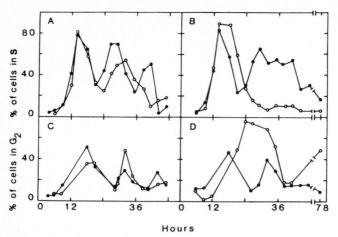

Hours

Fig. 4A–D. Effect of MBA on cell cycle progressions in synchronized cells. Cells were treated 4 h after release from isoleucine deprivation with 6.8 μM MNNG (**B, D**) or control medium (**A, C**). At the indicated times the percent of cells in S (**A, B**) was determined by autoradiography and the percent of cells in G_2 (**C, D**) was determined by flow cytometry. *Open symbols* indicate control medium and *closed symbols* represent the presence of 1 mM MBA

the first G_2 or the subsequent G_2 was dependent on the dose of MNNG (data not shown). These results suggest that ADP-ribosylation reactions, which do not seem to be rate limiting for DNA excision repair in quiescent cells, are essential for progression through S, G_2, and M and, thus, for subsequent rounds of DNA replication following DNA damage.

Effect of ADP-Ribosylation Inhibitors on Induction of Mutation and Malignant Transformation

Mutagenesis and malignant transformation are cellular responses to DNA damage which require DNA replication and cell division for expression. However, we have shown that in cycling cells inhibitors of ADP-ribosylation alter S-phase and block cell division following DNA damage. This effect can complicate studies designed to evaluate the effect of ADP-ribodylation inhibitors on mutagenesis and transformation. Figure 5 shows that even at very low doses of MNNG, MBA is cocytotoxic. Our approach to this potential problem has been to control the dose of MNNG and the number of survivors in a given assay so as to allow equal numbers of cell divisions to occur during expression in control- and inhibitor-treated populations. We found that following treatment with approx. 2 μM MNNG, mutation frequency is inhibited 50% (Fig. 6), while transformation frequency was enhanced 1200% (Table 2).

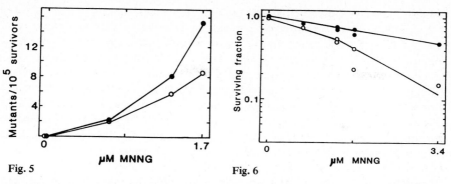

Fig. 5

Fig. 6

Fig. 5. Effect of MBA on MNNG-induced mutation at the ouabain loci. After treatment and expression performed as described in Fig. 5, cells were reseeded for determination of ouabain-resistant colonies in medium containing 3 mM ouabain. After 16 days the mutant colonies were fixed, stained, and counted. Mutants/10^5 viable cells for cells treated with MNNG (●) and MNNG + MBA (○)

Fig. 6. Effect of MBA on cell survival following MNNG in dividing cells. Cells were treated with MNNG ± MBA for 48 h and allowed to grow in control medium for an additional 48 h. The cells were reseeded for colony-forming ability. (●) MNNG, (○) MNNG + MBA

Table 2. Effect of ADP-ribosylation inhibitors on malignant transformation

Treatment	Survivors/ dish	Total survivors	Foci/dish	Transformation frequency $\times 10^{-4}$	Number of pop. doub. during ex- pression
Control	730	35,040	0/48	0	10.5
MBA	904	43,392	0/48	0	10.1
MNNG	813	68,312	3/84	0.44	10.2
MNNG + MBA	287	28,374	15/99	5.29	11.7

Discussion

In an attempt to segregate direct and indirect effects of ADP-ribosylation reactions related to DNA excision repair, we have studied the effect of MBA on PLDR in quiescent cells. The data presented here (Table 1 and Fig. 1) raise the question as to whether ADP-ribosylation reactions are involved directly in DNA excision repair or are necessary for some other processes required for cellular recovery from DNA damage that are coincident with DNA excision repair. We have observed a good correlation between the appearance and removal of DNA strand breaks and the time course of PLDR. However, in these studies we found no direct effect of MBA on DNA excision repair in quiescent cells (PLDR) and no correlation of MBA-induced DNA strand breaks and MBA-induced cytotoxicity following MNNG treatment. We previously observed in dividing cells that MBA is not cocytotoxic when added at 36 h or later [3] following MNNG treatment. Yet we still observed the appearance of an additional number of DNA strand breaks when MBA was added at 36 h (unpublished observations). Previous postulations that ADP-ribosylation inhibitors exert their cocytotoxic effects via effects on DNA ligase have been supported by correlations between cocytotoxic effects and increased number of DNA strand breaks [9]. Thus, our studies reported here do not lend support to a direct involvement of ADP-ribosylation in DNA excision repair.

The data presented here demonstrate that processes sensitive to MBA are required for dividing cells, but we have been unable to demonstrate a similar requirement in quiescent C3H10T1/2 cells during recovery from MNNG treatment. These data further suggest that following DNA damage, dividing cells require ADP-ribosylation in order to complete S at a normal rate (Fig. 4), to progress through G_2 and M (Fig. 4), and to initiate subsequent rounds of DNA replication (Fig. 3). These findings are in close agreement with those from studies conducted in human fibroblasts [11].

Attempts to assess the role of ADP-ribosylation reactions in other cellular responses to damage, such as mutagenesis and malignant transformation by means of ADP-ribosylation inhibitors, are complicated by the cytotoxic and cell cycle effects reported here. It is known that expression of the transformed phenotype in the C3H10T1/2 assay is a function of the number of population doublings completed when the culture reaches confluence [12]. Therefore, it is necessary to match the number of survivors per dish in the assay for both control and test populations. Employing a dose of carcinogen and number of cells which results in a high number of cells blocked in G_2 per dish can artifactually yield an apparent decrease in transformation frequency. After carefully establishing the appropriate dose of MNNG and appropriate number of cells to treat in order to obtain a similar number of survivors per dish, we observed large increases (12-fold) in transformation frequency and only small effects in mutagenesis. Our mutagenesis assay involves approximately four population doublings for expression and under these conditions it might still be argued that cells expressed in the presence of MBA, which undergo one less population doubling, may be showing lowered mutagenesis rates for this reason.

Clearly, ADP-ribosylation reactions are important following DNA damage in dividing cells. Large changes in chromatin structure accompany cell cycle progression. DNA excision repair must be coordinated with these events in dividing cells. We propose that poly(ADP-ribose) metabolism is involved in cellular recovery from DNA damage because it is required to either stabilize chromatin structure and/or mediate changes in chromatin structure that are necessary for dividing cells to effect normal cell cycle progression following DNA damage. Thus, ADP-ribosylation limits events which lead to transformation. It is likely that this function is closely coordinated with DNA excision repair.

Acknowledgments. This work was supported in part by NIH grant CA23994 and the TWU faculty research fund. We thank Lynne Gracy for assistance in manuscript preparation, Dr. J. Measel for flow cytometric analyses and Rene Meadows for assistance with cell synchrony experiments. We also thank Dr. K. Wielckens and Dr. H. Hilz for stimulating discussions during the course of this work.

References

1. Jacobson EL, Antol KM, Juarez-Salinas H, Jacobson MK (1983) Poly(ADP-ribose) metabolism in ultraviolet irradiated human fibroblasts. J Biol Chem 258:103–107
2. Wielckens K, Schmidt A, George E, Bredehorst R, Hilz H (1982) DNA fragmentation and NAD depletion. J Biol Chem 257:12872–12877
3. Jacobson EL, Smith JY, Mingmuang M, Meadows R, Sims JL, Jacobson MK (1984) Effect of nicotinamide analogues on recovery from DNA damage in C3H10T1/2 cells. Cancer Res 44: 2485–2492
4. Nudka N, Skidmore CJ, Chall S (1980) The enhancement of cytotoxicity of N-methyl-N-nitrosourea and of γ-radiation by inhibitors of poly(ADP-ribose) polymerase. Eur J Biochem 105:525–530
5. Durrant LG, Boyle JM (1982) Potentiation of cell killing by inhibitors of poly(ADP-ribose) polymerase in four rodent cell lines exposed to N-methyl-N-nitrosourea or UV light. Chem-Biol Interact 38:325–338
6. James MR, Lehman AR (1982) Role of poly(adenosine diphosphate ribose) in deoxyribonucleic acid repair in human fibroblasts. Biochemistry 21:4007–4013
7. Durkacz BW, Omidiji O, Gray DA, Shall S (1980) (ADP-ribose)$_n$ participates in DNA excision repair. Nature (London) 283:593–596
8. Morgan WF, Cleaver JE (1983) Effect f 3-aminobenzamide on the rate of ligation during repair of alkylated DNA in human fibroblasts. Cancer Res 43:3104–3107
9. Creissen D, Shall S (1982) Regulation of DNA ligase activity by poly(ADP-ribose). Nature (London) 296:271–272
10. Anhstrom G, Erixon K (1981) Measurement of strand breaks by alkaline denturation and hydroxyapatite chromatography. In: Friedberg EC, Hanawalt PC (eds) DNA repair. A laboratory manual of research procedures, vol 1, part B. Marcel Dekker, New York Basel, pp 403–418
11. Boorstein RJ, Pardee AB (1984) 3-Aminobenzamide is lethal to MMS-damaged human fibroblasts primarily during S phase. J Cell Physiol 120:345–353
12. Mordan LJ, Martner JE, Bertram JS (1983) Quantitative neoplastic transformation of C3H10T1/2 fibroblasts: Dependence upon the size of the initiated cell colony at confluence. Cancer Res 43:4062–4067

DNA Ligase II of Mammalian Cells

DEBBIE CREISSEN and SYDNEY SHALL[1]

Two Forms of DNA Ligase

A number of early reports on the purification of DNA ligase activities from mammalian sources suggested the possibility that more than one enzyme existed in these cells. The main indication arose from molecular weight estimates which revealed wide variations in different tissues. The DNA ligase from rabbit tissue had a mol. wt. of 95,000 [1] compared to 240,000 for the mouse enzyme [2]. Pedrali Noy et al. [3] purified the enzyme from a human heteroploid cell line, EUE. The enzyme was fractionated into two forms, one with a mol. wt. of 190,000, the other 95,000. In freshly prepared cell extracts the DNA ligase was present as the high molecular weight form, prolonged incubation, or enzyme purification lead to the appearance of the low molecular weight form, without variations in the total activity. A similar observation was made during purification of the calf thymus enzyme [4]. The smaller form of DNA ligase, in these instances, is therefore present as a result of dissociation of a high molecular weight form of the enzyme into a low molecular weight form (possibly a dimer into a monomer), or to proteolytic cleavage of the ligase.

The fact that the major ligase activity from mammalian cells can be converted to a smaller, active form, complicated the issue as to whether two distinct enzymes actually existed. In 1973, Söderhall and Lindahl [5] reported the presence of two DNA ligases in cell extracts from calf thymus tissue. These activities were separable by hydroxylapatite column chromatography and by gel filtration and were termed DNA ligase I (M_r 175,000) and DNA ligase II (M_r 95,000). Two DNA ligases isolated from cytoplasmic and nuclear fractions of rat liver [6–8] and from baby hamster kidney cells [9], appear to correspond to DNA ligase I and DNA ligase II described from calf thymus. We have succeeded in separating two distinct activities, on hydroxylapatite columns, from human tonsil extracts, from mouse leukaemic (L1210) cells and from a number of cell strains derived from human tumours (Fig. 1A).

These two forms of DNA ligase may be distinguished by a number of criteria, including molecular weight, sedimentation coefficient, heat lability, pH optimum and K_m for ATP. They have also been shown to be serologically unrelated [10] and to

---

1 Cell and Molecular Biology Laboratory, University of Sussex, Brighton, East Sussex, BN1 9QG, Great Britain

ADP-Ribosylation of Proteins
(ed. by F.R. Althaus, H. Hilz, and S. Shall)
© Springer-Verlag Berlin Heidelberg 1985

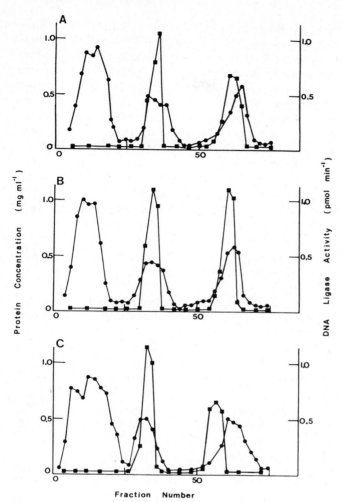

Fig. 1A–C. DNA ligase activity in L1210 cells. ● protein concentration; ■ DNA ligase activity. **A** Separation of DNA ligase activities from L1210 cells by hydroxylapatite column chromatography. Cells were harvested, salt extracts prepared and partial purification of DNA ligases I and II was carried out [23]. **B** Increase in DNA ligase II activity after exposure of L1210 cells to DMS. Cells were incubated for 20 min with 100 μM DMS before harvesting. **C** 3-aminobenzamide prevents the increase in DNA ligase II activity observed after exposure of L1210 cells to DMS. Cells were treated with 3 mM 3-aminobenzamide together with 100 μM DMS before harvesting

respond differently to cell proliferation [11]. This latter observation has led to the proposal of specific in vivo roles for the two enzymes. DNA ligase I increases, in parallel with polymerase α, when cells which are normally in a resting state are stimulated to divide, by various methods [11, 12] and is thus believed to be active in DNA replication. DNA ligase II, like polymerase β, does not respond in this fashion and these enzymes are thought to be involved in DNA repair. Further evidence for the existence of two DNA ligases with distinctive roles emerges from work on developmental systems. It has been shown that levels of two enzymes which can be separated on sucrose gradients, fluctuate independently according to the stage of development, probably reflecting differences in the requirement of cells for DNA replication and repair during maturation [13, 14]. An association of DNA ligase II with repair was demonstrated by a study of ligase activities during ram spermatogenesis [15]. DNA ligase II activity was closely associated with the germ cells which do exhibit DNA repair and was undetectable in elongated spermatids and spermatozoa which do not [16, 17]. We have added further weight to support a role for DNA ligase II in the repair process. Firstly, it was shown that total DNA ligase activity increased more than twofold when L1210 cells were treated with the DNA damaging agent, dimethyl sulphate (DMS), and secondly, it was demonstrated that the increase in activity was due entirely to an increase in levels of DNA ligase II, ligase I activity remaining essentially unchanged (Fig. 1B). This result has since been repeated in this laboratory with a number of human tumour cells. It has been confirmed by Chan et al. [18], who measured DNA ligase activities in a mutant strain (EM9) of Chinese hamster ovary cells; DNA ligase II increased 2.0- and 2.6-fold in the parental cells and EM9, respectively, following treatment with MMS. The increase in DNA ligase II activity in the experiments so far cited, occurs a short while after treatment with the DNA damaging agent (20–60 min) and does not therefore involve synthesis of new enzyme, but rather, activation of enzyme protein already present. In contrast, Mezzina et al. have published observations consistent with biosynthesis of ligase II after DNA damage [19]. An increase in enzyme activity was observed 24–48 h after treatment and was inhibited by the protein synthesis inhibitor, cycloheximide.

It should be noted that despite the many reports from several laboratories of two distinct DNA ligase activities, there remains some debate over this issue. It has been suggested that a single species of DNA ligase is present in mammalian cells and that the low molecular weight DNA ligase is an artifact arising from dissociation of the larger enzyme during the purification procedure [20, 21]. We would dispute this proposition on the basis of clearly distinguishable properties of the two activities and the absence of a decrease in ligase I activity, concomitant with the increase in ligase II following damage.

ADPRT and Ligase II Activation

The involvement of (ADP-ribose)$_n$ in the repair of DNA damage has been established through a number of experimental observations (review [22]). The activity of the enzyme responsible for the biosynthesis of (ADP-ribose)$_n$, nuclear ADPRT, is absolutely dependent on DNA ends and is stimulated by DNA damaging agents. A number of inhibitors of ADPRT are known and studies using these have provided the most conclusive evidence that ADP-ribosylation is required for the efficient repair of DNA damage. These inhibitors (including methylxanthines, nicotinamide analogues and thymidine) act synergistically with DNA-damaging agents in causing lethality in mammalian cells. The question of which step in DNA repair requires ADPRT activity now arises.

A requirement for ADPRT at the incision stage can be discounted since the release of methylated bases from DNA is not inhibited by ADPRT inhibitors and the appearance of strand breaks in DMS-treated DNA is similarly unaffected. The second step in the process, repair synthesis, is also unaffected by inhibition of ADPRT activity. A probable involvement of ADP-ribosylation at a late stage in DNA excision repair is therefore indicated.

We have demonstrated an involvement of $(\text{ADP-ribose})_n$ in the ligation step of repair by studying the effects of ADPRT inhibitors on DNA ligase activity in mammalian cells [23]. In crude cell extracts, the observed increase in total DNA ligase activity was completely abolished by the addition of 3-aminobenzamide, or other ADPRT inhibitors, to the growth medium, before DMS was added. The non-inhibitory analogues 3-aminobenzoic acid and nicotinic acid had no effect. Separation of DNA ligase I and II on hydroxylapatite columns showed that 3-aminobenzamide specifically prevents the increase in activity of ligase II and has no detectable effect on ligase I activity (Fig. 1C). This result has been observed in L1210 cells and in the human tumour cell strains, U251MG and U118MG, and has led us to propose that the activity of DNA ligase II is specifically increased during excision repair, and that this increase is mediated through the activity of ADP-ribosyl transferase.

References

1. Lindahl T, Edelman GM (1968) Polynucleotide ligase from myeloid and lymphoid tissues. Proc Natl Acad Sci USA 61:680–687
2. Beard F (1972) Polynucleotide ligase in mouse cells infected by polyoma virus. Biochim Biophys Acta 269:385–396
3. Pedrali Noy GCF, Spadari S, Ciarrocchi G, Pedrini AH, Palaschi A (1973) Two forms of the DNA ligase of human cells. Eur J Biochem 39:343–351
4. Bertazzoni U, Mathelet M, Campagnari F (1972) Purification and properties of a polynucleotide ligase from calf thymus glands. Biochim Biophys Acta 287:404–414
5. Söderhall S, Lindahl T (1973) Two DNA ligase activities from calf thymus. Biochem Biophys Res Commun 53:910–916
6. Zimmerman SB, Levin CJ (1975) A deoxyribonucleic acid ligase from nuclei of rat liver. J Biol Chem 250:149–155
7. Teraoka H, Simoyachi M, Tsukada K (1975) Two distinct polynucleotide ligases from rat liver. FEBS Lett 54:217–220
8. Teraoka H, Simoyachi M, Tsukada K (1977) Purification and properties of deoxyribonucleic acid ligases from rat liver. J Biochem (Tokyo) 81:1235–1260
9. Evans DG, Ton SH, Kier HM (1975) DNA ligase activities in BHK-21/C13 cells. Biochem Soc Trans 3:1131–1132
10. Söderhall S, Lindahl T (1975) Mammalian DNA ligases. J Biol Chem 250:8438–8444
11. Söderhall S (1976) DNA ligases during rat liver regeneration. Nature (London) 260:640–642
12. Teraoka H, Okamoto N, Tamura S, Tsukada K (1981) Induction of DNA ligase during stimulation of DNA synthesis in intact rat liver by a dietary manipulation. Biochim Biophys Acta 653:408–411
13. David JC, Vinson D, Lefresne J, Signoret J (1979) Evidence for a DNA ligase change related to early cleavage in axolotl egg. Cell Differentiation 8:451–459
14. David JC, Vinson D (1979) Duality and developmental changes of chicken thymus. Exp Cell Res 119:69–74

15. David JC, Vinson D, Loir M (1982) Developmental changes of DNA ligase during ram sperma-
    togenesis. Exp Cell Res 141:357–364
16. Kofman-Alfaro S, Chandley AC (1971) Radiation initiated DNA synthesis in spermatogenic
    cells of the mouse. Exp Cell Res 69:33–44
17. Gledhill BL, Darzynkiewicz Z (1973) Unscheduled synthesis of DNA during mammalian sper-
    matogenesis in response to UV irradiation. J Exp Zool 183:375–382
18. Chan JYH, Thompson LH, Becker FF (1984) DNA ligase activities appear normal in the CHO
    mutant EM9. Mutat Res 131:209–214
19. Mezzina M, Nocentini S, Sarasin A (1982) DNA ligase activity in carcinogen treated human
    fibrobalsts. Biochimie 64:743–748
20. Teraoka H, Tamura S, Tsukada K (1979) Evidence for a single species of DNA ligase localized
    in nuclei of rat liver. Biochim Biophys Acta 563:535–539
21. Teraoka H, Tsukada K (1982) Eukaryotic DNA ligase. J Biol Chem 257:4758–4763
22. Shall S (1984) ADP ribose in DNA repair. Adv Radiat Biol 11:1–69
23. Creissen D, Shall S (1982) Regulation of DNA ligase activity by poly(ADP-ribose). Nature
    (London) 296:271–272

Isolation of a DNA Ligase Mutant from L1210 Cells

BARBARA A. MURRAY, JUDY IRWIN, DEBBIE CREISSEN,
MANOOCHEHR TAVASSOLI, BARBARA W. DURKACZ, and SYDNEY SHALL[1]

Introduction

Current evidence suggests that the ADP-ribosylation of chromatin proteins is involved in DNA repair [2]. Much of this evidence comes from the use of inhibitors of ADPRT, such as 3-aminobenzamide (3AB), which retards DNA strand rejoining following DNA damage and potentiates the cytotoxicity of DNA damaging agents. Alkylation damage increases DNA ligase activity, predominantly that of DNA ligase II. This increase is prevented by ADPRT inhibitors. We have previously suggested that the requirement for ADPRT in DNA repair is at the ligation step [1].

To provide further independent evidence of ADPRT involvement in DNA repair and to investigate the molecular mechanisms of this involvement we have isolated cell variants altered in (ADP-ribose)$_n$ metabolism from mammalian cells. After mutagenesis variants were selected for resistance to the combined effect of 3AB and dimethyl sulphate (DMS). It was presumed that some variants would arise as a result of alterations in (ADP-ribose)$_n$ metabolism and/or in enzymes involved in DNA repair.

Isolation of Variants

Mouse leukaemia L1210 cells were treated with methylnitrosourea (to aout 10% survival) and after suspension in fresh medium and growth for 2 days to allow expression of the mutations, a recycling enrichment method of selection was carried out. This involved exposing the cells to DMS and 3AB simultaneously. The DMS concentration was chosen such that there was more than 90% survival in the absence of 3-aminobenzamide, but less than 10% survival in its presence. After eight cycles of this treatment (30 μM DMS, 2 mM 3AB), cells were treated with 60 μM DMS and 2 mM 3-aminobenzamide (1.0% survival of mutagenised cells, 0.4% survival of parental cells). The cells were then cloned.

---

1 Cell and Molecular Biology Laboratory, School of Biological Sciences, University of Sussex, Brighton, East Sussex, BN1 9QG, Great Britain

ADP-Ribosylation of Proteins
(ed. by F.R. Althaus, H. Hilz, and S. Shall)
© Springer-Verlag Berlin Heidelberg 1985

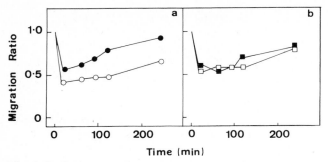

Fig. 1a,b. Sedimentation of nucleoids from cells treated with 10 μM DMS in the presence or absence of 3AB. Cells were incubated with or without 5 mM 3AB for 30 min before addition of 10 μM DMS. Samples were removed at various times after DMS addition and nucleoids prepared and run (Cook and Brazell). Results are expressed as the ratio of migration of DMS treated to untreated cells. **a** L1210 cells: ● control, ○ plus 3AB. **b** M3 cells: ■ control, □ plus 3AB

Non-toxic concentrations of 3AB enhance the cytotoxicity of DMS for wild-type (wt) L1210 cells quite markedly. One of the variant clones isolated, variant 3, showed no such potentiation of cytotoxicity of DMS by 3AB although its survival to DMS did not differ from wild-type L1210 cells.

Repair of DNA Strand Breaks

Nucleoids, which are structures containing all of the nuclear DNA but depleted in protein, from undamaged cells sediment rapidly in neutral sucrose gradients. Relaxation of supercoiled DNA by endonucleolytic incision during DNA repair results in nucleoids sedimenting with lower velocities — subsequent DNA repair leads to restoration of the rapidly sedimenting nucleoids. In wild-type L1210 cells the restoration of supercoiling following DMS treatment is partially inhibited by 3AB and 3-acetoamino-benzamide (3AAB). However, in variant 3 cells treated with 10 μM DMS the same loss and restoration of supercoiling is seen in the presence and absence of 3AB (Fig. 1) and 3-AAB.

Nuclear ADPRT Activity

Nuclear ADPRT activity is present in the variant 3 cells and is stimulated by the addition of deoxyribonuclease. 3-aminobenzamide inhibits this stimulation in both the variant 3 and in the wild-type L1210 cells. The variant cells have a twofold higher V_{max} than the wild-type cells. The K_m (for NAD) and K_i (for 3-aminobenzamide) values are the same for both wild-type L1210 and variant 3 cell enzymes; (K_i 13.2 μM for 3AB).

Table 1. DNA ligase I and II activities in parental and variant 3 cells[a]

	Controls				Exposed to DMS			
	I		II		I		II	
L1210	1.0	0.2 (6)	0.29	0.09 (6)	1.07	0.05 (3)	0.91	0.07 (3)
Variant 3	1.0	0.13 (6)	0.16	0.07 (6)	1.17	0.09 (2)	0.91	0.05 (2)
Variant 3[b]	1.66	0.22	0.26	0.12	1.96	0.14	0.32	0.08

[a] The enzyme activities are expressed relative to the activity in cells not exposed to DMS, SD (n)
[b] The relative enzyme activities of variant 3 compared to wild-type cells

DNA Ligase Activity

Variant 3 cells were found to have 70% higher DNA ligase I activity but the same DNA ligase II activity as the parental cells (Table 1). After treatment with 100 μM DMS the DNA ligase II activity in parental cells increased threefold, whereas the enzyme activity in variant 3 cells showed no such increase. In wild-type L1210 cells the increase in DNA ligase II activity after DMS treatment can be prevented by 3AB. In variant 3 cells no such increase in ligase activity is observed and 3AB has no effect on either ligase I or ligase II activities.

Discussion

We have isolated a variant of L1210 cells which survives the combined cytotoxicity of DMS and 3AB better than do the parental cells. Nucleoid and alkaline sucrose gradients show that variant 3 cells repair DNA damaged by DMS normally. At low DMS doses no inhibition of repair by ADPRT inhibitors was observed in the variant cells. However, with high doses of DMS and consequently greater amounts of damage, DNA repair was partially inhibited by 3AB in the variant cells. It was possible that an alteration in the ADPRT enzyme could account for the insensitivity of DNA repair to 3AB at low doses. However, ADPRT activity was present in the variant cells and was inhibitable by 3AB.

Variant 3 cells had higher ligase I levels but the same ligase II levels as wild-type cells. The most striking change observed in the variant 3 cells was the complete absence of the usual increase in DNA ligase II activity after treatment with DMS.

A mutation in the DNA ligase II enzyme of the variant cells such that it can no longer be activated by ADP-ribosylation could account for the lack of stimulation of ligase II activity after DNA damage. However, we must also account for the increase in DNA ligase I activity. It seems to us most likely that a single mutation has caused the pleiotropic effect of loss of activation of ligase II and the increased ligase I activity. The probability of these alterations arising from two independent mutations is very low, probably about 10^{-8}. Furthermore, an independent DNA ligase II alteration of the type observed in variant 3 does not seem to have any selective advantage for growth in our selective conditions of DMS plus 3AB. The inference is then that there is a close physiological connection between the DNA ligase I and ligase II activities. Many different hypotheses are available; including the possibility that a regulatory polypeptide which normally interacts with ligase II is now altered and now interacts with ligase I. Alternatively, there is regulation of the biosynthesis of DNA ligase I which is sensitive to the deficiency of ligase II and responds by raising the level of DNA ligase I.

It might be suggested that ligase II is a metabolic (or proteolytic) product of ligase I; however, this is unlikely because activation of ligase II is not associated with a decrease in ligase I activity. Thus it is very unlikely that the mutation in these cells is an inability to convert DNA ligase I to DNA ligase II. 3AB alone has no effect on cytotoxicity or nucleotide metabolism at the concentrations and duration of exposure used in these experiments.

In conclusion, we have isolated a variant (mutant) from wild-type L1210 cells which is insensitive to 3AB potentiation of DMS cytotoxicity. This variant is independent confirmatory evidence of ADPRT involvement in DNA repair and indicates that 3-aminobenzamide sensitivity is related to DNA ligase activity.

References

1. Creissen D, Shall S (1982) Regulation of DNA ligase activity by poly(ADP-ribose). Nature (London) 296:271–272
2. Shall S (1984) ADP ribose in DNA repair. Adv Radiat Biol 11:1–69

Environmental Stress and the Regulation of Poly(ADP-Ribose) Metabolism

MYRON K. JACOBSON[1], GILBERTO DURAN-TORRES[1],
HECTOR JUAREZ-SALINAS[1], and ELAINE L. JACOBSON[2]

Poly(ADP-Ribose) Metabolism in Vivo Following DNA Damage

The nuclear metabolism of poly(ADP-ribose) involves at least three enzymatic activities, poly(ADP-ribose) polymerase, poly(ADP-ribose) glycohydrolase, and protein-mono(ADP-ribose) lyase. Although each of these enzymes has been purified and studied, we still have a poor understanding of poly(ADP-ribose) metabolism and its regulation in intact cells. Two features of poly(ADP-ribose) metabolism in intact cells are illustrated in Fig. 1. First, the occurrence of environmentally-induced DNA damage results in a rapid elevation of the intracellular levels of polymer. Current evidence from both in vitro and in vivo studies argues that this alteration is regulated at the level of poly(ADP-ribose) polymerase and that the activating factor is the appearance of DNA strand breaks [1, 2]. A second feature is that polymers are rapidly

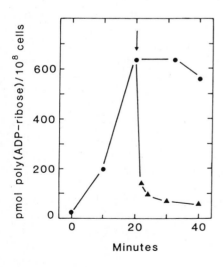

Fig. 1. Intracellular levels of poly(ADP-ribose) in cultured mouse cells (SVT2) following treatment with 400 μM MNNG. Polymer levels were determined as described elsewhere [6]. Benzamide was added to the culture medium to final concentration of 10 mM as indicated by the *arrow*. Polymer measurements made following benzamide addition are indicated by the *triangles*

1 Department of Biochemistry, North Texas State University/Texas College of Osteopathic Medicine, Denton, TX 76201, USA
2 Department of Biology, Texas Woman's University, Denton, TX 76204, USA

ADP-Ribosylation of Proteins
(ed. by F.R. Althaus, H. Hilz, and S. Shall)
© Springer-Verlag Berlin Heidelberg 1985

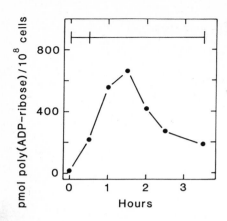

Fig. 2. Poly(ADP-ribose) levels in SVT2 cells subjected to a 30 min treatment at 45°C and returned to 37°C. The hyperthermic treatment was from 0 to 0.5 h

turning over in vitro. This indicates that individual polymers are required only transiently to fulfill their function or that the coordinated synthesis and degradation of the polymer results in a longer term change in chromatin structure.

Poly(ADP-Ribose) Metabolism Following Hyperthermia

The observation that inhibitors of poly(ADP-ribose) polymerase are cocytotoxic with DNA alkylating agents argues that poly(ADP-ribose) metabolism is necessary for cellular recovery from the cytotoxic effects of these agents. It is well known that cells respond to a wide variety of stressful conditions in a remarkably similar manner which involves rapid changes in both transcription and translation [3]. Our recent observation that other environmental stresses, such as hyperthermia or ethanol, also alter poly(ADP-ribose) metabolism suggested the possibility that poly(ADP-ribose) metabolism may be involved in a general response to environmental stress [4]. Thus, we have studied the mechanism of alteration of poly(ADP-ribose) metabolism by hyperthermia in cultured mouse cells (SVT2). For the studies described here, we have utilized an acute hyperthermic exposure, namely, 45°C for 30 min. Figure 2 shows the intracellular levels of poly(ADP-ribose) following an acute hyperthermic treatment. This treatment not only caused an elevation of poly(ADP-ribose), but it also potentiated by over 100-fold the accumulation of polymer induced by N-methyl-N'-nitro-N-nitrosoguanidine (MNNG) as shown in Table 1.

Hyperthermia is known to result in the appearance of a number of heat shock proteins whose synthesis is controlled at the level of both transcription and translation [3]. Thus, we have examined for effects of cycloheximide and actinomycin D on the alteration of poly(ADP-ribose) metabolism by hyperthermia. The results of Table 2 show that the presence of cyclohexamide resulted in over 80% inhibition of accumulation of poly(ADP-ribose) following hyperthermia, but did not affect accumulation in control cells. In contrast, the presence of actinomycin D, which inhibited uridine incorporation by over 90%, had no appreciable effect (data not shown).

Table 1. Effect of acute heat shock on accumulation of poly(ADP-ribose) following MNNG treatment

Treatment	No additions	34 μM MNNG[a]
	pmol/10⁸ cells	
Control (37°C)	24.5 ± 7.2 (n = 13)	61.2 ± 17.3 (n = 3)
Post-heat shock (37°C, 30 min following 45°C, 30 min)	302 ± 152 (n = 10)	4219 ± 1518 (n = 10)

[a] Present during the past 30 min prior to harvest

We have examined for effects of hyperthermia on both poly(ADP-ribose) polymerase activity and poly(ADP-ribose) glycohydrolase activity in vivo. An increase in polymerase activity should be reflected by an increased NAD consumption, while a decrease in glycohydrolase activity should result in a decreased rate of turnover of polymer. Table 3 shows that hyperthermic treatment resulted in an approximate fourfold increase in the amount of NAD consumed. Figure 3 shows the turnover of polymer in heat-shocked cells following MNNG treatment. In contrast to an estimated polymer half-life of 1 min in control cells, a half-life of 30 min was estimated in cells previously exposed to hyperthermia. Thus, these studies show that the alteration of poly(ADP-ribose) metabolism by hyperthermia involves effects on both poly(ADP-ribose) polymerase and poly(ADP-ribose) glycohydrolase.

It is now well established that DNA strand breaks are activators of poly(ADP-ribose) polymerase. By alkaline sucrose gradient analysis, we have been unable to attribute the activation of polymerase due to hyperthermia to an accumulation of DNA strand breaks [4]. However, it is possible that hyperthermia results in a small number of DNA strand breaks that are undetectable by alkaline sucrose gradient analysis yet produce a powerful activation of the polymerase. Conversely, it is also

Table 2. Effect of cyclohexamide (CH) on poly(ADP-ribose) accumulation

Treatment	Control	+ CH	% inhibition
	pmol/10⁸ cells		
Heat shock (45°C, 30 min)			
No additions	25	< 0.5	>95
34 μM MNNG[a]	945	196	79
Post-heat shock (37°C, 30 min)			
No additions	260	16	94
34 μM MNNG[a]	2979	487	84
Control (37°C)			
400 μM MNNG[a]	1310	1355	0

[a] Present during the last 30 min prior to harvest

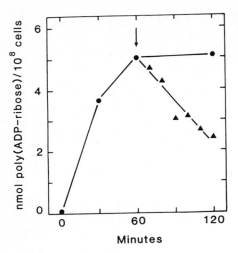

Fig. 3. Poly(ADP-ribose) levels and turnover in SVT2 cells subjected to an acute hyperthermic treatment. Cells were treated for 30 min at 45°C prior to return to 37°C (time 0). MNNG (34 μM) was added at time 0. Benzamide was added to the culture medium to a final concentration of 10 mM as indicated by the *arrow*. Polymer measurements made following benzamide addition are indicated by the *triangles*

Table 3. Effect of acute heat shock on NAD depletion

Treatment	No addition	34 μM MNNG[b]
	NAD depletion[a] pmol/10^8 cells	
Control (37°C)	0	8,800
Heat shock (45°C, 30 min)	4,300	38,000
Post-heat shock (37°C, 30 min)	5,500	45,900

[a] Initial NAD content 119,400 pmol/10^8 cells
[b] Present during the last 30 min prior to harvest

possible that the polymerase is also regulated by factors other than DNA strand breaks.

The major alteration of poly(ADP-ribose) metabolism by hyperthermia involves regulation of poly(ADP-ribose) glycohydrolase activity. We are not aware of other reports of physiological regulation of this enzyme. While cyclohexamide had no effect on NAD depletion (data not shown), it did effectively inhibit polymer accumulation (Table 2). Thus, the regulation of glycohydrolase activity appears to involve translationally-controlled protein synthesis caused by hyperthermia. It is interesting that a group of translationally-controlled heat-shock proteins recently have been reported to be associated with the nuclear matrix [5]. This raises the possibility that these proteins regulate glycohydrolase activity. It also raises the interesting possibility that poly-(ADP-ribose) metabolism is associated with the nuclear matrix.

A Possible General Role of Poly(ADP-Ribose) Metabolism in Cellular Responses to Stress

Studies have demonstrated interesting interactions of different types of environmental stress with regard to the consequences of stress, cell killing, or adaptive responses that promote survival [3]. Combinations of different types of stress, such as hyperthermia, ethanol, and DNA damaging agents act synergistically to promote cell killing. In addition, both ethanol and hyperthermia induce the acquisition of thermotolerance and heat-shock proteins. These interactions have led to the proposal that many types of stress are mediated by a common mechanism. In view of the known synergistic effects of different stresses on both biological responses to stress and on poly(ADP-ribose) metabolism, the possibility that poly(ADP-ribose) metabolism is part of a general response to stress should be considered. Further, the studies described here show that alteration of poly(ADP-ribose) metabolism by stress may involve regulation of either poly(ADP-ribose) polymerase or poly(ADP-ribose) glycohydrolase or both.

Acknowledgments. This work was supported in part by NIH grant CA23994, the North Texas State University Faculty Research Fund and by the Texas Ladies Auxillary to the Veterans of Foreign Wars.

References

1. Benjamin RC, Gill DM (1980) Poly(ADP-ribose) synthesis in vitro programmed by damaged DNA: A comparison of DNA molecules containing different types of strand breaks. J Biol Chem 255:10501–10508
2. Jacobson EL, Entol KM, Juarez-Salinas H, Jacobson MK (1983) Poly(ADP-ribose) metabolism in UV irradiated human fibroblasts. J Biol Chem 258:103–107
3. Schlesinger MJ, Ashburner M, Tissieres A (eds) (1982) Heat shock from bacteria to man. Cold Spring Harbor Lab, Cold Spring Harbor
4. Juarez-Salinas H, Duran-Torres G, Jacobson MK (1984) Alteration of poly(ADP-ribose) metabolism by hyperthermia. Biochem Biophys Res Commun 122:1381–1388
5. Reiter T, Penman S (1983) "Prompt" heat shock proteins: Translationally regulated synthesis of new proteins associated with the nuclear matrix-intermediate filaments as an early response to heat shock. Proc Natl Acad Sci USA 80:4737–4741
6. Jacobson MK, Payne DM, Juarez-Salinas H, Alvarez-Gonzales R, Sims JL, Jacobson EL (1983) Determination of in vivo levels of polymeric and monomeric ADP-ribose by fluorescence methods. In: Miwa M, Hayaishi O, Shall S, Smulson M, Sugimura T (eds) Methods in enzymology, posttranslational modifications. Jpn Sci Soc Press, Tokyo, pp 165–174

Tumor Promoter Phorbol-Myristate-Acetate Induces a Prooxidant State which Causes the Accumulation of Poly(ADP-Ribose) in Fibroblasts

NEETA SINGH, GUY G. POIRIER, and PETER A. CERUTTI[1]

Many tumor promoters induce a cellular prooxidant state, i.e., they increase the concentrations of active oxygen, organic-hydroperoxides, and -radicals [1]. The question arises, therefore, whether these highly reactive metabolic intermediates play a role in the promotor-induced modulation of the expression of growth- and differentiation-related genes. Posttranslational modification of chromosomal proteins by poly(ADPR) could be involved in this process because it is intimately related to the cellular redox-state, DNA strand breakage, and chromatin conformation. Poly(ADPR) appears to play a role in various aspects of chromatin metabolism. Its inhibition has been shown to affect DNA repair [2, 3], the formation of sister chromatid exchanges and chromosomal aberrations, cell differentiation [4, 5], and malignant transformation [6, 7]. We report our studies of the effect of the mouse skin tumor promoter phorbol-12-myristate-13-acetate (PMA) in cultured rodent and human fibroblasts.

Monolayer cultures of mouse embryo fibroblasts C3H10T1/2 and human fibroblasts 3229 were treated with PMA at a dose of 25 ng ml^{-1} and the poly(ADPR) content determined by the fluorescence assay of Jacobson et al. [8]. Parallel cultures were treated with the methylating agent N-methyl-N'-nitro-nitrosoguanidine (MNNG, 136 μM) and served as a reference system which has already been studied in detail by other investigators. PMA caused a 10–15-fold increase in poly(ADPR) concentration within 2–3 h. Even after prolonged incubation for 14 h the poly(ADPR) levels had not completely returned to those of untreated controls. As expected from previous experiments the increase in poly(ADPR) concentration after MNNG treatment was much more rapid [9]. A 35-fold stimulation was reached in 20 min. NAD represents the substrate for poly(ADPR) synthetase. As shown in Fig. 1, 136 μM MNNG caused a 33% decrease in cellular NAD levels within 20 min, while the moderate accumulation of poly(ADPR) caused by PMA had no effect on NAD even after 5 h incubation.

Further evidence for a fundamental difference in the mechanism of poly(ADPR) accumulation following PMA and MNNG treatment derives from experiments with inhibitors of macromolecular synthesis. As shown in Fig. 2 for human fibroblasts, suppression of RNA synthesis with actinomycin D (2 μg ml^{-1}) reduced the PMA-induced accumulation of poly(ADPR) by 90%. A similar result was obtained with cycloheximide (5 μg ml^{-1}) which inhibits de novo protein synthesis. In contrast, these

---

1   Department of Carcinogenesis, Swiss Institute for Experimental Cancer Research, 1066 Epalinges, Switzerland

ADP-Ribosylation of Proteins
(ed. by F.R. Althaus, H. Hilz, and S. Shall)
© Springer-Verlag Berlin Heidelberg 1985

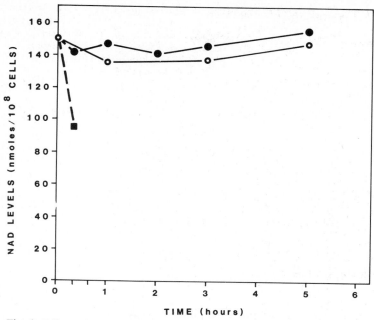

Fig. 1. Effect of PMA on NAD levels in C3H10T1/2 cells. Rapidly growing monolayer cultures were treated with 25 ng ml^{-1} PMA and NAD concentrations determined according to the method of Jacobson and Jacobson [16] (●—●); treatment with 136 μM MNNG (■); untreated control (○—○)

drugs had no effect on the stimulation of poly(ADPR) synthesis by MNNG. On the basis of these results it was conceivable that de novo synthesis of poly(ADPR) synthetase was necessary for the increase in poly(ADPR) induced by PMA. However, DNAse I treatment of permeabilized 10T1/2 cells which had been exposed to PMA resulted only in a twofold increase in poly(ADPR) synthetase activity relative to controls which had not been treated with PMA. This increase may not suffice for the 10—15-fold increase in poly(ADPR) levels induced by PMA and may merely result from the mitogenic action of PMA.

Poly(ADPR) residues attached to chromosomal proteins are subject to rapid turnover. Indeed, polymer half-life times (t1/2) of 1 min or less have been estimated following treatment of cells with alkylating agents [10]. Degradation of poly(ADPR) occurs by several enzymes most importantly poly(ADPR) glycohydrolase, phosphodiesterase, and ADPR-lyase. Inhibition of poly(ADPR) degradation rather than the stimulation of its synthesis could lead to an increase in the stationary poly(ADPR) concentration. The present methodology for the estimation of t1/2 uses an inhibitor of poly(ADPR) synthetase, such as 3-aminobenzamide. We measured a t1/2 of poly-(ADPR) in PMA-treated 10T1/2 cells of 2 min. Because the time required for the inhibitor 3-aminobenzamide to reach the site of its action in the cell may become rate limiting, t1/2 values below 1—2 min cannot be measured. Therefore, our data does not exclude the possibility that PMA induced an increase in polymer half-life from a few seconds to 2 min.

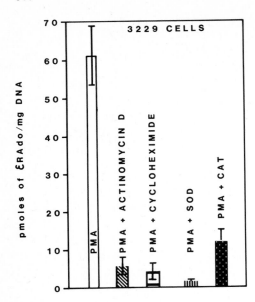

Fig. 2. Effect of inhibitors of de novo RNA and protein synthesis and of the antioxidant enzymes CuZn-superoxide dismutase and catalase on PMA-induced accumulation of poly(ADPR) in human fibroblasts 3229 (for experimental conditions see text)

As mentioned above many tumor promoters induce a cellular prooxidant state. Because poly(ADP) ribosylation of chromosomal proteins could play a role in active oxygen-induced modulation of gene expression by PMA we studied the effect of antioxidants. As shown in Fig. 2 for human fibroblasts 3229 the extracellular addition of moderate concentrations of CuZn-superoxide dismutase and catalase suppressed the PMA-induced accumulation of poly(ADPR) by 80—100%. Heated catalase was inactive. The low molecular weight antioxidant butylated-hydroxytoluene also suppressed the increase in poly(ADPR). These results suggest that active oxygen produced in a superoxide driven Fenton reaction represents an intermediate in the mechanism of action of PMA. The fact that the same antioxidants had no effect on poly(ADPR) synthesis induced by MNNG further emphasizes the fundamental difference between the two agents.

There is convincing evidence that DNA containing breaks stimulates poly(ADPR) synthetase [11]. For the case of MNNG, DNA breaks are induced as a consequence of repair processes. In our recent work with SV40 infected CV-1 monkey cells MNNG induced a proportionate increase in strand breakage and poly(ADP) ribosylation in SV40 minichromosomes. Because of the low molecular weight of SV40 DNA artifactual induction of breakage during cell permeabilization and DNA isolation can be avoided (strand breakage in SV 40 molecules can be measured with high sensitivity by the determination of the fraction of relaxed circles produced from superhelical molecules). These results suggest that DNA breakage is the cause for the increase in poly-(ADPR) of SV40 minichromosomes. In analogy we speculated that active oxygen induced DNA strand breakage might be the cause for the observed increase in poly-(ADPR) levels [12, 13] following PMA treatment. We attempted to measure DNA by the alkaline unwinding method of Birnboim [14]. For this purpose 10T1/2 cells or 3229 fibroblasts were grown in monolayers on the bottom of small test tubes and treated with PMA. Their DNA was then analyzed for strand breakage without the necessity to detach the cells from the glass surface. Despite the high sensitivity and reproducibility of this assay no convincing evidence was obtained for a PMA-induced increase in the stationary concentration of DNA breaks.

From the results described above the PMA-induced accumulation of poly(ADPR) in rodent and human fibroblasts can be characterized in the following manner (1) maximal poly(ADPR) levels are reached after 2–3 h; (2) the PMA-induced increase in poly(ADPR) levels does not result in a decrease in the cellular NAD concentration; (3) de novo RNA and protein synthesis are required, but poly(ADPR) synthetase activity increases only slightly; (4) antioxidants prevent poly(ADPR) accumulation; (5) there is no convincing evidence for the induction of DNA breakage by PMA in fibroblasts. These characteristics clearly distinguish the mechanism of action of PMA from that of alkylating agents and ionizing radiation. They are reminiscent of changes in poly(ADPR) in response to starvation of Ehrlich ascites cells, heat shock, or picolinic acid treatment of mouse cells and glycerol treatment of HeLa cells [10, 15]. In all these cases no measurable change in the stationary concentration of DNA strand breaks was observed. We can only speculate about the mechanism of PMA-induced accumulation of poly(ADPR). The following model appears attractive to us: a regulatory protein is produced as a consequence of the PMA-induced cellular prooxidant state which inhibits degradation of poly(ADPR), e.g., by inhibiting poly-(ADPR) glycohydrolase.

Acknowledgments. We thank Dr. M. Jacobson for valuable advice, Ms I. Zbinden for excellent technical assistance, the Swiss National Science Foundation and the Swiss Association of Cigarette Manufacturers for research support, the International Union against Cancer for an Eleanor Roosevelt fellowship to one of us (G.P.) and the Swiss Government for a postdoctoral fellowship (N.S.).

References

1. Cerutti PA (1985) Prooxidant states and tumor promotion. Science 227:375–381
2. Jacobson EL, Antol KM, Juarez-Salinas, Jacboson MK (1983) Poly(ADP-ribose) metabolisms in ultraviolet irradiated human fibroblasts. J Biol Chem 258:103–110
3. Durkacsz B, Omidiji O, Gray D, Shall S (1980) (ADP-ribose) participates in DNA excision repair. Nature (London) 283:593–596
4. Althaus FR, Lawrence SD, He YZ, Sattler GL, Tsukada Y, Pitot HC (1982) Effects of altered (ADP-ribose) metabolisms on expression of fetal functions by adult hepatocytes. Nature (London) 300:366–368
5. Farzaneh F, Zalin R, Brill D, Shall S (1982) DNA strand breaks and ADP-ribosyl transferase activation during cell cycle differentiation. Nature (London) 300:362–366
6. Borek C, Morgan WF, Ong A, Cleaver JE (1984) Inhibition of malignant transformation in vitro by inhibitors ofpoly(ADPR-ribose) synthesis. Proc Natl Acad Sci USA 81:243–247
7. Kun E. Kirsten E, Milo G, Kurian P, Kumari H (1983) Cell cycle dependent intervention by benzamide of carcinogen induced neoplastic transformation and in vitro poly(ADP ribosyl)-ation of nuclear proteins in human fibroblasts. Proc Natl Acad Sci USA 80:7219–7223
8. Jacobson MK, Payne DM, Juarez-Salinas H, Sims JL, Jacobson EL (1984) In: Wold F, Moldave K (eds) Post translational modifications. Methods in enzymology, vol 106. Academic Press, London New York, pp 483–494

9. Jacobson MK, Levi V, Juarez-Salinas H, Barton RA, Jacobson EL (1980) Effect of carcinogenic N-alkyl-N-nitroso compounds on nicotinamide adenine dinucleotide metabolisms. Cancer Res 40:1797–1802

10. Wielkens K, George E, Pless T, Hilz H (1983) Stimulation of poly(ADP-ribosyl)ation during Ehrlich ascites tumor cell "starvation" and suppression of concomitant DNA fragmentation by benzamide. J Biol Chem 258:4098–4104

11. Lubet RA, Kiss E, Gallagher MM, Dively C, Kouri RE, Schechtman LM (1983) Induction of neoplastic transformation and DNA single-strand breaks in C3H/10T1/2 clone 8 cells by polycyclic hydrocarbons and alkylating agents. J Natl Cancer Inst 71:991–997

12. Hollenberg M, Ghani P (1982) In: Hayaishi O, Ueda K (eds) ADP-ribosylation reactions. Academic Press, London New York, pp 439–450

13. Uchigata Y, Yamamoto H, Kawamura A, Okamoto H (1982) Protection by superoxide dismutase, catalase, and poly(ADP-ribose) synthetase inhibitors against alloxan and streptouotocin-induced islet DNA strand breaks and against the inhibition of proinsulin synthesis. J Biol Chem 257:6084–6088

14. Birnboim HC (1982) DNA strand breakage in human luekocytes exposed to a tumor promoter, phorbol myristate acetate. Science 215:1247–1249

15. Kidwell WR, Purnell MR (1983) In: Miwa M, Hayaishi O, Shall S, Smulson M, Sugimura T (eds) ADP-ribosylation, DNA repair and cancer. Jpn Sci Soc Press, Tokyo, pp 243–252

16. Jacobson EL, Jacobson MK (1979) Pyridine nucleotide synthesis in 3T3 cells. J Cell Physiol 99:417–426

Kinetic Studies of the ADP-Ribosylation of HSP-83 in Chicken Embryo Fibroblast Cells

LARS CARLSSON[1]

Abbreviations:

IEF/SDS-PAGE	Isoelectric focusing/sodium dodecyl sulphate polyacrylamide gel electrophoresis
SDS	Sodium dodecyl sulphate
HSP-83	Heat-shock inducible protein with M_r 83,000
CEF	Chicken embryo fibroblast cells

1. Introduction

ADP-ribosylation of proteins has been detected and characterized to occur as a post-translational modification both on cytoplasmic and nuclear polypeptides. Usually, protein acceptors have been characterized by in vitro reactions, utilizing $[^{32}P]$-NAD$^+$ with high specific activity as substrate for enzyme-catalyzed transfer of ADP-ribose. Several cytoplasmic acceptors have in this way been identified, including such diverse proteins as the protein synthesis elongation factor 2 [1, 2] and several cytoplasmic structural proteins [3, 4]. However, few studies have tried to answer the question of which proteins act as in vivo acceptors, mainly because of a lack of a suitable radioactively labeled precursor. Recently I showed that the by far most abundant intracellular acceptor for mono(ADP-ribose) in vivo is a polypeptide with a M_r of 83,000 and identified by several criteria as identical with the stress-inducible and glucose-regulated HSP-83 [5]. This was made possible by utilizing $[^3H]$-adenosine as a precursor to intracellular $[^3H]$-ATP and thus also $[^3H]$-NAD$^+$, and separating the total cell homogenate by two-dimensional isoelectric focusing/SDS polyacrylamide gel electrophoresis (2D IEF/SDS-PAGE) followed by fluorography. It was further shown that both heat shock and glucose starvation can induce drastic changes in the incorporation of tritiated ADP-ribose into HSP-83, suggesting that this modification of the protein plays an important physiological function. However, so far no enzymatic activity or other specific function has been assigned to HSP-83, other than its enhanced transcription during stress situations induced in a variety of ways [6]. With this technique,

1 Department of Medical and Physiological Chemistry, Biomedical Center, Uppsala University, Box 575, S-751 23 Uppsala, Sweden

ADP-Ribosylation of Proteins
(ed. by F.R. Althaus, H. Hilz, and S. Shall)
© Springer-Verlag Berlin Heidelberg 1985

I detected some minor ADP-ribose acceptors with apparent mol. wts. of 100,000, 140,000, and 150,000 [5]. The identity of these polypeptides have not been clarified.

I have now further investigated the kinetics of incorporation of tritiated ADP-ribose into HSP-83 and the disappearance of the label during heat shock, as well as its subcellular distribution.

2. Materials and Methods

2.1 Cell Culture and Labeling of Cells

Chicken embryo fibroblast cells (CEF) were prepared from the carcasses of 10-day-old embryos and maintained in complete growth medium as described [5]. The cells were used between the third and seventh subcultivation after initiation of the cultures. Cells were labeled in vivo with 125 μCi of [2,8,5'-^{3}H]adenosine per ml of culture media (50.5 Ci mmol^{-1}; 1 Ci = 3.7 $\times$ 10^{10} Bq; New England Nuclear) as described [5].

2.2 Subcellular Fractionation

Cells were fractionated into a soluble fraction, a nuclear fraction, and a plasma membrane fraction in a two-phase system utilizing a mixture of buffered polyethylenglycol 6000 (Sigma) and Dextrane T500 (Pharmacia, Uppsala, Sweden) as described by Albertsson [8]. The three distinct fractions obtained were analyzed on two-dimensional IEF/SDS-PAGE.

2.3 Gel Electrophoretic Analysis of Total Cell Homogenates and Fluorography

Cells or subcellular fractions were harvested in SDS and urea containing lysis buffer exactly as described [5], and total cellular homogenates analyzed by one- and two-dimensional gel electrophoresis [7]. The gels were both stained with Coomassie brilliant blue-R and processed for fluorography as described [5].

2.4 Quantitation of Tritiated ADP-ribose and Tritiated RNA

Quantitation of incorporation of radioactivity into individual polypeptides or RNA molecules were done by cutting out the appropriate part of the one- or two-dimensional gel, incubating the gel piece with 750 μl of 0.5 M NaOH at 65°C for 3 h in order to hydrolyze and elute the radioactivity from the polyacrylamide piece. This was followed by neutralization and quantitation of the radioactivity by liquid scintillation spectroscopy in 20 ml of Aquasol (New England Nuclear).

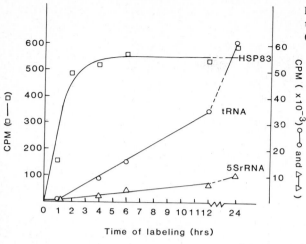

Fig. 1. Time course of [³H]-incorporation into HSP-83 (□ - -□), tRNA (○- - -○), and 5S rRNA (△- - -△)

3. Results

3.1 Kinetics of Labeling of HSP-83

Chicken embryo fibroblasts were labeled with [³H]-adenosine for various lengths of time, the cells harvested, and the homogenates analyzed by one-dimensional SDS-PAGE and fluorography. Thereafter the amount of radioactivity incorporated into HSP-83 and various RNA species were directly quantitated as described in Materials and Methods. As illustrated in Fig. 1, the incorporation of tritium into HSP-83 is half-maximal saturated within 1 h of labeling of cells, while incorporation of [³H]-adenosine into tRNA and ribosomal 5S RNA continue to increase linearly for more than 24 h. The time it takes to reach saturation of labeling of HSP-83 resembles the time course for saturation of the ATP and NAD⁺ pools with radioactive precursor. This indicates that the actual modification of HSP-83 is not the rate limiting step and, thus, the turnover rate of the ADP-ribose moiety is very rapid.

3.2 Kinetics of Disappearance of Labeled HSP-83 After Heat Shock

During the initial investigation it was oberved that the ADP-ribosylation of HSP-83 was reversed after incubating cells for a short period of time at elevated temperature [5]. This suggested that the modification might play a physiologically important regulatory role for the function of HSP-83. To determine more precisely what importance the ADP-ribosylation might have for the heat-shock response, it was of interest to follow the time course more closely of the removal of the ADP-ribose from HSP-83 during elevated temperature. Therefore, cells were prelabeled with [³H]-adenosine for 5 h at 37°C and then transferred to 42°C for various length of time. As is illustrated in Fig. 2 it takes up to 30 min after the heat shock is initiated before a significant decrease in radioactivity is detected in HSP-83. This result argues against the pos-

Time of heat shock (min)

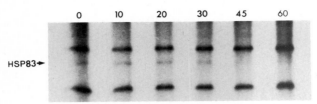

Fig. 2. Time course of disappearance of [³H]-ADP-ribose from HSP-83 during heat shock. CEF were labeled and heat shocked as described and total cellular homogenates analyzed on one-dimensional SDS-PAGE and fluorography. Only the portion of the autoradiogram corresponding to HSP-83 is shown in the figure

sibility of an "early" function of the ADP-ribose in the heat-shock response, since stimulation of transcription of heat-shock genes are achieved within minutes after transfer to the elevated temperature ([6], see further Discussion).

3.3 Subcellular Distribution of HSP-83

To further understand the nature and possible function of HSP-83 and its ADP-ribo-sylated variants. I have investigated its subcellular distribution by subjecting homo-genates of CEF to a rapid centrifugal fractionation procedure, utilizing a two-phase system (see Materials and Methods). In preliminary experiments it was observed that the result of the fractionation very much depended on how the cells were harvested from the tissue culture plates prior to homogenization and fractionation. Therefore, it was warranted to make a direct comparison between cells removed from the sub-strate either by trypsin treatment or by scraping the cells with a rubber policeman. Figure 3 illustrates the result from this kind of experiment. As can be seen, the major fraction of HSP-83 is found in a soluble form when cells are harvested by trypsiniza-tion (Fig. 3A). On the other hand, in homogenates of cells mechanically removed from the tissue culture plate, most of the HSP-83 is found in the plasma membrane fraction (Fig. 3F). A significant, but smaller part of HSP-83 is also found in the nuclear fractions (Fig. 3B, E).

4. Discussion

Labeling of HSP-83 with [³H]-adenosine as precursor rapidly reaches saturation, with a time course that most likely reflects the time it takes to saturate the intracellular ATP and NAD⁺ pools with radioactivity. There is further evidence that effective glyco-hydrolases are present in the cell and which can act on the ADP-ribose moiety of HSP-83. Direct comparison of cell homogenates made in SDS and urea (as described in this paper) with homogenates made with the nonionic detergent Nonidet-P40 show

TRYPINIZED CELLS   SCRAPED CELLS

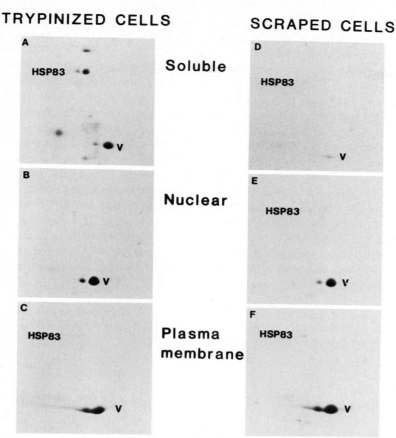

Fig. 3A–F. Subcellular distribution of HSP-83. Cells were either trypsinized (**A–C**) or scraped (**D–F**) from the culture plates before fractionation by the two-phase system described in [8]. The three fractions obtained were analyzed by two-dimensional IEF/SDS-PAGE. Only the portion of the gels including HSP-83 are shown. **A, D** Soluble fractions; **B, E** nuclear fractions; **C, F** plasma membrane fractions. *Arrows* point to HSP-83 and vimentin (*V*)

that within 2 min of incubation of the extracts at room temperature or 37°C, almost all of the isoelectric variants of HSP-83 containing ADP-ribose is converted to the unmodified form (unpublished observations). Furthermore, the in vitro labeling of HSP-83 with [^{32}H]-NAD$^+$ as substrate does not result in any efficient incorporation of radioactivity into this protein compared with other in vitro cytoplasmic acceptors [5]. Taken together these results show that in the presence of nonionic detergents, and thus, nondenaturing conditions, HSP-83 is efficiently demodified. These results also indicate that the turnover of ADP-ribose in HSP-83 is very high, and that there are effective regulatory mechanisms in the cell for determining the level of ADP-ribose modification.

The removal of ADP-ribose from HSP-83 during incubation of cells at elevated temperature proceeds with a relatively slow rate in relation to the induction of the expression of heat-shock genes. Therefore, one can conclude that the demodification of HSP-83 does not participate in the initial phases of the heat-shock response, but is a secondary reaction to the stress situation.

To further advance the knowledge and search for a possible physiological function for HSP-83 in the cell, it is of importance to analyze its subcellular distribution. It has been suggested, using antibodies towards HSP-83, that its main localization is in the cytoplasm in cells, contrary to most of the other HSP-proteins which are found accumulated in the nucleus during heat shock [6]. In the studies presented here I see an extensive dependence of the subcellular distribution depending on how the cells have been harvested prior to homogenization. This is not only reflected in the sub-cellular distribution of HSP-83, but can be detected for several cytoplasmic poly-peptides and even the main cyto-matrix proteins. Therefore, it is reasonable to believe that the results obtained by mechanical removal of cells from their support more closely reflects the normal intracellular distribution. In conclusion, it means that HSP-83 to a great extent, but not exclusively, is found in close proximity with the plasma membrane. However, since it is easily solubilized in intact cells, it is not an integral membrane protein, but most likely associated with the periferal protein layer present just beneath the plasma membrane in most cell types [9, 10].

Acknowledgment. I thank Dr. Ingrid Blikstad for continued discussion during the course of this work and participation in the subcellular fractionation studies described here. This work was supported by a grant from the Swedish Natural Science Research Council.

References

1. Honjo T, Nishizuka Y, Hayaishi O, Kato I (1968) Diphtheria toxin-dependent adenosine diphosphate ribosylation of aminoacyl transferase II and inhibition of protein synthesis. J Biol Chem 243:3553–3555
2. Raeburn S, Collins JF, Moon HM, Maxwell ES (1971) Aminoacyltransferase II from rat liver I. Purification and enzymatic properties. J Biol Chem 246:1041–1048
3. Kaslow HR, Groppi VE, Abood ME, Bourne HR (1981) Cholera toxin can catalyze ADP-ribosylation of cytoskeletal proteins. J Cell Biol 91:410–413
4. Amir-Zaltsman Y, Ezra E, Scherson T, Zutra A, Littauer US, Salomon Y (1982) ADP-ribosylation of microtubule proteins as catalyzed by cholera toxin. EMBO J 1:181–186
5. Carlsson L, Lazarides E (1983) ADP-ribosylation of the M_r 83,000 stress-inducible and glucose-regulated protein in avian and mammalian cells: Modulation by heat shock and glucose starvation. Proc Natl Acad Sci USA 80:4664–4668
6. Schlesinger MJ, Ashburner M, Tissières A (eds) Heat shock. From bacteria to man. Cold Spring Harbor Lab, Cold Spring Harbor
7. Hubbard BD, Lazarides E (1979) Copurification of actin and desmin from chicken smooth muscle and their copolymerization in vitro to intermediate filaments. J Cell Biol 80:166–182
8. Albertsson P-Å (1970) Partition of cell particles and macromolecules in polymer two-phase systems. Adv Protein Chem 24:309–341
9. Lindberg U, Carlsson L, Markey F, Nyström L-E (1979) The unpolymerized form of actin in non-muscle cells. In: Jasmin G, Cantin M (eds) Methods and achievements in experimental pathology, vol VIII. Karger, Basel, pp 143–170
10. Lindberg U, Höglund AS, Karlsson R (1981) On the ultrastructural organization of the micro-filament system and the possible role of profilactin. Biochemie 63:307–323

Effects of 3-Aminobenzamide on Cell-Cycle Traverse and Viability of Human Cells Exposed to Agents which Induce DNA Strand-Breakage

PAUL J. SMITH, CATHERINE O. ANDERSON, and STEVEN H. CHAMBERS[1]

1. Introduction

There is evidence that poly(ADP-ribosyl)ation, in response to DNA damage (review [1]), permits changes in chromatin structure [2] which may facilitate the activity of repair enzymes [3] and become a major factor in the control of cell-cycle traverse [4]. This paper explores the role of chromatin structure and poly(ADP-ribosyl)ation (using the inhibitor, 3-aminobenzamide; 3AB; [5]) in the responses of human cells to either a DNA specific ligand (Hoechst dye 33341; Ho33342) or X-radiation. Cells derived from an ataxia telangiectasia (A-T) patient have been included in the study to provide a hypersensitive control (reviews [6, 7]) in which anomalous cell survival and cell-cycle responses [8–10] to radiation may reflect a primary defect in chromatin structure [11, 12].

The study of chromatin structure commonly involves the use of nuclease digestion techniques (review [13]) which have previously revealed that transcriptionally active or potentially active regions of chromatin (representing perhaps only 5–10% chromatin) demonstrate enhanced sensitivity or accessibility to endonucleases. Chromatin accessibility has also been studied by the use of various intercalating and cationic probes to nuclear DNA (review [14]). In the present study both a deoxribonuclease and a bis-benzimidazole dye have been used as probes for chromatin organisation.

The dye, Ho33342, shows fluorescence enhancement on binding to DNA [15] and has been reported [16] to cause cell killing, inhibition of DNA synthesis, mutation and DNA strand-breakage in rodent cells. It was hypothesized that Ho33342 would induce DNA damage (and cell killing) in a manner which would reflect the innate sensitivity of chromatin to the ligand.

2. Materials and Methods

2.1 Cell Culture

The two fibroblast lines, MRC5CVI and AT5BIVA, are SV_{40} transformed derivatives of the normal and A-T parental strains MRC5 and AT5BI respectively. The trans-

---

1 MRC Clinical Oncology and Radiotherapeutics Unit, Hills Road, Cambridge, Great Britain

ADP-Ribosylation of Proteins
(ed. by F.R. Althaus, H. Hilz, and S. Shall)
© Springer-Verlag Berlin Heidelberg 1985

formed fibroblasts were kindly supplied by Dr. C. Arlett, MRC Cell Mutation Unit, Sussex. The human colon adenocarcinoma cell strain (HT29) was supplied to the Unit by Dr. J. Fogh (Sloan Kettering Institute, N.Y.). Cells were maintained in monolayer culture using standard procedures.

2.2 Analysis of Ho33342 Uptake and Binding by Flow Cytometry

The uptake and binding of Ho33342 (CP Laboratories, Bishops Stortford, U.K.) to cellular DNA was determined by fluorescence intensity measurements using preparations of permeabilised cells (see below) in LS buffer. Initial studies on HT29 cells used a flow cytometer incorporating a Spectra Physics (Mountain View, CA) 164-05 argon laser, tuned to the 356 and 363 UV lines. Median fluorescence (515−560 nm) distributions were determined for populations of single cells. Comparative studies on all three cell lines involved the use of an Ortho Diagnostics Instruments (Westwood, MA) system 50-H cell sorter with an associated 2150 computer system. Bound intranuclear Ho33342 was excited using a Coherent (Paolo Alto, CA) Innova 90 argon ion laser, emitting 200 mW in the UV region (wavelengths of 351.1 to 363.8 nm). Both fluorescence (390−440 nm) and scatter (at 5−15°; for the identification of cell debris) measurements were made on pulse area.

2.3 Cell Survival

Trypsin/versene detached cells were X-irradiated (using a 250 kV, 15 mA, Pantak X-irradiator; filtration of 2.32 mm copper half-value thickness) in suspension in complete medium (10^5 cells ml) with subsequent dilution of cells in growth medium for te measurement of clonogenic potential. In Ho33342 or 3AB toxicity experiments, cells were collected from logarithmic phase cultures and allowed to attach at 2.5 × $10^2 - 5 × 10^4$ cells per 9 cm diameter plastic dish (Sterilin) containing growth medium at 37° for a period of 16 h. Ho33342 or 3AB was then added to the indicated dilution and cultures incubated for the given period at 37° (pH being controlled in dye experiments by the addition 10 mM Hepes). Following treatment, cultures were: washed with phosphate buffered saline, growth medium replaced and incubated for the measurement of viability by clonogenic potential. In experiments involving 3AB, the inhibitor was present for 15 min before treatment, during treatment and for the specified recovery period.

2.4 Assay for DNA Strand Breaks

2.4.1 Preparation of Cells.
Assays for DNA damage (and Ho33342 uptake) were carried out on cultures which had been subjected to a standard protocol for the generation of freeze/thawed (permeabilised) cells directly from monolayer, adapted [17, 18] from the method described by Ganesan et al. [19]. Cells were resuspended in a low salt (LS) buffer (10 mM Tris-HCl, pH 8; 100 mM NaCl; 10 mM EDTA; 1 mg ml^{-1} bovine serum albumin).

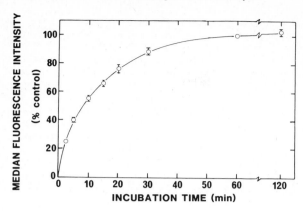

Fig. 1. The time course of uptake of Ho33342 by HT29 cells. The median of the distribution of fluorescence intensity is expressed as a percentage of the value at 1 h for 8.7 μM dye. *Data points* represent arithmetic means (± SE) for 2–8 independent experiments

2.4.2 DNase II Nicking. Permeabilised cells (8×10^6) were pelleted in a series of 1.5 ml ampoules by a short (1–2 s) centrifugation in an Eppendorf bench centrifuge. Drained pellets were then resuspended in 250 μl of prewarmed LS buffer (pH 7.75) containing DNase II (20 Kunitz U ml^{-1}; bovine spleen type V; Sigma). Samples were incubated in a water bath at 37° for 30 min and the reaction stopped by dilution of the cell suspension into 2.4 ml ice-cold LS buffer (pH 8.0) in preparation for the standard FADU assay described below.

2.4.3 DNA Unwinding Assay. DNA strand-breaks were measured by a modification of the method developed by Birnboim and Jevcak (FADU; fluorimetric analysis of DNA unwinding; [20]) involving the time-dependent partial unwinding of cellular DNA in crude cell lysates exposed to alkaline solutions. Treatment-induced enhancement of DNA unwinding (due to strand-breaks or alkali-labile damage) was determined by the expression; log (%D control)-log (%D treated), where %D was the % double-stranded DNA, and was found to give a linear relationship for X-irradiated cells up to at least 12 Gy. The X-ray calibration of the assay (see Sect. 3) was used to determine breaks 10^{-9} daltons DNA assuming 0.5 breaks 10^{-9} daltons mol. wt. Gy^{-1} [21].

2.5 Cell-Cycle Distribution Analysis

Cell-cycle phase distributions were determined by flow cytometry using an ethidium bromide/Triton technique [17].

3. Results

3.1 Characteristics of Ho33342-Induced DNA Strand-Breaks in Human Cell Lines

Figure 1 shows that the uptake of Ho33342 quantified by flow cytometry, into human adenocarcinoma cells (HT29) is initially rapid with a fluorescence maximum achieved after a 1 h exposure (for 8.7 μm). Figure 2 compares the dose-dependent

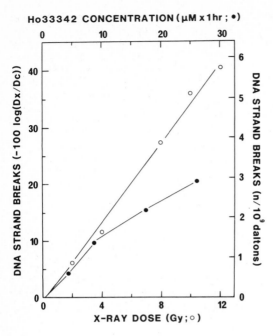

Fig. 2. Comparison of the induction of DNA strand-breaks (including those arising from alkali-labile lesions) in HT29 cells as a function of dye concentration (1 h exposure) or X-ray dose (irradiated monolayer cultures). *Symbols:* ○ X-ray data (n = 3); ● Ho33342 data (n = 2–7). *Data points* represent arithmetic means (SE < 15%) for the number (n) of determinations indicated above in *parentheses*. (Data from [17]; by permission of Taylor Francis Ltd.)

induction of DNA strand-breaks in HT29 cells as a consequence of ligand binding or X-irradiation. The induction response for radiogenic damage was linear whereas the ligand gave a biphasic response, the break relating to the concentration at which fluorescence quenching occurred. Figure 3 compares the fate of ligand- or X-radiation-induced DNA strand-breaks and shows that Ho33342-associated lesions disappear relatively slowly.

In order to monitor Ho33342-DNA binding in normal and A-T transformed fibroblasts the mean fluorescence intensity measurements were corrected for the mean cellular DNA content. The data shown in Fig. 4 indicate that less ligand is bound per unit of cellular DNA in the HT29 cells than by the transformed fibroblast lines.

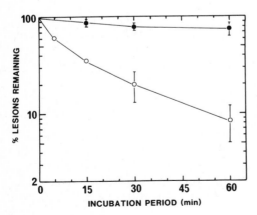

Fig. 3. Disappearance of strand-breaks from the DNA of HT29 cells following Ho33342 exposure (8.7 μM for 1 h) or X-irradiation (8 Gy; in air). *Data points* (with associated SE values) represent arithmetic means for multiple determinations: ● Ho33342 (four experiments); ○ X-rays (two experiments). (Data from [18]; by permission of IRL Press)

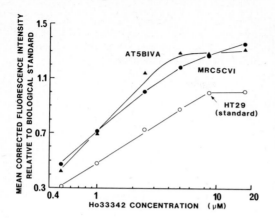

Fig. 4. Comparison of the concentration-dependent uptake of Ho33342 (1 h exposure) by three human cell lines as measured by flow cytometry. Ho33342-DNA associated fluorescence intensity measurements were corrected for the mean amount of DNA within a given cell population and related to a biological standard (i.e. HT29 cells exposed to 8.7 μM Ho33342 for 1 h at 37°). *Symbols:* ○ HT29; ● MRC5CVI; ▲ AT5BIVA. *Data points* represent mean of at least two determinations (SE < 10%). (Data from [18]; by permission of IRL Press)

The results shown in Fig. 4 define a standard treatment of 8.7 μM for 1 h as the minimal exposure condition under which plateau levels of fluorescence enhancement are achieved in the three cell lines. The levels of DNA strand-breakage induced under such conditions were determined and the results are shown in Table 1. In keeping with the increased ligand-binding capacity, the level of DNA damage is elevated in the fibroblast lines compared to the HT29 control. However, the A-T cell line shows a significantly greater number of lesions than the normal fibroblast line.

To assess whether the results obtained with Ho33342 related to some innate feature of cellular chromatin, namely accessibility, permeabilised cells were exposed to a low level of DNase II (under conditions which produced a linear relationship between enzyme concentration and break frequency) and accessibility monitored as the frequency of enzyme-induced DNA strand-breaks. The results (Table 1) show that nuclease sensitivity correlated with the frequency of ligand-induced breaks.

Table 1. Comparison of the sensitivity of three human cell lines to Ho33342 and DNase II-induced DNA strand-beakage. (Data taken from [18])

Cell line	DNA strand-breaks (N 10^{-9} daltons)	
	Ho33342-induced (8.7 μM × 1 h)	DNase II-induced (20 u ml^{-1} × 0.5 h)
HT29	1.4 ± 0.1[a] (7)	3.5 ± 0.2[a] (6)
MRC5CVI	2.8 ± 0.3 (5)	6.5 ± 0.8 (6)
AT5BIVA	4.6 ± 0.3 (5)	10.5 ± 0.9 (4)

[a] Values represent arithmetic means (± SE) for the number of determinations indicated in parentheses

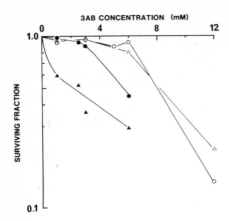

Fig. 5. Representative toxicity curves for the response of MRC5CVI (○ ●) and AT5BIVA (△ ▲) to 3AB for a 24 h (*open symbols*) or continuous (*closed symbols*) exposure period

3.2 Cell Survival and Cell-Cycle Effects of 3-Aminobenzamide (3AB) on X-Irradiated Cells

The toxic effects of a continuous or 24 h exposure to various concentrations of 3AB were determined for the two transformed cell lines (Fig. 5). Continuous exposure to this inhibitor was more toxic to the A-T cell line. A standard inhibitor dose of 6 mM was selected and found to enhance the sensitivity of normal cells to low doses of X-radiation only when the drug was present continuously during post-irradiation incubation (Fig. 6). The X-ray sensitivity of A-T cells was not affected by the short term (24 h) or continuous presence of 3AB.

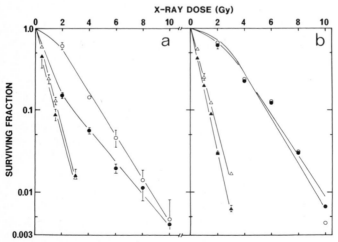

Fig. 6a,b. Comparison of the effects of the post-irradiation presence of 3AB, for continuous (a) or a 24 h (b) exposure, on the X-ray survival responses of MRC5CVI (○ ●) and AT5BIVA (△ ▲) cells. *Open symbols* represent untreated controls and *closed symbols* indicate 3AB exposure. *Data points* (with associated SE) represent arithmetic means for 2–4 determinations. (Data from Smith et al., Int. J. Radiat. Biol. 47:701–712, 1985)

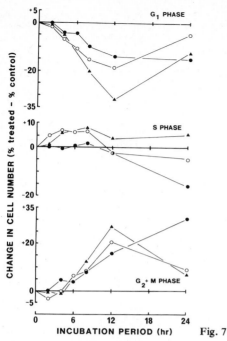

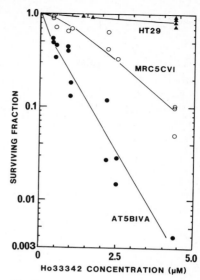

Fig. 7

Fig. 8

Fig. 7. Perturbations of cell-cycle phase distributions for cells exposed to 4 Gy X-radiation. *Data points* represent arithmetic means (SE < 3%) of results from two (HT29) and three (fibroblasts) experiments. The data represent the loss or gain of cells, with respect to the number of untreated control cells within a given cell-cycle compartment. *Symbols:* ▲ HT29; ○ MRC5CVI; ● AT5BIVA

Fig. 8. Concentration-dependent Ho33342 survival responses of human cell lines (1 h exposure). *Data points* represent accumulated results for three experiments for each cell line

Figure 7 shows the X-ray-induced (4 Gy) changes in cell-cycle distributions for the three cell lines. In the normal fibroblast and HT29 cell lines, irradiation produces a transient accumulation of cells in S phase, in keeping with the inhibitory effect on de novo DNA synthesis. The A-T cell line shows no initial effect of X-radiation on the proportion of cells in S phase which is consistent with the relative resistance of de novo DNA synthesis to radiogenic inhibition in A-T-derived cells [7, 11, 22]. All of the cell lines show progressive accumulations of cells in $G_2 + M$ phase from 2 to 12 h following irradiation. Between 12 and 24 h post-irradiation, HT29 and MRC5CVI cells demonstrate recovery from retention in $G_2 + M$ phase whereas the A-T cells continue to accumulate leading to an absence of replenishment of the G_1 fraction. Hence, the number of cells retained in G_2 phase at 24 h post-treatment reflects the efficiency of the late recovery process.

The results in Table 2 indicate that 3AB does not modify the entry of irradiated cells into $G_2 + M$ phase for either cell line (c.f. values at 12 h post-irradiation; Table 2) but does appear to inhibit (or at least delay) the expression of the recovery phenomenon in cells in $G_2 + M$ phase. Furthermore, the effectiveness of 3AB is dependent

Table 2. Effect of 3AB on the retention of X-irradiated cells in G_2+M phase

Cell line	X-ray dose (Gy)	3AB (6 mM)	Retention of cells in G_2 phase at t(h) after irradiation (% treated − % control)[a]			
			t = 12	Ratio (+3AB/−3AB)	t = 24	Ratio (+3AB/−3AB)
MRC5CVI	0	−	Control		Control	
	0	+	ns[b]		ns	
	4.0	−	17.7 ± 0.8	1.0	7.6 ± 1.1	2.0
	4.0	+	18.5 ± 0.1		15.3 ± 2.4	
	6.75	−	nd[c]		11.9 ± 3.1	1.8
	6.75	+	nd		20.9 ± 1.1	
	8.0	−	nd		19.0 ± 1.4	1.4
	8.0	+	nd		26.7 ± 1.4	
AT5BIVA	0	−	Control		Control	
	0	+	ns		ns	
	1.0	−	nd		5.8 ± 1.3	2.2
	1.0	+	nd		12.8 ± 1.9	
	4.0	−	15.0 ± 3.5	1.1	31.8 ± 2.0	0.9
	4.0	+	16.3 ± 4.5		29.7 ± 2.6	

[a] Values represent arithmetic means (± SE) for 2 to 11 (majority 6) determinations
[b] ns = not significantly different from control
[c] nd = not determined

upon the X-ray dose for each strain, there being fewer cells available for 3AB-induced delay at higher doses. The A-T cells are responsive to 3AB at a low dose of X-rays, but this effect is overshadowed by the radiogenic perturbation as the X-ray dose is increased.

3.3 Cell Survival and Cell-Cycle Effects of 3AB on Ho33342-Treated Cells

Figure 8 compares the Ho33342-survival responses of the three human lines. The HT29 cell line shows resistance to the ligand whereas the A-T cell line shows enhanced cell killing compared to the MRC5CVI line. The resistance of the HT29 cell line is consistent with the reduced level of ligand uptake/binding whereas the sensitivity of the A-T line presumably reflects the elevated levels of DNA damage.

Table 3 presents the results of a cytotoxicity and cell-cycle study on the effects of the ligand either alone or in the presence (for treatment and recovery) of 3AB. The results in Table 3 show that normal cells are sensitized to Ho33342 by the continuous presence of 3AB, whereas the A-T cells are not. When analysed at 24 h after Ho33342 exposure, the predominant difference in cell-cycle distribution for the two fibroblast lines is the increase in the retention of A-T cells in G_2 phase with the consequential reduction of the G_1 compartment. The presence of 3AB for 24 h elevated the reten-

Table 3. Comparison of the effects of 3-aminobenzamide on Ho33342-induced cytotoxicity and cell-cycle phase perturbations in human adenocarcinoma cells and transformed fibroblasts

Cell line	Treatment[a]	Cytotoxicity[b] (% Survival)	Expt	G_1	S	G_2+M	$\dfrac{(G_2+M)^{+3AB}}{(G_2+M)^{-3AB}}$
HT29	8.7 μM dye	nd[d]	1	+7.1	−11.8	+4.7	
			2	+4.2	−16.0	+11.8	
	8.7 μM dye + 3AB	nd	1	+1.9	−10.4	+8.5	1.8
			2	−3.2	−16.8	+20.0	1.7
MRC5CVI	1 μM dye	54 ± 10	1	−3.3	−4.9	+8.2	
			2	−0.7	−10.8	+10.1	
	1 μM dye + 3AB	8 ± 3 (6.5 ± 0.2)	1	−9.4	−5.1	+14.4	1.8
			2	−13.0	−5.0	+18.0	1.8
AT5BIBVA	1 μM dye	25 ± 9	1	−10.5	−10.7	+21.3	
			2	−6.3	−9.5	+15.7	
	1 μM dye + 3AB	20 ± 2 (5.0 ± 2.0)	1	−12.1	−9.3	+23.3	1.1
			2	−8.5	−10.7	+19.0	1.2

[a] 3AB treatment at 6 mM (continuous exposure)
[b] Values represent arithmetic means (± SE) for three determinations and are presented as data either corrected or uncorrected (numbers in parentheses), for 3AB toxicity
[c] Cell cycle distribution determined by flow cytometric analysis; each value represents the mean of duplicate determinations within a given experiment (SE approx. 2%). A 24 h exposure to 6 mM 3AB had no significant effect on cell-cycle phase distributions of MRC5CVI cells (G_1 = 48.6%, S = 33.7%, G_2+M = 17.7%) or AT5BIVA cells (G_1 = 40.5%, S = 34.2%, G_2+M = 25.3%) or HT29 cells (G_1 = 50.9%, S = 33.9%, G_2+M = 15.2%). Fibroblast data taken from [18]
[d] nd = not determined

tion of Ho33342-treated normal (factor of 1.8) but not A-T (factor of 1.1−1.2) cells in G_2+M phase. The HT29 cell line responds in a similar fashion to the MRC5CVI cell line except that, as expected, a higher ligand concentration was required.

4. Discussion

It has been proposed [18] that bis-benzimidazole dye molecules, when bound to the nuclear DNA of intact cells, elicit conformational changes which are enzymatically converted (with an efficiency dependent upon enzyme accessibility) to strand-breaks of alkali-labile sites. In agreement with this model, the three lines studied ranked in order of both the expressed frequency of ligand-induced breaks and the feature of chromatin accessibility as monitored by DNase II sensitivity.

The abnormal cell killing and cell-cycle perturbation effects of Ho33342 on A-T cells can be attributed to the dose-modifying effect (in terms of DNA damage) of the chromatin anomaly. Experiments to determine the effects of inhibition of poly(ADP-ribosyl)ation revealed that the ligand-induced accumulation of cells in G_2 phase (measured at 24 h post-treatment) was enhanced by the inhibitor 3AB in normal but not A-T cells. This differential effect is probably due to the fact that the number of ligand-treated A-T cells retained in G_2 is already elevated, thus removing a potentially inhibitor-responsive population. The corresponding effect of 3AB on ligand toxicity is in keeping with the cell-cycle effects but more difficult to interpret. Continuous exposure to the inhibitor was required to increase ligand cytotoxicity in normal cells whereas the A-T cells were non-responsive. This refractory state of A-T cells may relate to the increased toxicity of 3AB alone.

All three cell lines studied demonstrated a similar and progressive accumulation in G_2 +M phase at $2-12$ h following X-irradiation. In the normal controls, at $12-24$ h there was an overall release of cells from the delayed state whereas in the A-T line, cells continued to accumulate. The continuous presence of 3AB increased the number of normal cells retained in G_2 +M phase. In A-T cells exposed to low levels of radiation, 3AB increased the retention of cells in G_2 +M phase to an extent commensurate with the responses of a normal cell undergoing the same absolute level of radiogenic delay. This apparently normal capacity (in relation to the magnitude of the G_2 +M delay) of irradiated A-T cells to respond to 3AB is compatible with the observations of Zwelling et al. [23] that the levels of X-ray induced poly(ADP-ribosyl)ation are normal in A-T cells. However, the predicted capacity of 3AB to modify A-T survival at low X-ray doses is not observed probably due to the elevated toxicity of 3AB alone when present continuously.

Zwelling et al. have also reported [23] that unirradiated A-T cells demonstrate elevated levels of poly(ADP-ribosyl)ation activity. Since this function is dependent upon the presence of DNA strand-breaks [1], such findings may indicate that A-T cells exhibit a modified steady state of enzymatically-induced breaks in chromatin which would be consistent with a state of enhanced nuclease susceptibility.

Ho33342 appears to induce the type of DNA damage which can activate poly(ADP-ribosyl)ation-dependent events, and that inhibition in normal cells enhances ligand toxicity. In A-T cells the critical targets for ligand attack may have already undergone extensive ribosylation (as discussed above) with subsequent chromatin relaxation [2] providing unusually accessible regions for Ho33342 and nucleases, resulting in relatively refractory substrates for the DNA damage/repair modifying effects of subsequent exposure to 3AB. In conclusion, identification of the genomic distribution of preferred sites for ligand and nuclease attack is of obvious importance and may reveal minor fractions of the genome which are unusually dependent upon poly(ADP-ribosyl)ation events for the control of cell-cycle traverse following exposure to DNA damaging agents.

Acknowledgments. We thank Professor N.M. Bleehen for encouragement and support; and Dr. J.V. Watson for the provision of the flow cytometry facility.

References

1. Shall S (1982) ADP-ribose in DNA repair. In: Hayaishi O, Ueda K (eds) ADP-ribosylation reactions. Academic Press, London New York, p 472
2. Poirer GC, De Murcia G, Jongstra-Billen J, Niedergang C, Mandel P (1982) Poly(ADP-ribosyl)-ation of polynucleosomes causes relaxation of chromatin structure. Proc Natl Acad Sci USA 79:3423–3427
3. Creissen D, Shall S (1982) Regulation of DNA ligase activity by poly(ADP-ribose). Nature (London) 296:271–272
4. Rowley R, Zorch M, Leeper DB (1984) Effect of caffeine on radiation – induced mitotic delay: delayed expression of G_2 arrest. Radiat Res 97:178–185
5. Purnell MR, Whish WJD (1980) Novel inhibitors of poly(ADP-ribose) synthetase. Biochem J 185:775–777
6. Paterson MC, Smith PJ (1979) Ataxia telangiectasia: an inherited human disorder involving hypersensitivity to ionizing radiation and related DNA-damaging chemicals. Annu Rev Genet 13:291–318
7. Bridges BA, Harnden DG (1982) Ataxia-telangiectasia – a cellular and molecular link between cancer, neuropathology and immune deficiency. Wiley, Chichester
8. Zampetti-Bosseler F, Scott D (1981) Cell death, chromosome damage and mitotic delay in normal human, ataxia telangiectasia and retinoblastoma fibroblasts after X-radiation. Int J Radiat Biol 39:547–558
9. Imray FP, Kidson C (1983) Perturbations of cell-cycle progression in gamma-irradiated ataxia telangiectasia and Huntington's disease cells detected by DNA flow cytometric analysis. Mutat Res 112:369–382
10. Ford MD, Martin L, Lavin MF (1984) The effects of ionizing radiation on cell cycle progression in ataxia telangiectasia. Mutat Res 125:115–122
11. Painter RB, Young BR (1980) Radiosensitivity in ataxia telangiectasia: A new explanation. Proc Natl Acad Sci USA 77:7315–7317
12. Jaspers NGJ, de Wit J, Regulski MR, Bootsma D (1982) Abnormal regulation of DNA replication and increased lethality in ataxia telangiectasia cells exposed to carcinogenic agents. Cancer Res 42:335–341
13. Igo-Kemenes T, Horz W, Zachau HG (1982) Chromatin. Annu Rev Biochem 51:89–121
14. Darzynkiewicz Z (1979) Acridine orange as a molecular probe in studies of nucleic acids in situ. In: Melamed MR, Mullaney PF, Mendelsohn ML (eds) Flow cytometry and sorting. Wiley, New York, p 285
15. Latt SA (1979) Fluorescent probes of DNA microstructure and synthesis. In: Melamed MR, Mullaney PF, Mendelsohn ML (eds) Flow cytometry and sorting. Wiley, New York, p 139
16. Durand RE, Olive PL (1982) Cytotoxicity mutagenicity and DNA damage by Hoechst 33342. J Histochem Cytochem 30:111–116
17. Smith PJ, Anderson CO (1984) Modification of the radiation sensitivity of human tumour cells by a bis-benzimidazole derivative. Int J Radiat Biol 46:331–344
18. Smith PJ (1984) Relationship between a chromatin anomaly in ataxia telangiectasia cells and enhanced sensitivity to DNA damage. Carcinogenesis 5:1345–1350
19. Ganesan AK, Smith CA, Van Zeeland AA (1981) Measurement of the pyrimidine dimer content in DNA in permeabilized bacterial or mammalian cells with endonuclease V of bacteriophage T_4. In: Friedberg EC, Hanawalt PC (eds) DNA repair. A laboratory manual of research techniques, vol I. Marcel Dekker, New York, p 89
20. Birnboim HC, Jevcak JJ (1981) Fluorometric method for rapid detection of DNA strand-breaks in human white blood cells produced by low doses of radiation. Cancer Res 41:1889–1892
21. Kanter PM, Schwartz HS (1982) A fluorescence enhancement assay for cellular DNA damage. Mol Pharmacol 22:145–151

22. Smith PJ, Paterson MC (1983) Effect of aphidicolin on de novo DNA synthesis, DNA repair and cytotoxicity in gamma-irradiated human fibroblasts: Implications for the enhanced radio-sensitivity in ataxia telangiectasia. Biochim Biophys Acta 739:17–26
23. Zwelling LA, Kerrigan D, Mattern MR (1983) Ataxia-telangiectasia cells are not uniformly deficient in poly(ADP-ribose) synthesis following X-irradiation. Mutat Res 120:69–78

Potentiation of the Effects of Bleomycin and 1,3-Bis(2-Chloroethyl)-1-Nitrosourea on A549 Cell Cultures by Inhibitors of Poly(ADP-Ribose) Synthetase

DORIS A. GRAY, JOHN M. LUNN, MICHAEL R. PURNELL, and ADRIAN L. HARRIS[1]

Introduction

The observation of Miller [1] that single-strand breaks in DNA stimulate poly(ADP-ribose) synthetase has led to a large amount of research exploring a possible role for this enzyme in DNA repair [2]. Inhibitors of the enzyme have been shown to enhance the cytotoxicity of DNA damaging agents [3]. The majority of work has focussed on monofunctional alkylating agents in conjunction with inhibitors of poly(ADP-ribose) synthetase, but because this class of compounds has extremely little clinical use we have explored the interaction of the inhibitor 3-acetoamidobenzamide (3AAB) with the glycopeptide antibiotic, bleomycin (BLM), or with the bifunctional alkylating agent, 1,3-bis(2-chloroethyl)-1-nitrosourea (BCNU). Both of these drugs are widely used in cancer chemotherapy, either singly or in combination with other treatments.

Cytotoxicity

In these investigations, we have used A549 cells, derived from a human lung carcinoma, grown as adherent cultures in RPMI 1640 medium supplemented with 10% foetal bovine serum. The effect of either BLM or BCNU on cell proliferation was assessed by measuring the increase in the number of cells in cultures 3 days after drug treatment. This is expressed as "percentage cell number increase", defined as

$$\frac{[(\text{cell number) day 3} - (\text{cell number) day 0}]\ \text{drug}}{[(\text{cell number) day 3} - (\text{cell number) day 0}]\ \text{control}} \times 100.$$

When 3AAB was present in drug-treated cultures, the appropriate control cultures were treated with 3AAB.

1   Cancer Research Unit, University of Newcastle upon Tyne, Royal Victoria Infirmary, Newcastle upon Tyne NE1 4LP, Great Britain

ADP-Ribosylation of Proteins
(ed. by F.R. Althaus, H. Hilz, and S. Shall)
© Springer-Verlag Berlin Heidelberg 1985

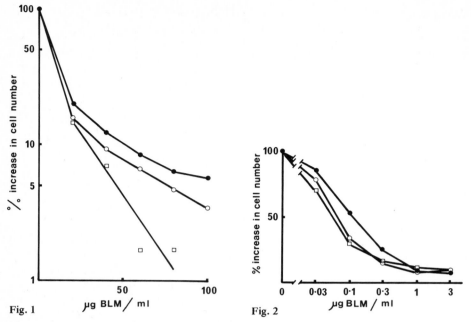

Fig. 1. Effect of 3AAB on the retardation of proliferation of A549 cells 3 days after a 30 min exposure to BLM. ● 0 mM 3AAB; ○ 2 mM 3AAB; □ 4 mM 3AAB. Triplicate cultures were counted for each determination. Results ● and ○ are the means of three independent experiments

Fig. 2. Effect of 3AAB on the retardation of proliferation of A549 cells exposed continuously to BLM for 3 days. Triplicate cultures were counted for each determination. ● 0 mM 3AAB; ○ 1 mM 3AAB; □ 2 mM 3AAB

Bleomycin

The proliferation of A549 cells in 72 h following a 30 min exposure to increasing concentrations of BLM is shown in Fig. 1. The dose response curve is biphasic, with apparently sensitive and resistant populations of cells. Cytotoxicity was enhanced by 3AAB only at BLM concentrations greater than 20 μg ml^{-1}, i.e. only in "resistant" cells. A more detailed study in the range 0–10 μg BLM ml^{-1} ("sensitive" cells) failed to reveal any potentiation by 3AAB (results not shown).

Pulmonary toxicity, the major side effect of BLM therapy in cancer patients, is minimized by continuous infusion of relatively low concentrations of BLM, rather than by injection of the drug as a bolus. An analogous approach showed that BLM applied continuously over 72 h was toxic to A549 cells at much reduced concentrations (Fig. 2). Further, even at low BLM concentrations, cytotoxicity was enhanced by 3AAB. Thus, for example, the BLM concentration required to reduce proliferation by 50% (ID$_{50}$) fell from 0.116 μg ml^{-1} in the absence of 3AAB to 0.071 or 0.062 μg ml^{-1} in the presence of 1 or 2 mM 3AAB respectively (Fig. 3).

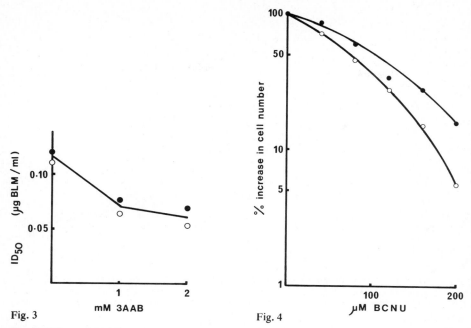

Fig. 3. Effect of 3AAB on the dose of BLM, applied continuously for 3 days, required to reduce proliferation of A549 cells by 50%. Results from two independent experiments are shown

Fig. 4. Effect of 3AAB on the retardation of proliferation of A549 cells 3 days after a 30 min exposure to BCNU. Triplicate cultures were counted for each determination. Results are the means from two independent experiments. ● 0 mM 3AAB; ○ 2 mM 3AAB

BCNU

Exposure of A549 cells to increasing concentrations of BCNU for 30 min gave rise to a broad shoulder type dose response curve (Fig. 4). Potentiation of BCNU cytotoxicity by 3AAB was observed throughout the concentration range of BCNU tested (40–200 μM).

DNA Strand Breaks

Following treatment of A549 cell suspensions with either BLM or BCNU, DNA strand breaks were measured by alkaline sucrose gradient centrifugation. Exposure of cells to BLM (500 μg ml^{-1}) produced considerable DNA strand scission (Fig. 5), but much of this was repaired 2 h after drug removal. Inclusion of 3AAB in the medium during the recovery phase had only a slight effect on DNA strand rejoining (Fig. 5). Immediately after a similar 30 min treatment of A549 cells with BCNU (50 μM), essentially no

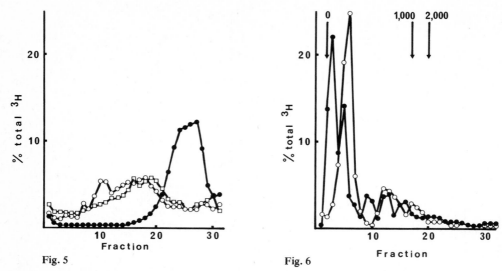

Fig. 5   **Fig. 6**

Fig. 5. DNA strand breaks produced by a 30 min exposure of A549 cells to BLM (500 μg ml⁻¹). Strand breaks were assayed by alkaline sucrose gradient centrifugation [14]. Sedimentation was from *right to left.* ● Immediately after BLM treatment; ○ 2 h recovery, 0 m*M* 3AAB; □ 2 h recovery, 5 m*M* 3AAB

Fig. 6. DNA strand breaks produced by a 30 min exposure of A549 cells to BCNU (50 μ*M*). Strand breaks were assayed by alkaline sucrose gradient centrifugation [14]. ● Immediately after BCNU treatment; ○ BCNU treatment followed by 500 rad X-irradiation. *Arrows* indicate peak positions for DNA of cells exposed to 0, 1 000 and 2 000 rad γ-irradiation

DNA strand breaks could be detected (Fig. 6). However, since X-irradiation (500 rad) subsequent to BCNU treatment also failed to produce detectable DNA strand breaks (Fig. 6), it seems likely that strand breaks have been masked by cross-linking effects.

Poly(ADP-Ribose) Synthetase

Poly(ADP-Ribose) synthetase activity was measured by the incorporation (TCA-insoluble) of [³H]-NAD in permeabilized cells immediately following treatment of A549 cell suspensions for 30 min with either BLM (100 μg ml⁻¹) or BCNU (500 μ*M*). Whereas BLM produced an eightfold stimulation of activity, no increase could be detected after BCNU treatment (Fig. 7).

Discussion

The biphasic appearance of the BLM cytotoxicity dose response curve (Fig. 1) resembles that described by Barranco and Humphrey [4] for CHO cells. The selective sensitivity

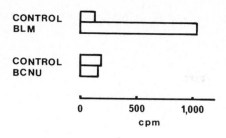

Fig. 7. Stimulation of poly(ADP-ribose) synthetase activity in A549 cells treated for 30 min with either BLM (100 μg ml^{-1}) or BCNU (500 μM). Synthetase was assayed in permeabilized cells [15] by measuring the amount of [^{3}H]-NAD incorporated in 30 min into TCA-insoluble material

of M-phase cells may, in part, explain the presence of sensitive and resistant populations of cells [4]. Further, the lack of enhancement of BLM cytotoxicity by 3AAB in "sensitive" cells could be due to the absence of repair of potentially lethal damage in M-phase cells [5].

The dose response curve for the proliferation of BCNU-treated A549 cells (Fig. 4) resembles in shape the reported survival curve for CHO cells similarly treated [6]. With CHO cells, the initial shoulder on the curve is followed by a linear region where cell survival decreased exponentially with increasing BCNU concentrations. However, cell number increases of much less than 5% are experimentally difficult to measure accurately, and so the BCNU dose response curve for A549 cell survival (Fig. 4) could not be extended to the linear region. Consequently it is not possible to identify, from the data of Fig. 4, whether 3AAB is affecting the shoulder or slope of the dose response curve, or both. Work is in progress to clarify this point using colony forming ability to assess cytocoxicity. In addition, preliminary experiments suggest that potentiation of cytotoxicity by 3AAB is less effective with BCNU than with monofunctional alkylating agents.

Our finding of potentiation by 3AAB of BCNU cytotoxicity contrasts with other recent reports. Berger et al. [7] failed to detect any 3-aminobenzamide (3AB)-mediated potentiation of BCNU toxicity to L1210 mouse leukemic cells, and likewise Boorstein and Pardee [8] detected no enhancement by the same inhibitor of BCNU toxicity to human fibroblasts. This may reflect differences in the cell systems. Alternatively, since 3AAB is some 6—8 times more effective as a poly(ADP-ribose) synthetase inhibitor than is 3AB [9], it could be that the concentrations of 3AB used by Berger et al. [7] and Boorstein and Pardee [8], 10 mM and 4 mM respectively, were not sufficiently high.

Detection of DNA strand breaks after treating cells with the levels of drugs used to produce cytotoxicity probably requires techniques more sensitive than alkaline sucrose gradient centrifugation. Indeed, preliminary experiments suggest that the alkali-unwinding technique of Birnboim and Jevcak [10] may detect BLM-induced DNA strand breaks. BCNU-induced DNA strand breaks have been reported [11]. However, assessment of these is complicated by DNA cross-linking [12]. Also, since BCNU produces DNA strand breaks for a number of hours after drug withdrawal [13], it may be that breaks should be measured at a later time. Such a delay in DNA repair could explain why no increase in poly(ADP-ribose) synthetase activity was detected immediately following BCNU treatment of A549 cells (Fig. 7). Indeed, Dornish and Smith-Kielland [13] have found a modest increase in poly(ADP-ribose) synthetase activity in BCNU-treated Ehrlich ascites tumour cells 2—4 h after withdrawal of the drug, corrresponding to the time of maximal DNA repair.

Thus, it would appear that cytotoxicity of the clinically useful drugs BLM and BCNU can be potentiated using 3AAB, an inhibitor of poly(ADP-ribose) synthetase.

Acknowledgment. This work was supported by the North of England Cancer Research Campaign.

References

1. Miller EG (1975) Stimulation of nuclear poly (adenosine diphosphate ribose) polymerase activity from HeLa cells by endonucleases. Biochim Biophys Acta 395:191—200
2. Shall S (1982) ADP-ribose in DNA repair. In: Hayaishi U, Ueda K (eds) ADP-ribosylation reactions. Academic Press, London New York, pp 477—520
3. Nduka N, Skidmore CJ, Shall S (1980) The enhancement of cytotoxicity of N-methyl-N-nitrosourea and of γ-radiation by inhibitors of poly(ADP-ribose) polymerase. Eur J Biochem 105:525—530
4. Barranco SC, Humphrey RM (1971) The effects of bleomycin on survival and cell progression in Chinese hamster cells in vitro. Cancer Res 31:1218—1233
5. Barranco SC, Bolton ME (1977) Cell cycle phase recovery from bleomycin-induced potentially lethal damage. Cancer Res 37:2589—2591
6. Barranco SC, Humphrey RM (1971) The effects of 1,3-bis(2-chloroethyl)-1-nitrosourea on survival and cell progression in Chinese hamster cells. Cancer Res 31:191—195
7. Berger NA, Catino DM, Vietti TJ (1982) Synergistic antileukemic effect of 6-aminonicotinamide and 1,3-bis(2-chloroethyl)-1-nitrosourea on L1210 cells in vitro and in vivo. Cancer Res 42:4382—4386
8. Boorstein RJ, Pardee AB (1984) Factors modifying 3-aminobenzamide cytotoxicity in normal and repair-deficient human fibroblasts. J Cell Physiol 120:335—344
9. Purnell MR, Whish JD (1980) Novel inhibitors of poly (ADP-ribose) synthetase. Biochem J 185:775—777
10. Birnboim HC, Jevcak JJ (1981) Fluorometric method for rapid detection of DNA strand breaks in human white blood cells produced by low doses of radiation. Cancer Res 41:1889—1892
11. Erickson LC, Bradley MO, Kohn KW (1977) Strand breaks in DNA from normal and transformed human cells treated with 1,3-bis(2-chloroethyl)-1-nitrosourea. Cancer Res 37:3744—3750
12. Erickson LC, Bradley MO, Ducore JM, Ewig RAG, Kohn KW (1980) DNA crosslinking and cytotoxicity in normal and transformed human cells treated with antitumour nitrosoureas. Proc Natl Acad Sci USA 77:467—471
13. Dornish JM, Smith-Kielland I (1981) Alkylation of Ehrlich ascites tumor cells elicits a response from poly (ADP-ribose) polymerase. In: Seeberg E, Kleppe K (eds) Chromosome damage and repair. Plenum Press, New York London, pp 497—501
14. Gray DA, Durkacz BW, Shall S (1981) Inhibitors of nuclear ADP-ribosyl transferase retard DNA repair after N-methyl-N-nitroso-urea. FEBS Lett 131:173—177
15. Benjamin RC, Gill DM (1980) ADP-ribosylation in mammalian cell ghosts. J Biol Chem 255:10493—10501

Cell Death Caused by the Anti-Cancer Drug Methotrexate:
Does ADP-Ribosyl Transferase Have a Role to Play?

KEVIN M. PRISE, JOZSEF C. GAAL, and COLIN K. PEARSON[1]

Introduction

Thymidylate residues required for cellular DNA synthesis are obtained exogenously via the salvage pathway and/or are synthesised de novo via dUMP in a reaction catalysed by thymidylate synthase (dUMP → dTMP). The thymidine monophosphate can then be phosphorylated by kinases to dTDP and dTTP before being incorporated into DNA.

The formation of thymidylate from dUMP has long been a major target for cancer chemotherapeutic agents which either inhibit thymidylate synthase directly (for example, 5-fluorodeoxyuridine) or dihydrofolate reductase (the folate antagonists such as methotrexate), the activity of which is essential for dTMP formation. The consequences of such inhibition are that the intracellular concentration of dTTP falls and that of dUMP increases [1]. Studies using cultured lymphoid cells report that these changes are accompanied by an increase in intracellular dUTP concentration of more than 1000-fold [2]. Taken with the reduction in dTTP to about 2% of the original amount, this brought the ratio of intracellular dUTP/dTTP from less than 10^{-5} to less than 10^{-1}. A consequence of this is that dUMP is incorporated into cellular DNA in place of dTMP.

Subsequent excision-repair of the DNA at the sites of uracil incorporation may only result in the re-incorporation of uracil in drug-treated cells because of the high dUTP: dTTP ratio. This may lead to a futile cycle of uracil incorporation and excision resulting in an increase in the steady-state level of DNA strand breaks, a possible contributory factor leading ultimately to cell death.

This hypothesis has been investigated together with studies on ADP-ribosylation reactions expected to be affected by this putative DNA damage and/or perturbations in chromatin structure caused by the methotrexate.

---

1 Department of Biochemistry, University of Aberdeen, Marischal College, Aberdeen AB9 1AS, Scotland, Great Britain

ADP-Ribosylation of Proteins
(ed. by F.R. Althaus, H. Hilz, and S. Shall)
© Springer-Verlag Berlin Heidelberg 1985

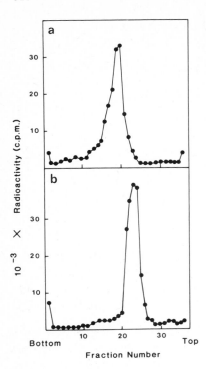

Fig. 1a,b. Nucleoid sedimentation analysis. HeLa cells were grown for 24 h at 37°C in the presence of [6-^{3}H] thymidine at 1 μCi ml^{-1} (sp.act. 26 Ci mmol^{-1}). The medium was then removed and replaced with fresh medium (a) or with medium containing 100 μM methotrexate (b) and incubation continued for 1 h. Subsequent nucleoid sedimentation analysis (6000 rpm for 30 min at 20°C, SW50.1 rotor) was as described by Cook and Brazell [3]. Fractions were collected from the bottom of gradients and precipitated onto GF/C filters for radioactivity determination. Sedimentation is from *right to left*

Do DNA Strand Breaks Occur in Cells Treated with Methotrexate?

Neutral sucrose gradient sedimentation analysis under high salt conditions [3] shows (Fig. 1) that nucleoids from cells exposed to 100 μM methotrexate for 1 h sediment more slowly than controls [ratio 0.856 $\pm$ 0.04 (SD), n = 3]. If we accept that this reduction in sedimentation rate is due to DNA strand breaks this represents about 300 breaks per genome [4].

Since methotrexate diminishes tetrahydrofolate (FH$_4$) levels it effectively inhibits not only thymidylate synthesis but that of purines and some amino acids also requiring FH$_4$. In order to isolate the effect of the methotrexate to that on dihydrofolate reductase, and hence thymidylate synthesis, cells were grown with hypoxanthine (25 μM) present also. Subsequent analysis revealed that nucleoids still sedimented more slowly than controls [ratio 0.879 $\pm$ 0.055 (SD), n = 4].

Activation of ADPR-Transferase in Methotrexate-Treated Cells

Transferase activity [5] increased rapidly reaching some 300% stimulation 3 h after exposure of cells to the methotrexate and remained higher than control activities for 24 h (Fig. 2). However, when the growth medium also contained hypoxanthine, for

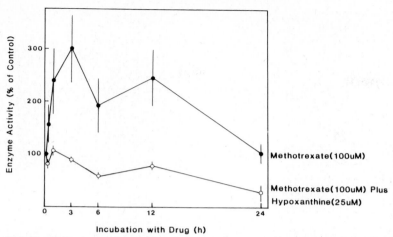

Fig. 2. ADPR-transferase activity in permeabilized cells. Conditions for growing cells and adding drugs was as described for Fig. 1 except that [^{3}H]-thymidine was not present. At the times indicated cells were recovered, permeabilized and assayed for transferase activity [5]. The cell growth media contained either 100 μM methotrexate alone (●) or 100 μM methotrexate and 25 μM hypoxanthine (○)

reasons explained above, no activation of the transferase was observed but the nucleoid sedimentation rate from these cells was still slower than that from controls, as stated above, showing that DNA strand breaks were still present. (Neither methotrexate nor hypoxanthine affected enzyme activity when added to a transferase assay.)

It appears from this that the DNA strand breaks are not the cause of the increased transferase activity, unless the hypoxanthine in some way inhibits only the activation of the enzyme caused by the methotrexate, without affecting the basal level of activity.

Methotrexate had no affect on ADP-ribose degradation, showing that we are observing enhanced ADP-ribose synthesis. This was not due to gross increases in cellular NAD$^+$ content, since these were similar in control and drug-treated cells over 24 h except for a small transient decrease in the first 2 h.

Protein ADP-Ribosylation in Drug-Treated Cells

Three bands appear darker on autoradiographs of acrylamide gels of ADP-ribosylated proteins from cells exposed to methotrexate for 24 h (Fig. 3, arrows). This is more evident in the permeabilized cells incubated at the higher NAD$^+$ concentration in the ADP-ribosyl transferase reaction (compare track 2 with 5 in Fig. 3A). We have not yet examined proteins from cells exposed to methotrexate for a shorter duration. The three proteins that the bands represent have mol. wts. of about 96 kD, 17 kD and 14 kD. Possible identities include a topoisomerase subunit or a peptide fragment derived from the transferase (96 kD) and core histones (17 kD, 14 kD), although

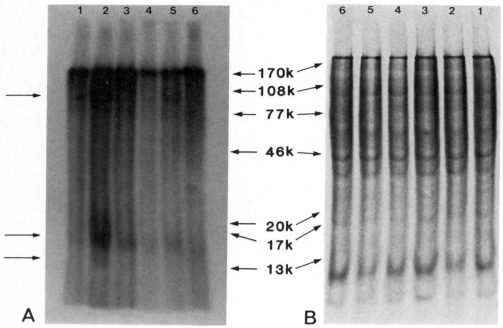

Fig. 3A,B. SDS gradient gel electrophoresis and autoradiography of ADP-ribosylated proteins. Permeabilized HeLa cells (5×10^6) were incubated for 20 min at 26°C with 0.133 μM [^{32}P]-NAD$^+$, sp.act. 245.8 Ci mmol^{-1}, made up to 1 μM or 100 μM with non-radioactive NAD$^+$. Each assay contained about 5.9×10^6 cpm. Reactions (75 μl) were stopped by adding 75 μl of solution containing 2% (w/v) SDS, 50 mM 2-mercaptoethanol, 6 M urea, 10 mM sodium bisulphite, 5 mM 4-aminobenzamidine, 0.5 mM PMSF, 0.15 μg of pepstatin and 0.15 μg of leupeptin in 100 mM Tris-HCl pH 5.4 and 40% (v/v) glycerol. The mixture was immediately boiled for 4 min. Slab-gel electrophoresis was carried out in a 6–18% linear gradient polyacrylamide gel made in 25 mM sodium phosphate pH 6.8, 1% (w/v) SDS and 3 M urea. A 4% stacking gel was used made in 10 mM sodium phosphate pH 6.0. The electrode buffer was 25 mM sodium phosphate pH 6.8, 1% SDS. Gels were subjected to electrophoresis for 6–7 h at a constant current of 40 mA.

Gels were prepared for autoradiography (**A**) after staining with Coomassie blue (**B**). A standard curve of $\log_{10}$ mol. wt. against $\log_{10}$ acrylamide concentration was constructed using 11 marker proteins. The same amount of protein was applied to each well (about 350 μg).

Key: Tracks 1, 3, 4, and 6, proteins from control cells; tracks 2 and 5, protein from cells grown for 24 h in the presence of 100 μM methotrexate before permeabilization. In the subsequent transferase reaction NAD$^+$ was present at either 100 μM (tracks 1–3) or 1 μM (tracks 4–6)

the 14 kD protein is unlikely to be a histone other than H$_4$ modified by about five ADP-ribose moieties.

The migration pattern of all the radioactive proteins from the drug-treated cells appears to be identical to that from control cell proteins.

These results suggest that the three particular proteins identified are modified by only short ADP-ribose chains and that their greater radioactive content is because a larger proportion of the molecules in each population becomes modified in the metho-trexate-treated cells. Longer ADP-ribose chains would lead to clearly detectable changes in electrophoretic mobility as a result of the increased mol. wt.

Concluding Remarks

DNA strand breaks are detectable within the first hour of exposing HeLa cells to the anti-cancer drug methotrexate but not after this time. Cell viability (from Trypan Blue exclusion) does not change much until after about 12 h, but declines continuously thereafter. Although the DNA breaks are repaired quickly, nevertheless it remains possible that they have an early triggering affect on other biochemical processes which only much later result in cell death.

The enhancement of ADPR-transferase activity is also an early consequence of drug-exposure and leads to defined changes in protein ADP-ribosylation, although the significance of these changes remains unclear at this time. The kinetics of cell death during exposure to methotrexate are similar when hypoxanthine is also present (when the activation of transferase is prevented but the basal activity is still present; Fig. 2), or when 3AB is present with the methotrexate.

We note, however, that when ADPR-transferase is experimentally inhibited an alternative, faster operating DNA repair pathway that does not require ADP-ribosylation reactions may operate [6, 7]. A single experiment of ours supports this idea. We did not detect DNA strand breaks when cells were exposed to methotrexate and 3AB (5 mM) together for 1 h.

We conclude then by considering that the activation of ADP-ribosylation reactions is a normal consequence of the DNA damage and/or chromatin perturbation caused by methotrexate, but that despite this, or the possible activation of an alternative DNA repair pathway when the transferase is inhibited, the methotrexate successfully kills cells on prolonged exposure. This may point to a different factor(s) as the major cause of cell death in these drug-treated cells.

Acknowledgments. We wish to thank SERC and MRC for their financial support and Professor H.M. Keir for his encouragement and provision of facilities.

References

1. Fridland A (1974) Effect of methotrexate on deoxyribonucleotide pools and DNA synthesis in human lymphocytic cells. Cancer Res 34:1883–1888
2. Goulian M, Bleile B, Tseng BY (1980) Methotrexate-induced misincorporation of uracil into DNA. Proc Natl Acad Sci USA 77:1956–1960
3. Cook PR, Brazell IA (1975) Supercoils in human DNA. J Cell Sci 19:261–279
4. Durkacz BW, Irwin J, Shall S (1981) The effect of inhibition of (ADP-ribose)$_n$ biosynthesis on DNA repair assayed by the nucleoid technique. Eur J Biochem 121:65–69
5. Wallace HM, Gordon AM, Keir HM, Pearson CK (1984) Activation of ADP-ribosyltransferase in polyamine-depleted mammalian cells. Biochem J 219:211–221
6. Bohr V, Klenow H. (1981) 3-aminobenzamide stimulates unscheduled DNA synthesis and rejoining of strand breaks in human lymphocytes. Biochem Biophys Res Commun 102:1254–1261
7. Zwelling LA, Kerrigan D, Pommier Y (1982) Inhibitors of poly(adenosine diphosphoribose) synthesis slow the resealing rate of X-ray-induced DNA strand breaks. Biochim Biophys Res Commun 104:897–902

Cytogenetic, Cytotoxic, and Cell-Cycle Effects of 3-Aminobenzamide in Human Lymphoblastoid Cells Proficient or Deficient in Nucleotide Biosynthesis

JEFFREY L. SCHWARTZ and RALPH R. WEICHSELBAUM[1]

Abbreviations:

3AB	3-aminobenzamide
CHO	Chinese hamster ovary
HGPRT	hypoxanthine-guanine phosphoribosyl transferase
$6TG^R$	6-thioguanine-resistant
TK	thymidine kinase
TFT^R	trifluorothymidine-resistant
SCE	sister chromatid exchange

Introduction

In some recent studies it has been reported that 3-aminobenzamide (3AB), an inhibitor of poly(ADP-ribose) polymerase [7, 10] also inhibits the pathway for de novo synthesis of DNA precursors. Specifically, 3AB inhibits the metabolism of glucose and methionine into precursors of DNA in WI-L2 lymphoid cells [6], C3H mouse fibroblasts [1], and Chinese hamster ovary (CHO) cells [4]. It has been suggested that some of the reported effects of 3AB, such as its effect on cell-cycle progression [8], may be due to a disturbance in nucleotide precursor pathways rather than an inhibition of poly(ADP-ribose) synthesis [1–4, 6]. To test this hypothesis, we have examined the effects of 3AB on SCE induction, cytotoxicity, and cell-cycle progression in three different human lymphoblastoid cell lines; two deficient in salvage nucleotide synthesis pathways and one competent in both salvage and de novo nucleotide synthesis. We hypothesized that if 3AB inhibits de novo nucleotide synthesis pathways, then cells deficient in salvage nucleotide synthesis pathways should be more sensitive to the various effects of 3AB, especially its cytotoxic and cell-cycle effects.

---

1 Present address: Department of Radiation Oncology, University of Chicago Medical Center, 5841 South Maryland Ave., Chicago, IL 60637, USA
Department of Cancer Biology, Harvard School of Public Health, 665 Huntington Ave., Boston, MA 02115, USA

ADP-Ribosylation of Proteins
(ed. by F.R. Althaus, H. Hilz, and S. Shall)
© Springer-Verlag Berlin Heidelberg 1985

Materials and Methods

The human diploid lymphoblastoid cell line, TK6, isolated from WI-L2 cells and first characterized by Liber and Thilly [5] served as the parental line. It has both hypoxanthine-guanine phosphoribosyl transferase (HGPRT) and thymidine kinase (TK) activity. Two spontaneously occurring clones were isolated from this line on the basis of their ability to grow in either 2 μg ml^{-1} trifluorothymidine (TFT^R; TK deficient) or 5 μg ml^{-1} 6-thioguanine (6TG^R; HGPRT deficient). All three lymphoblastoid cell lines were maintained under exponential growth conditions in RPMI-1640 medium supplemented with 10% (v/v) heat-inactivated (2 h at 56°C) horse serum, 100 U ml^{-1} penicilin, and 100 μg ml^{-1} streptomycin. The cells were grown in suspension and the cell density kept between 10^4 and 10^6 cell ml^{-1}.

Survival, as measured by cloning efficiency of cells grown in different concentrations of 3AB, was determined as described previously [5]. For each treatment, two to four 96-well microtiter plates were used to determine the cloning efficiency.

For the determination of growth characteristics, 10 ml cell cultures containing 1—20 mM 3AB were initiated with 2 × 10^4 cells. Daily cell counts were made for up to 7 days. Cell cultures were diluted 1:1 with fresh medium containing the appropriate drug after each cell count. The cell concentration never exceeded 10^6 cells ml^{-1}. Cell cycle doubling times were determined from the slopes of the growth curves. Slopes were calculated from least squares regression analysis of the data.

Bromodeoxyuridine (final concentration 10 μM) was added to the cells along with different concentrations of 3AB for two cell cycles (28 h) for SCE analysis. Two hours before culture termination, Colcemid (2 × 10^{-7} M, final concentration), was added. Cells were then treated with 0.075 KCl for 5 min, and fixed in 3:1 methanol: acetic acid. SCEs were differentiated by a modified fluorescence-plus-Giemsa technique as previously described [9]. Twenty-five cells were scored for each treatment.

Results

All three human lymphoblastoid cell lines responded to 3AB in a similar manner. They were equally sensitive to the cytotoxic effects of 3AB (Fig. 1) and to the effects of 3AB on cell cycle progression (Fig. 2). Five to 10 mM concentrations of 3AB increased cell doubling times in a dose-dependent manner and decreased the plating efficiencies in the three cell lines. There was no apparent increase sensitivity to 3AB in the 6TG^R or TFT^R cells. Both 15 and 20 mM 3AB blocked cell cycling. In some experiments, 1 mM 3AB appeared to increase plating efficiency slightly. SCE induction, measured in TK6 and 6TG^R cells, increased at a similar rate (Fig. 3) in both cell lines.

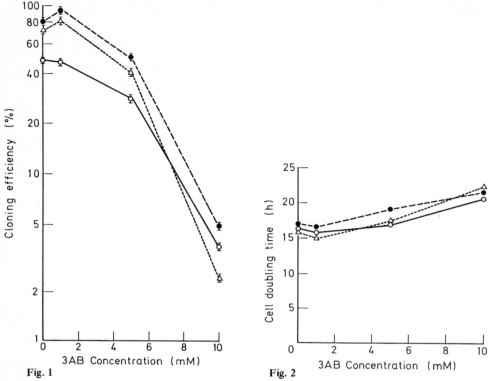

Fig. 1. Survival of TK6 (●), 6TG[R] (○), and TFT[R] (x) cells exposed to different concentrations of 3AB. Mean and SE of three experiments

Fig. 2. Cell doubling time of TK6 (●), 6TG[R] (○), and TFT[R] (x) cells exposed to different concentrations of 3AB. Mean of two determinations

Discussion

In some recent studies [1–4, 6], Cleaver et al. have questioned the validity of poly-(ADP-ribose) polymerase studies that are based on the use of inhibitors of ribosylation such as 3AB. The high concentrations of 3AB commonly used might cause other cellular effects unrelated to poly(ADP-ribose) synthesis. For example, 3AB has been reported to inhibit de novo nucleotide synthesis [1–4, 6]. Inhibition of de novo nucleotide synthesis could slow cell growth and lead to imbalances of nucleotide pools. Cleaver [4] has recently reported that a 6TG[R]-CHO line, unable to incorporate exogenous purines and therefore HGPRT-deficient, was more sensitive to the cytotoxic effects of 3AB as compared to a CHO line proficient in HGPRT activity. The increased sensitivity is believed to be due to the reported 3AB-mediated inhibition of de novo nucleotide synthesis [1–4, 6].

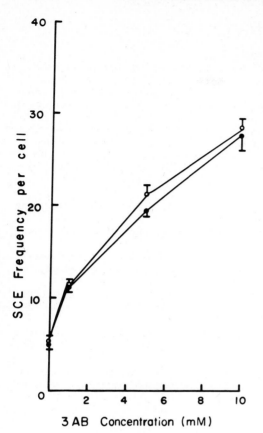

Fig. 3. SCE frequency per cell of TK6 (●) and 6TG^R (○) cells exposed to different concentrations of 3AB. 25 cells per treatment counted. Mean and SE of three experiments

In contrast to this report [4], are the results presented here. Whereas Milam and Cleaver [6] reported that 3AB inhibited de novo nucleotide synthesis in WI-L2 lymphoid cells, the degree of inhibition must not be great enough to affect cell viability or cell-cycle progression because in WI-L2-derived cells that are deficient in TK or HGPRT activity, enzymes involved in the salvage pathway of nucleotide synthesis, there is no increased sensitivity to 3AB as compared to lymphoid cells competent in these salvage pathways. In contrast, aminopterin, which is a potent inhibitor of de novo nucleotide synthesis, blocked growth in the 6TG^R and TFT^R cells and inhibited TK6 growth substantially (data not presented). Thus, the reported sensitivity of the 6TG^R CHO line to 3AB [4] might not be due to its defect in nucleotide synthesis and the reported inhibition of nucleotide synthesis by 3AB and related compounds might not be of great importance in the cytotoxic, cytogenetic, or cell-cycle response of exposed human lymphoblastoid cells.

Acknowledgments. This work was supported by Grant No. CA-29883 from the National Institutes of Health.

References

1. Borek C, Morgan WF, Ong A, Cleaver JE (1984) Inhibition of malignant transformation in vitro by inhibitors of poly(ADP-ribose) synthesis. Proc Natl Acad Sci USA 81:243–247
2. Cleaver JE, Bodell WJ, Morgan WF, Zelle B (1983) Differences in the regulation of poly(ADP-ribose) of repair of DNA damage from alkylating agents and ultraviolet light according to cell type. J Biol Chem 258:9059–9068
3. Cleaver JE, Bodell WJ, Borek C, Morgan WF, Schwartz JL (1983) Poly(ADP-ribose): spectator or participant in excision repair of DNA damage. In: Miwa M et al. (eds) ADP-ribosylation, DNA repair and cancer. Jpn Sci Soc Press, Tokyo/VNU Science Press, Utrecht, pp 195–207
4. Cleaver JE (1984) Differential toxicity of 3-aminobenzamide to wild-type and 6-thioguanine-resistant Chinese hamster cells by interference with pathways of purine biosynthesis. Mutat Res 131:123–127
5. Liber HL, Thilly WG (1982) Mutation assay at the thymidine kinase locus in diploid human lymphoblasts. Mutat Res 94:467–485
6. Milam KM, Cleaver JE (1984) Inhibitors of poly(ADP-ribose) synthesis also affect other metabolic processes. Science 223:589–591
7. Oikawa A, Tohda H, Kanai M, Miwa M, Sugimura T (1980) Inhibitors of poly(adenosine diphosphate ribose) polymerase induce sister chromatid exchanges. Biochem Biophys Res Commun 97:1311–1316
8. Schwartz JL, Morgan WF, Kapp LN, Wolff S (1983) Effects of 3-aminobenzamide on DNA synthesis and cell cycle progression in Chinese hamster ovary cells. Exp Cell Res 143:377–382
9. Schwartz JL, Banda MJ, Wolff S (1982) 12-O-Tetradecanoylphorbol-13-acetate (TPA) induces sister-chromatid exchanges and delays in cell cycle progression in Chinese hamster and human cell lines. Mutat Res 92:393–409
10. Sims JL, Sikorski GW, Catino DM, Berger SJ, Berger N (1982) Poly(adenosine diphosphoribose)polymerase inhibitors stimulate unscheduled deoxyribonucleic acid synthesis in normal human lymphoblasts. Biochemistry 21:1813–1821

Poly(ADP-Ribose) Polymerase Inhibition and the Induction of Mutation in V79 Cells

NITAIPADA BHATTACHARYYA and SUKHENDU B. BHATTACHARJEE[1]

Introduction

Synthesis of poly(ADP-ribose), a nuclear polymer, has been found to increase in response to the DNA damage by various agents like alkylating agents [8, 11], UV [6, 10], and ionizing radiations [17]. The biological significance of poly(ADP-ribose) is not clear. But activity of the polymer has been related to the regulation of DNA synthesis [12], RNA synthesis [15], sister chromatid exchanges [14], cell differentiation [9], cell transformation [5], and repair of damaged DNA [6—8]. However, there is no definite experimental evidence of requirement for poly(ADP-ribose) in any of the cellular processes mentioned above. Most of the evidence for roles of poly(ADP-ribose) is based on the inhibition of the enzyme poly(ADP-ribose) polymerase by different inhibitors.

We have studied the influence of the two inhibitors of the enzyme poly(ADP-ribose) polymerase, benzamide (BZA) and nicotinamide (NA) on N-methyl-N'-nitro-N-nitroso-guanidine (MNNG), X-ray, and UV-induced killing and 8-azaguanine (AZ) resistant mutation in V79 Chinese hamster cells.

Methods

Details of culture conditions, media, etc., procedures for survival and mutational studies have been published earlier [2—4]. In short, after attachment for 4 h, cells in the order of 10^2 to 10^4 were exposed to MNNG, X-ray, or UV. Cells were then grown in the absence or presence of inhibitors. Medium containing the inhibitors were withdrawn after 24 h, fresh medium added, and incubated till colony formation.

For mutation study, about 10^6 cells were treated and allowed 7—9 days of expression, with 3—4 subcultures. Cells were then replated at concentration of 5×10^5 per 100 mm Corning dishes in selection medium containing 3 μg ml^{-1} of azaguanine in complete growth medium. About 95% of the azaguanine resistant clones were sensitive

---

1 Crystallography and Molecular Biology Division, Saha Institute of Nuclear Physics, Sector 1, Block 'AF', Bidhan Nagar, Calcutta, 700 064, India

ADP-Ribosylation of Proteins
(ed. by F.R. Althaus, H. Hilz, and S. Shall)
© Springer-Verlag Berlin Heidelberg 1985

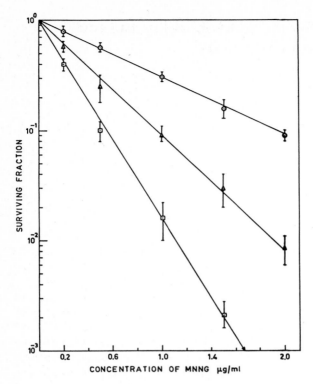

Fig. 1. Survival of V79 Chinese hamster cells exposed to different concentrations MNNG for 1 h, without (○) or with subsequent 5 mM benzamide (□) or 15 mM nicotinamide (△) treatment for 24 h

to HAT medium (complete medium containing hypoxanthine 10^{-4} M, aminopterin 3×10^{-7} M, and thymidine 10^{-5} M). Experimental data shown represent the average of 2–5 independent observations with the standard deviations.

Results

To avoid the toxicity of the inhibitors themselves nontoxic doses of the inhibitors were used. Treatment for 24 h with 5 mM BZA or 15 mM NA did not affect the colonogenic property of the undamaged cells and were used in the following studies. When MNNG-exposed cells were treated with BZA or NA for 24 h, the rate of cell killing increased more than three times by BZA treatment and about two times by NA treatment. The results are shown in the Fig. 1.

Figure 2 shows the influence of 1 to 5 mM BZA on X-ray-induced killing. It can be seen that BZA treatment for 24 h after irradiation, did not change the slope of the survival curve. There was only a small reduction in the extrapolation number. A similar observation was also made with NA treatment (data not shown).

Figure 3 shows the effect of 1 to 5 mM BZA on UV-induced killing. It can be seen from Fig. 3 that BZA influenced the extrapolation number of the UV-survival curve,

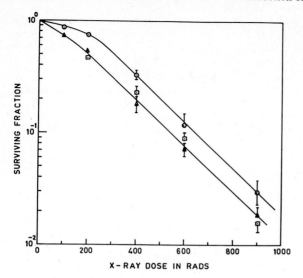

Fig. 2. Survival of V79 cells exposed to different doses of X-ray without (○) or with 5 mM (△) or 1 mM benzamide (□) treatment for 24 h

the rate of killing remaining unchanged. Similar results have also been obtained with NA (data not shown).

Effect of the inhibition of poly(ADP-ribose) polymerase activity by NA or BZA treatment after MNNG, X-ray, or UV-exposures on the induction of AZ resistant mutants is shown in Table 1. It can be seen from Table 1 that the background mutant yield was not influenced by BZA or NA treatment, but MNNG-induced yield was reduced significantly. However, the yield of mutants induced by X-ray or UV remained unaffected by such treatments.

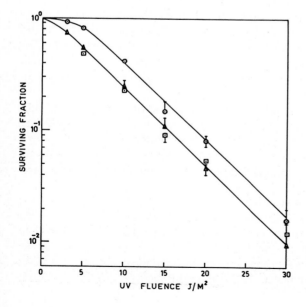

Fig. 3. Survival of V79 Chinese hamster cells exposed to different doses of UV without (○) or with 5 mM (△) or 1 mM (□) benzamide

Table 1. Influence of inhibition of poly(ADP-ribose) polymerase on MNNG X-ray and UV-induced 8-azaguanine resistant mutant in V79 cells

Agents	Dose	Mutation frequency (m.f.) $\times 10^{-5}$	m.f. $\times 10^{-5}$ 24 h treatment with	
			Nicotinamide (15 mM)	Benzamide (5 mM)
MNNG	0	0.3 ± 0.2	0.3 ± 0.2	0.5 ± 0.3
	0.2 μg ml^{-1}	10.5 ± 2.2	6.0 ± 1.1	2.6 ± 0.9
	0.5 μg ml^{-1}	23.5 ± 2.6	16.6 ± 3.0	5.8 ± 1.0
	1.0 μg ml^{-1}	57.7 ± 4.5	28.4 ± 6.0	17.0 ± 5.8
X-ray	200 rad	1.9 ± 0.5	2.2 ± 0.6	1.9 ± 0.4
	400 rad	3.4 ± 0.8	3.8 ± 0.8	3.4 ± 0.7
	600 rad	6.1 ± 1.1	6.4 ± 1.1	6.4 ± 0.7
UV	8 J m^{-2}	4.1 ± 1.1	4.8 ± 2.0	4.3 ± 1.5
	15 J m^{-2}	8.5 ± 1.4	8.1 ± 2.5	9.2 ± 2.0
	20 J m^{-2}	12.6 ± 2.0	12.7 ± 2.2	11.3 ± 1.8

Discussion

The results presented above clearly show that poly(ADP-ribose) polymerase may have different roles in the recovery of MNNG-induced damage and X-ray or UV-induced damage. Similar conclusions were also arrived at from other studies with alkylating agents and UV with different biological end points [6, 14].

Since the colonogenic properties of undamaged cells were unaffected by inhibitors in 24 h, it could be assumed that there were no other significant metabolic effects on undamaged cells. Recently it has been shown that BZA and 3ABA at concentrations higher than 2 mM had some adverse effects on other metabolic activities [13]. We had shown that lower concentrations of BZA could also increase the MNNG-induced killing [4], indicating that the effect might not be due to the influence on other metabolic process(es). It could be due to the influence of poly(ADP-ribose) polymerase on the cellular repair activity.

However, these studies do not clarify the role of poly(ADP-ribose) in any particular step of the repair process(es). But it is evident that such inhibition suppressed an error prone component of repair. It is likely that poly(ADP-ribose) might have a regulatory role in the postreplication repair in MNNG-treated cells. The possibility of modulation of excision of 0−6 methylguanine which is supposed to promutagenic lesion cannot be ruled out.

The insignificant role of poly(ADP-ribose) in UV- and X-ray-induced killing and mutation could be attributed to the amount and duration of stimulation of the polymer synthesis [6, 17]. Since it had been reported that the types of the DNA breaks could specify the amount of stimulation of polymer [1] and all the breaks do not stimulate the synthesis [16] the mechanism by which the repair enzymes act at the damaged site of the DNA in mammalian cells may be different for different damaging agents. Thus, the stimulation of the polymer synthesis would be different and the inhibition would be expected to give rise to differential response. Thus, the requirement of poly(ADP-ribose) may not be as essential for the recovery after UV- or X-ray-induced damage, as for the recovery with alkylating agents.

References

1. Benjamin RC, Gill DM (1980) Poly(ADP-ribose) synthesis in vitro programmed by damaged DNA. J Biol Chem 255:10502–10508
2. Bhattacharjee SB, Pal B (1982) Tetracycline induced mutation in cultured Chinese hamster cells. Mutat Res 101:329–338
3. Bhattacharjee SB, Chatterjee S, Pal B (1982) Survival and mutation of Chinese hamster cells after ultraviolet irradition and caffeine treatment. Mutat Res 106:137–146
4. Bhattacharyya N, Bhattacharjee SB (1983) Suppression of N-methyl-N′-nitro-N-nitrosoguanidine induced mutation in Chinese hamster V79 cells by inhibition of poly(ADP-ribose) polymerase activity. Mutat Res 121:287–292
5. Borek C, Morgan WF, Ong A, Cleaver JE (1984) Inhibition of malignant transformation in vitro by inhibitors of poly(ADP-ribose) synthesis. Proc Natl Acad Sci USA 81:243–247
6. Cleaver JE, Bodell WJ, Morgan WF, Zelle B (1983) Differences in regulation by poly(ADP-ribose) of repair of DNA damage from alkylating agents and ultraviolet light according to cell type. J Biol Chem 258:9059–9068
7. Creissen D, Shall S (1982) Regulation of DNA ligase activity by poly(ADP-ribose). Nature (London) 296:271–272
8. Durkacz BW, Omidiji O, Gray DA, Shall S (1980) (ADP-ribose)$_n$ biosynthesis participates in DNA excision repair. Nature (London) 283:593–596
9. Farzaneh F, Zalin R, Bill D, Shall S (1982) DNA strand breaks and ADP-ribosyl transferase activation during cell differentiation. Nature (London) 300:362–366
10. Jacobson EL, Antol KM, Juarez-Salinas H, Jacobson MK (1983) Poly(ADP-ribose) metabolism in ultraviolet irradiated human fibroblasts. J Biol Chem 258:103–107
11. Juarez-Salinas H, Sims JL, Jacobson MK (1979) Poly(ADP-ribose) levels in carcinogen treated cells. Nature (London) 282:740–741
12. Kitamura A, Tanigawa Y, Yamamoto T, Kawamura M, Doi S, Shimoyama M (1979) Glucocorticoid induced reduction of poly(ADP-ribose) synthetase in nuclei from chickembryo liver. Biochem Biophys Res Commun 87:725–733
13. Milam KM, Cleaver JE (1984) Inhibitors of poly(adenosine diphosphate-ribose) synthesis: Effect on other metabolic processes. Science 223:589–591
14. Morgan WF, Cleaver JE (1982) 3 aminobenzamide synergistically increases sister chromalid exchanges in cells exposed to methylmethane sulphonate but not to ultraviolet light. Mutat Res 104:361–365
15. Taniguchi T, Agemore M, Kameshita I, Nishikimi M, Shizuta Y (1982) Participation of poly-(ADP-ribosyl)ation in the depression of RNA synthesis caused by treatment of mouse lymphoma cells with methylnitrosourea. J Biol Chem 257:4027–4030
16. Walker IG (1984) Lack of 4-nitroquinoline 1-oxide on cellular NAD levels. Mutat Res 139:155–159
17. Zwelling LS, Kerrigan D, Mattern MR (1983) Ataxia-telangiectasia cells are not uniformly defficient in poly(ADP-ribose) synthesis following X-irradiation. Mutat Res 120:69–78

Effect of 3-Aminobenzamide on CHO Cells Replicating in Medium Containing Chloro-Deoxyuridine

T.S.B. ZWANENBURG[1] and A.T. NATARAJAN[1,2]

1. Introduction

Sister chromatid exchanges (SCEs) can be visualized in several ways, mainly by allowing cells to incorporate thymidine analogues, such as bromo- or chloro-deoxyuridine (or radioactive thymidine) which finally results in sister chromatids with different staining properties. Labeling schemes are principally: (1) growing cells for two cell cycles in the presence of the analogues or for one cycle in the analogue followed by a second one in thymidine or a different analogue: (2) incubation for at least four generations in bromodeoxyuridine (BrdUrd) followed by two cell cycles in thymidine or one cycle in thymidine and the second one in BrdUrd: (3) incubation for three cell cycles in different concentrations of BrdUrd which allows a discrimination of the frequencies of SCEs induced in the first, second, and third S-phase.

DNA damaging agents, specifically those of the S-dependent type, induce SCEs at fairly low concentrations and, therefore, scoring of SCEs in the screening of genotoxic agents has become popular. However, care has to be exercised because data exist which indicate that BrdUrd can be involved in the induction of SCEs by DNA damaging agents [1, 2] as well as by inhibitors of the nuclear enzyme poly(ADP-ribose) synthetase [3]. These inhibitors, such as 3-aminobenzamide (3AB) and benzamide, comprise a group of agents that significantly increases SCE frequencies, without damaging DNA directly [3, 4]. It has been shown that 3AB is primarily effective when it is present during the second cell cycle after the initial addition of BrdUrd, when BrdUrd containing DNA is used as the template for replication [3]. When cells are grown for four generations in BrdUrd containing medium, most of the SCEs are induced by 3AB in the first S-phase after return to normal, thymidine containing medium [3]. The extent of the increase in the frequency of SCEs depends on the concentration of 3AB used as well as the amount of BrdUrd incorporated and is independent of the type of nucleoside present in the medium during the second cell cycle [5].

When CldUrd is incorporated instead of BrdUrd the baseline frequency of SCEs is increased dramatically [6] and most SCEs are formed during the second S-phase, inde-

1 Department of Radiation Genetics and Chemical Mutagenesis, State University of Leiden, Wassenaarseweg 72, 2333 AL Leiden, The Netherlands
2 J.A. Cohen Institute, Interuniversity Institute for Radiopathology and Radiation Protection, Leiden, The Netherlands

ADP-Ribosylation of Proteins
(ed. by F.R. Althaus, H. Hilz, and S. Shall)
© Springer-Verlag Berlin Heidelberg 1985

pendent of the analogue present during the second cycle [7]. This striking similarity between the effect of CldUrd incorporation and 3AB-treatment of BrdUrd-substituted cells warranted further extension of the investigations on CldUrd-substituted cells.

Data from experiments concerning the effect of inhibition of poly(ADP-ribose) synthetase in CldUrd-substituted cells on cell cycle progression (as determined by flow cytometric analysis), cytotoxicity, and induction of mutations and chromosomal aberrations are reported here.

2. Materials and Methods

2.1 Cell Culture Conditions and Substitution with Analogues

Chinese hamster ovary cells (CHO) were cultured as monolayers in Ham's F10 medium supplemented with 15% newborn calf serum (Gibco), penicillin (100 IU ml^{-1}) and streptomycin (100 μg ml^{-1}) at 37°C and an atmosphere of 5% CO_2. The generation time was about 12 h. In all experiments exponentially growing cells were used.

Substitution with BrdUrd and CldUrd was obtained by growing cells for one or two cell cycles in medium containing 1 μM 5-fluorodeoxyuridine and 100 μM deoxycytidine, together with the analogue and thymidine in a ratio of either 3:7 or 7:3 or with analogue or thymidine alone. The sum of the concentration of analogue and thymidine was kept constant at 10 μM and in the text, the base-analogue concentrations are expressed as percentages of the total analogue + thymidine concentration.

2.2 Flow Cytometry

Cells were trypsinized, washed with PBS, and permabilized by resuspending them at 1.10^6 cells ml^{-1} in a hypotonic buffer containing 10 mM Tris-HCl (pH 7.8), 1 mM EDTA, 4 mM MgCl$_2$, 30 mM mercaptoethanol, and 0.05% Triton X-100. They were kept on ice for 10 min. Then an equal volume of a solution containing 100 mM Tris-HCl (pH 7.4), 75 mM MgCl$_2$, 10% ethanol, 25 μg ml^{-1} ethidium bromide, and 50 μg ml^{-1} mitramycin was added. After 5 min on ice the samples were analyzed using a FACS IV fluorescent-activated cell sorter (Becton Dickinson, Sunny vale, CA, USA) equipped with an argon laser tuned to 457 nm with an intensity of 100 MW. Fluorescence histograms were recorded for about 25,000 cells per sample. The fluorescence was measured using a long-pass 580 nm filter.

2.3 Cell Killing and Induction of Mutations (HGPRT$^-$)

Cells were incubated with various concentrations CldUrd and 3AB for 24 h. Then survival was determined by incubation of five petri dishes per point, each with 150 cells. Expression was started by inoculation of five petri dishes of 14.5 cm with 3.5×10^5 unsubstituted cells. In order to ascertain the subculturing of sufficient numbers of viable mutants, 2×10^6 substituted cells per dish were seeded. The cells were incubated

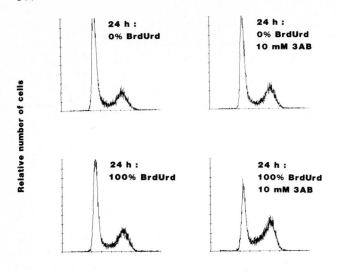

Relative fluorescence intensity

Fig. 1. Influence of 3AB on the distribution of BrdUrd-substituted cells over the cell cycle stages

for 9 days and subcultured regularly during this period. Selection was carried out by incubation of the cells at 10^5 cells per dish (ten dishes) together with 5 μg ml^{-1} thioguanine. Cloning efficiency was determined by incubation of 100 cells per dish (five dishes). All colonies were fixed after an incubation period of 8 days, stained with 0.1% methylene blue, and counted by eye.

2.4 Chromosomal Aberrations and Premature Chromosome Condensation (PCC)

2.4.1 Chromosomal Aberrations. Cells were incubated with 0% or 100% CldUrd together with various concentrations of 3AB for 24 h, the last 2 h in the presence of 10^{-5}% Colcemid. Chromosomal preparations were made by routine procedures.

2.4.2 PCC. A series of 14.5 cm petri dishes was incubated with 6×10^5 CHO cells for 24 h. Then 0% or 100% CldUrd was added together with various concentrations of 3AB for another 24 h.

Mitosis were obtained by incubation of a series of bottles (75 cm^2) with 6×10^5 CHO cells per bottle for 42 h. Then 20 ml of fresh medium with 10^{-5}% Colcemid was added for another 6 h and mitotic cells were collected by shaking the bottles manually.

Fusion was obtained principally the same as described [8]. Interphase cells were fused with the mitotic cells at a ratio of 5 to 1, using 55% PEG (MW 1000), and incubated for 1 h in F10 medium.

All samples were fixed in 3:1 methanol/acetic acid after hypotonic treatment for 10 min in 1% sodium citrate at 37°C. Air-dried preparations were made and stained with 2% Giemsa for 8 min.

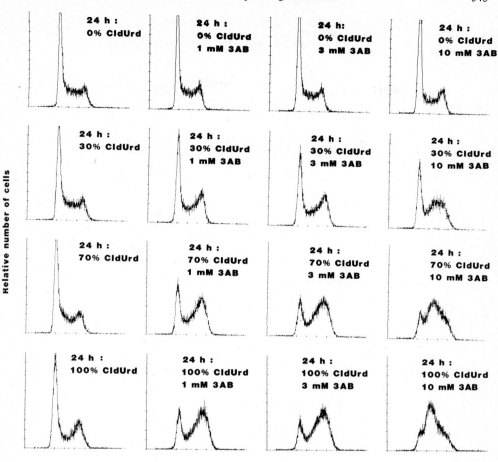

Relative fluorescence intensity

Fig. 2. Influence of 3AB on the distribution of CldUrd-substituted cells over the cell cycle stages

3. Results

3.1 Flow Cytometry

Treatment of cells with 3AB for 24 h disturbs the progression of cells through the cell cycle only slightly (Figs. 1 and 2). Relatively more cells appear in the S and $G_2 + M$ peak when cells are incubated with 100% BrdUrd and simultaneously treated with 10 mM 3AB (Fig. 1). Incorporation of CldUrd during two cell cycles induces G_2 delay (Fig. 2) as has been reported previously [9]. When these cells are treated simultaneously with 3AB, progression is affected very heavily, when CldUrd and 3AB are present at high concentrations (Fig. 2). However, this delay is abolished when 3AB is omitted from the second cell cycle (Fig. 4). Moreover, it overcomes the G_2 delay

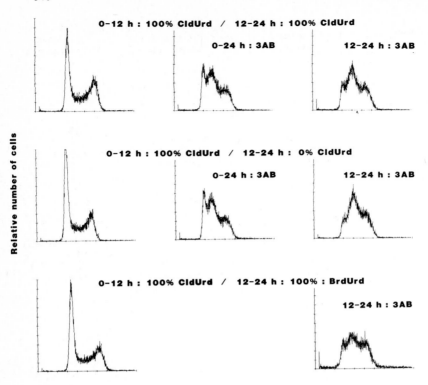

Relative fluorescence intensity

Fig. 3. Influence of the type of nucleoside present in the second S-phase on the distribution of cells over the cell cycle stages. Concentration of 3AB is 10 mM

induced by CldUrd alone when 3AB is present only during the first cycle of incorporation of CldUrd (Fig. 4). Replacement of CldUrd in the second cell cycle by BrdUrd or thymidine does not alter the magnitude of the delay (Fig. 3).

A difference exists whether 3AB is present during the entire period of 24 h or only for the last 12 h of incubation. Relatively more cells are in G_1 and early S when the cells are treated for 24 h with 3AB (Figs. 3 and 4).

3.2 Cell Killing and Induction of Mutations

Incorporation of CldUrd into DNA reduces colony-forming ability to about 10% when cells are grown for 24 h in medium containing 100% CldUrd. Simultaneous treatment with 3AB reduces cell survival, depending on the amount of CldUrd incorporated and concentration of 3AB (Fig. 5).

However, under these conditions 3AB does not increase the frequency of gene mutations at the HGPRT locus (Table 1).

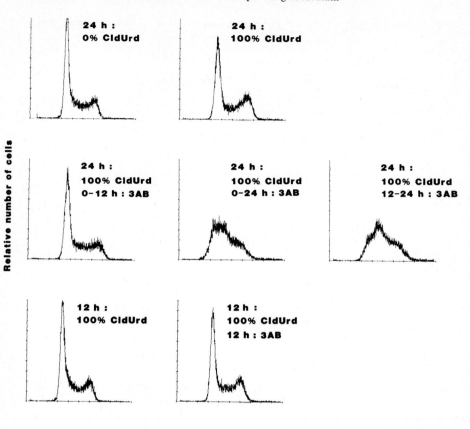

Relative fluorescence intensity

Fig. 4. Cell cycle delay induced by 3AB (10 m*M*)

Table 1. Mutation frequencies in CldUrd-substituted cells treated with various concentrations of 3AB

CldUrd conc. (%)	Mutation frequency[a]			
	Dose of 3AB (m*M*)			
	0	1	3	10
0	0.1	0.1	0.3	0.3
30	3.1	2.9	2.9	3.0
70	12.0	11.4	nd[b]	nd

[a] Mutations/10^5 surviving cells
[b] nd = not determined

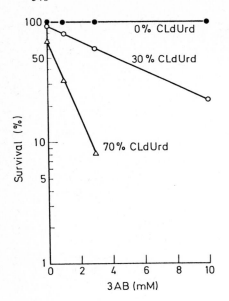

Fig. 5. Influence of 3AB on colony-forming ability of CldUrd-substituted cells

3.3 Chromosomal Aberrations and Premature Chromosome Condensation

CldUrd does not induce chromosomal aberrations when cells are grown for only 12 h in the analogue [9]. However, incubation for 24 h in medium containing 100% CldUrd results in numerous aberrations, which are all of the chromatid type (Table 2). Treatment with a low dose of 3AB (1 mM) increases this frequency. At 10 mM the mitotic index is drastically decreased and less aberrations are observed.

Scoring for aberrations in substituted G_2 cells with premature condensed chromosomes reveals that the appearance of G_2 delay induced by the CldUrd coincides with an enhanced number of aberrations. Treatment with 1 mM 3AB increases this frequency and with 10 mM 3AB, similar to metaphase cells, no significant increase can be observed.

4. Discussion

The present results show that 3AB induces a shift in the relative number of cells in the different cell cycle phases. The extent of the effect depends on several conditions: (1) 3AB is only slightly effective in unsubstituted cells during treatment for two cell cycles. This effect on unsubstituted cells correlates with the observation that 3AB induces a small number of SCEs in the first S-phase during replication of unsubstituted DNA [5]. (2) Incorporation of high amounts of BrdUrd and simultaneous treatment with 3AB for two cell cycles induces more cells to be in the S and G_2 + M peak.

Table 2. Influence of 3AB on the induction of chromosomal aberrations by CldUrd

CldUrd conc. (%)	3AB (mM)	Mitotic index	Cells analyzed	Cells abnormal	Aberrations/100 metaphases[a]			
					B^1	B^{11}	RB^1	Total
0	–	49	100	2	2	1		3
0	1	51	100	2	2			2
0	10	37	100	2	2	1		3
100	–	39	100	66	60	24	79	163
100	1	14	100	80	75	37	148	260
100	10	4	100	68	56	28	62	146
Premature chromosome condensation:								
0	–		25	3	12			12
0	1		25	4	8	8		16
0	10		25	3	8	4		12
100	–		25	23	152	36	236	424
100	1		25	24	272	96	372	740
100	10		25	22	144	68	260	472

[a] B^1, chromatid break; B^{11}, iso-chromatid break; RB^1, chromatid exchange

(3) 3AB is very effective in retarding progression of cells through the cell cycle when CldUrd is incorporated into DNA. The number of cells in the S and G_2 + M peak increases, depending on the amount of CldUrd incorporated and concentration of 3AB applied for two cell cycles. (4) The type of nucleoside present during the second cell cycle (CldUrd, BrdUrd, or thymidine) is without any influence on the effect of 3AB. This situation is similar to the reported increase in SCEs in BrdUrd substituted cells by inhibitors of poly(ADP-ribose) synthetase [3] as well as to the high baseline level of SCEs after incorporation of CldUrd into DNA [7]. (5) The data show that 3AB is most effective when it is present during the second cell cycle when CldUrd-substituted DNA becomes replicated. This is in accordance with the observation that most of the induced SCEs are formed when 3AB is present during the second cell cycle after the initial addition of BrdUrd [3]. However, a difference exists whether 3AB is present during both cell cycles of incorporation of CldUrd or only during the last 12 h. Less delay is present when the cells are treated for 24 h, simultaneously with the incorporation of CldUrd. The additional presence of 3AB during the first cell cycle apparently induces a significant delayed start of the second S-phase. This is supported by the findings that 3AB is slightly effective in unsubstituted cells and that it reduces G_2 delay induced by CldUrd (incorporated during two S-phases) when it is present during the first 12 h.

In contrast to the finding that 3AB does not induce chromosomal aberrations in CHO cells when BrdUrd is incorporated (to moderate levels) [10] treatment of cells with 3AB simultaneously with the incorporation of CldUrd results in an increased frequency of aberrations. When high doses of 3AB are used this increase is levelled down; most probably it reflects the selection of less damaged cells due to the immense delay during S-phase of the damaged cells. Numerous aberrations are present in G_2

cells when visualized as premature condensed chromosomes. Similarly, the number of aberrations in metaphase cells after treatment with a high 3AB concentration is reduced to control level.

An interesting observation of the present study is that 3AB effectively reduces colony-forming ability of CldUrd-substituted cells in a dose-dependent way, without affecting mutation frequency.

A conclusive explanation cannot be offered at the moment as to why 3AB induces SCEs and delay in cell cycle progression in BrdU substituted cells and in addition chromosomal aberrations and cell killing in CldUrd-substituted cells. The presented results indicate a correlation between the ability of 3AB to induce SCEs and to delay the progression of cells through the cell cycle. Since ADP-ribosylation is stimulated in response to the introduction of DNA strand breaks [11, 12] and treatment of cells with the inhibitors results in an inhibited repair of these breaks [13, 14], it is likely to assume an involvement of DNA strand breaks in the effects observed in CldUrd- and BrdUrd-substituted cells. These breaks might be very specific with regard to their nature, to the site in the genome where they occur, and to the timing of appearance. Earlier investigations failed to show the presence of an enhanced number of DNA strand breaks due to the incorporation of BrdUrd and CldUrd [9]. However, treatment with inhibitors of poly(ADP-ribose) synthetase increases the number of breaks in CldUrd-substituted cells as measured with an alkaline [15] and a neutral technique [16]. Therefore, biochemical investigations concerning the origin of the DNA strand breaks involved are in progress in our laboratory and will probably throw light on the mechanism of formation of SCEs as well as on the action of inhibitors of poly(ADP-ribose) synthetase.

Acknowledgments. We thank Dr. H.J. Tanke, Dr. P. Oljans, Dr. F. Darroudi, and Ms. I. Bussman for their assistance. This work was supported by the Foundation for Basic Medical Research (FUNGO) and the Queen Wilhelmina Fund (KWF).

References

1. Ockey CA (1981) Methyl methanesulphonate (MMS) induced SCEs are reduced by BrdU used to visualize them. Chromosoma 84:243–256
2. Natarajan AT, Tates AD, Meijers M, Neuteboom I, Vogel N de (1983) Induction of sister-chromatid exchanges (SCEs) and chromosomal aberrations by mitomycin C and methyl methanesulfonate in Chinese hamster ovary cells. Mutat Res 121:211–223
3. Natarajan AT, Csukas I, Zeeland AA van (1981) Contribution of incorporated 5-bromodeoxyuridine in DNA to the frequencies of sister-chromatid exchanges induced by inhibitors of poly(ADP-ribose)polymerase. Mutat Res 84:125–132
4. Oikawa A, Tohda H, Kaina M, Miwa M, Sugimura T (1980) Inhibitors of poly(adenosine diphosphate ribose) polymerase induce sister chromatid exchanges. Biochem Biophys Res Commun 97:1311–1316
5. Zwanenburg TSB, Natarajan AT (1984) 3-Aminobenzamide-induced sister chromatid exchange are dependent on incorporated bromodeoxyuridine in DNA. Cytogenet Cell Genet 38:278–281
6. DuFrain RJ, and Garrand TJ (1981) The influence of incorporated halogenated analogues of thymidine on the sister-chromatid exchange frequency in human lymphocytes. Mutat Res 91: 233–238

7. O'Neill JP, Heartlein MW, Preston RJ (1983) Sister chromatid exchanges and gene mutations are induced by the replication of 5-bromo- and 5-chloro-deoxyuridine substituted DNA. Mutat Res 109:259—270

8. Pantelias GE, Maillie HD (1983) A simple method for premature chromosome condensation induction in primary human and rodent cells using polyethylene glycol. Som Cell Genet 9: 533—547

9. Zwanenburg TSB, Mullenders LHF, Natarajan AT, Zeeland AA van (1984) DNA-lesions, chromosomal aberrations and G_2 delay in CHO cells cultured in medium containing bromo- or chloro-deoxyuridine. Mutat Res 127:155—168

10. Natarajan AT, Csukas I, Degrassi F, Zeeland AA, Palitti F, Tanzarella C, Salvia R de, Fiore M (1982) Influence of inhibition of repair enzymes on the induction of chromosomal aberrations by physical and chemical agents. In: Natarajan AT, Obe G, Altmann H (eds) DNA repair, chromosome alterations and chromatin structure. Prog Mutat Res 4:47—59

11. Miller EG (1975) Stimulation of nuclear poly(adenosine diphosphateribose)polymerase activity from HeLa cells by endonucleases. Biochem Biophys Acta 395:191—200

12. Benjamin RC, Gill DM (1980) ADP-ribosylation in mammalian cell ghosts. Dependence of poly(ADP-ribose) synthesis on strand breakage. J Biochem (Tokyo) 255:10493—10501

13. Durkacz BW, Omidiji O, Gray DA, Shall S (1981) (ADP-ribose)$_n$ participates in DNA excision repair. Nature (London) 283:593—596

14. Zwelling LA, Kerrigan D, Pommier Y (1982) Inhibitors of poly(adenosine diphosphoribose) synthesis slow the resealing rate of X-ray-induced DNA strand breaks. Biochem Biophys Res Commun 104:879—902

15. Dillehay LE, Thompson LH, Carrone AV (1984) DNA-strand breaks associated with halogenated pyrimidine incorporation. Mutat Res 131:129—136

16. Zwanenburg et al. (unpublished)

Poly(ADP-Ribose) Synthesis in Lymphocytes of Persons Occupationally Exposed to Low Levels of Ionizing Radiation

HANS ALTMANN[1], ALEXANDER TOPALOGLOU[1], DIETMAR EGG[2], and ROBERT GÜNTHER[2]

1. Introduction

The fact that ionizing radiation decreased the cellular NAD^+ level was discovered more than 20 years ago. Ten years later it was published that 90% of the NAD^+ synthesized in the cell is used for the synthesis of poly(ADP-ribose) [1]. DNA strand-breaks seem to be the most important step responsible for the increase in poly(ADP-ribose) polymerase activity that mediates the NAD^+ effect [2]. ADP-ribosylation is required both for normal cell functions and as a chromatin modification factor following DNA damage [3].

DNA damage produced by high doses of ionizing radiation (21 Krad) consumes the entire cellular pool of NAD^+ within 5 min and transiently produces a mass of poly(ADP-ribose) equivalent to about 10% that of the cell DNA [4].

Irradiation of cancer patients by relatively high doses caused an increase of UV-induced UDS in peripheral lymphocytes at a time when repair processes according to the foregoing radiation therapy were no more detectable [5], and in parallel an increase in poly(ADP-ribose) synthesis could be detected [6]. UDS was also increased in lymphocytes of people receiving elevated background levels of chronic low dose exposure of high LET radiation by inhalation of ^{222}Rn and its daughters [7, 8] and in persons occupationally exposed to low levels of γ-irradiation [9, 10]. Sister chromatid exchanges (SCE) were determined in lymphocytes of the same population groups and no significant difference could be found in spontaneously occurring SCEs, while mitomycin C induced SCEs were significantly reduced in persons exposed to radiation [10]. In animal experiments it was also found that after an applied X-ray dose of 1 Gy the increase in UDS was accompanied by an increase in poly(ADP-ribose) synthesis [11]. By the present investigations we wanted to find out whether, in addition to an increase of UDS, also poly(ADP-ribose) synthesis was changed in lymphocytes of chronically exposed persons.

---

1 Institut für Biologie, Forschungszentrum Seibersdorf, A-2444 Seibersdorf, Austria
2 L. Boltzmann-Institut für Balneologie und Klimaheilkunde, Bad Gastein, Austria

ADP-Ribosylation of Proteins
(ed. by F.R. Althaus, H. Hilz, and S. Shall)
© Springer-Verlag Berlin Heidelberg 1985

2. Materials and Methods

2.1 Probands

Test persons for the present lymphocyte experiments were the same who had been used for UDS studies. This personnel is working in a therapy station for treatment of rheumatic diseases, mainly ankylosing spondilytis. The treatment room was a former gold mine with a temperature of about 41°C and a relative humidity of about 95%. The main radioactivity in this mine results from ^{222}Rn and its daughters and reaches about 3 nCi l^{-1} of air. The occupationally exposed test persons (miners) received a mean blood dose calculated from α-emission and external γ-radiation between 800 and 1600 mrad yr^{-1} [7]. The second group of test persons were employees of the spa treatment hospital. The external radiation in these rooms reaches 120—300 mrad yr^{-1}, the α-blood dose was 0.06—2.3 mrad mo^{-1} and the γ-blood dose 8—13 mrad mo^{-1}. Controls were medical students and teachers from the University of Innsbruck.

2.2 Mono- and Poly(ADP-Ribose) Synthesis ([^{3}H]-NAD$^+$ Incorporation)

For studies on mono- and poly(ADP-ribose) synthesis, lymphocytes were made permeable to NAD$^+$ by hypoosmotic cold shock at 4°C for 15 min. 1 μCi [^{3}H]-NAD$^+$ was added to the incubation buffer and the radioactivity counted in a PCA precipitate [12].

2.3 Poly(ADP-Ribose) Synthesis (P^{32} Incorporation)

Lymphocytes were suspended in 2 ml incubation system without inorganic phosphate, and treated with 0.1 mCi P^{32} for 30 min at 37°C. The reaction was terminated by adding cold trichloroacetic acid (6% final concentration). The precipitate was washed three times with cold TCA (6%) and lipids extracted with diethyl ether. In 1 N KOH, RNA was hydrolized at room temperature overnight. (NH$_4$)$_2$CO$_3$ solution was used to obtain pH 8.9 and proteins were removed by phenol extraction. The aqueous extract was passed through a column containing Affi-gel 601, a boronate resin. The elution and determination of poly(ADP-ribose) was done according to Hakam et al. [13].

Purity estimation was based on HPCL chromatography of phosphodiesterase digests. The P^{32} incorporation method was used in some probands, parallel to the [^{3}H]-NAD incorporation in permeabilized cells. Both methods used have given comparable results.

3. Results

3.1 Noninduced Poly(ADP-Ribose) Synthesis in Lymphocytes of Chronically Exposed People

In contrast to the activation of the poly(ADP-ribose) polymerase by DNA strand-breaks after ionizing radiation, the poly(ADP-ribose) synthesis in lymphocytes of

Table 1. Poly(ADP-ribose) synthesis (see text)

	Controls	Miners (3 nCi ^{222}Rn l^{-1} air)	Spa personnel (200 pCi ^{222}Rn l^{-1} air)
n	16	16	16
$\bar{x}$	14.8	10.7	11.4
s	5.5	3.2	3.1

[a] Controls to miners: 1.4% significant
Controls to spa personnel: 3.7% significant
Spa personnel to miners: not significant

persons occupationally exposed to chronic low dose radiation over years showed opposite results (Table 1).

3.2 UV-Induced Poly(ADP-Ribose) Synthesis

To obtain comparable results to UDS, UV-induced poly(ADP-ribose) synthesis was also determined. Table 2 shows induced poly(ADP-ribose) synthesis, 90 min after UV-irradiation (20 J m^{-2}, 254 nm max.).

Table 2. Induced poly(ADP-ribose) synthesis (see text)

	Student t-Test[a]		
	Controls	Miners	Spa personnel
n	16	16	16
$\bar{x}$	36.9	23.3	24.2
S.D.	14.2	9.9	9.1
	Variance analysis (r-Test)[b]		
	Controls	Miners	Spa personnel
n	16	14	13
$\bar{x}$	36.9	20.3	22.5
S.D.	14.1	5.4	3.7
r		18.1	47.6
		41.5	39.4
			8.2

[a] Controls to miners: 0.4% significant
Controls to spa personnel: 0.5% significant
Miners to spa personnel: not significant
[b] Controls to miners: 0.03% significant
Controls to spa personnel: 0.14% significant
Miners to spa personnel: not significant

UV-irradiation-dependent DNA strand-breaks (incision during DNA repair) are no potent inducers of poly(ADP-ribose) polymerase, but also in this case significant differences to controls could be obtained.

4. Discussion

In previous experiments we could find in lymphocytes of persons occupationally exposed for a short time to ionizing radiation and also in cancer patients after radiation therapy, higher poly(ADP-ribose) synthesis compared to unexposed controls. Also the chromatin structure determined by micrococcal nuclease (MNase) digestion of DNA within chromatin was changed in lymphocytes of exposed persons [14]. Labeled poly(ADP-ribose) (from [3H]-NAD) was much more increased in the MNase sensitive region, compared to the MNase resistant region within chromatin of lymphocytes (unpublished results) after short-term exposure of persons to ionizing radiation.

It was reported that UDS after UV irradiation is enhanced in the presence of poly-(ADP-ribose) inhibitors [15]. On the basis of these results we could explain the correlation between the lower levels of poly(ADP-ribose) synthesis in miners and the increase in UDS in their lymphocytes. But this cannot explain the enhanced UDS in lymphocytes of persons after short-term exposure parallel to increased poly(ADP-ribose) synthesis. Factors which influence DNA repair incorporation are also the precursor pool and shifts in lymphocyte subpopulations. But in our own investigations we did not find different UDS rates caused by shifts of B- and T-cell ratios [9] and also no changes in the effective thymidine precursor pool in lymphocytes of the two population groups [7]. The observed increase in UDS in lymphocytes of miners could be an induction of de novo synthesis of repair enzymes, similar to the induction of error free and/or error prone (SOS) repair known in microorganisms. Many publications show that some steps in DNA repair may be regulated by the poly(ADP-ribose) polymerase system. Imbalances in UDS can generate mutations, but on the other hand, there is no evidence that the poly(ADP-ribose) polymerase system is participating in induction of mutations [16]. ADP-ribosylation of RNA polymerase is also involved in the regulation of transcription [17]. The fact that UDS is enhanced in both groups, people exposed to acute and chronically applied radiation, seems to be dependent on more overlapping mechanisms.

The induction of the synthesis of repair enzymes, poly(ADP-ribose) dependent changes in chromatin structure, and the regulation of nuclease activity by ADP-ribosylation could take part in the reactions described.

Since poly(ADP-ribose) polymerase decreased in lymphocytes of aged people [18], the lower activity of poly(ADP-ribose) synthesis in persons chronically irradiated could be a molecular biological sign of early aging.

Acknowledgments. This work was supported in part by IAEA Research Contract No. 3315 and AFOSR Grant-Nor. 84-0390.

References

1. Rechsteiner M, Hillyard D, Olivery BM (1976) Magnitude and significance of NAD turnover in human cell line D98/AH2. Nature (London) 259:695–696
2. Skidmore CH, Davies MI, Goodwin PM, Omidiji O, Zia'ee A, Shall S (1980) Poly(ADP-ribose) polymerase – an enzyme sensitive to DNA damage. In: Smulson ME, Sugimura T (eds) Novel ADP-ribosylation of regulatory enzymes and proteins. Elsevier/North Holland, Amsterdam New York, pp 197–205
3. Altmann G, Dolejs I, Topaloglou A, Sooki-Toth A (1979) Faktoren, die die DNA-Reparatur in "spacer and core DNA" von Chromatin menschlicher Zellen beeinflussen. Stud Biophys 76: 195–203
4. Benjamin RC, Gill M (1980) A connection between poly(ADP-ribose) synthesis and DNA strand breakage. In: Smulson ME, Sugimura T (eds) Novel ADP-ribosylation of regulatory enzymes and proteins. Elsevier/North Holland, Amsterdam New York, pp 227–237
5. Tuschl H, Kovac R (1980) The effect of radiation therapy on unscheduled DNA synthesis in lymphocytes of tumor patients. In: Altmann H, Riklis E, Slor H (eds) DNA repair and late effects. NRCN, Israel, pp 305–309
6. Altmann H (unpublished)
7. Tuschl H, Altmann H, Kovacs R, Topaloglou A, Egg D, Günther R (1980) Effects of low dose radiation on repair processes in human lymphocytes. Radiat Res 81:1–9
8. Altmann H, Tuschl H (1978) DNA repair investigations in lymphocytes of persons living in elevated natural background radiation areas (Bad Gastein). In: Late biological effects of ionizing radiation, vol I. IAEA, Vienna, pp 437–445
9. Tuschl H, Kovac R, Altmann H (1983) UDS and SCE in lymphocytes of persons occupationally exposed to low levels of ionizing radiation. Health Phys 45:1–7
10. Tuschl H, Altmann H, Kovac R, Topaloglou A (1983) Effects of chronic low dose exposure on DNA repair processes and sister chromatid exchanges. In: Biological effects of low level radiation. IAEA, Vienna, pp 185–190
11. Gueth L, Weniger P, Altmann H, Vincze I (1984) Induction of DNA repair in spleen cells by in vivo treatments. In: Abstr 16th meeting of FEBS, June 25–30, Moscow, USSR, p 305
12. Altmann H, Dolejs I (1982) Poly(ADP-ribose) synthesis, chromatin structure and DNA repair in cells of patients with different diseases. In: Natarajan AT, Obe G, Altmann H (eds) DNA repair, chromosome alterations and chromatin structure. Alsevier, Amsterdam, pp 167–175
13. Hakam A, McLick J, Kun E (1984) Separation of poly(ADP-ribose) by high performance liquid chromatography. J Chromatography 296:369–377
14. Klein W, Kocsis F, Altmann H, Günther R, Egg D, Sandri B (1983) Chromatinstruktur von Lymphozyten unter erhöhter Rn-Exposition. OEFZS-Ber No A0379, BL-409/83, pp 1–9
15. Miwa M, Kanai M, Kondo T, Hoshino H, Ishihara K, Sugimura T (1981) Inhibitors of poly-(ADP-ribose) polymerase enhance unscheduled DNA synthesis in human peripheral lymphocytes. Biochem Biophys Res Commun 100:463–470
16. Sugimura T, Miwa M (1983) Poly(ADP-ribose) and cancer research. Carcinogenesis 4:1503–1506
17. Slattery E, Digman JD, Matsui T, Roeder RG (1983) Purification and analysis of a factor which suppresses nick-induced transcription by RNA polymerase II and its identity with poly-(ADP-ribose)polymerase. J Biol Chem 258:5955–5959
18. Klein W, Dolejs I, Kocsis F, Altmann H, Sagaster P, Gruber F (1983) Molekularbiologische Untersuchungen zum Alterungsprozeß von Lymphozyten unterschiedlicher Personengruppen. OEFZS-Ber No 4229, BL-430/83, pp 1–23

ADP-Ribosylation and Cellular Differentiation

ADP-Ribosyltransferase in Protozoan Differentiation

GWYN T. WILLIAMS[1]

Investigation of Differentiation in Protozoan Experimental Systems

Cell differentiation in higher eukaryotes involves the selective expression of the information contained in the genome to produce different cell types. In multicellular organisms these cell types develop in parallel along numerous distinct pathways, whereas protozoa, being unicellular, can only follow one sequence of transformations at a time. Nevertheless, the problem of the selective expression of genetic information is also at the centre of the differentiation of protozoa, which involves the production of a series of morphologically and biochemically different types of cells. Because of the large difference, in evolutionary terms, between protozoa and the cells of vertebrates, studies of the mechanism of differentiation of these two types of cells should allow the identification of the common elements of the process which form the basis of eukaryotic cell differentiation as a whole.

The life-cycles of protozoa generally include several transformations between different cell types, and many of these transformations can be reproduced in vitro. Protozoan differentiation can often therefore be investigated in experimental systems which are well suited to detailed analysis at a number of levels. Unlike untransformed vertebrate cell lines which normally have a limited life-span, protozoan cells can be grown and manipulated indefinitely in vitro and often induced to differentiate by simple stimuli such as changes in temperature.

Some protozoa offer uniquely suitable experimental systems for the investigation of phenomena of general importance e.g. African trypanosomes for the analysis of gene rearrangements. Other species have proved valuable in the analysis of other aspects of cell differentiation. These include the differentiation of *Naegleria gruberi* amoebae to flagellates [1] and the various morphological transformations of *Trypanosoma cruzi* [2].

Investigation of *Trypanosoma cruzi* differentiation is also important for practical reasons. *T. cruzi* infection of man results in Chagas' disease, a widespread and often fatal infection occurring in Southern and Central America [3]. Since this parasite differentiates between two interdependent stages in its mammalian host, investigation of

---

1 Anatomy Department, University of Birmingham Medical School, Vincent Drive, Birmingham BI5 2TJ, Great Britain

ADP-Ribosylation of Proteins
(ed. by F.R. Althaus, H. Hilz, and S. Shall)
© Springer-Verlag Berlin Heidelberg 1985

the mechanism of differentiation may allow new targets for chemotherapy to be identified. Such an approach may also be useful for several other protozoan infections, notably malaria. Divergence in the mechanisms of differentiation between protozoan and mammalian cells is of particular relevance in this context. In view of the continuing importance of differentiation for several cell types in adult mammals, the optimal effect on a protozoan infection would only be obtained with compounds which disrupt protozoan differentiation much more than mammalian differentiation.

ADP-Ribosyltransferase Activity in Protozoa

Nuclear ADP-ribosyltransferase activity is distributed very widely, if not universally, in the nuclei of eukaryotic cells [4]. The enzyme has been detected in the free-living protozoan *Tetrahymena pyriformis* by Tsopanakis and colleagues [5], using assay conditions very similar to those used for the enzymes from higher eukaryotes.

The ADP-ribosyltransferase present in the nuclei of the parasites causing a form of malaria in mice, *Plasmodium yoelii nigeriensis,* has been characterised in detail [6]. The enzyme seems to be similar in most respects to those found in higher eukaryotes. The enzyme may well be found in all protozoa, as it has also been detected in *Trypanosoma brucei* (E.E. Okolie, pers. comm., F. Farzaneh, pers. comm.), and in all other species so far examined.

The Effect of Inhibitors of ADP-Ribosyltransferase on Protozoan Differentiation

Several of the Kinetoplastida (protozoa which possess a characteristic organelle containing extra-chromosomal DNA, the kinetoplast) can be induced to differentiate in vitro. Both *Trypanosoma* species and *Leishmania* species belong to this category, and have proved to be most suitable for studies of the effects of enzyme inhibitors.

The differentiation of *Trypanosoma cruzi* has been studied in both extracellular and intracellular experimental systems [2, 7, 8]. This parasite normally proliferates inside mammalian cells as the non-motile amastigote, and differentiates to the flagellated trypomastigote form towards the end of the intracellular infection. Observation of the accumulation of motile trypomastigotes in the culture medium after rupture of host cells therefore provides a convenient method of assessing the success of the infection cycle.

Normally, most trypomastigotes are released 4 or 5 days after the original infection (Fig. 1). However, in the presence of the ADP-ribosyltransferase inhibitor, 3-aminobenzamide [9], release was significantly and reproducibly delayed (Fig. 1A). The presence of 3-aminobenzoic acid produced no significant effect. Stronger inhibitory effects were produced by other enzyme inhibitors. 5-Methylnicotinamide [10] (Fig. 1B) and 3-methoxybenzamide [9] (Fig. 1C) almost totally blocked the appearance of the

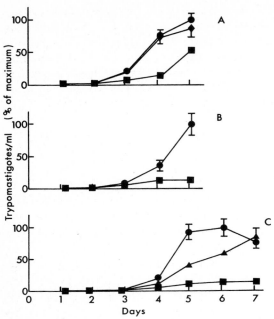

Fig. 1A–C. Inhibition of the intracellular infection cycle of *T. cruzi*. ◆ 5 m*M* 3-aminobenzoic acid (Aldrich) present continuously. ■ (A) 5 m*M* 3-aminobenzamide present continuously. ■ (B) 5 m*M* 5-methylnicotinamide present continuously. ■ (C) 2 m*M* 3-methoxybenzamide present from day 1 onwards. ▲ 2 m*M* 3-methoxybenzamide present day 1–3. ● no inhibitor. Data taken from [8], by copyright permission of the Rockefeller University

trypomastigotes. Removal of the inhibitor resulted in the appearance of trypomastigotes in the supernatant (Fig. 1C), indicating that the effect was reversible and not the result of cell death.

However, a complete cycle of infection requires both the proliferation of amastigotes and the release of trypomastigotes in addition to differentiation between these two forms, so that a number of other possible explanations for the effects observed were examined. One such possible explanation was that the proliferation of the parasite, rather than its differentiation, was inhibited under the conditions of the experiment. This possibility was examined directly by monitoring the accumulation of *T. cruzi* amastigotes in the presence and absence of the inhibitors. No decrease in the rate of accumulation was detected with any of the inhibitors (Table 1).

A second explanation considered was a direct inhibitory effect on the release of the trypomastigotes from the mammalian cells. If this were correct, differentiation from the non-motile amastigote to the flagellated trypomastigote would occur normally, and cells treated with the inhibitor would contain trapped trypomastigotes. Staining of infected cells treated with inhibitor showed that the parasites were present as amastigotes, i.e. that differentiation to the trypomastigote form was inhibited [2].

Further analysis of the effects of inhibitors of ADP-ribosyltransferase on the differentiation of *T. cruzi* required the development of an extracellular differentiation

Table 1. Accumulation of intracellular *T. cruzi* amastigotes in the presence of inhibitors of ADP-ribosyltransferase

Inhibitor (m*M*)	Parasites/ field[b]	Parasites/ infected cell[b]
–	7.0 ± 1.6	7.0 ± 0.5
3-aminobenzamide[a] (5)	9.3 ± 2.6	10.6 ± 1.2
3-methoxybenzamide[a] (2)	6.3 ± 1.9	8.4 ± 1.2
5-methylnicotinamide (5)	6.9 ± 1.2	9.0 ± 1.1
Nicotinamide (10)	6.0 ± 2.0	8.0 ± 0.8

[a] Data from [7]

[b] Numbers of parasites 94 h after initiating infection. Means and SE of four cultures are shown for each inhibitor

system [2, 7, 8]. This system allowed indirect effects on the host cells to be excluded. The experimental system developed used non-motile amastigotes grown extracellularly at 37°C and induced morphological transformation simply by transfer to appropriate medium at 27°C. Both of the flagellated forms of *T. cruzi*, the epimastigote and the trypomastigote, were produced under these conditions. The rate of differentiation was therefore monitored by counting the total number of motile parasites (Fig. 2). These made up about 40% of the total after 5 days.

In this extracellular system, the effect of ADP-ribosyltransferase inhibitors was even more marked (Fig. 2) [7, 8]. Once again, control experiments showed that the inhibition was reversible and could not be explained by an effect on the proliferation of either motile or non-motile cells (Fig. 2 and Table 2 respectively). As in the intracellular system, ADP-ribosyltransferase inhibitors seemed to affect specifically some process involved only in differentiation which was not required for the proliferation of the cells. The observations exclude any overwhelming effect on cellular metabolism not related to differentiation.

A number of different ADP-ribosyltransferase inhibitors were also examined for effects on this system [7]. 3-aminobenzamide, theophylline, benzamide and nicotinamide all inhibited the appearance of motile parasites. 10 m*M* nicotinamide, for example, produced greater than 95% inhibition of differentiation. In contrast, cultures containing 10 m*M* nicotinic acid, which does not inhibit the enzyme, were indistinguishable from controls.

If the compounds tested produced their effects by inhibition of a specific process, or a small group of related processes, they should only work when added before the events concerned occur. This hypothesis was tested by adding 5-methylnicotinamide at different times to amastigotes differentiating extracellularly (Fig. 3). Despite the significant asynchrony of differentiation in this system, which would reduce the possibility of observing such an effect, it was found that optimal inhibition of differentiation was only produced when the 5-methylnicotinamide was added in the first 15 h of the incubation. As most motile parasites did not appear before 90 h, the inhibitor seemed to affect a very early event in the differentiation process. This observation is clearly not consistent with inhibition of proliferation or non-specific cytotoxicity in this system.

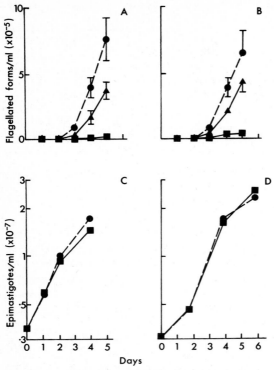

Fig. 2A–D. Effect of ADP-ribosyltransferase inhibitors on differentiation (**A, B**) and proliferation (**C, D**) of *T. cruzi*. ■ 2 m*M* 3-methoxybenzamide (**A, C**) or 5 m*M* 5-methylnicotinamide (**B, D**) continuously present. ▲ 2 m*M* 3-methoxybenzamide (**A**) or 5 m*M* 5-methylnicotinamide (**B**) present up to 24 h. ● no inhibitor. Data taken from [8], by copyright permission of the Rockefeller University

Table 2. Effect of ADP-ribosyltransferase inhibitors on the extracellular proliferation of *T. cruzi* amastigotes

Inhibitor (m*M*)	Proliferation rate (% of control)
–	100 ± 7^a
5-Methylnicotinamide (5)	100 ± 5
3-Aminobenzamide (5)[b]	116 ± 12
Benzamide (2.5)[b]	96 ± 6
3-Methoxybenzamide (2)	74 ± 4
Nicotinamide (10)[b]	70 ± 15
Theophylline (5)[b]	69 ± 4

[a] Mean and SE of four cultures
[b] Data from [7]

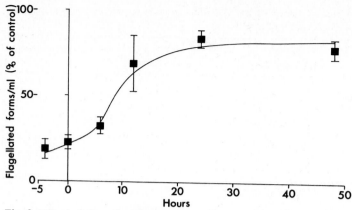

Fig. 3. Effect of time of addition of 5-methylnicotinamide on *T. cruzi* differentiation. Amastigotes were transferred to Warren's medium at 27°C at time zero. 5-Methylnicotinamide was added, to a final concentration of 5 m*M*, to separate cultures at the times indicated and the numbers of flagellated cells counted with a haemocytometer at 96 h. This was expressed as a percentage of the flagellated cells in control cultures without 5-methylnicotinamide. Data taken from [8], by copyright permission of the Rockefeller University

Morphological differentiation of the African trypanosome, *T. brucei* can also be studied in vitro, and the inclusion of benzamide inhibitors of ADP-ribosyltransferase also inhibited differentiation in this system (J.D. Barry, pers. comm.).

The *Leishmania,* close relatives of the *Trypanosoma,* also undergo differentiation in vitro. Morphological transformation in these protozoa can be studied in an experimental system which is similar to that used to study extracellular differentiation in *T. cruzi,* (e.g. [11]). Non-motile amastigotes are harvested from the macrophages in which they grow and transferred to medium at 27°C where they transform to flagellated pro-

Table 3. Effect of 3-methoxybenzamide on the differentiation of *Leishmania mexicana mexicana* amastigotes[a]

	Differentiating forms at 24 h (% of total parasites in control)
Control[b]	93
3-Methoxybenzamide[c] (from 0 h)	22
3-Methoxybenzamide[c] (from 8 h)	50
3-Methoxybenzamide[c] (from 16 h)	63

[a] Methoxybenzamide was present at 2.5 m*M*, a concentration which had no effect on promastigote proliferation. Parasite numbers were significantly less in methoxybenzamide treated cultures than in controls. This may be due to the inability of the undifferentiated parasites to survive or proliferate without differentiating, or to selective toxicity in *Leishmania* amastigotes (unpublished work of J. Capaldo and G.H. Coombs)

[b] Mean of four experiments

[c] Mean of two experiments

Table 4. Antimalarial activity of inhibitors of ADP-ribosyltransferase and chloroquine against *P. yoelii* infection in mice. (From Okolie et al. [13])

Inhibitor/drug	Total PD[a] Inhibitor/drug	Total PD Saline control	Inhibitor (drug)/ saline ratio
Nicotinamide	34	35	0.97
5-Methylnicotinamide	26	35	0.74
3-Aminobenzamide	36	35	1.03
M-methoxybenzamide	18	35	0.51
Theophylline	24	35	0.69
Theobromine	22	35	0.63
Chloroquine sulphate	5	35	0.14

[a] PD = Parasite Density

mastigotes. Addition of inhibitors of ADP-ribosyltransferase to this system inhibited this differentiation but did not affect the proliferation of promastigotes (Table 3, [12]). A requirement for ADP-ribosyltransferase in differentiation but not proliferation therefore appears widely distributed among kinetoplastid protozoa.

Potential of ADP-Ribosyltransferase Inhibitors in Chemotherapy

The potential use of inhibitors of ADP-ribosyltransferase for blocking the differentiation of pathogenic protozoa in human disease probably rests on the development of compounds which preferentially inhibit the protozoan enzymes. Currently available inhibitors were originally identified by their effects on ADP-ribosyltransferases from the cells of vertebrates and cannot therefore be specific for the protozoan enzymes.

There do appear to be significant quantitative differences between the effects of known inhibitors of ADP-ribosyltransferase on the enzyme from *Plasmodium* when compared to their effects on mammalian enzymes [6]. In addition, the degree of inhibition observed differs considerably between different protozoa, since the enzyme from *Tetrahymena pyriformis* [5] was much more susceptible to inhibition both by nicotinamide and by thymidine than was the *Plasmodium* enzyme.

Since the observed inhibition of differentiation was reversible, effective treatment of protozoan infections would probably involve using ADP-ribosyltransferase inhibitors to block the progression of the infection and increase the effectiveness of conventional cytotoxic agents in eliminating the parasites.

The effect of currently available inhibitors on the development of malaria (*Plasmodium yoelii nigeriensis* infection) in mice has been investigated (Table 4, [13]). Perhaps surprisingly even some of these compounds, which would also be expected to inhibit the immune response against the parasites [14, 15], inhibited the progress of the disease. However, the mode of action of these compounds in whole animals could be different from that observed with other parasites in vitro.

Conclusion

ADP-ribosyltransferase therefore seems to be involved in the differentiation of a very wide range of eukaryotes, from mammals and other vertebrates to protozoa. This suggests that the enzyme is an essential part of a general mechanism of differentiation which has been conserved at least since the divergence of the evolutionary pathways leading to vertebrates and protozoa.

ADP-ribosyltransferases are found in the plasma membranes as well as the nuclei of eukaryotic cells. Although the question is far from settled, what evidence there is suggests that the nuclear enzyme is the one needed for differentiation [17] (although it is, of course, possible that both are involved). The involvement of the nuclear enzyme may be due to a requirement for changes in the integrity of DNA to allow conformational changes or genetic rearrangement to occur [14, 16, 17]. Protozoan systems may well prove very valuable in investigating the role which ADP-ribosyltransferase plays in this process.

Acknowledgments. I thank the U.K. Medical Research Council and the Wellcome Trust for financial support. I thank J. Capaldo and G.H. Coombs for permission to include some of their unpublished work.

References

1. Fulton C (1983) Macromolecular synthesis in the quickchange act of *Naegleria*. J Protozool 30:192–198
2. Williams GT (in press) Control of differentiation in *Trypanosoma cruzi*. Curr Top Microbiol Immunol
3. Brener Z (1973) Biology of *Trypanosoma cruzi*. Annu Rev Microbiol 27:347–383
4. Ueda K, Hayaishi O (1982) Poly(ADP-ribose) synthetase. In: Hayaishi O, Ueda K (eds) ADP-ribosylation reactions. Academic Press, London New York, p 117
5. Tsopanakis C, Leer JC, Nielsen OF, Gocke E, Shall S, Westergaard O (1978) Poly(ADP-ribose) metabolizing enzymes in nuclei and nucleoli of *Tetrahymena pyriformis*. FEBS Lett 93:297–300
6. Okolie EE, Onyezili NI (1983) ADP-ribosyl transferase in *Plasmodium* (malaria parasites). Biochem J 209:687–693
7. Williams GT (1983) *Trypanosoma cruzi:* Inhibition of intracellular and extracellular differentiation by ADP-ribosyl transferase antagonists. Exp Parsitol 56:409–415
8. Williams GT (1984) Specific inhibition of the differentiation of *Trypanosoma cruzi*. J Cell Biol 99:79–82
9. Purnell MR, Whish WJD (1980) Novel inhibitors of poly(ADP-ribose) synthetase. Biochem J 185:775–777
10. Clark JB, Ferris GM, Pinder S (1971) Inhibition of nuclear NAD nucleosidase and poly ADP-ribose polymerase activity from rat liver by nicotinamide and 5-methylnicotinamide. Biochim Biophys Acta 238:82–85
11. Simpson L (1968) The leishmania-leptomonad transformation of *Leishmania donovani:* Nutritional requirements, respiration changes and antigenic changes. J Protozool 15:201–207
12. Capaldo J, Coombs GH (1983) New approaches to the chemotherapy of leishmaniases. Parasitology 87:xli

13. Okolie EE, Adewunmi AI, Enwonwu CO (1982) Antimalarial activity of inhibitors of ADP-ribosyl transferase against *Plasmodium yoelii* infection in mice. Acta Trop 39:285–289
14. Johnstone AP, Williams GT (1982) Role of DNA breaks and ADP-ribosyl transferase activity in eukaryotic differentiation demonstrated in human lymphocytes. Nature (London) 300:368–370
15. Johnstone AP (1984) Rejoining of DNA strand breaks is an early event during the stimulation of quiescent lymphocytes. Eur J Biochem 140:401–406
16. Farzaneh F, Zalin R, Brill D, Shall S (1982) DNA strand breaks and ADP-ribosyl transferase activation during cell differentiation. Nature (London) 300:362–366
17. Williams GT, Johnstone AP (1983) ADP-ribosyl transferase, rearrangement of DNA, and cell differentiation. Biosci Rep 3:815–830

ADP-Ribosyltransferase Activity in
Trypanosoma brucei

FARZIN FARZANEH[1], SYDNEY SHALL[2], PAUL MICHELS[3],
and PIET BORST[4]

Introduction

The nuclear enzyme ADPRT (EC: 2.4.2.30), which is entirely dependent on the presence of DNA-containing strand breaks for its activity [1], is required for efficient DNA excision repair [2]. ADPRT activity is also an obligatory requirement for the expression of the differentiated phenotype in a number of eukaryotic cells (see [3]). Williams [4] has shown that inhibition of ADPRT activity blocks the differentiation, but not proliferation, of *Trypanosoma cruzi* amastigotes to epimastigotes and trypomastigotes. Although the presence of ADPRT activity in *Plasmodium yoelii* has been demonstrated [6], the presence of this enzyme in trypanosomes has not been directly investigated. In addition, it is not clear whether the protozoan ADPRT activity can be stimulated by DNA strand breaks and whether the inhibitors of ADPRT activity in higher eukaryotes can also block the activity of this enzyme in trypanosomes. Here we repor the detection of ADPRT activity in *T. brucei,* its activation by DNA damage and its inhibition by benzamide and its analogues. We also show that intact trypanosomes are readily permeable to the ADPRT inhibitor 3-aminobenzamide.

Presence of ADPRT Activity in *Trypanosoma brucei*

In freshly isolated blood-stream trypanosomes, which are permeabilised to NAD by a mild hypotonic shock, there is a low but detectable level of basal ADPRT activity (Fig. 1 ■). Induction of DNA strand breaks by exogenous DNase I increases this

1 Harris Birthright Research Center for Fetal Medicine, Department of Obstetrics and Gynaecology, Kings College School of Medicine, Denmark Hill, London SE5 8RX, Great Britain
2 Cell and Molecular Biology Laboratory, University of Sussex, Brighton BN1 9QG, Great Britain
3 Section for Medical Enzymology, University of Amsterdam, Jan Swamerdam Institute, 1005 GA Amsterdam, The Netherlands. Present address: International Institute of Cellular and Molecular Pathology, Avenue Hippocrat 75, 1200 Bruxelles, Belgium
4 Section for Medical Enzymology, University of Amsterdam, Jan Swamerdam Institute, 1005 GA Amsterdam, The Netherlands. Present address: The Netherlands Cancer Institute, Antoni van Leeuwenhoekhuis, Plesmanlaan 121, 1066 CX Amsterdam, The Netherlands

ADP-Ribosylation of Proteins
(ed. by F.R. Althaus, H. Hilz, and S. Shall)
© Springer-Verlag Berlin Heidelberg 1985

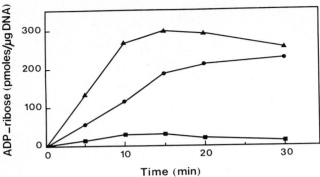

Time (min)

Fig. 1. ADPRT activity in permeabilised *T. brucei*. Trypanosomes were permeabilised to NAD by a mild hypotonic shock. ADPRT activity was measured either in the absence of exogenous nucleases (■), or in the presence of 10 U ml^{-1} of DNase I (▲) for the estimation of basal and potential activities, respectively. The basal ADPRT activity in trypanosomes pre-treated with 100 μM dimethylsulphate, prior to permeabilisation (●)

activity approximately tenfold (Fig. 1 ▲). The basal ADPRT activity can also be stimulated fourfold by exposure of the trypanosomes to 100 μM DMS for 15 min, prior to permeabilisation (Fig. 1 ●).

In order to confirm that the acid-insoluble radioactive material synthesised during the ADPRT assay is in fact (ADP-ribose)$_n$, this material was digested with a number of hydrolytic enzymes. RNase A, DNase I, Micrococcal nuclease and proteinase K failed to degrade this material. However, snake venome phosphodiesterase degraded 96% of the acid-insoluble radioactive material (Table 1). This strongly suggests that the incorporated radioactive material was (ADP-ribose)$_n$.

Inhibition of ADPRT Activity

The ADPRT activity in permeabilised *T. brucei* is completely inhibited by 100 μM 3-aminobenzamide (equimolar concentration to the substrate) (Fig. 2 □). At this concentration, 3-aminobenzoic acid did not block the synthesis of (ADP-ribose)$_n$

Table 1. Sensitivity of ADPRT assay product to enzyme hydrolysis

Enzyme used	pmol [^{14}C]-ADP-ribose remaining	%(ADP-ribose)$_n$ degraded
Control	754	0
RNase A	741	2
DNase I	735	3
Micrococcal nuclease	739	2
Proteinase K	689	9
Snake venom phosphodiesterase	34	96

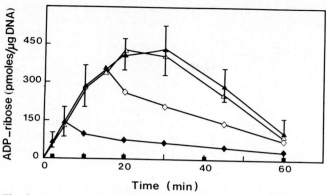

Fig. 2. Inhibition of ADPRT activity. The potential ADPRT activity was measured either in the presence (■), or absence of either 100 μM 3-aminobenzamide (▲), or 100 μM 3-aminobenzoic acid (△). Alternatively 100 μM 3-aminobenzamide was added either 5 min (◆), or 15 min (◇) after the start of the assay

(Fig. 2 △). The addition of 3-aminobenzamide after either 5 or 15 min of incubation with [^{14}C]-NAD, was followed by a reduction in the level of already synthesized (ADP-ribose)$_n$ (Fig. 2 ◆ and ◇). Therefore, a degrading enzyme, possibly poly(ADP-ribose) glycohydrolase, is present and active in the permeabilised trypanosomes.

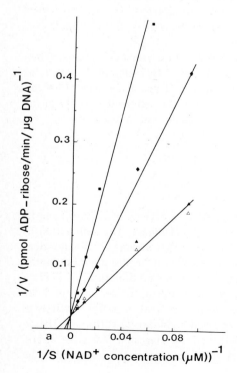

Fig. 3. Double-reciprocal plot of ADPRT activity. The total ADPRT activity was measued in the absence (▲) or presence of 5 μM (◆) or 10 μM (■) 3-aminobenzamide, or 10 μM 3-aminobenzoic acid (△)

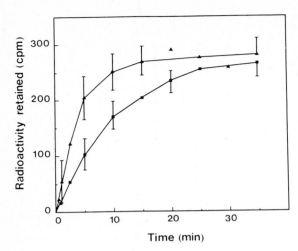

Radioactivity retained (cpm)

Time (min)

Fig. 4. Permeability of *T. brucei* to 3-aminobenzamide. Intact or permeabilised trypanosomes were incubated at either 4°C (■) or 37°C (▲) in a physiological saline containing 1 mM [³H]-3-aminobenzamide. After three very rapid washes, the retained radioactivity was measured. Results are expressed as the difference between intact and permeabilised trypanosomes which served as control

The K_m and V_{max} of the potential ADPRT activity (measured in the presence of 10 U ml^{-1} of exogenous DNase I), estimated from a double-reciprocal plot, are 110 ± 17 μM NAD and 39 ± 5 pmol ADP-ribose min^{-1} μg^{-1} DNA, respectively. (Fig. 3 ▲). 3-Aminobenzamide is a competitive inhibitor of ADPRT activity in *T. brucei*. The K_i, measured at 5 μM and 10 μM, is 4.3 ± 0.5 μM (Fig. 3 ♦ and ■). The estimated K_i values for two other competitive inhibitors, benzamide and 3-methoxybenzamide, are 2.6 ± 0.4 μM and 2.9 ± 0.5 μM, respectively (results not shown). The acid analogues of these compounds, 3-aminobenzoic acid (Fig. 3 △), benzoic acid and 3-methoxybenzoic acid, do not inhibit the ADPRT activity.

Freshly isolated intact trypanosomes are permeable to 3-aminobenzamide. When incubated in a physiological saline in the presence of [³H]-3-aminobenzamide, at either 4°C or 37°C, this compound readily permeates the trypanosomes (Fig. 4 ■ and ▲). The faster rate of permeation at 37°C may be due to the increased mobility of the trypanosomes and better mixing of the incubation mixture at this temperature.

Discussion

Studies reported here demonstrate the presence of ADPRT activity in *T. brucei* and its inhibition by benzamide and its amide, but not acid, analogues. The stimulation of ADPRT activity by DNA damage suggests that the role of this enzyme in trypanosomes may be similar to that in higher eukaryotic cells. It has already been shown that inhibition of ADPRT activity blocks the differentiation, but not proliferation, in *T. cruzi* [5] and *T. brucei* (J.D. Barry, pers. comm.). We have previously demonstrated the formation of DNA strand breaks and the obligatory involvement of ADPRT activity in the cytodifferentiation, but not proliferation, of avian myoblasts in culture [3]. Based on this and other similar observations (see [4]) we have suggested that DNA strand break formation and rejoining, modulated by ADPRT activity, may be involved in the developmental regulation of cellular differentiation. We have postulated that this may be due to the possible involvement of gene amplification, mobility in the genome and/or regional chromatin relaxation mediated by the action of topoisomerases in this process.

African trypanosomes, like *T. brucei*, depend on antigenic variation to evade the immune response of their vertebrate hosts. Antigenic variation is caused by repeated switches in the composition of the surface coat (variable surface glycoproteins, VSGs) and most switches require the duplicative transposition of a VSG gene [7]. We are currently investigating the possible involvement of ADPRT activity in VSG gene switching in *T. brucei*.

References

1. Benjamin RG, Gill DM (1980) Poly(ADP-ribose) synthesis in vitro programed by damaged DNA. J Biol Chem 255:10943–10508
2. Durkacz WD, Omidiji O, Gray DA, Shall S (1980) (ADP-ribose)$_n$ participates in DNA excision repair. Nature (London) 283:593–596
3. Farzaneh F, Zalin R, Brill D, Shall S (1982) DNA strand breaks and ADP-ribosyl transferase activation during cell differentiation. Nature (London) 300:362–366
4. Williams GT, Johnstone AP (1983) ADP-ribosyl transferase, rearrangement of DNA, and cell differentiation. Biosci Rep 3:815–830
5. Williams GT (1983) *Trypanosoma cruzi:* Inhibition by ADP-ribosyl transferase antagonists of intracellular and extracellular differentiation. Exp Parasitol 56:409–415
6. Okolie EE, Onyezili NI (1983) ADP-ribosyl transferase in *Plasmodium* (malaria Parasite). Biochem J 209:687–693
7. Borst P, Cross GAM (1982) The molecular basis for trypanosome antigenic variation. Cell 29:291–303

ADP-Ribose: A Novel Aspect of Steroid Action

GEORGE S. JOHNSON[1] and SEI-ICHI TANUMA[2]

1. Introduction

Glucocorticoid hormones have multiple effects on cellular physiology involving interactions with many types of cells. The majority, if not all, of the effects of glucocorticoids involve changes in gene expression, hence, studies with these steroids have been
useful not only to understand the biochemical basis of hormone action, but also as
model systems to understand the control of gene expression [2].

Understanding the molecular basis for glucocorticoid action and that of other
steroids has been the object of much research attention for the past several years.
Although the mechanistic details have not yet been firmly established, several general
features have been developed. After entering the cell (via a process which is not clearly
understood) steroids bind to a specific receptor protein. The resultant steroid-receptor
complex undergoes an "activation" characterized by changes in several biochemical
parameters of the receptor protein [17]. Perhaps the most prominent characteristics
of the activated receptor most easily identifiable with gene expression are its strong
affinity for DNA and its nuclear location. The cellular location of the unactivated
receptor is a matter of some controversy. Glucocorticoid and other steroid receptors
are believed to be cytoplasmic and translocate to the nucleus upon activation. However, this model for the estrogen receptor has recently been challenged. Evidence has
been obtained using monoclonal antibody localization [4] or cytochalasin B-induced
enucleation [22] that the receptor is normally present in the nucleus. To explain
the discrepancy of these results with those of previous experiments, it was proposed
that the nuclear association of the unactivated receptor is weak so that the receptor
is released into the cytoplasmic compartment following cell disruption. Similar localization of the glucocorticoid receptor may prove to be true, but this idea must await
additional experimentation.

In prokaryotes certain proteins have been shown to affect gene expression by binding to highly specific DNA sequences [8]. Ongoing research on steroid activation of

1 Laboratory of Molecular Biology, National Cancer Institute, National Institutes of Health,
  Bethesda, MD 20205, USA
2 Present address: Department of Physiological Chemistry, Faculty of Pharmaceutical Sciences,
  Teikyo University, Sagamiko, Kanagawa 199-01, Japan

ADP-Ribosylation of Proteins
(ed. by F.R. Althaus, H. Hilz, and S. Shall)
© Springer-Verlag Berlin Heidelberg 1985

genes suggests the existence of similar mechanisms in eukaryotes. In general, DNA sequences essential for gene regulation are located in the 5' region of the genome, upstream from the RNA initiation start site. Biochemical and genetic analyses have established that specific DNA sequences in this region of glucocorticoid sensitive genomes are required for induction and that steroid-receptor binding to these regions is required for initiation of RNA polymerase activity [9, 10, 16].

How steroid-receptor binding to these regions leads to gene activation is not known. Active genes are generally in a more open configuration characterized by hypersensitivity to nuclease digestion [21]. Chromatin structural changes are possibly instrumental for the steroid activation as steroids have been demonstrated to cause changes in nuclease sensitivity in responsive genes [1]. The biochemical basis for these steroid-induced changes in structure is the object of much speculation, but a covalent modification of chromosomal proteins is a likely candidate. One such modification, ADP-ribosylation, has been clearly shown to cause global alterations in chromatin structure in vitro [11]. We reasoned that similiar localized changes in chromatin structure in the steroid-receptor binding region or in the promoter region of steroid sensitive genomes could be instrumental in glucocorticoid activation of genes. The two biological systems we have chosen to analyze this possibility are growth hormone (GH) synthesis in rat pituitary GH_3 cells and mouse mammary tumor virus (MMTV) expression in mouse mammary tumor cells, two systems commonly used to study steroid action.

ADP-ribosylation of proteins has been associated with gene expression and steroid action in other systems. These results have been discussed elsewhere [3, 18, 19] and will not be mentioned in detail here.

2. Inhibitors of ADP-Ribosylation Increase Expression of Glucocorticoid Sensitive Genes

2.1 Mouse Mammary Tumor Virus Expression

Two species of MMTV RNA are present in cultured mouse mammary tumor cells, a full length 35S and a spliced 24S species [15]. When cells are treated with glucocorticoids, the cellular content of both RNA species increases considerably. The increase is due to stimulation of RNA synthesis, a rapid action reaching a maximum within 10–30 min after steroid addition [14, 24]. Activation is insensitive to inhibitors of protein synthesis demonstrating that it is a primary action of the steroids on the gene and not dependent upon prior activation of another gene. When the established cell line of mouse mammary tumor cells (34I line) is treated with 3-aminobenzamide (3-ABm), an inhibitor of $(ADP\text{-}ribose)_n$ synthetase [12], the content of both RNA species increases (Fig. 1; [19]). In this experiment induction is about the same as with dexamethasone, a synthetic glucocorticoid. But it should be emphasized that the kinetics of accumulation is considerably different with the two agents. The time required for half maximal accumulation of both RNA species is about 2 h for dexamethasone and about 25 h for 3-ABm (Fig. 2). 3-Aminobenzoic acid is ineffective as an inducing agent [19].

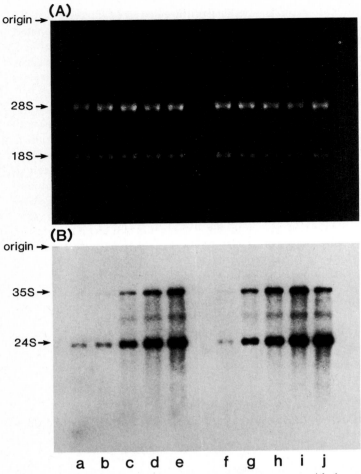

Fig. 1A,B. Accumulation of MMTV RNA during treatment with dexamethasone or 3-ABm. 34I cells were grown and treated, and RNA was extracted and analyzed as described [19]. A Ethidium bromide staining pattern of the gel. Identical staining of the 18S and 28S rRNA demonstrates that an equivalent amount of RNA was added to each gel lane. B Hybridization of RNA to [^{32}P]-labeled MMTV (env) DNA. *a–e* 3-ABm treatment: *a* 0 h, *b* 4 h, *c* 16 h, *d* 32 h, *e* 64 h. *f–j* Dexamethasone treatment: *f* 0 h, *g* 1 h, *h* 2 h, *i* 4 h, *j* 16 h

This large increase in MMTV RNA during 3-ABm treatment is not always observed for reasons which are not understood. In other studies we found a much decreased response to benzamide (BAm) (Fig. 3) or to 3-ABm (data not shown). Interestingly, dexamethasone is a much more effective agonist in cells treated with BAm than it is in control cells (Fig. 3).

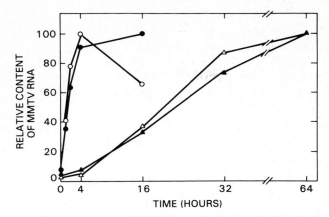

Fig. 2. Kinetics of MMTV RNA accumulation. The intensity of the bands shown in Fig. 1 were determined by densitometry. The maximum value for each group was defined as 100%. ▲ and △ 24S and 35S RNA from 3-ABm treated cells; ● and ○ 24S and 35S RNA from dexamethasone-treated cells

2.2 Growth Hormone Expression

Treatment of GH_3 cells with nicotinamide and related compounds increases synthesis of GH as measured by radioimmunoassay [3]. This induction is associated with a larger cellular content of GH mRNA indicating that the effect is due, at least in part, to changes in RNA. GH synthesis in the GH_3 cells is activated by glucocorticoids and thyroid hormone (T_3) [20]; moreover, activation is synergistic when both hormones are used. The nicotinamide-related drugs also activate GH synthesis synergistically with T_3 demonstrating that they may be mimicking glucocorticoid action in these cells. The action of the nicotinamide durgs is complex and they may interact with more than one control mechanism. For example, nicotinic acid, which does not inhibit

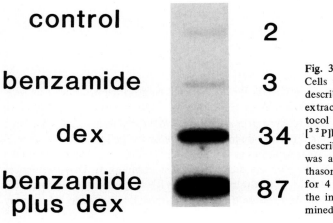

Fig. 3. MMTV RNA in 34I cells. Cells were grown and treated as described [19]. MMTV RNA was extracted by the "cytodot" protocol [23]. Hybridization to [^{32}P]labeled MMTV DNA was as described [19]. BAm treatment was at 8 mM for 48 h; dexamethasone treatment was a 10^{-7} M for 4 h. The numbers represent the intensity of the bands determined by densitometry

ADP-ribosylation, also increases GH synthesis; however, activation with this agent is not synergistic with T_3, in contrast to the observations with nicotinamide. The mechanism of action of nicotinic acid in thse cells remains to be investigated. It should be noted that prolactin synthesis is also increased by N'-methylnicotinamide, the most effective agent in inducing GH synthesis [3]. The prolactin genome is not under glucocorticoid regulation, but it is stimulated by estrogens [20]. Possibly ADP-ribose may be important in estrogen action.

3. Endogenous ADP-Ribosylation

The results described suggest that inhibition of ADP-ribosylation leads to an increase in gene expression; i.e., loss of (ADP-ribose)$_n$ from chromosomal protein(s) is associated with gene activation. It follows that (ADP-ribose)$_n$ could be a negative regulator of gene expression. To more clearly establish this point and to learn more about how (ADP-ribose)$_n$ may influence gene activity, it is essential to know which proteins are ADP-ribosylated in intact cells and if loss of (ADP-ribose)$_n$ from certain of these proteins is associated with gene activation. Numerous proteins can be ADP-ribosylated in broken cell homogenates. Among these, the high mobility group (HMG) proteins [13], which are associated with at least some active genes [21], are possible candidates for regulation of gene expression. We observed that HMG proteins 14 and 17 are ADP-ribosylated in intact 34I cells and that endogenous ADP-ribose is lost from these proteins with a half-time of about 15 min [19], a time span associated with induction of the MMTV genome [14, 24]. With 3-ABm treatment the loss of (ADP-ribose)$_n$ is much slower, presumably a result of attrition and dilution of the modified proteins by cell growth and division. The slow loss is associated with the slow accumulation of MMTV RNA (Fig. 2). The results we have obtained are consistent with the idea that ADP-ribosylation of the HMG proteins is important for steroid activation of genes. However, the evidence linking these proteins to the activation is only indirect. Possibly ADP-ribosylation of other proteins not included in our protein fraction is important. Proof for this involvement will come with direct studies in in vitro experiments.

4. Conclusions

Our results suggest that a loss of (ADP-ribose)$_n$ from chromosomal proteins is an integral component in steroid activation of genes. We propose the following model to explain our ideas. The model is based primarily on the MMTV genome which has been the most thoroughly studied of the glucocorticoid sensitive genes, but the general features are applicable to other genomes as well. Genetic and biochemical experiments have demonstrated that binding of the steroid-receptor complex to DNA sequences 150 to 200 base pairs upstream from the start site for RNA synthesis is essential for gene activation [5, 9, 10, 16]. We propose that an ADP-ribosylated protein, which may be one of the HMG proteins, binds to this region or between this region and the cap site and thereby prevents RNA polymerase initiation. One consequence of the binding of the activated steroid-receptor complex is the removal of the $(ADP-ribose)_n$ from this protein possibly by activation of an ADP-ribosyl protein lyase [6]. Alternatively, the ADP-ribosylated protein may be degraded by receptor activation of a specific protease. The question arises if the removal of the ADP-ribosylated protein is sufficient to activate the gene. The induction with 3-ABm treatment argues that this may be the case; however, the variability of the response to 3 ABm indicates that other factors are involved.

Obviously, this model is quite speculative. To test these ideas it will be necessary to learn which proteins are associated with the regulatory regions of the steroid-sensitive genes, which of these proteins are ADP-ribosylated, and if ADP-ribosylation of these proteins influences steroid-induced RNA initiation in in vitro assays. These analyses are difficult and beyond the present "state of the art" in molecular biology.

The 34I cells used in our analysis contain several copies of the MMTV genome stably integrated within chromosomes. Thus, analysis with this system is very difficult. A more pure genetic system would be necessary for analysis of our model. Recently, Ostrowski et al. [7] have introduced the regulatory region of the MMTV genome (the long terminal repeat) into cells as a chimeric bovine papilloma virus plasmid which replicates as an extrachromosomal episome. Expression of this episome is regulated by glucocorticoids. In collaboration with this group we have learned that expression from this episomal system is stimulated by nicotinamide and curiously and unexpectedly by nicotinic acid (M. Ostrowski, S. Tanuma, and G. Johnson 1984, unpublished observations). We hope that future studies with this system will be useful to learn more about the role of ADP-ribosylation in the complex mechanisms of gene regulation.

References

1. Burch BE, Weintraub H (1983) Temporal order of chromatin structural changes associated with activation of the major chicken vitellogenin gene. Cell 33:65–76
2. Higgins SJ, Gehring U (1978) Molecular mechanisms of steroid hormone action. Adv Cancer Res 28:313–397
3. Kimura N, Kimura N, Cathala G, Baxter JD, Johnson GS (1983) Nicotinamide and its derivatives increase growth hormone and prolactin synthesis in cultured GH_3 cells: role for ADP-ribosylation in modulating specific gene expression. DNA 2:195–203
4. King WJ, Greene GL (1984) Monoclonal antibodies localize oestrogen receptor in the nuclei of target cells. Nature (London) 307:745–747
5. Majors J, Varmus HE (1983) A small region of the mouse mammary tumor virus long terminal repeat confers glucocorticoid hormone regulation on a linked heterologous gene. Proc Natl Acad Sci USA 80:5866–5870
6. Oka J, Ueda K, Hayaishi O, Komura H, Nakanishi K (1984) ADP-ribosyl protein lyase. J Biol Chem 259:986–995
7. Ostrowski MC, Richard-Foy H, Wolford RG, Berard DS, Hager GL (1983) Glucocorticoid regulation of transcription at an amplified episomal promoter. Mol Cell Biol 3:2045–2057
8. Pabo CO, Sauer RT (1984) Protein-DNA recognition. Annu Rev Biochem 53:293–321

9. Payvar F, Wrange Ö, Carlstedt-Duke J, Okret S, Gustafsson J-Å, Yamamoto KR (1981) Purified glucocorticoid receptors bind selectively in vitro to a cloned DNA fragment whose transcription is regulated by glucocorticoids in vivo. Proc Natl Acad Sci USA 78:6628–6632
10. Pfahl M (1982) Specific binding of the glucocorticoid-receptor complex to the mouse mammary tumor proviral promoter region. Cell 31:475–482
11. Poirier GG, De Murcia G, Jongstra-Bilen J, Niedergang C, Mandel P (1982) Poly (ADP-ribosyl)-ation of polynucleosomes causes relaxation of chromatin structure. Proc Natl Acad Sci USA 79:3423–3427
12. Purnell MR, Whish WJD (1980) Novel inhibitors of poly(ADP-ribose) synthetase. Biochem J 185:775–777
13. Reeves R, Chang D, Chung S-C (1981) Carbohydrate modifications of the high mobility group proteins. Proc Natl Acad Sci USA 78:6704–6708
14. Ringold GM, Yamamoto KR, Bishop JM, Varmus HE (1977) Glucocorticoid-stimulated accumulation of mouse mammary tumor virus RNA: increased rate of synthesis of viral RNA. Proc Natl Acad Sci USA 74:2879–2883
15. Robertson DL, Varmus HE (1979) Structural analysis of the intracellular RNAs of murine mammary tumor virus. J Virol 30:576–589
16. Scheidereit C, Geisse S, Westphal HM, Beato M (1983) The glucocorticoid receptor binds to defined nucleotide sequences near the promoter of mouse mammary tumour virus. Nature (London) 304:749–752
17. Schmidt TJ, Litwack G (1982) Activation of the glucocorticoid-receptor complex. Physiol Rev 62:1131–1192
18. Shimoyama M, Kitamura A, Tanigawa Y (1982) Glucocorticoid effects on poly (ADP-ribose) metabolism. In: Hayaishi O, Ueda K (eds) ADP-ribosylation reactions. Academic Press, London New York, pp 465–475
19. Tanuma S-I, Johnson LD, Johnson GS (1983) ADP-ribosylation of chromosomal proteins and mouse mammary tumor virus gene expression. J Biol Chem 258:15371–15375
20. Tashjiam AH Jr (1979) Clonal strains of hormone-producing pituitary cells. Methods Enzymol 58:527–535
21. Weisbrod S (1982) Active chromatin. Nature (London) 297:289–295
22. Welshons WV, Lieberman ME, Gorski J (1984) Nuclear localization of unoccupied receptors. Nature (London) 307:747–479
23. White BA, Bancroft FC (1982) Cytoplasmic dot hybridization. J Biol Chem 257:8569–8572
24. Young HA, Shih TY, Scolnick EM (1977) Steroid induction of mouse mammary tumor virus: effect upon synthesis and degradation of viral RNA. J Virol 21:139–146

ADP-Ribosylation and Glucocorticoid-Regulated Gene Expression

SEI-ICHI TANUMA[1] and GEORGE S. JOHNSON[2]

1. Introduction

ADP-ribosylation of chromosomal proteins has been studied primarily in broken cell systems, such as permeabilized cell, isolated nuclei, and chromatin, where labeled NAD^+ can be used as a substrate. Although the modification has been suggested to be related to several cellular functions, the physiological significance is not well understood [1−3]. This is largely due to a lack of suitable methods for analysis of ADP-ribosylation in intact cells in which to probe precise relationships.

The total $(ADP\text{-ribose})_n$ incorporation in the broken cell systems is much greater than the amount of $(ADP\text{-ribose})_n$ in intact cells [4, 5], and there is no assurance that proteins which are ADP-ribosylated in the broken cell systems contain $(ADP\text{-ribose})_n$ in intact cells. Hence, before the relevance of observations in the broken cell systems can be evaluated, it must be established which proteins are ADP-ribosylated in intact cells and how much $(ADP\text{-ribose})_n$ they contain. Furthermore, to obtain a better understanding of functions for this modification, it is essential to know if changes in *endogenous* ADP-ribosylation of these proteins accompany alterations in cellular functions.

This paper describes our simple analytical methods for determination of changes in endogenous $(ADP\text{-ribose})_n$ on some acceptor molecules and presents evidence of the involvement of ADP-ribosylation of chromosomal proteins in the regulation of gene expression.

2. Analytical Method for Endogenous ADP-Ribosylation

2.1 Isolation of ADP-Ribosylated Proteins

To analyze endogenous ADP-ribosylation, [^{3}H]adenosine was used as a precursor for $(ADP\text{-ribose})_n$ [6]. Continuous labeling for 16 h, approximately one cell doubling time

---

1 Department of Physiological Chemistry, Faculty of Pharmaceutical Sciences, Teikyo University, Sagamiko, Kanagawa 199-01, Japan
2 Laboratory of Molecular Biology, National Cancer Institute, National Institutes of Health, Bethesda, MD 20205, USA

ADP-Ribosylation of Proteins
(ed. by F.R. Althaus, H. Hilz, and S. Shall)
© Springer-Verlag Berlin Heidelberg 1985

Table 1. Isolation of ADP-ribosylated HMG proteins and histone H1 from control and 3-ABm treated cells

Fraction	Radiactivity Control		3-ABm	
	cpm × 10^{-3}	(%)	cpm × 10^{-3}	(%)
I. Whole cells	59,792	(100)	56,421	(100)
II. Nuclei	42,811	(72)	37,939	(67)
III. CsCl density gradient	1,198	(2.0)	1,051	(1.9)
IV. 5% PCA extraction	109	(0.18)	87	(0.15)
V. CM-Sephadex C-50 column chromatography	9.3	(0.016)	4.7	(0.008)

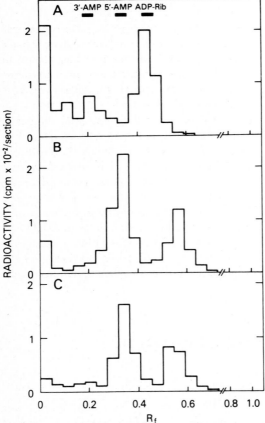

Fig. 1A–C. Chromatographic analysis of (ADP-ribose)$_n$ residues associated with HMG proteins and histone H1 isolated from control and 3-ABm treated cells. Radioactive material (fraction V, control) was treated with 0.3 N NaOH for 1 h at 0°C (**A**). Fraction V [control (**B**) and 3-ABm (**C**)] was digested with snake venome phosphodiesterase for 3 h at 37°C. The hydrolytic product was analyzed by cellulose thin layer chromatography, a solvent system containing 0.1 *M* sodium phosphate buffer (pH 6.8) ammonium sulfate, and *n*-propanol (100:60:2, v/w/v)

of mouse mammary tumor cells (34I cell line), was performed in the absence or presence of 10 mM 3-aminobenzamide (3-ABm), an inhibitor of (ADP-ribose)$_n$ synthetase [7]. Under these conditions, 3-ABm had no effect on cell growth. The majority of incorporated adenosine was present in DNA and RNA. Thus, ADP-ribosylated proteins had to be separated from these nuclei acids. This isolation procedure involved nuclei isolation, CsCl density gradient centrifugation, 5% perchloric acid extraction, and CM-Sephadex C-50 column chromatography (Table 1). More than 99.9% of the total radioactivity was in the nucleic acid fractions and only about 0.02% in the protein fraction (fraction V). Radioactivity (fraction V) in the 3-ABm treated cells was about half that of the control.

2.2 Identification of (ADP-Ribose)$_n$

In order to establish that the radioactive component in fraction V was (ADP-ribose)$_n$, cold alkaline hydrolysis and phosphodiesterase treatment were applied. Thin layer chromatographic analysis of the product following alkaline hydrolysis in control cells demonstrated three radioactive components (Fig. 1A); one at the origin indicating the presence of poly- or oligo(ADP-ribose), a second at the position of ADP-ribose, and a third migrating with 3'-AMP. The latter material was probably derived from a

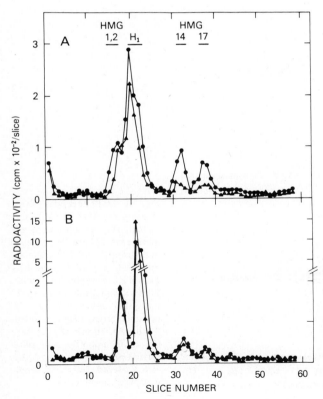

Fig. 2A,B. Acetic acid/urea polyacrylamide gel electrophoresis of ADP-ribosylated HMG proteins and histone H1. The proteins [Table 1, fraction V, control (●) and 3-ABm (▲)] labeled with [^{3}H]adenosine (A) or [^{14}C]lysine (B) were separated by electrophoresis in acetic acid/urea 15% acrylamide slab gels at 100 V for 4 h

small amount of contaminating RNA. The digestion products with phosphodiesterase from control and 3-ABm treated cells are shown in Fig. 1B,C, respectively. 5'-AMP and phosphoribosyl-AMP were apparent, and an average chain length of 1.5 to 1.8 was calculated, suggesting that the modification was primarily mono(ADP-ribose).

2.3 Acceptor Molecules for (ADP-Ribose)$_n$

To determine which proteins contained (ADP-ribose)$_n$, fraction V was analyzed by acetic acid/urea polyacrylamide gel electrophoresis. As shown in Fig. 2A, the major portion of radioactivity migrated with histone H1 and radioactivity was also found with staining bands of the four HMG proteins. Following 3-ABm treatment for 16 h, ADP-ribosylation of HMG 14 and 17 was almost completely inhibited, while that of HMG 1 and 2 and histone H1 decreased less. The reduced ADP-ribosylation cannot be attributed to differences in protein extraction or proteolysis since Coomassie blue staining patterns were very similar and 3-ABm treatment did not affect the incorporation of labeled lysine into HMG proteins and histone H1 (Fig. 2B). ADP-ribosylation

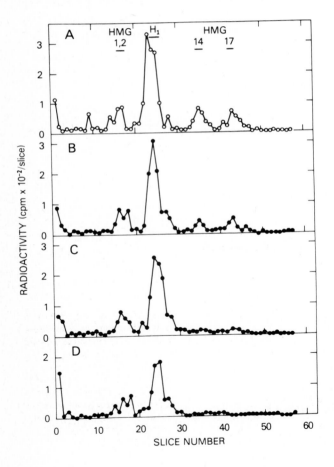

Fig. 3A–D. Endogenous (ADP-ribose)$_n$ on HMG proteins and histone H1 treated with dexamethasone. 341 cell were labeled with for 16 h with [^{3}H]adenosine (100 μCi ml^{-1}). The cultures were rinsed five times with PBS(–), and fresh media without label containing 10^{-7} M dexamethasone were added for 0 min (A), 10 min (B), 30 min (C), or 60 min (D). ADP-ribosylated proteins were extracted and analyzed by acetic acid/urea gel electrophoresis

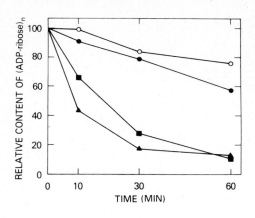

Fig. 4. Kinetics of the loss of $(ADP\text{-}ribose)_n$ from chromosomal proteins. The radioactivity in the protein peaks shown in Fig. 3 was tabulated and plotted as a function of time. The initial value (zero time) was defined as 100% for each peak. ● histone H1; ○ HMG 1 and 2; ▲ HMG 14; ■ HMG 17

of HMG proteins is of particular interest since HMG proteins are associated with actively transcribing genes [8, 9].

To determine the percentage of HMG molecules which were modified, cells were labeled with [^{14}C]lysine and [^{3}H]adenosine of known specific activity and HMG proteins were purified as described above. Assuming 20 lysine residues and 1.5 ADP-ribose residues/HMG molecule, it was calculated that only 0.01–0.03% of HMG proteins contained $(ADP\text{-}ribose)_n$ in control cells. These results suggest that ADP-ribosylation of HMG proteins may have a very specific role in chromatin functions.

3. ADP-Ribosylation and Glucocorticoid-Regulated Mouse Mammary Tumor Virus Gene Expression

3.1 DeADP-Ribosylation in Glucocorticoid-Treated Cells

Using this method, we have analyzed endogenous ADP-ribosylation in relation to mouse mammary tumor virus (MMTV) gene expression in the 34I cell line. This cell line provides a suitable system to study the relationship between ADP-ribosylation of chromosomal proteins and hormone-regulated gene expression since it contains stably integrated MMTV genes that are under glucocorticoid regulation [10, 11]. Activation of MMTV RNA transcription in 34I cells by glucocorticoids is maximal within 10–30 min [10, 11]. If $(ADP\text{-}ribose)_n$ metabolism is involved in this gene expression, alterations in ADP-ribosylation of chromosomal proteins must be observed within this time frame. 34I cells were first labeled with [^{3}H]adenosine for 16 h and then endogenous $(ADP\text{-}ribose)_n$ was determined at different times after addition of dexamethasone [12]. Without steroid addition (zero time), HMG proteins and histone H1 contained $(ADP\text{-}ribose)_n$ (Fig. 3A). Following addition of dexamethasone (Fig. 3B–D), the $(ADP\text{-}ribose)_n$ on HMG 14 and 17 was rapidly lost with a half-time of 8 and 17 min, respectively, while $(ADP\text{-}ribose)_n$ on histone H1 and HMG 1 and 2 was much more stable over this time period (Fig. 4). A similar loss of $(ADP\text{-}ribose)_n$ from HMG 14 and 17, but not from HMG 1 and 2 and histone H1, was observed by treatment

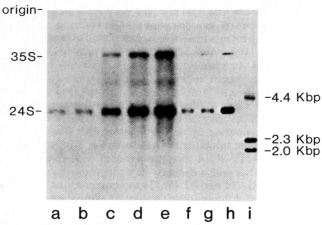

origin-

35S-

24S-

-4.4 Kbp

-2.3 Kbp
-2.0 Kbp

a b c d e f g h i

Fig. 5. Analysis of RNA complementary to MMTV DNA in 34I cells treated with (ADP-ribose)$_n$ synthetase inhibitors. Total RNA from cells treated with 10 mM 3-ABm for 0 (*lanes a, f*), 4 (*lane b*), 16 (*lanes c, h*), 32 (*lane d*), and 64 h (*lane e*), or with 10 mM 3-aminobenzoic acid for 16 h (*lane g*), was subjected to electrophoresis on 1.2% agarose gels, transferred to a nitrocellulose paper, and hybridized with nick-translated [^{32}P]-labeled MMTV DNA specific for the *env* region. The size of the RNA were extrapolated from the migration of radiolabeled λ *Hind* III DNA standards (*lane i*) (4.4, 2.3, and 2.0) kilobase pair (Kbp) and 18S and 28S rRNA from the ethidium bromide staining pattern

with another glucocorticoid, triamcinolone (10^{-7} M). Dexamethasone 21-mesylate, an analogue of dexamethasone without glucocorticoid action [13], did not apparently stimulate loss of (ADP-ribose)$_n$ from HMG 14 and 17 during a 3 h treatment at 10^{-7} M.

3.2 Increased Expression of MMTV Gene by Inhibition of ADP-Ribosylation

A decrease in (ADP-ribose)$_n$ on HMG 14 and 17 by glucocorticoids suggested the possible involvement of ADP-ribosylation of these proteins in regulation of MMTV RNA synthesis. If so, a decrease in ADP-ribosylation by other mechanisms might induce MMTV gene. To evaluate this possibility, we analyzed the effect of 3-ABm on MMTV RNA synthesis [12]. RNA isolated from 34I cells contains two major species of MMTV RNA, a 35S genomic RNA which contains genetic information for *gag*, *pol*, and *env* proteins and a 24S spliced RNA which codes for *env* protein [10, 11]. As shown in Fig. 5, increases in the levels of both RNA species were observed after 16 h treatment with 10 mM 3-ABm. The time required to reach half maximal levels was 20–25 h, and the stimulation of each RNA was approximately 15-fold after 64 h. The related analogue, 3-aminobenzoic acid (10 mM), which did not inhibit ADP-ribosylation in isolated nuclei and intact cells had no effect on MMTV RNA levels after 16 h treatment (Fig. 5). 3-ABm had essentially no effect on actin gene expression. Endogenous ADP-ribosylation of HMG 14 and 17 was reduced by a 16 h treatment with 3-ABm, while that of HMG 1 and 2 and histone H1 decreased much less (Fig. 2A). It should be noted that following a 4 h treatment with 3-ABm, there was essentially no increase

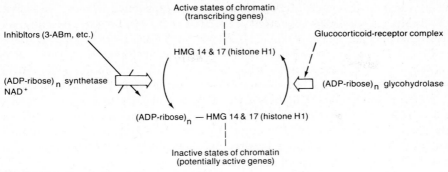

Fig. 6. ADP-ribosylation and gene expression

in MMTV RNA levels (Fig. 5) and only a small decrease in endogenous $(ADP\text{-ribose})_n$, further demonstrating the correlation of loss of $(ADP\text{-ribose})_n$ from chromosomal proteins, especially from HMG 14 and 17, with expression of MMTV gene.

4. Discussion

Glucocorticoid hormones regulate syntheses of specific mRNA species [14, 15]. Interaction of the steroid-receptor complex with highly localized regions of chromatin is essential for this induction, but little of the molecular basis for this activation mechanism is understood. The primary action of glucocortiocoids is very rapid; thus, any alterations in chromatin which are essential for this activation would be observed within a few minutes after steroid addition. Also, it is possible that proteins known to be associated with transcriptionally active genes, such as the HMG 14 and 17 [8, 9], are important.

Our results clearly demonstrate that a loss of $(ADP\text{-ribose})_n$ from HMG 14 and 17 is a consequence of glucocorticoid action. The rapid loss (hydrolysis) of $(ADP\text{-ribose})_n$ on HMG 14 and 17 occurred in the same time frame as the induction of MMTV RNA synthesis by glucocorticoids, suggesting that deADP-ribosylation may be an important event in MMTV gene expression, and implying that the glucocorticoid-receptor complex may stimulate $(ADP\text{-ribose})_n$ removal, a reaction catalyzed by $(ADP\text{-ribose})_n$ glycohydrolase [16, 17] or ADP-ribosyl protein lyase [18]. This proposal is supported by the observation that 3-ABm increases MMTV RNA levels; associated with this increase in MMTV RNA was a decrease in endogenous ADP-ribosylation of HMG 14 and 17, which is presumably a result of dilution of the modified molecules by cell division.

Glucocorticoids affect the expression of only a few genes, whereas HMG proteins are associated with possibly all active genes since no specificity has been observed. Thus, it is not obvious why ADP-ribosylation of HMG proteins should be specific for glucocorticoid-regulated genes. One possible explanation is that ADP-ribosylated HMG 14 and 17 are not uniformly distributed in chromatin but that most of HMG proteins which contain (ADP-ribose)$_n$ are associated with potentially active chromatin which includes glucocorticoid-regulated genes, such as the MMTV gene. Recently, we observed the preferential association of ADP-ribosylated HMG 14 and 17 with chromatin which are sensitive to digestion with micrococcal nuclease, an enzyme which selectively digests transcriptionally active chromatin [19, 20]. Furthermore, (ADP-ribose)$_n$ on HMG 14 and 17 and histone H1, but not on HMG 1 and 2, in nuclease-sensitive chromatin was rapidly hydrolyzed by the action of glucocorticoids (S. Tanuma and G. Johnson 1984, unpublished observation). We speculate that ADP-ribosylated HMG 14 and 17 (and histone H1) locate on regulatory regions in the long terminal repeat (LTR) of MMTV genome, possibly on binding sequences of glucocorticoid-receptor complex [21] or a cap site for transcription.

The observations with glucocorticoids and 3-ABm are consistent with the idea that (ADP-ribose)$_n$ on chromosomal proteins serves as a negative regulator for glucocorticoid-regulated gene expression. In general, deADP-ribosylation of chromosomal proteins may be an integral component in induction of potentially active genes (Fig. 6). A hydrolysis of (ADP-ribose)$_n$ on HMG 14 and 17 (and histone H1) could cause subtle localized changes in chromatin structure making specific binding sequences and promotor sequences more accessible to the glucocorticoid-receptor complex and RNA polymerase II, respectively.

References

1. Hayaishi O, Ueda K (1977) Poly(ADP-ribose) and ADP-ribosylation of proteins. Annu Rev Biochem 46:95–116
2. Purnell MR, Stone PR, Whish WJD (1980) ADP-ribosylation of nuclear proteins. Biochem Soc Trans 8:215–227
3. Mandel P, Okazaki H, Nidergang C (1982) Poly(adenosine diphosphate ribose). Prog Nucleic Acid Res Mol Biol 27:1–51
4. Kanai M, Miwa M, Kuchino Y, Sugimura T (1982) Presence of branched portion in poly-(adenosine diphosphate ribose) in vivo. J Biol Chem 257:6217–6223
5. Kreimeyer A, Wielckens K, Adamietz P, Hilz H (1984) DNA repair-associated ADP-ribosylation in vivo. J Biol Chem 259:890–896
6. Tanuma S, Johnson GS (1983) ADP-ribosylation of nonhistone high mobility group proteins in intact cells. J Biol Chem 259:4067–4070
7. Purnell MR, Whish WJD (1980) Novel inhibitors of poly(ADP-ribose) synthetase. Biochem J 185:775–777
8. Weisbrod S, Weintraub H (1979) Isolation of subclass of nuclear proteins responsible for conferring a DNase I-sensitive structure on globin chromatin. Proc Natl Acad Sci USA 76:630–635
9. Goodwin GH, Mathew CGP (1982) Role in gene structure and function. In: Johns EW (ed) HMG chromosomal proteins. Academic Press, London New York, pp 193–221
10. Young HA, Shih TY, Scolnick EM, Parks WF (1977) Steroid induction of mouse mammary tumor virus: effect upon synthesis and degradation of viral RNA. J Virol 21:139–146
11. Ringold GM, Yamamoto KR, Bishop JM, Varmus HE (1977) Glucocorticoid-stimulated accumulation of mouse mammary tumor virus RNA: increased rate of synthesis of viral RNA. Proc Natl Acad Sci USA 74:2879–2882
12. Tanuma S, Johnson LD, Johnson GS (1983) ADP-ribosylation of chromosomal proteins and mouse mammary tumor virus gene expression. J Biol Chem 258:15371–15375

13. Simons SS Jr, Thompson EB (1981) Dexamethasone 21-methylate: an affinity label of gluco-corticoid receptors from rat hepatoma tissue culture cells. Proc Natl Acad Sci USA 78:3541–3545

14. Yamamoto KR, Alberts BM (1976) Steroid receptor: elements for modulation of eukaryotic transcription. Annu Rev Biochem 45:731–746

15. Higgins SJ, Gehring U (1978) Molecular mechanism of steroid hormone action. Adv Cancer Res 28:313–397

16. Miwa M, Tanaka M, Matsushima T, Sugimura T (1974) Purification and properties of glyco-hydrolase from calf thymus splitting ribose-ribose linkage of poly(adenosine diphosphate ribose). J Biol Chem 135:449–455

17. Tavassoli M, Tavassoli MH, Shall S (1983) Isolation and purification of poly(ADP-ribose) glycohydrolase from pig thymus. Eur J Biochem 135:449–455

18. Oka J, Ueda K, Hayaishi O, Komura H, Nakanishi K (1984) ADP-ribosyl protein lyase. J Biol Chem 259:986–995

19. Tata JR, Baker B (1978) Enzymatic fractionation of nuclei: polynucleosomes and RNA poly-merase II as endogenous transcriptional complexes. J Mol Biol 118:249–272

20. Levy-Wilson B, Cannor W, Dixon GH (1979) A subset of testis nucleosomes enriched in tran-scribed DNA sequences contains high mobility group proteins as major structural components. J Biol Chem 254:609–620

21. Scheidereit C, Geisse S, Westphal HM, Beato M (1983) The glucocorticoid receptor binds to defined nucleotide sequences near the promoter of mouse mammary tumor virus. Nature (London) 304:749–752

Poly(ADP-Ribose) Synthetase and Cell Differentiation

ARNOLD I. CAPLAN[1]

Introduction

Developing chick limb mesenchymal cells differentiate into cartilage, bone, muscle, and other connective tissue phenotypes [1–3]. These differentiation events are cued by local signals which stimulate, inhibit, or provide permissive conditions to allow one of these expressional pathways to become available. Such local cuing can arise from a number of sources, including the dispersion of nutrients and other molecules via the differentiating vascular system [4–6]. The vascular system sets up vascular-rich and vascular-poor domains within the limb prior to overt differentiation of limb cells into particular phenotypes. For example, nicotinamide is stored in the yolk of the developing embryo and may be nonuniformly dispersed throughout the limb because of this distinctive vascular pattern. I have previously proposed the "NAD hypothesis" which stated that this local cuing was responsible for the control of cellular NAD levels which, in turn, communicated with chromatin-associated machinery to influence these complex differentiation events [1, 4, 7–9]. Two observations gave support to this hypothesis: First, exogenously added nicotinamide caused increased cellular NAD levels concomitant with an inhibition of cartilage formation [7, 9]; and second, ás cells differentiate into chondrocytes, a threefold increase in the synthesis of poly(ADP-ribose) was observed [8].

Recent experiments clearly indicate that this hypothesis is *not* correct and must be modified [10, 11]. Central to our new hypothesis is the observation that *both* nicotinamide and nicotinic acid cause increases in cellular NAD levels of limb mesenchymal cells, but only nicotinamide causes an inhibition of chondrogenesis. This observation alone indicates that nicotinamide, itself, and not NAD, is the bioactive agent which can affect chondrogenesis. Also, in chick limb mesenchymal cell cultures treated with the poly(ADP-ribose)synthetase (pADPRS)inhibitors, 3-aminobenzamide or benzamide, reduced chondrogenesis was observed (3-aminobenzoic acid or benzoic acid which do not inhibit pADPRS had no effect). In these cases, reduced chondrogenesis was due to a reduction in cell proliferation and a decrease in chondrogenic expression independent of cell proliferation. Importantly, nicotinamide was temporally more effective in inhibiting chondrogenesis than 3-aminobenzamide or benzamide; nicotinamide was

1 Biology Department, Case Western Reserve University, Cleveland, OH 44106, USA

ADP-Ribosylation of Proteins
(ed. by F.R. Althaus, H. Hilz, and S. Shall)
© Springer-Verlag Berlin Heidelberg 1985

effective at inhibiting chondrogenesis when introduced at any time during the 4-day-period following introduction of cells into culture, while the other inhibitors had to be present early during the culture period (i.e., during the first 2 days) in order to be effective [11, 12].

These observations strongly support the view that nicotinamide is the bioactive agent responsible for the inhibition of chondrogenesis. Low internal pools of nico-tinamide would enhance chondrogenesis, while high levels would inhibit this process. Also, based on both the inhibition produced by nicotinamide and 3-aminobenzamide or benzamide, it is likely that a prominent cellular target for the action of nicotin-amide is chromatin-associated pADPRS. If cellular levels of nicotinamide are, indeed, an important driving force in the control of chondrogenesis in developing chick limbs, then the "nicotinamide hypothesis" could explain this function. The nicotinamide hypothesis states that cellular nicotinamide levels directly control the activity of pADPRS which, itself, must be associated with chromatin transitions involved in the differentiation phenomena.

The chromatin-associated effects involving nicotinamide and pADPRS are not directly amenable to experimental verification at this time [12, 13]. This is due to the unique binding properties of pADPRS for DNA [14, 15]. In addition, the levels of nicotinamide necessary to affect cells in culture may not be comparable to in vivo levels. The use of nicotinamide in the experimental manipulations of the chick limb mesenchymal cell system in culture has clearly implicated pADPRS in control of developmentally related events. The data presented below show that levels of pADRPS increase during the decisional phases of differentiation events. These increases may, again, be interpreted to indicate a role of this enzyme in chromatin rearrangement reactions associated with phenotypic differentiation. However, another possibility can be put forth which indicates that pADPRS acts as a sentinel for DNA strand breaks prevalent during these periods of profound developmental transitions. Further-more, pADPRS can act in a "fail-safe" mechanism to eliminate cells by exhausting NAD levels in cells which do not appropriately rejoin these DNA strand breaks or in cases where excess breaks occur. This fail-safe function would require an increase in pADPRS at or just before the time of major chromatin rearrangements during deci-sional phases of cell differentiation.

pADPRS and Development

Tissues from various stages of embryonic development and from cell cultures were analyzed for pADPRS activity by sonicating homogenates and incubating them with [^{3}H]NAD at 25°C for 1 min followed by ice-cold TCA precipitation of newly formed polymer. Assays were linear with time, optimized for Mg^{++}, pH, substrate, and enzyme concentration [12]. Lineweaver-Burke plots of all samples showed that the K_m for NAD was between 76 and 86 μM with a variable V_{max}. The product of the reaction was poly(ADP-ribose) by several criteria: resistance to DNase-I, RNase-B, but complete degradation by snake venom phosphodiesterase; chain lengths of 9.5 to 11.5; chroma-tographic, electrophoretic, and immunoprecipitation analysis of the reaction product

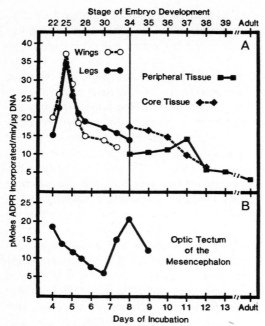

Fig. 1A,B. pADPRS levels as a function of in vivo limb and mesencephalon development. **A** Whole limbs, either wings or legs, were dissected from staged embryos (stages 22–34) and assayed for pADPRS activity. After stage 34 (day 8), only legs were denuded of ectoderm, the feet discarded, and the remaining tissues dissected into core (chondrogenic/osteogenic) and peripheral (myogenic/fibrogenic) compartments. These tissues were then analyzed for pADPRS activity [12] and expressed as radioisotopically-labeled ADP-ribose units, derived from [^{3}H]NAD, incorporated into poly(ADP-ribose) min^{-1} μg^{-1} DNA. The means from three separate experiments with at least four reactions per sample are presented. **B** The optic tectum of the embryonic mesenchephalon was dissected from staged embryos (stages 22–35), processed, and analyzed for pADPRS activity μg^{-1} DNA. The means from two separate experiments with at least four reactions per sample are presented

showed that greater than 90% of the radiolabel migrated or was associated with auto-modified pADPRS; and 0.4 M NaOH-treatment shifted the [^{3}H] from a position coincident with protein to one coincident with nucleic acid on CsCl equilibrium density centrifugation. Taken together, these results were interpreted to indicate that the enzyme activity is a reflection of the number of enzyme units or molecules present in a tissue at the time of assay [12].

In Vivo Limb Development

Since cartilage formation is a predominant feature of early (stages 22–29) in vivo limb development [16], whole limbs were used to study the relationship of pADPRS activity to chondrogenic differentiation and are depicted in Fig. 1A, left panel. The observed changes can be separated into three distinct temporal phases: in the first phase (stages 22–25), intermediate levels of pADPRS activity rapidly rise and become maximal. During this time period, undifferentiated limb mesenchymal cells begin the process which leads to the initiation of chondrogenic expression and the first signs of cyto-differentiation can be detected in histological preparations as metochromatic staining indicative of the deposition of cartilage proteoglycans [1, 16, 17]. In the second phase (stages 22–29), a rapid decrease in the level of poly(ADP-ribosylation) is temporally correlated with massive and overt chondrogenic expression [18]. The third phase (stages 29–34) is characterized by a slow, progressive decrease in the capacity of the tissue to synthesize poly(ADP-ribose) and is coincident with cartilage maturation and the initiation of osteogenesis [19, 20]. These assays show that a twofold transient peak of pADPRS activity occurs in developing limbs during a 24 h period that is temporally correlated with the initiation of cartilage differentiation.

Subsequent differentiation events during limb development are characterized by a continuation of the chondrogenesis to osteogenesis transition in the core region and importantly, the initiation of muscle formation in the peripheral regions. Since the older limbs are larger and more amenable to separation of the core (hard tissue) and peripheral (soft tissue) compartments, legs from day 8 embryos through adult were dissected and tissues segregated into core and peripheral regions and separately assayed for pADPRS activity. By day 8, when the core and peripheral regions can first be cleanly dissected, the core compartment contains 1.6 times more pADPRS activity per cell than the myogenic compartment. As leg development proceeds, the pADPRS activity associated with the core gradually decreases. In contrast, the pADPRS activity present in the peripheral tissues during muscle differentiation exhibits a pattern of gradual increase in the level of activity with a small, but significant and reproducible peak associated with day 11 myogenesis. On day 11, the majority of myoblasts have withdrawn from the cell cycle, but have not yet fused to form multinucleated myo-tubes [21]. After day 11, coincident with the expressional events involved in terminal differentiation, including the biosynthesis of muscle-specific proteins, a sharp decrease in the pADPRS activity is observed. Adult leg muscles maintained this low level of pADPRS activity. These data show that there is a transient increase in pADPRS activity which is correlated with the initiation of the terminal myogenic differentiation events.

Because these transient peaks of activity at stage 25 and on day 11 of limb development were observed, the question can be asked if these are specific to the tissues involved or a reflection of events involving the whole embryo. In this respect, tissues from the optic tectum of the mesencephalon were dissected and assayed under conditions comparable to those for timb tissue. As seen in Fig. 1B, this brain tissue exhibits transient increases and decreases in activity which are not correlated with those increases seen in limb tissue. These data strongly suggest that the transient increase in pADPRS activity at stage 25 and day 11 is specifically involved with molecular and cellular events associated with limb development and differentiation. In addition, if one extrapolates the pADPRS activity from the separated core and peripheral tissues (Fig. 1A) back to an earlier time (stage 25), it appears that the peripheral region probably contains low levels of pADPRS activity, while the core region contains high levels of activity. This extrapolation may be interpreted to suggest that a transient

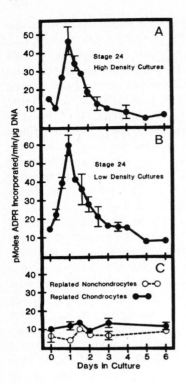

Fig. 2A–C. pADPRT activity as a function of in vitro limb mesenchyme development. **A** Mesenchymal cells from stage 24 limbs were isolated and seeded at 5×10^7 cells/100 mm culture plate in serum-containing nutrient medium [22]. On a specific day in culture, the cell layers are rinsed with saline, scraped off the culture dish, processed, and assayed for pADPRS activity μg^{-1} DNA [12]. Standard deviations represent results from three separate experiments with duplicate plates. **B** Mesenchymal cells from stage 24 limbs were isolated, seeded at 7.5×10^6 cells/100 mm culture dish in serum-containing nutrient medium and were treated as in **A**. **C** Chondrocytes and nonchondrocytes were isolated from day 8 high density stage 24 mesenchymal cultures and seeded at an initial density of 7.5×10^6 cells/100 mm culture dish [25, 26]. At specific times, cells were isolated, processed, and analyzed for pADPRS activity μg^{-1} DNA [12]. The results were derived from two separate replating preparations with four separate rate determinations per experiment

increase in pADPRS activity occurs exclusively in the chondrogenic region of the limb during early limb development.

pADPRS Activity During in Vitro Chondrogenesis and Myogenesis

The in vivo studies above indicated that a temporal correlation exists between the differentiation events and peaks of pADPRS activity. With this in mind, we have analyzed the pADPRS activity as a function of the differentiation and development events of stage 24 limb mesenchymal cells and committed day 11 myoblasts in culture, since these systems are more amenable to experimental manipulation. As illustrated in Fig. 2A, freshly seeded cultures of stage 24 limb mesenchymal cells plated under conditions which optimize chondrogenesis [1, 22] contain basal levels of pADPRS activity. By day 1 under these high density conditions in culture, pADPRS activity has increased fourfold and subsequently decreases to the basal levels observed in freshly isolated cells. The transient increase in pADPRS activity in culture occurs just prior to the initiation of the synthesis of cartilage-specific molecules [1, 23] and is followed by a period of overt chondrogenic expression which is correlated with the return to basal levels of pADPRS activity.

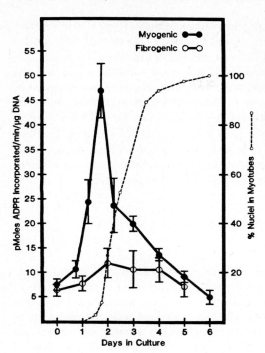

Fig. 3. pADPRS activity as a function of in vitro myogenesis and fibrogenesis. Myoblast and muscle-associated fibroblast cultures were rinsed, scraped, and assayed for their pADPRS activity μg^{-1} DNA [12]. Standard deviations were derived from at least three separate tissue preparations with duplicate samples. At least four reactions are performed on each samples. The extent of myoblast fusion [12, 27, 28] was determined and plotted

Stage 24 limb mesenchymal cells when plated at low initial densities do not exhibit evidence of chondrogenesis [24] but, nonetheless, commitment and differentiation events proceed and are well underway during the first few days of tissue culture. Analysis of the pADPRS activity in low density cultures show that there is an identical pattern of increase and then decrease of pADPRS activity as was observed in high density chondrogenic cultures. These data indicate that the transient increase and then decrease in pADPRS activity is related to the differentiation event itself regardless of the phenotype involved. In both the in vivo and in vitro studies described above, it is clear that peaks of the high level of pADPRS activity are related to initiation of differentiation of chondrogenic, nonchondrogenic, myogenic, and neural tissues.

To further test this conclusion, expressive chondrocytes and nonchondrocytes from day 8 high density cultures can be separately isolated and introduced back into cultures [25, 26]. These replated chondrocytes and cultures of nonchondrocytes contain homogeneous populations of terminally differentiated cells which continue their expressional program. Thus, such replated cells do not undergo events specific to phenotypic acquisition, but rather maintain their current differentiated state. However, and importantly, replated chondrocytes and nonchondrocytes proliferated in culture in the manner showing identical kinetics to that of freshly isolated primary cultures of stage 24 limb mesenchymal cells. As evidenced by Fig. 2B, the levels of pADPRS activity are constant and at basal levels throughout the culture period of such replated chondrocytes and nonchondrocytes. These data further strengthen the conclusion that pADPRS activity increases during the differentiation event and such increases are not related to the expression of differentiated cells.

The analysis of pADPRS activity during in vitro myogenesis (Fig. 3) demonstrates that a rapid, fourfold increase in activity occurs by day 2. This increase occurs prior to and overlaps the period of myoblast fusion as plotted in Fig. 3 and temporally corresponds to the initiation of the appearance of new muscle-specific RNAs and the initiation of their translation [27]. Coincident with overt and massive expression of these muscle transcripts is a rapid decrease in the pADPRS activity. When fibroblasts are isolated and purified from such cultures and then assayed for pADPRS activity, these muscle-associated fibroblasts exhibit basal levels of pADPRS activity [12]. Thus, the increase in pADPRS activity seems to be associated with the differentiation events of myogenic cells in culture, while the decrease in activity to basal levels is an inherent characteristic of cells which are expressing their differentiated phenotype.

Discussion

The results summarized above show that levels of pADPRS activity undergo dramatic changes as limb mesenchymal cells differentiated into muscle and cartilage both in intact limbs and in cell cultures. These changes are characterized by a transient increase in the level of pADPRS activity which is correlated with the initiation of differentiation of these particular phenotypes. After the expressional events are initiated, there is a sharp decline of pADPRS activity to basal levels. This pattern of pADPRS activity is characteristic of differentiating cells whether they express muscle cartilage, noncartilage, or neural tissue phenotypes, but is not observed in cells or tissues which have already progressed through these differentiation events.

As mentioned in the Introduction, the nicotinamide hypothesis suggests that nicotinamide can be a major effector involved with the differentiation of limb mesenchymal cells into particular phenotypes. If unusually high levels of nicotinamide are encountered, these levels of nicotinamide could serve to affect the transient increase in pADPRS activity which has been correlated with decisional events. There are two implications from this hypothesis with regard to pADPRS: The first would suggest that differentiation of limb mesenchymal cells and/or myoblasts into chondrogenesis and/or into the terminal phase of myogenesis involves alterations in chromatin structure assisted or involving pADPRS activity. Under these conditions nicotinamide could inhibit the enzyme from its chromatin-associated function and, therefore, affect the differentiation events.

An alternative explanation links the DNA repair process with differentiation. As cells differentiate into specific phenotypic pathways, an increase in DNA strand breaks could be sustained [28]. Since pADPRS has a high binding affinity for such DNA strand breaks, this enzyme could act as a fail-safe system or "sentinel" to monitor DNA integrity during the dynamic structural transitions associated with terminal cell differentiation. The enzyme may act to ensure the fidelity of tissue differentiation either by assisting in the repair of DNA strand breaks or by elimination of cells with extensive DNA damage. The elimination would occur when the pADPRS remained active as it was bound to DNA strand breaks; this continued activity would deplete the cell's NAD levels to the point where the cell would not be able to function and, thus, cause the cell's suicide [29]. With the "sentinel, fail-safe" proposal in mind, it may be that at the time of phenotypic commitment, stage 24 limb mesenchymal cells, both prechondrocytes and nonchondrocytes, are provided with increased amounts of pADPRS in anticipation of developmentally associated DNA strand breaks involved with chromatin rearrangement. Such a model would require that the pADPRS increase is a programmed part of the differentiation phenomenon, but does not drive the process. In this case, one would have to speculate that pADPRS is intimately associated with the differentiation machinary as a support element and if pADPRS activity does not increase in anticipation of differentiation, it may be that the differentiation events will not proceed. Inhibition of pADPRS by nicotinamide and other agents may, thus, affect the differentiation process by indirect means. These speculations form the basis for future experimentation.

Acknowledgments. The data presented here were obtained by B.W. Cherney and are a part of his thesis [12, 13]. These studies were supported by grants from the National Institutes of Health.

References

1. Caplan AI (1981) The molecular control of muscle and cartilage development. Liss, New York, pp 37–68
2. Fallon JF, Caplan AI (eds) (1983) Limb development and regeneration, part A. Liss, New York
3. Kelley RO, Goetinck PF, MacCabe JA (eds) (1983) Limb development and regeneration, part B. Liss, New York
4. Caplan AI, Koutroupas S (1973) The control of muscle and cartilage development in the chick limb: The role of differential vascularization. J Embryol Exp Morphol 29:571–583
5. Jargiello DM, Caplan AI (1983) The establishment of vascular derived microenvironments in the developing chick wing. Dev Biol 97:364–374
6. Caplan AI (1985) The vasculature and limb development. Cell Differ 16:1–11
7. Caplan AI, Rosenberg MJ (1975) Interrelationship between poly(adenosine diphosphoribose) synthesis, intracellular NAD levels and muscle or cartilage differentiation from embryonic chick limb mesodermal cells. Proc Natl Acad Sci USA 72:1852–1857
8. Caplan AI, Neidergang H, Okazaki H, Mandel P (1979) Poly(ADP-ribose) levels as a function of chick limb mesenchymal cell development as studied in vitro and in vivo. Dev Biol 72:102–109
9. Cherney BW, Midura RJ, Caplan AI (1982) Poly(ADP-ribose) and the differentiation of embryonic tissue. In: Hayaishi EO, Ueda K (eds) ADP-ribosylation reactions. Academic Press, London New York, pp 389–406
10. Midura RJ (1984) The relationship of nicotinamide adenine denucleotide to the cytodifferentiation of embryonic limb mesecnhyme. Thesis, Case Western Reserve Univ, Cleveland, Ohio, USA
11. Midura RJ, Cherney BW, Caplan AI (in press) The relationship of nicotinamide adenine dinucleotide to the chondrogenic differentiation of limb mesenchymal cells. Dev Biol 110
12. Cherney BW (1984) ADP-ribosylation reactions in embryonic differentiation. Thesis, Case Western Reserve Univ, Cleveland, Ohio, USA
13. Cherney BW, Midura RJ, Caplan AI (in press) Poly(ADP-ribose)synthetase and chick limb mesenchymal cell differentiation. Dev Biol
14. Yoshihara K, Kamiya T (1982) Poly(ADP-ribose)synthetase-DNA interaction. In: Hayashi O, Ueda K (eds) ADP-ribosylation reactions. Academic Press, London New York, pp 157–171

15. Benjamin RC, Gill DM (1980) Poly(ADP-ribose)synthetase in vitro programmed by damaged DNA: a comparison of DNA molecules containing different types of strand breaks. J Biol Chem 255:10502–10508
16. Searls RL, Janners M (1969) The stabilization of cartilage properties in the cartilage forming region of the embryonic chick limb. J Exp Zool 170:365–376
17. Royal PD, Sparks KJ, Goetinck PF (1980) Physical and immunochemical characterization of proteoglycans synthesized during chondrogenesis in the chick embryo. J Biol Chem 255: 9870–9878
18. Dessau W, von der Mark H, von der Mark K, Fischer S (1980) Changes in the pattern of collagens and fibronectin during limb bud chondrogenesis. J Embryol Exp Morphol 57:51–63
19. Osdoby P, Caplan AI (1981) First bone formation in embryonic chick limbs. Dev Biol 86: 147–156
20. Caplan AI, Syftestad G, Osdoby P (1983) The development of bone and cartilage in tissue culture. J Clin Orthoped Rel Res 174:342–363
21. Herrmann H, Heywood SM, Marchok AC (1970) Reconstruction of muscle development as a sequence of macromolecular synthesis. In: Moscona AA, Monroy A (eds) Current topics in developmental biology. Academic Press, London New York, pp 181–342
22. Caplan AI (1970) Effects of the nicotinamide-sensitive teratogen 3-acetylpyridine on chick limb cells in culture. Exp Cell Res 62:341–355
23. Hascall VC, Oegema, Brown M, Caplan AI (1976) Isolation and characterization of proteoglycans from avian embryonic limb bud chondrocytes grown in vitro. J Biol Chem 251:3511–3519
24. Osdoby P, Caplan AI (1979) Osteogenesis in cultures of limb mesenchymal cells. Dev Biol 72: 102–109
25. Lennon DP, Osdoby P, Carrino DA, Vertel BM, Caplan AI (1983) Isolation and characterization of chondrocytes and non-chondrocytes from high density chick limb bud cultures. J Craniofac Genet Dev Biol 3:235–351
26. Carrino DA, Lennon DP, Caplan AI (1983) Extracellular matrix and the maintenance of the differentiated state: proteoglycans synthesized by replated chondrocytes and non-chondrocytes. Dev Biol 99:132–144
27. Devlin R, Emergson CP (1979) Coordinate accumulation of contractile protein mRNAs during myoblast differentiation. Dev Biol 69:202–216
28. Farzaneh F, Zalin R, Brell D, Shall S (1982) DNA strand breaks and ADP-ribosyl transferase activation during cell differentiation. Nature (London) 300:362–366
29. Berger NA, Weber G, Kaichi AS, Petzold SJ (1978) Relation of poly(adenosine diphosphoribose) synthesis to DNA synthesis and cell growth. Biochim Biophys Acta 519:105–117

Interaction of Cyclic AMP and ADP-Ribosylation in a Hormonally Responsive Cell Line

SUSAN GEAR CARTER[1]

Introduction

The adenylate cyclase system is a complex of proteins which responds to several stimuli with a single, integrated response. A hormonal signal is translated by multiple systems, including the stimulation of cAMP, to the nucleus [1, 2]. Regulation of cAMP synthesis in response to a hormone requires GTP as a cofactor for hormonal stimulation of the adenylate cyclase system [3, 4]. GTP appears critical for the action of various toxins, such as cholera and pertussis toxin, as well as for hormonal activity [5, 6].

ADP-ribosylation is a posttranslational covalent modification of cellular proteins. In studies with both whole cell and purified proteins, several investigators have shown that the guanine-nucleotide-binding regulatory proteins of the adenylate cyclase complex are ADP-ribosylated [5, 7]. Toxins, such as cholera toxin and pertussis toxin, stimulate the ADP-ribosylation of these regulatory proteins and modify cellular responses by either stimulating or inhibiting the adenylate cyclase system. The similarity between hormonal stimulation of the adenylate cyclase system and toxin-induced change in function of this same system suggests a possible role of posttranslational cellular protein modification by ADP-ribosylation in response to hormonal stimuli.

The MCF-7 cell line is a human breast cancer cell line with differentiation potential [8, 9] that provides a system to study in vitro the interaction of the process of ADP-ribosylation with reagents that may modify the adenylate cyclase system and a system to study the effect of hormonal manipulation of cellular growth and differentiation on these same processes.

Results and Discussion

The interaction of adenyl nucleotides with the process of ADP-ribosylation was evaluated using a permeable cell assay system to allow cellular uptake of the substrate NAD. The system used was one in which the MCF-7 human breast cancer cells were

1 Cleveland Metropolitan General Hospital, Division of Hematology/Oncology, Case Western Reserve University, School of Medicine, 3395 Scranton Road, Cleveland, OH 44109, USA

ADP-Ribosylation of Proteins
(ed. by F.R. Althaus, H. Hilz, and S. Shall)
© Springer-Verlag Berlin Heidelberg 1985

Table 1. Effect of nucleotides and specific inhibitors on [^{32}P]-ADP-ribose incorporation

Sample	DPM [^{32}P]-ADP-ribose incorporated	% Control
Control	45,507	
AMP 5 mM	81,002	178
ADP 5 mM	13,295	29
ATP 5 mM	9,052	20
Cyclic AMP 5 mM	108,822	239
Dibutyryl cyclic AMP 5 mM	72,904	160
GMP 5 mM	44,597	98
GDP 5 mM	12,287	27
GTP 5 mM	3,641	8
cGMP 5 mM	50,968	112
Nicotinamide 10 mM	17,741	39
Methyl-nicotinamide 10 mM	28,716	63
Theophylline 1 mM	16,836	37

rendered permeable to exogenously supplied nucleotides by a cold shock with near isotonic buffer [10]. The synthesis of ADP-ribose was then measured as the amount of [^{32}P]-ADP-ribose incorporated into acid-precipitable material [11]. The MCF-7 cells were grown under standard tissue culture conditions and were studied in log-phase growth.

As shown in Table 1, both the adenyl and guanyl triphosphates and diphosphates markedly inhibited ADP-ribosylation as measured by incorporation of radioactive label into acid-precipitable material. In contrast, AMP and cAMP stimulated ADP-ribosylation with a greater than twofold increase in labeling occurring in the presence of cAMP. GMP and cGMP had no effect. This degree of inhibitory and stimulatory effects of the adenyl nucleotides on ADP-ribosylation has not been previously shown. A study using a nonhormone responsive cell line showed only minimal inhibition by all the adenyl nucleotides [11]. Since the actions of many hormones are mediated in a major part by the intracellular second messenger cAMP [12], one can conjecture that the difference in response between a hormone-responsive and a hormone-nonresponsive cell line is related to the activity of this second messenger system.

The MCF-7 cell line has been shown to have a variable response to estradiol. One group demonstrated a serum requirement for estrogen responsiveness that was unrelated to estrogen receptor number [9]. Another group showed estrogen regulation of net DNA synthesis in this cell line [13]. In our system there was no difference in response to cAMP or ATP among cells grown in serum-containing media, charcoal-extracted media, or charcoal extracted media with added estradiol, nor were there differences in growth at the time of assay.

The stimulation of ADP-ribosylation of proteins associated with both pertussis toxin and cholera toxin involves modification of plasma membrane proteins [5–7]. To localize the site of maximal ADP-ribosylation of cellular proteins after exposure to cAMP, isolated nuclear preparations were simultaneously compared to whole cell preparations. Nuclei were isolated by a nonionic detergent method [14].

Table 2. Comparison of whole cell and isolated nuclei preparations

Sample	DPM [^{32}P]-ADP-ribose incorporated	% Control
Whole cell		
Control	39,988	
Cyclic AMP 5 mM	82,775	207
ADP-ribose 5 mM	17,625	55
ADP-ribose 5 mM + cyclic AMP 5 mM	16,393	51
Nuclei		
Control	2,329	
Cyclic AMP 5 mM	2,269	97
ADP-ribose 5 mM	844	36
ADP-ribose 5 mM + cyclic AMP 5 mM	891	38

As shown in Table 2, cAMP increased [^{32}P]-ADP-ribose incorporation above control level only in the whole cell preparation. Autoradiography of SDS-polyacrylamide gel electrophoretically separated proteins from this experiment showed [^{32}P]-ADP-ribose incorporation in many cellular proteins, including proteins identified as nuclear histones and probably ADP-ribose polymerase.

The NAD analogue ADP-ribose and the benzamide analogue 3-aminobenzamide are inhibitors of ADP-ribosylation [11, 15]. Both inhibited ADP-ribosylation in the MCF-7 permeabilized cell assay. The 50% inhibition by ADP-ribose 5 mM could not be reversed by the addition of cAMP 5 mM. In contrast, 3-aminobenzamide 50 μM inhibited ADP-ribosylation by 80%, but this inhibition was partially reversible by the addition of cAMP 5 mM. This suggests that there might be a different site of inhibition for each analogue with cAMP reacting more competitively with 3-aminobenzamide than with ADP-ribose.

ADP-ribose has also been shown to be an inhibitor of the purified poly(ADP-ribose) glycohydrolase [16]. Like ADP-ribose, cAMP is an inhibitor of this enzyme. To determine if the stimulatory effect of cAMP on ADP-ribosylation was secondary to inhibition of the poly(ADP-ribose) glycohydrolase, other inhibitors of the enzyme, such as CaCl$_2$ and NaCl, were studied.

As shown in Table 3, in the whole cell system, both salts inhibited [^{32}P]-ADP-ribose incorporation both in the absence and presence of cAMP in the assay. In contrast, in the isolated nuclei assay, CaCl$_2$ had a stimulatory effect on [^{32}P]-ADP-ribose incorporation which was only slightly decreased by cAMP. This suggests that the effect of CaCl$_2$ on ADP-ribosylation is a complex one, but that the stimulatory effect of cAMP on ADP-ribosylation of proteins cannot be explained solely on the basis of its inhibition of poly(ADP-ribose) glycohydrolase.

The cycloalkyl lactamimide, MDL 12330A, is an inhibitor of the guanyl nucleotide stimulatory protein [17]. This compound has several inhibitory effects besides inhibiting ADP-ribosylation of the guanyl nucleotide stimulatory protein. It also inhibits adenylate cyclase and NAD-glycohydrolase. This compound was used to further clarify whether the stimulatory effect of cAMP on [^{32}P]-ADP-ribose incorporation was

Table 3. Reactants that influence the rate of ADP-ribosylation. Each reactant is present in the reaction mix at the final concentration noted below

Sample	DPM [^{32}P]-ADP-ribose incorporated	% Control
Whole cell		
Control	53,863	
CaCl$_2$ 10 mM	26,066	48
NaCl 125 mM	33,670	63
Cyclic AMP 1 mM	86,279	160
Cyclic AMP 1 mM + CaCl$_2$ 10 mM	45,208	84
Cyclic AMP 1 mM + NaCl 125 mM	46,960	87
Nuclei		
Control	1,153	
CaCl$_2$ 10 mM	1,968	171
NaCl 125 mM	1,042	90
Cyclic AMP 1 mM	1,030	89
Cyclic AMP 1 mM + CaCl$_2$ 10 mM	1,767	153
Cyclic AMP 1 mM + NaCl 125 mM	692	60

primarily due to mono(ADP-ribosylation) of nonnuclear proteins by the ADP-ribose transferase. In the absence of cAMP, MDL 12330A 500 μM inhibited [^{32}P]-ADP-ribose incorporation by 80%. When cAMP 5 mM was added to the assay system, this inhibition persisted, suggesting that the effect of exogenous cAMP in the MCF-7 cell system was mediated through the process of mono(ADP-ribosylation).

To determine whether culture conditions which increased endogenous cAMP would show the same stimulation of [^{32}P]-ADP-ribose incorporation, cells were grown in cAMP 100 μM, isoproterenol 2.5 mM, or propranolol 250 μM. Cells grown in cAMP showed no stimulation of ADP-ribosylation. Although tumor cell lines have been shown to transport adenine nucleotides across cell membranes [18], this lack of response may reflect lack of cAMP uptake by intact cells. Cells grown in the presence of the beta-receptor agonist isoproterenol showed a complete inhibition of ADP-ribosylation that was not corrected by the addition of cAMP to the culture system. In contrast, the beta-receptor antagonist propranolol had a 50% stimulatory effect on ADP-ribosylation that was not enhanced with the addition of cAMP to the culture media. These results are contrary to what would be expected from these beta-receptor agents since one would expect the beta agonist to increase intracellular cAMP. While these findings are contrary to the stimulatory effect on ADP-ribosylation that one sees with acutely added cAMP, they may suggest an inhibitory role for chronically elevated levels of cAMP on the process of cellular ADP-ribosylation.

In summary, the hormone-responsive MCF-7 human breast cancer cell line has been used to study the interaction of cAMP and ADP-ribosylation. In the premeabilized cell assay using this cell line, added cAMP markedly stimulated ADP-ribosylation, whereas the inhibitory effect of ATP is retained. This effect of cAMP has not been seen in other cell lines studied. Further studies of this system show that various poly(ADP-ribose) glycohydrolase inhibitors do not mimic the action of cAMP suggesting that cAMP works by mechanisms other than the inhibition of this enzyme. The data from isolated nuclei and from studies using a mono(ADP-ribose) inhibitor, cycloalkyl lactamimide, suggest that this stimulatory effect of cAMP occurs primarily in association with mono(ADP-ribosylation) of cytoplasmic and cell membrane proteins.

Acknowledgments. We thank the American Cancer Society of Ohio, Cuyahoga County Unit, for research support and the Merrell Dow Research Institute, Cincinnati, Ohio, for the gift of the chemical MDL 12330A.

References

1. Freidman DL (1976) Role of cyclic nucleotides in cell growth and differentiation. Physiol Rev 56:652–708
2. Murdoch GH, Rosenfeld MG, Evans RM (1982) Eukaryotic transcriptional regulation and chromatin-associated protein phosphorylation by cyclic AMP. Science 218:1315–1317
3. Pfeuffer T (1977) GTP-binding proteins in membranes and the control of adenylate cyclase activity. J Biol Chem 252:7224–7234
4. Northrup JK, Smigel MD, Sternweis PC, Gilman AG (1983) The subunits of the stimulatory regulatory component of adenylate cyclase. J Biol Chem 258:11369–11376
5. Hsia JA, Moss J, Hewlett EL, Vaughn M (1984) ADP-ribosylation of adenylate cyclase by pertussis toxin. J Biol Chem 259:1086–1090
6. Moss J, Vaughn M (1977) Mechanism of action of choleragen. J Biol Chem 252:2455–2457
7. Gill DM, Meren R (1978) ADP-ribosylation of membrane proteins catalyzed by cholera toxin: basis of the activation of adenylate cyclase. Proc Natl Acad Sci USA 75:3050–3054
8. Horwitz KB, Costlow ME, McGuire WL (1975) MCF-7: A human breast cancer cell line with estrogen, androgen, progesterone, and glucocorticoid receptors. Steroids 26:785–795
9. Page MJ, Field JK, Everett NP, Green CD (1983) Serum regulation of the estrogen responsiveness of the human breast cancer cell line MCF-7. Cancer Res 43:1244–1250
10. Berger NA, Johnson ES (1976) DNA synthesis in permeabilized mouse L cells. Biochim Biophys Acta 425:1–17
11. Berger NA, Weber G, Kaichi AS (1978) Characterization and comparison of poly(adenosine diphosphoribose) synthesis and DNA synthesis in nucleotide-permeable cells. Biochim Biophys Acta 519:87–104
12. Greengard P (1979) Cyclic nucleotides, phosphorylated proteins and the nervous system. Fed Proc 38:2208–2217
13. Aitken SC, Lippman ME (1982) Hormonal regulation of net DNA synthesis in MCF-7 human breast cancer cells in tissue culture. Cancer Res 42:1727–1735
14. Jackson V, Chalkey R (1974) Separation of newly synthesized nucleohistone by equilibrium centrifugation in cesium chloride. Biochemistry 13:3952–3957
15. Sims JL, Sikorski GW, Catino DM, Berger SJ, Berger NA (1982) Poly(adenosine-diphospho-ribose) polymerase inhibitors stimulate unscheduled deoxyribonucleic acid synthesis in normal human lymphocytes. Biochemistry 21:1813–1821
16. Tavassoli M, Tavassoli MH, Shall S (1983) Isolation and purification of poly(ADP-ribose) glycohydrolase from pig thymus. Eur J Biochem 135:449–455
17. Bitonti AJ (1984) Inhibition of ADP-ribosylation-transferase activity of cholera toxin by MDL 12330A and chorpromazine. Biochem Biophys Res Commun 120:700–706
18. Rapaport E, Fishman RF, Gercel C (1983) Growth inhibition of human tumor cells in soft-agar cultures by treatment with low levels of adenosine-5'-triphosphate. Cancer Res 43:4402–4406

Poly(ADP-Ribose) Synthetase Inhibitor Effects on Cellular Functions

WILLIAM R. KIDWELL[1], PHILIP D. NOGUCHI[2], and MICHAEL R. PURNELL[3]

1. Introduction

The original report that certain poly(ADP-ribose) synthetase inhibitors block the repair of DNA chain breaks [1] has been confirmed by a number of laboratories [2–4]. In spite of this, a systematic approach to answering the fundamental question of whether synthetase plays an obligatory role in the repair process has not been undertaken. We have attempted to evaluate this possibility by comparing the potency of six synthetase inhibitors for inhibiting synthetase and for blocking the repair of γ-ray-induced DNA chain breaks. The results of this analysis were decidely mixed — some inhibitors being more portent in blocking repair than predicted on the basis of their K_i for synthetase and some being less effective than anticipated. Consequently, assays were made of the potency of the inhibitors for reducing cell viability or for affecting RNA synthesis. Cell viability was reduced by the synthetase inhibitors in direct relationship to their potency as synthetase inhibitors, with one exception. The effects of the inhibitors on the uptake of RNA precursors were, however, not related to their K_i's for synthetase.

2. Effects of Synthetase Inhibitors on the Repair of γ-Ray-Induced DNA Chain Breaks

2.1 Theoretical Considerations

Most of the compounds that have been found to be good synthetase inhibitors are structurally related to the nicotinamide moiety of the enzyme substrate, NAD. Additionally, these synthetase inhibitors have been found to act competitively with respect to NAD. Rearrangement of the Michaelis-Menten equation for a competitive inhibitor yields the following:

1   Laboratory of Pathophysiology, National Cancer Institute, Bethesda, MD 20205, USA
2   Cell Biology Branch, FDA, NIH, Bethesda, MD 20205, USA
3   Cancer Research Unit, The Royal Victoria Infirmary, Newcastle upon Tyne, NEI 4LP, Great Britain

ADP-Ribosylation of Proteins
(ed. by F.R. Althaus, H. Hilz, and S. Shall)
© Springer-Verlag Berlin Heidelberg 1985

a) $1/V = K_m/V_{max} \cdot S(1 + I/K_i) + 1/V_{max}$ or

b) $1/V - 1/V_{max} = K_m/V_{max} \cdot S(1 + I/K_i)$,

where V = the reaction rate, V_{max} = the maximum velocity, S = substrate concentration, K_m = apparent substrate affinity constant, and K_i = apparent affinity constant of the inhibitor.

A plot of $1/V$ vs I/K_i will be linear in a cell if the substrate concentration is unchanged by the inhibitors. The plot will be colinear for a family of inhibitors if they all act on the same enzyme in a cell and if the inhibitors all have equal access to the enzyme. If one makes the assumption that poly(ADP-ribose) synthesis is rate limiting for the repair of DNA chain breaks, then the K_i for a particular synthetase inhibitor will be directly proportional to its K_i for blocking repair. In this case, a plot of the reciprocal of the repair rate vs inhibitor concentration normalized against its K_i for synthetase would be linear and similar plots for a variety of compounds that block repair via blocking synthetase would be colinear.

2.2 Observed Effects

To test the above premise, it was necessary to establish conditions under which the rate of repair could be properly judged. We utilized the alkaline elution method essentially as outlined by Zwelling et al. [5]. In this method, the first order rate constant for elution of [^{14}C]thymidine-labeled DNA is proportional to the number of DNA chain breaks, within limits [5].

Since quantitation of the absolute number of DNA chain breaks is difficult, we chose to express the number of breaks in terms of rad equivalents of DNA breaks [5]. A standard curve was constructed of the first order DNA elution rate through filters under alkaline conditions [5]. Cells were irradiated at doses of γ-rays from 0 to 1000 rad. A plot of the DNA elution rate constant was linear up to 800 rad. Next, the amount of time necessary to repair 400 rad equivalents of DNA damage was determined. The repair rate was linear for about 7.5 min with completion of repair taking 10 min. We chose a time of 5 min at which time about 50% of the repair of 400 rad was accomplished in cells not exposed to synthetase inhibitors. HeLa cells were irradiated for 400 rad then incubated for 5 min in the presence or absence of synthetase inhibitors to permit DNA repair to be accomplished. Inhibitors utilized were 3-acetamidobenzamide (AAB, K_i = 0.25 μM), 3-aminobenzamide (AB, K_i = 2.8 μM), 3-methoxybenzamide (MB, K_i = 0.41 μM), 3-hydroxybenzamide (HB, K_i = 1.05 μM), benzamide (B, K_i = 0.91 μM), and 3-nitrobenzamide (NB, K_i = 1.03 μM) as determined in a poly(ADP-ribose) synthetase assay with HeLa cell nuclei. Inhibitors were tested for their ability to reduce the number of rad equivalents of DNA breaks repaired in 5 min at 37°C over a concentration range of 0 to 40 mM, where solubilities permitted.

The results of our analyses are presented in Fig. 1. As indicated, the plots of 1/rad equivalents of breaks repaired vs inhibitor concentration, I, normalized against its K_i for synthetase, was approximately colinear at low inhibitor concentrations for AAB, AB, HB, and B. NB was more inhibitory for repair than anticipated based on its K_i

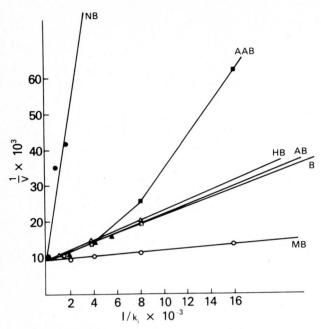

Fig. 1. Relationship between poly(ADP-ribose) synthesis inhibitory potential and DNA repair inhibitory potential of benzamides

for synthetase, whereas MB was less potent than expected. Also, AAB was found to be more potent at blocking repair at high concentrations than anticipated, suggesting that at least two different modes of inhibition of repair by AAB were involved.

2.3 Conclusions

The above results suggest that a causal relationship between the inhibitory potential of the benzamides for blocking DNA repair and inhibiting synthetase might exist for four of six of the compounds, but only at low inhibitor concentrations. However, because there are two exceptions, namely, NB and MB, the apparently causal relationship may be fortuitous. These two compounds might penetrate cells differently. They may affect cellular processes other than synthetase or the premise regarding synthetase as a rate limiting step in repair might not be tenable. In fact, the latter conclusion was reached using *Drosophila melanogaster* cells because the synthetase was found to be inactivated by temperature shock conditions that did not eliminate the repair of γ-ray induced DNA strand breaks [6]. From the data presented here, a convincing case either for or against synthetase operating as a rate limiting enzyme in the repair process, cannot be made. It is possible that synthetase might facilitate repair, but not be essential for it. If that were the case, a certain basal repair rate that was insensitive to synthetase inhibitors or temperature shock would exist. This possibility cannot be ruled out.

3. Effects of 3-Acetamidobenzamide on the DNA Size in EGF-Treated Cells

Recent reports have suggested that when epidermal growth factor (EGF) interacts with its receptor, the receptor or a fragment thereof, might relocate to the nucleus of cells and, via an ATP-dependent nuclease activity associated with the receptor, activate nuclear genes by inducing DNA chain breaks [7]. Consistent with this report are the observations that A431 cells, which possess large numbers of EGF receptors, are adversely affected by super stimulation with high concentrations of EGF [8]. A431 cells were labeled for 16 h with 0.05 μCi ml^{-1} [^{14}C]thymidine. The cells were then treated with or without EGF (20 ng ml^{-1}) and/or AAB (10 mM) for 4 h at 37°C and then DNA integrity was analyzed by the nucleoid sedimentation method [9]. The results of the experiment were as follows. AAB treatment had no effect on the apparent nucleoid size in the absence of EGF. EGF alone had no effect on the nucleoid sedimentation rate. However, a combination of the two agents reduced the apparent nucleoid size dramatically. It is concluded that the steady state level of DNA chain breaks is not markedly affected by EGF but if chain rejoining is prevented by the inclusion of AAB in the culture, an accumulation of chain breaks in response to EGF can be discerned. These results offer no proof of the direct or indirect effects of poly-(ADP-ribose) synthetase on the repair process. They only show that repair rate can be reduced by AAB.

The inhibition of the repair process is probably not due to a trivial effect of AAB on cellular nucleotide triphosphate levels, however. When CHO-K1 cells were treated with 10 mM AAB for 1 h, there was no effect on the cellular concentration of ribonucleotides. The amounts of the ribonucleotides were 1.31, 0.97, 7.51, and 0.88 ng 10^{-6} cells for UTP, CTP, ATP, and GTP for the control cultures, respectively. For the AAB treated cells the values were 1.34, 1.01, 7.99, and 0.96 for UTP, CTP, ATP, and GTP, respectively. Whatever the mechanism, the use of AAB for slowing DNA strand rejoining may offer a novel approach for analyzing the effects of a variety of growth factors, steroids, etc., on DNA strand integrity. AAB use for this purpose may be particularly appropriate since its effects on cell growth are completely reversible for up to one cell doubling time exposure (see Sect. 6).

4. Effects of Synthetase Inhibitors on Cell Viability

As an alternative to the proposal that benzamides block DNA repair by inhibiting synthetase, the possibility of indirect effects must be considered. One approach to detecting such "indirect effects" is to evaluate synthetase inhibitor action on cell viability in the absence of any applied DNA damage. It is quite clear that viability is affected. However, viability itself might be decreased via effects on synthetase. We have considered this possibility by comparing the potency of the benzamides for reducing cell viability in relation to their potency as synthetase inhibitors. When CHO-K1 cells were treated with the benzamides for 24 h, followed by determining

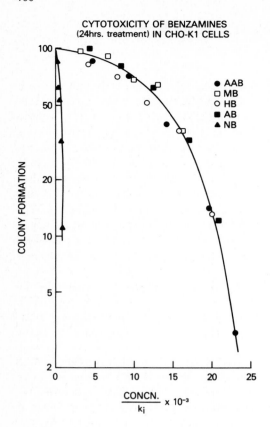

Fig. 2. Correlation between the poly(ADP-ribose) synthetase inhibition potential and cell killing potential of benzamides

cell cloning efficiency as a test for viability, we observed that benzamide treatment reduced cell viability in relation to the potency of the compounds as synthetase inhibitors (Fig. 2).

In fact, the loss of viability is at least as clearly related to the synthetase potency of the compounds as is the relationship to inhibition of γ-ray induced chain break repair — with five of six compounds giving superimposable plots of cloning efficiency vs inhibitor concentration normalized against its K_i for synthetase. As was true for the results in Fig. 1, NB was more effective in reducing cloning efficiency than predicted based on its K_i for synthetase.

5. Synthetase Inhibitor Effects on RNA Precursor Uptake

An analysis of the effects of benzamides on [³H]uridine incorporation into RNA was also performed. The cells were pretreated with various concentrations of the inhibitors for 20 min, then pulse labeled for 30 min with 2 μCi ml⁻¹ of [³H]uridine. The incorporation into RNA was performed by standard techniques. When the effects of the

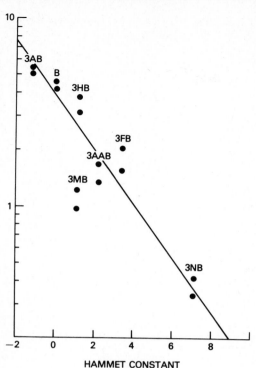

Fig. 3. Correlation between ability of benzamides to inhibit uridine uptake into RNA and the Hammett constant

benzamides were evaluated in terms of their inhibition of RNA labeling in relation to their potency as synthetase inhibitors, no correlation was apparent (not shown). However, a direct correlation between the electron withdrawing or electron donating potential of the substituent at the meta position of the benzamide and its potency in blocking [³H]uridine incorporation was found. This is shown in Fig. 3. The log of the relative amount of [³H]uridine incorporation in the presence of each benzamide species was plotted vs the Hammet constant for the specific compound. The plot was linear. Thus, the effect of the benzamides on RNA labeling is unlikely to be directly attributable to effects on poly(ADP-ribose) synthetase.

6. Reversible Cell Cycle Arrest by 3-Acetamidobenzamide

From Fig. 1 it is apparent that the effects of 3-acetamidobenzamide on DNA repair are attributable to more than one cellular target. Thus, plots of the reciprocal of the repair rate vs AAB concentration normalized against the K_i of AAB for synthetase are not linear except at very low AAB concentrations. Although it might be anticipated that an inhibition of RNA synthesis might indirectly lead to effects on DNA repair, this was judged unlikely because the repair of single strand breaks is extremely rapid and also because equivalent effects of NB and AAB on [³H]uridine incorporation

Fig. 4. Arrest of cell growth in G_1 and G_2 by AAB

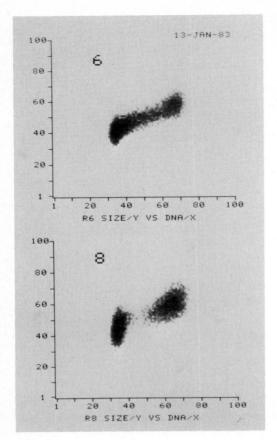

into RNA were seen (5 mM inhibitor concentration) even though the effectiveness of the two compounds in blocking DNA repair were quite different. Consequently, we considered the possibility that AAB might act on some additional cellular process. One possibility was an effect on the cell cycle, since a cell growth arrest at a specific cell cycle position had previously been observed with other synthetase inhibitors [10].

Both AAB and NB were tested for their effects on the cell cycle arrest position of CHO-K1 cells using flow microfluorimetry. NB and AAB treatment (8 h, 5 or 10 mM concentration) produced a growth arrest of the cells. However, NB arrested cell growth in all phases of the cell cycle, while AAB produced a growth arrest in G_1 and G_2 without affecting cell transit through S. This is illustrated in Fig. 4. AAB and NB effects on the cells differed in one other way. Provided the inhibitor treatment was maintained for no more than one cell doubling time, the growth arrest by AAB was reversible, while that produced by NB was not.

7. Summary

Several points emerge from the results that have been presented. First, it is apparent that synthetase inhibitors are inherently toxic to cells, even in the absence of exposure to DNA damaging agents. Second, there are probably multiple targets for the compounds in cells. Third, the effects of the inhibitors on repair cannot be clearly ascribed to effects on synthetase because of the multiplicity of cellular targets. Fourth, no increase in the number of DNA chain breaks is discernable in cells exposed to the most potent synthetase inhibitor unless the cells are treated with some agent potentially capable of inducing DNA chain breaks (γ-rays, EGF, etc.). Some of the inhibitors, AAB in particular, may be useful for assessing the effects of agents that activate the cellular genetic apparatus via gene rearrangements in which DNA chain breaks must necessarily be introduced. AAB may be a useful tool for obtaining a reversible growth arrest of cells at G_1 and/or G_2 irregardless of whether or not cell growth arrest is mediated via an inhibition of poly(ADP-ribose) synthetase. Finally, the results of the inhibitor effects on the nucleotide triphosphate pools of cells make it unlikely that the inhibition of the repair of single strand DNA breaks is trivially produced by a cellular depletion of macromolecular precursors.

References

1. Skidmore CJ, Davies MI, Goodwin PM, Halldorsson H, Lewis PJ, Shall S, Zaiee AA (1979) The involvement of poly(ADP-ribose) polymerase in the degradation of NAD caused by γ-irradiation and N-methyl-N-nitrosourea. Eur J Biochem 101:135–142
2. Juarez-Salinas H, Sims JL, Jacobson MK (1979) Poly(ADP-ribose) levels in carcinogen-treated cells. Nature (London) 282:740–741
3. James MR, Lehman AR (1982) Role of poly(ADP-ribose) in DNA repair in human fibroblasts. Biochemistry 21:4007–4013
4. Althaus FR, Lawrence SO, Sattler GL, Pitot HC (1982) ADP-ribosyltransferase activity in cultured hepatocytes. Interactions with DNA repair. J Biol Chem 257:5528–5535
5. Zwelling LA, Kohn KW, Ross WE, Ewig RA, Anderson T (1978) Kinetics of formation and disappearance of DNA cross links in mouse L1210 cells treated with cis and trans diaminodichloroplatinum (II). Cancer Res 38:1762–1768
6. Nolan N, Kidwell WR (1982) Effect of heat shock on poly(ADP-ribose) synthetase and DNA repair in Drosophilia cells. Radiat Res 90:187–203
7. Carpenter G (1984) Properties of the receptor for epidermal growth factor. Cell 37:357–358
8. Kawamoto T, Sato JD, Polikoff J, Sato GH, Mendelsohn J (1983) Growth stimulation of by EGF; identification of high affinity receptors for EGF by an anti-receptor monoclonal antibody. Proc Natl Acad Sci USA 80:1337–1341
9. Cook PR, Brazell IA, Pawsley SA, Gianelli F (1978) Changes induced by UV light in the superhelical DNA of lymphocytes from subjects with Xeroderma Pigmentasum and normal controls. J Cell Sci 29:117–127
10. Kidwell WR, Mage M (1976) Changes in poly(ADP-ribose) and poly(ADP-ribose) synthetase in synchronously dividing HeLa Cells. Biochemistry 15:213–217

Poly(ADP-Ribose) Synthetase Inhibitors Induce Islet B-Cell Regeneration in Partially Depancreatized Rats

HIROSHI OKAMOTO[1,2], HOROSHI YAMAMOTO[2], and YUTAKA YONEMURA[3]

Introduction

The removal of the insulin-producing pancreatic B-cells from an animal induces a syndrome which features many of the acute metabolic derangements of diabetes as they are known in man. Von Mering and Minkowski [1] in 1890 were the first to produce this form of diabetes by surgically removing the pancreata of dogs. Alloxan in 1943 [2] and streptozotocin in 1963 [3] were found to be highly selective B-cytotoxins in animals and to be extremely potent diabetogenic substances. Since then, alloxan and streptozotocin have been widely used to produce diabetes in experimental animals. Recently, Okamoto and his colleagues [4–10] have clarified that alloxan and streptozotocin cause DNA strand breaks in pancreatic B-cells to stimulate nuclear poly(ADP-ribose) synthetase, thereby depleting the intracellular NAD level and inhibiting B-cell functions, including proinsulin synthesis. We also reported that alloxan- and streptozotocin-induced diabetes can be prevented by inhibiting poly(ADP-ribose) synthetase of B-cells [8–10] and that after the combined administration of the diabetogenic substances with poly(ADP-ribose) synthetase inhibitors, some B-cells may be converted to the tumor cells [7, 11]. More recently, we have obtained evidence that poly(ADP-ribose) synthetase inhibitors prevent experimental diabetes caused by subtotally removing the pancreata of rats [12]. Morphological evidence indicates that regeneration or proliferation of the pancreatic B-cells occurs in the poly(ADP-ribose) synthetase inhibitor-treated rats.

Administration of Poly(ADP-Ribose) Synthetase Inhibitors to 90% Depancreatized Rats

Male Wistar rats weighing 180–200 g were 90% depancreatized according to the technique described by Foglia [13] and maintained on standard rat chow. As shown in

---

1 Department of Biochemistry, Tohoku University School of Medicine, Sendai 980, Japan
2 Department of Biochemistry, Toyama Medical and Pharaceutical University School of Medicine, Toyama 930-01, Japan
3 Department of Surgery, Kanazawa University School of Medicine, Kanazawa 920, Japan

ADP-Ribosylation of Proteins
(ed. by F.R. Althaus, H. Hilz, and S. Shall)
© Springer-Verlag Berlin Heidelberg 1985

90% Pancreatectomy

Fig. 1. Administration of poly(ADP-ribose) synthetase inhibitors to 90% depancreatized rats

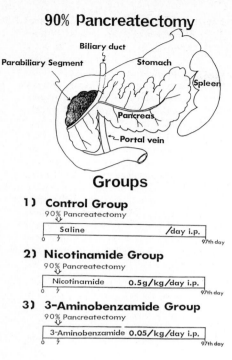

Groups

1) Control Group

90% Pancreatectomy

Saline	/day i.p.

0 7   97th day

2) Nicotinamide Group

90% Pancreatectomy

Nicotinamide	0.5g/kg/day i.p.

0 7   97th day

3) 3-Aminobenzamide Group

90% Pancreatectomy

3-Aminobenzamide 0.05/kg/day i.p.

0 7   97th day

Fig. 1, from 7 days before the partial pancreatectomy, nicotinamide, at a dose of 0.5 g kg^{-1} body weight in 2 ml saline, and 3-aminobenzamide·HCl, at a dose of 0.05 g kg^{-1} body weight in 2 ml saline, were injected intraperitoneally into each group of five rats every day. Nicotinamide and 3-aminobenzamide are inhibitors of pancreatic B-cell poly(ADP-ribose) synthetase; the inhibitory ability of the latter is about tenfold as potent as that of the former [9]. The injection was continued until the 90th postoperative day. Five control rats received saline daily before and after the partial pancreatectomy. The residual pancreatic tissue, referred to as the remaining pancreas, was removed 3 months after the partial pancreatectomy and fixed in Bouin's solution. Hydrated 5 μm sections of paraffin-embedded pancreatic tissues were stained for insulin, glucagon, and somatostatin by the peroxidase antiperoxidase method [14].

Effect of Poly(ADP-Ribose) Synthetase Inhibitor Administration on Urinary and Plasma Glucose Levels

As shown in Fig. 2, the control rats exhibited glucosuria 1 to 3 months after the 90% pancreatectomy. However, in rats receiving 0.5 g kg^{-1} nicotinamide or 0.05 g kg^{-1} 3-aminobenzamide daily, the urinary glucose excretion level decreased markedly. Plasma glucose levels, before and after an intravenous glucose load, in the rats receiving the poly(ADP-ribose) synthetase inhibitors were also significantly decreased in

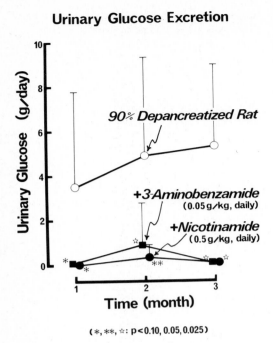

Urinary Glucose Excretion

Fig. 2. Urinary glucose excretion in partially depancreatized rats with or without poly(ADP-ribose) synthetase inhibitor injection. 1, 2, and 3 months after the 90% pancreatectomy, the urine of each rat was collected for 24 h. Urinary glucose levels were measured by the glucose oxidase method. ○ Control rats; ● nicotinamide-injected rats; ■ 3-aminobenzamide-injected rats. Statistical significance of differences between rats treated with and without poly(ADP-ribose) synthetase inhibitors was analyzed using Student's *t* test. Each point is the mean for five different rats; *vertical bars* show SD when larger than the symbol indicating the mean value. *, ** and ☆ = p<0.10, p<0.05, and p<0.025 vs control rats. The time after the partial pancreatectomy is shown on the *abscissa*

comparison with those in the control rats [12]. These results indicate that poly(ADP-ribose) synthetase inhibitors can prevent or improve diabetes mellitus in partially depancreatized rats.

Morphological Examinations of Pancreatic Tissues

We next examined morphologically the remaining pancreas stained with hematoxylin and eosin of the partially depancreatized rats with and without poly(ADP-ribose) synthetase inhibitor injection. Three months after the partial pancreatectomy, the islets of control rats without poly(ADP-ribose) synthetase inhibitors were decreased in number [18.4 ± 9.8 islets per cm^2 (mean ± SD, n = 7); 47.3 ± 15.7 islets cm^{-2} (mean ± SD, n = 18) in normal untreated rats], small in size and had irregular contours (Fig. 3A). Fibrotic degeneration and degranulation were frequently encountered. These observations were in agreement with those reported by others [15, 16]. However, the islets in the remaining pancreas of rats which received the nicotinamide and 3-aminobenzamide injection were extremely large (Fig. 3B); their diameters were 0.221 ± 0.141 mm (mean ± SD, n = 105) and 0.143 ± 0.114 mm (mean ± SD, n = 89), respectively, while the value of normal untreated rat islets was 0.112 ± 0.090 mm (mean ± SD, n = 28). The number of islets per cm^2 was increased in nicotinamide- and 3-aminobenzamide-injected rats [56.0 ± 23.1 (mean ± SD, n = 6) and 52.4 ± 43.1 (mean ± SD, n = 5),

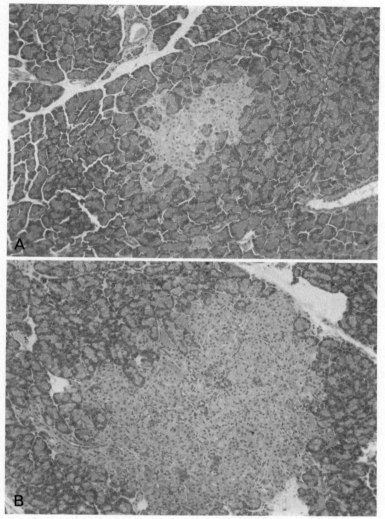

Fig. 3. Pancreatic islets in pancreata stained with hematoxylin and eosin. **A** Remaining pancreas from control 90% depancreatized rat. **B** Remaining pancreas from nicotinamide-treated rat. Magnification × 825

respectively]. The contours of the islets were somewhat irregular; loops and cords of insular cells pushed their way out between the adjacent acini. The size of each cell in these islets was normal. Therefore, an increase in cell number was considered to be responsible for the increase in islet size in partially depancreatized rats treated with poly(ADP-ribose) synthetase inhibitors. When the remaining pancreata were immuno-histochemically stained, almost all the area of the enlarged islets in nicotinamide- and 3-aminobenzamide-treated rats was densely stained for insulin, but a small number of islet cells were stained for insulin in the control rats (Table 1). On the other hand, cells

Table 1. Cell numbers in the parabiliary segment of rat pancreatic tissues[a]

Treatment of rats	B-cell	A-cell (no. cm^{-2})	D-cell
90% depancreatized	442 ± 482	359 ± 120	185 ± 72
+ 0.5 g kg^{-1} day^{-1} nicotinamide	3204 ± 1412[b]	310 ± 207	133 ± 119
+ 0.05 g kg^{-1} day^{-1} 3-aminobenzamide	2286 ± 1173[c]	310 ± 161	92 ± 49
Normal untreated	2086 ± 507	285 ± 85	191 ± 42

[a] The number was determined in the parabiliary segment of at least five rats and expressed as mean ± SD
[b] $p < 0.01$
[c] $p < 0.05$ vs control 90% depancreatized rats

stained for glucagon (A-cells) were localized only on the peripheries of enlarged islets in the remaining pancreata of rats treated with nicotinamide or 3-aminobenzamide, and their number was almost the same as that in islets of normal rats. The number of cells stained for somatostatin (D-cells) was also unchanged. These immunohisto-chemical findings indicate that it is the B-cell population that increased in the islets of the remaining pancreas of poly(ADP-ribose) synthetase inhibitor-treated rats.

A Possible Role of Poly(ADP-Ribose) Synthetase Inhibitors in B-Cell Regeneration

The present study demonstrated that poly(ADP-ribose) synthetase inhibitors can prevent diabetes in partially depancreatized rats. Morphological and immunohisto-chemical examinations of the remaining pancreas of poly(ADP-ribose) synthetase inhibitor-treated rats showed a marked increase in the B-cell population. Therefore, it is reasonable to assume that poly(ADP-ribose) synthetase inhibitors induce pancre-atic B-cell regeneration, thereby improving diabetes mellitus caused by partial pancre-atectomy. The fact that poly(ADP-ribose) synthetase inhibitors induce pancreatic B-cell regeneration seems to mean that the enzyme may play a role in restricting B-cell replication. In fact, our recent results from an autoradiographic approach indicates that the proportion of labeled B-cell nuclei in the poly(ADP-ribose) synthetase inhibitor-administered rats 2 to 4 days after the pancreatectomy was about three times larger than in the control 90% depancreatized rats.

We have already shown that alloxan and streptozotocin induce islet DNA strand breaks and that poly(ADP-ribose) synthetase acts to repair the DNA breaks, consuming islet NAD [4–7]. This rapid and marked depletion of islet NAD has been regarded as the primary molecular mechanism behind the B-cell necrosis. The B-cells seem to be making a suicide response to repair the damaged DNA. Therefore, poly(ADP-ribose) synthetase inhibitors can prophylactically prevent alloxan and streptozotocin diabetes by blocking the NAD consumption (Fig. 4).

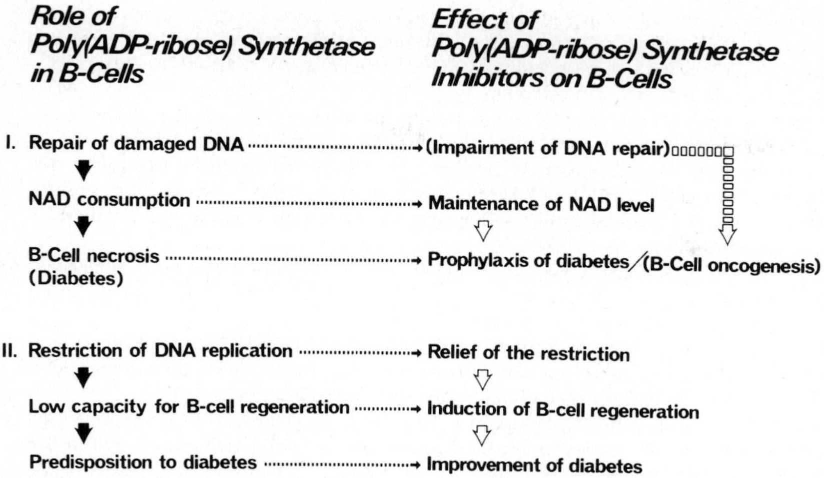

Role of
Poly(ADP-ribose) Synthetase
in B-Cells

Effect of
Poly(ADP-ribose) Synthetase
Inhibitors on B-Cells

I. Repair of damaged DNA ···→(Impairment of DNA repair)□□□□□□□□

   ↓

   NAD consumption ···→ Maintenance of NAD level

   ↓   ▽

   B-Cell necrosis ··→ Prophylaxis of diabetes ╱ (B-Cell oncogenesis)
   (Diabetes)

II. Restriction of DNA replication ··························→ Relief of the restriction

   ↓   ▽

   Low capacity for B-cell regeneration ··············→ Induction of B-cell regeneration

   ↓   ▽

   Predisposition to diabetes ·······························→ Improvement of diabetes

Fig. 4. A possible role of poly(ADP-ribose) synthetase and its inhibitors in DNA repair and replication of pancreatic B-cells [18]

Evidence in the present paper suggests an alternative function of poly(ADP-ribose) synthetase inhibitors in the improvement of surgical diabetes. In this case, poly(ADP-ribose) synthetase inhibitors may relieve restriction of DNA replication and so cause B-cell regeneration (Fig. 4). Since a low capacity for B-cell regeneration has been suggested as a predisposition for the development of human diabetes [17], the present study may provide a novel clue for the prevention and treatment of human diabetes.

Acknowledgments. This work was supported in part by grants-in-aid for Cancer Research and for Scientific Research from the Ministry of Education, Science and Culture, Japan.

References

1. Mering J von, Minkowski O (1890) Diabetes mellitus nach Pankreasexstirpation. Arch Exp Pathol Pharmakol 26:371–381
2. Dunn JS, Sheehan HL, McLetchie NGB (1943) Necrosis of islets of Langerhans produced experimentally. Lancet I:484–487
3. Rakieten N, Rakieten ML, Nadkarni MV (1963) Studies on the diabetogenic actions of streptozotocin. Cancer Chemother Rep 29:91–98
4. Okamoto H (1981) Regulation of proinsulin synthesis in pancreatic islets and a new aspect to insulin-dependent diabetes. Mol Cell Biochem 37:43–61
5. Yamamoto H, Uchigata Y, Okamoto H (1981) Streptozotocin and alloxan induce DNA strand breaks and poly(ADP-ribose) synthetase in pancreatic islets. Nature (London) 294:284–286
6. Yamamoto H, Uchigata Y, Okamato H (1981) DNA strand breaks in pancreatic islets by in vivo administration of alloxan or streptozotocin. Biochem Biophys Res Commun 103:1014–1020
7. Okamoto H, Yamamoto H (1983) DNA strand breaks and poly(ADP-ribose) synthetase activation in pancreatic islets — a new aspect to development of insulin-dependent diabetes and

pancreatic B-cell tumors. In: Miwa M, Hayaishi O, Shall S, Smulson M, Sugimura T (eds) ADP-ribosylation, DNA repair and cancer. Proc 13th Int Symp Princess Takamatsu Cancer Res Fund, Tokyo, 1982. Jpn Sci Soc Press, Tokyo/VNU Science Press BV, Utrecht, pp 297–308

8. Yamamoto H, Okamoto H (1980) Protection by picolinamide, a novel inhibitor of poly(ADP-ribose) synthetase, against both streptozotocin-induced depression of proinsulin synthesis and reduction of NAD content in pancreatic islets. Biochem Biophys Res Commun 95:474–481

9. Uchigata Y, Yamamoto H, Kawamura A, Okamoto H (1982) Protection by superoxide dismutase, catalase, and poly(ADP-ribose) synthetase inhibitors against alloxan- and streptozotocin-induced islet DNA strand breaks and against the inhibition of proinsulin synthesis. J Biol Chem 257:6084–6088

10. Uchigata Y, Yamamoto H, Nagai H, Okamoto H (1983) Effect of poly(ADP-ribose) synthetase inhibitor administration to rats before and after injection of alloxan and streptozotocin on islet proinsulin synthesis. Diabetes 32:316–318

11. Yamagami T, Miwa A, Takasawa S, Yamamoto H, Okamoto H (1985) Induction of rat pancreatic B-cell tumors by the combined administration of streptozotocin or alloxan and poly-(adenosine diphosphate ribose) synthetase inhibitors. Cancer Res 45:1845–1849

12. Yonemura Y, Takashima T, Miwa K, Miyazaki I, Yamamoto H, Okamoto H (1984) Amelioration of diabetes mellitus in partially depancreatized rats by poly(ADP-ribose) synthetase inhibitors. Diabetes 33:401–404

13. Foglia VG (1944) Caracteristicas de la diabetes en la rata. Rev Soc Argent Biol 20:21–37

14. Sternberger LA, Hardy PH Jr, Cuculis JJ, Meyer HG (1970) The unlabeled antibody enzyme method of immunohistochemistry. J Histochem Cytochem 18:315–333

15. Martin JM, Lacy PE (1963) The prediabetic period in partially pancreatectomized rats. Diabetes 12:238–242

16. Volk BW, Lazarus SS (1964) Ultrastructure of pancreatic B cells in severely diabetic dogs. Diabetes 13:60–70

17. Hellerström C, Andersson A, Gunnarsson R (1976) Regeneration of islet cells. Acta Endocrinol 83 (Suppl 205):145–160

18. Okamoto H (1985) Molecular basis of experimental diabetes: degeneration, oncogenesis and regeneration of pancreatic B-cells of islets of Langerhans. BioEssays (Cambridge University Press) 2:15–21

Regulation of Lymphocyte Proliferation by a Continuous Production and Repair of DNA Strand Breaks

WENDA L. GREER and J. GORDIN KAPLAN[1]

Introduction

Recent studies with chick myoblasts [1], Friend erythroleukemia cells [2], and human peripheral blood lymphocytes [3] have shown that an increase in ADP-ribosylation and in DNA strand breaks is associated with cell differentiation. Previous reports from this lab have shown that in mouse splenic lymphocytes there is repair of about 3200 strand breaks per diplod genome within 2 h of mitogenic stimulation [4, 5]. This repair did not occur in the presence of inhibitors of ADP-ribose polymerase; neither did the cells proliferate. On the other hand, rejoining of the breaks occurred in the absence of mitogen when nicotinamide at the appropriate concentration was added to the growth medium [4]. In spite of the nicotinamide-induced repair, cells did not proliferate. These observations indicated that NAD levels are rate limiting for ADP-ribosylation and thus for repair in resting lymphocytes, and show that repair of DNA strand breaks is necessary but not sufficient for lymphocyte activation. In previous reports from others [3, 6] and from us [4, 5], the assumption was made that the occurrence of strand breaks in the differentiated state and their repair in proliferation were unique, punctual events, an assumption we now believe to be false.

Cell Preparation and Culture

Balb/c male mice, 8–12 weeks old were killed by cervical dislocation and their spleens or thymus disrupted through a wire mesh screen. Red blood cells were removed by lysis with 0.83% ammonium chloride. Cells were cultured in RPMI1640 medium with 10% fetal calf serum, and 2 mM glutamine, at a density of 2×10^6 cells ml^{-1}.

1 Biochemistry Department, University of Alberta, Edmonton, Alberta, T6G 2H7, Canada

ADP-Ribosylation of Proteins
(ed. by F.R. Althaus, H. Hilz, and S. Shall)
© Springer–Verlag Berlin Heidelberg 1985

Table 1. Repair of DNA strand breaks after Con A stimulation in mouse and pig lymphocytes

Lymphocyte	Strand breaks repaired per diploid genome	Time after Con A stimulation of maximal repair
Mouse spleen	3200	2 h
Pig peripheral blood	2500	5 h

Detection of DNA Strand Breaks

DNA strand breaks were detected using the fluorometric analysis of DNA unwinding technique developed by Birnboim and Jevcak [7]. After denaturation at pH 12.8 for 1 h, the cell lysate was adjusted to pH 11, and ethidium bromide was added. Under these conditions, ethidium bromide binds to, and fluoresces selectively in double-stranded DNA. From the percentage of double-stranded DNA remaining after alkaline treatment, one can estimate the number of DNA strand breaks by using a calibration curve obtained from cells treated with various doses of gamma radiation (500 rad of γ-radiation produce 6000 single-strand breaks per diploid genome [8]).

Assay for ADP-Ribosylation

ADP-ribosylation was measured as incorporation of $[^3H]$-NAD$^+$ into the acid-insoluble fraction of permeabilized cells according to the method of Berger and Johnson [9].

Results and Discussion

Within 2 h of addition of concanavalin A (Con A) to mouse splenic lymphocytes, there was repair of 3200 strand breaks per diploid genome (Table 1); DNA strand breaks were assayed by means of fluorescence analysis of DNA unwinding [7]. This technique was used to demonstrate the same phenomenon in pig peripheral blood lymphocytes within 5 h (Table 1). This technique does not differentiate between alkali labile bonds and frank breaks; however, this argument does not apply to the work of Johnstone and Williams [3] who used the method of nucleoid sedimentation in neutral gradients. Thus in human lymphocytes, at least, frank breaks exist in resting cells.

Methoxybenzamide (MBA), an inhibitor of ADP-ribose polymerase prevented the Con A-induced repair in all three systems as well as the proliferative response; of all the parameters of activation examined, only the increase in transport of monovalent cations that follows soon after addition of mitogen [10] occurred in the presence of inhibitors of ADP-ribosylation.

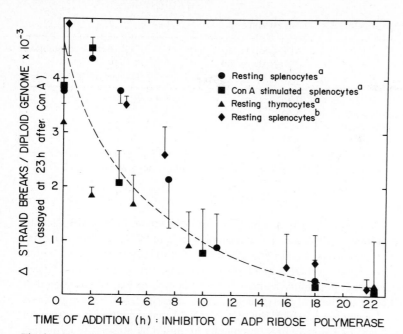

Fig. 1. Effect of methoxybenzamide or 3-aminobenzamide on number of DNA strand breaks, when added at various times after Con A stimulation. All samples were assayed for strand breaks at 23.5 h. (a) Cells were treated with 5 mM methoxybenzamide. (b) Cells were treated with 5 mM 3-aminobenzamide

It was shown in the human system [3] and in mouse [4] that inhibitors of ADP-ribosylation caused less inhibition of DNA synthesis when they were added at later times after mitogenic treatment; nevertheless, they still had considerable effect on proliferation of mouse splenic lymphocytes even when added at 26 h following initiation of culture with the mitogen Con A, which is 24 h after the time at which strand breaks reached a minimum. This suggested that either the inhibitors have some non-specific action which affects lymphocyte activation other than through DNA repair, or that there is a continuous production of DNA strand breaks which are normally promptly repaired by a system requiring ADP-ribosylation. Our data are consistent with the latter hypothesis which was tested in experiments in which MBA was added to cells at various times from 0 to 22 h from onset of culture. All samples were assayed at 23 h for DNA strand breaks. The data in Fig. 1 represent the number of strand breaks in MBA-treated cultures minus those in the untreated controls. In presence of MBA, there was an accumulation of breaks; the longer MBA was present, the greater the number of breaks accumulated. DNA strand breaks accumulated at about the same rate in both resting and Con A-stimulated splenocytes as well as in thymocytes. The same observations were made if alternative inhibitors of ADP-ribosylation were used, such as 3-aminobenzamide. The hyperbolic shape of the curve in Fig. 1 suggests that breaks may be produced at a progressively lower rate during culture with Con A, or perhaps that the cells have a low permeability for the inhibitors.

Table 2. Reversibility of MBA effects in mouse splenocytes[a]

Treatment	ADP-ribosylation at 10 h [^{3}H]-NAD bound 10^6 cells	[^{3}H] TdR incorporation at 48 h (cpm 10^6 cells $\times 10^{-3}$)
Con A 2 μg ml^{-1}	0.32	20
Con A 2 μg ml^{-1} + MBA 5 mM	0.09	2
Con A 2 μg ml^{-1} + MBA 5 mM washed out at 2 h	0.37	21

[a] ADP-ribosylation was assayed in permeabilized cells at 10 h after Con A stimulation, and DNA synthesis at 48 h

An alternative explanation for the accumulation of DNA strand breaks with increasing time of incubation with MBA and the continued effects of MBA at times well after the minimum number of strand breaks was attained at 2 h of culture, would be a non-specific effect of the inhibitor on DNA synthesis rather than that on ADP-ribosylation-mediated repair. One such possibility consistent with the data of Fig. 1, would be that MBA was itself directly or indirectly introducing breaks into the DNA. This hypothesis was excluded by experiments in which the effect of MBA was measured in cultures in which ADP-ribosylation had already been maximally inhibited by alternative inhibitors such as 3-aminobenzamide or high concentrations of nicotinamide, or heat shock [11] (data not shown). No additional breaks were introduced by MBA under these conditions. Furthermore, there was no effect of MBA on cellular levels of purine or pyrimidine nucleotides, and MBA treatment resulted in only 4% cell death. Table 2 shows that the inhibitory effects of MBA on ADP-ribosylation, and on cell proliferation were completely reversible. In addition, the data in Table 3 show a direct proportionality between concentration of MBA on the one hand and inhibition of ADP-ribosylation in permeabilized cells, increased number of DNA strand breaks and decreased cell proliferation at 48 h on the other hand. This, and the data cited above, permit us to conclude that MBA acts on this system by inhibiting the poly(ADP-ribosylation) reaction,

Table 3. Effect of increasing concentration of MBA on ADP-ribosylation, DNA strand breaks and cell proliferation[a]

MBA concentration (mM)	ADP-ribosylation (nmol [^{3}H]-NAD bound 10^6 cells)	DNA strand breaks accumulated per diploid genome ($\times 10^{-3}$)	[^{3}H] $\cdot$ TdR incorporation at 48 h (cpm $\times 10^{-3}$)
0	0.31	0	22
0.5	0.27	0.2	18
1.0	0.24	0.8	15
2.5	0.18	1.2	7
5.0	0.13	3.4	1

[a] MBA and Con A were added to all samples at 0 h. ADP-ribosylation and DNA strand breaks were assayed at 23 h. [^{3}H]-TdR incorporation was measured at 48 h

Table 4. Repair of DNA strand breaks induced by radiation in mouse lymphocytes

Treatment	Number of strand breaks induced or repaired	
Resting	0	
Con A 2 μg ml^{-1} 2 h	2240[a]	Repaired
Resting + 500 rad T = 0 h	6720[a]	Induced by radiation
Resting + 500 rad T = 2 h	5088[a]	Repaired after radiation treatment
Con A + 500 rad T = 0 h	8224[b]	Induced by radiation
Con A + 500 rad T = 2 h	6656[b]	Repaired after radiation treatment

[a] This refers to a change in number of strand breaks from those present in resting lymphocytes
[b] This refers to a change in number of breaks from those present in Con A-stimulated lymphocytes

whatever non-specific effects it may have in other system [12, 13]. Therefore we conclude that there is a continuous production of DNA strand breaks in both resting and Con A-stimulated lymphocytes, breaks that are normally repaired by a system that requires ADP-ribosylation of some acceptor protein(s). This continuous production of breaks after Con A stimulation may well be associated with the decondensation of the chromatin that normally takes place as lymphocytes undergo blast transformation [14].

It is obvious that there is something unique about the breaks that are present normally in resting lymphocytes; they are quite unlike breaks induced by gamma radiation or neutrons, in that these are readily and rapidly repaired [15]. Table 4 shows that resting lymphocytes efficiently repair DNA strand breaks after ^{60}Co gamma radiation treatment, but only to the level inherently present in these cells. These breaks are also different in their effect from those unrepaired breaks reported to be induced by 5-fluorouracil [16]. A 5 h treatment with 150 μM 5-fluorouracil introduced about 30% more breaks than those present in resting lymphocytes, yet the treated cells were still able to undergo the expected increase in protein and RNA synthesis as well as the complex morphological changes of blast transformation. Furthermore, cells with a comparable number of breaks induced by gamma radiation also underwent blast transformation [17]. It seems that the breaks per se may not be so significant in lymphocyte activation; what seems to be important is the continuous process of breaking and repair. It is possible that repair of certain of these naturally occurring breaks may be required because of their location with respect to DNA sequence or chromatin; for example, they might occur at sites of initiation of replication, or at transcription sites for key regulatory enzymes. It is also possible that the breaks allow the topological structure of the DNA to be altered in such a way as to regulate certain cellular events.

Table 5 compares DNA strand breaks and their repair in different lymphoid tissues. It is interesting that thymocytes contain fewer breaks than splenocytes and do not undergo further decrease in number of breaks after stimulation with Con A plus interleukin 2 (IL-2). This may give some support to the hypothesis that DNA strand breaks are associated with cell differentiation. Most of the cells in the thymus are immature and not yet fully differentiated [18]; they may not yet have acquired the breaks observed in splenocytes. Nevertheless, MBA inhibited thymocyte proliferation and also caused an accumulation of strand breaks proportional to the duration of incubation of inhibitor (Fig. 1), as in the case of splenocytes. This also suggests that

Table 5. Repair of DNA strand breaks and cell proliferation in splenic and thymic lymphocytes after mitogen treatment

Cell source	Treatment	% Double-stranded DNA[a]	% Blasts[b]	[³H] Incorporation 10^6 cells at 48 h (cpm)
Spleen	Resting	46 ± 3	5	2,400 ± 900
	Con A 2 μg ml^{-1}	71 ± 5	82	42,300 ± 6,000
Thymus	Resting Con A 3 μg ml^{-1} + interleukin 2	62 ± 4	3	1,200 ± 150
	8 U ml^{-1}	65 ± 4	65	28,400 ± 900

[a] This refers to the % double-stranded DNA remaining after 1 h denaturation in alkali: values are inversely proportional to the number of strand breaks. Using a calibration curve obtained from cells treated with various doses of γ-radiation, the difference in % double-stranded DNA observed in spleen compared to thymus was found to represent a difference of 3000 to 4000 strand breaks per diploid genome

[b] Blasts were defined as any cell with diameter larger than 10 μm

the presence of the breaks is less significant in regulation of proliferation than their constant repair, and indicates that repair of a small number of breaks occurs after stimulation of thymocytes but these are too few to be detectable. This would suggest that perhaps only a small proportion of the breaks repaired after stimulation of splenocytes may be important to activation of these cells.

We propose a model in which there is a continuous production and repair of DNA strand breaks in resting and Con A-stimulated mouse lymphocytes as well as in resting and Con A plus interleukin 2-stimulated thymocytes. A change in the steady state level of these breaks soon after stimulation occurs in the splenic but not thymic lymphocytes; this change is probably due to a transient increase in rate of repair, due to a rapid but transient increase in internal concentration of NAD^+, the substrate of the poly(ADP-ribose) polymerase. This shift in steady state is necessary but not sufficient for the initiation and maintenance of the proliferative response.

References

1. Farzaneh F, Zalin R, Brill D, Shall S (1982) DNA strand breaks and ADP-ribosyl transferase activation during cell differentiation. Nature (London) 300:362–366
2. Scher W, Friend C (1978) Breakage of DNA and alterations in folded genomes by inducers of differentiation in Friend Erythroleukemic cells. Cancer Res 38:841–849
3. Johnstone AP, Williams GT (1982) Role of DNA breaks and ADP ribosyltransferase activity in eukaryotic differentiation demonstrated in human lymphocytes. Nature (London) 300:368–370
4. Greer WL, Kaplan JG (1983) Regulation of repair of naturally occurring DNA strand breaks in lymphocytes. Biochem Biophys Res Commun 122:366–372

5. Greer WL, Kaplan JG (1984) A decrease in the steady state level of DNA strand breaks as a factor in the regulation of lymphocyte proliferation. In: Skehan P, Friedman S (eds) Growth, cancer and the cell cycle. Symp 16th Int Cell Cycle Soc. Humana Press, Clinton, NJ, pp 125–134

6. Williams GT, Johnstone AP (1983) ADP-ribosyl transferase, rearrangement of DNA, and cell differentiation. Biosci Rep 3:815–830

7. Birnboim HC, Jevcak JJ (1981) Fluorometric method for rapid detection of DNA strand breaks in human white blood cells produced by low doses of radiation. Cancer Res 41:1889–1892

8. Ormerod MR (1976) Radiation induced strand breaks in the DNA of mammalian cells. In: Yuhas JM, Tennant RW, Rogers JD (eds) Biology of radiation carcinogenesis. Raven Press, New York, pp 67–90

9. Berger NA, Johnson ES (1976) DNA synthesis in permeabilized mouse L cells. Biochim Biophys Acta 125:1–17

10. Kaplan JG (1978) Membrane cation transport and the control of proliferation of mammalian cells. Annu Rev Physiol 40:16–35

11. Nolan NL, Kidwell WR (1982) Effect of heat shock on poly(ADP-ribose) synthetase and DNA repair in Drosophila cells. Radiat Res 90:187–203

12. Cleaver JE, Bodell WJ, Morgan WF, Zelle B (1983) Differences in the regulation by poly (ADP-ribose) or repair of DNA damage from alkylating agents and ultraviolet light according to cell type. J Biol Chem 258:9059–9068

13. Milam KM, Cleaver JE (1984) Inhibitors of poly (adenosine diphosphate-ribose) synthesis: Effect on other metabolic processes. Science 223:589–591

14. Setterfield G, Hall R, Bladon T, Little J, Kaplan JG (1983) Changes in structure and composition of lymphocyte nuclei during mitogenic stimulation. J Ultrastruct Res 82:264–284

15. McWilliams RS, Cross WG, Kaplan JG, Birnboim HC (1983) Rapid rejoining of DNA strand breaks in resting human lymphocytes after irradiation by low doses of ^{60}Co γ rays or 14.6 MeU neutrons. Radiat Res 94:499–507

16. Greer WL, Kaplan JG (1983) DNA strand breaks in murine lymphocytes: Induction by purine and pyrimidine analogues. Biochem Biophys Res Commun 115:834–840

17. Roy C, Brown DL, Lapp WA, Kaplan JG (1982) Stimulation in the murine mixed lymphocyte reaction: Role of T cells. Immunobiology 163:383

18. Fudenberg HH, Stiles DP, Caldwell JL, Wells JV (1980) Basic and clinical immunology. Lang Medical Publ, Los Altos, Calif

Requirement for ADP-Ribosyltransferase Activity and Rejoining of DNA Strand Breaks During Lymphocyte Stimulation

ALAN P. JOHNSTONE[1]

Introduction

From work on chick embryo muscle cells the transient appearance of single-strand breaks in DNA, regulated by nuclear ADP-ribosyltransferase (ADPRT), has been proposed as part of a general mechanism for eukaryotic differentiation [1, 2]. Investigation of the stimulation of quiescent lymphocytes has provided concordant results supporting this hypothesis, and has allowed a preliminary dissection of the molecular events of lymphocyte activation underlying immune responses.

Mitogen Stimulation and Nuclear ADP-Ribosyltransferase

When added at the same time as mitogen, competitive inhibitors of ADPRT (3-amino-benzamide, 3-methoxybenzamide and 5-methylnicotinamide) inhibited the proliferative response of human peripheral lymphocytes and of purified T cells to an optimal dose of pokeweed mitogen, PHA or Con A [3, 4], and also of murine splenic lymphocytes to Con A or LPS (A.P. Johnstone, unpublished), each in a dose-dependent manner. The minimal doses showing total inhibition (2.5–5 mM) abrogated the response throughout the whole range of mitogen doses [4] and the cells did not even show a delayed response as long as the inhibitor was present [3]. The non-inhibitory acid analogues had much less effect at the same concentration (A.P. Johnstone, unpublished) and the inhibitors did not affect the proliferation of already activated lymphocytes or lymphoblastoid cell lines [4], thus demonstrating their specificity. Furthermore, addition of guanosine and adenosine did not overcome the inhibitory effect (A.P. Johnstone, unpublished) thus arguing against the proposal [5] that it is caused by blocking purine biosynthesis. Whilst I favour the interpretation that the effects are mediated through nuclear ADPRT [3, 4], the possibility that the cytoplasmic/membrane enzyme is involved cannot be excluded.

---

1 Department of Immunology, St. George's Hospital Medical School, London SW17 ORE, Great Britain

ADP-Ribosylation of Proteins
(ed. by F.R. Althaus, H. Hilz, and S. Shall)
© Springer-Verlag Berlin Heidelberg 1985

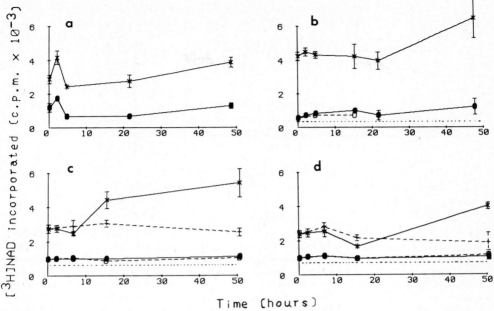

Fig. 1a–d. ADPRT activity of human lymphocytes following PHA stimulation. Cells (10^7 for each time point) were incubated at 10^6 ml^{-1} in RPMI 1640 containing 5% foetal calf serum in the presence or absence of 2.5 μg ml^{-1} PHA (Pharmacia). At the indicated times the cells were washed, permeabilised and ADPRT activity determined as described [14] except the filters were presoaked in 5 mM NAD and the acid wash also contained 1 mM NAD, 1% nicotinamide and 2% tetrasodium pyrophosphate. Results from four repeated experiments are expressed as incorporation of radioactivity (from [^{3}H]-NAD) into acid-insoluble material – 1000 cpm is equivalent to 17 pmol NAD$^+$ incorporated/10^6 cells. ●——● Intrinsic activity with PHA; ○– – –○ intrinsic activity without PHA; *——* DNAse-inducible, maximal activity with PHA; + – – – + DNAse-inducible, maximal activity without PHA. Background incorporation in the absence of cells is indicated by the *horizontal dotted line*

This requirement for ADPRT activity appears to be a very early and transient event. Addition of ADPRT inhibitors to the lymphocytes after stimulation had much less effect, such that within 1 h of PHA stimulation, about 20% of the cells detected as responding at day 2 by DNA synthesis had passed through the inhibitable stage. A plateau of 60–70% of normal response was obtained at about 16 h; for PHA on purified T cells and for Con A the escape from inhibition was slower but reached 100% at 24 and 32 h respectively [3, 4]. To investigate whether this early requirement for ADPRT reflects a need for increased enzyme activity, lymphocyte ADPRT was assayed at various times after stimulation (Fig. 1). By 48 h, a 1.5–2-fold increase in maximal (DNAse-inducible) activity was observed and also in some experiments a lesser increase in the intrinsic activity, in agreement with other workers [6–8]. Occasionally, early transient increases were detected (e.g. 2 h in Fig. 1a; 4–15 h in Fig. 1b) but not reproducibly (Fig. 1). Resting lymphocytes have a very low intrinsic ADPRT level, only 1.5–2 times assay background, and so small fluctuations are not

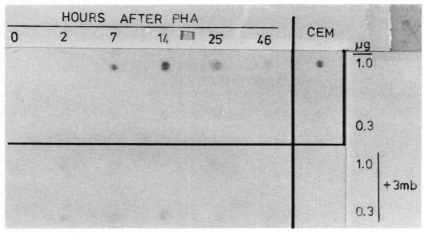

Fig. 2. Prevention by 3-methoxybenzamide of increased expression of c-myc following mitogen stimulation [11] of human lymphocytes. Cells were incubated with PHA as for Fig. 1 in the presence or absence of 2.5 mM 3-methoxybenzamide. At the indicated times the cells were lysed, total RNA prepared and 1 or 0.33 μg spotted onto PALL Biodyne nylon membrane. RNA from a lymphoblastoid cell line expressing c-myc, CEM, was similarly prepared and spotted. The membrane was hybridised to [^{32}P]-labelled c-myc probe [15] and the resultant autoradiograph is shown

detectable. Thus the enzyme can fulfil its function early during activation without a large change in its activity level. Greer and Kaplan [9] report an increase of about 50% in ADPRT activity at 20 and 40 min following Con A stimulation of mouse lymphocytes and returning to resting levels at 1 h. Whether they are referring to intrinsic or maximal activity is not stated, their cell preparation (disrupted spleen lysed with NH_4Cl) contains at least 20% dead cells [10], and their initial values for resting cells (440 pmol NAD incorporated/10^6 cells) is about 25 times higher than those of other workers ([7] and Fig. 1), and so their data are difficult to interpret. Neither can we detect in either murine or human lymphocytes these authors' rapid (30 min) increase in NAD levels following mitogen stimulation [16].

The process of lymphocyte activation can be investigated by determining whether or not a particular parameter of activation is prevented by inhibitors of ADPRT. Events that are not inhibited either occur before ADPRT is required or are independent of it. Such changes are the binding of mitogen to the cell surface, the early increase in phosphatidylinositol turnover in the plasma membrane [4], and the early increase in membrane chemiluminescence. Events that are inhibited presumably occur after ADPRT is required. These include the rejoining of DNA breaks (see below), subsequent protein and DNA synthesis [3, 4] and the increased expression of the c-myc gene [11] (Fig. 2). The later response of activated blasts to IL-2 does not appear to involve ADPRT [17].

Table 1. Nucleoid analysis of DNA strand break changes during stimulation: comparison of lymphocytes with fibroblasts[a]

Time (days)	Lymphocytes[b]	Fibroblasts[c]	
1	1.28 ± 0.16	n.d.	
2	1.59 ± 0.11	1.00 ± 0.05	
3	1.55 ± 0.13	1.03 ± 0.04	[d]1.05 ± 0.14

[a] Data are expressed as a ratio of migration of nucleoids from stimulated relative to quiescent cells [3, 4] — faster migration corresponds to less breaks
[b] Mouse splenic lymphocytes stimulated with 3 μg ml^{-1} Con A at day 0 compared with unstimulated cultures
[c] 3T3 cells grown in 10% serum compared with cultures with no serum from day 0
[d] Cultures with no serum from day 0 to 2 and then restimulated with 10% serum

DNA Strand Breaks

Concurrent with the requirement for ADPRT activity is the rejoining of DNA strand breaks detected by nucleoid analysis, with and without ethidium bromide, and by fluorimetric assay of alkali unwinding ([3, 4] and Table 1). The breaks have recently been quantitated as equivalent to 200 rad γ-irradiation (approx. 2000 breaks/diploid genome). The resting cells are as proficient as activated cells at repairing extra radiation-induced breaks (A.P. Johnstone, unpublished) and so the endogenous breaks are either especially maintained or are constantly turned over, as found for chick embryo myotubes [2]. Addition of 5 mM 3-aminobenzamide to quiescent lymphocytes for 4 h resulted in more strand breaks on nucleoid analysis (migration 0.81 ± 0.02 relative to non-inhibited control cultures) whereas this incubation had no effect on either 16 h PHA-activated lymphocytes (ratio, 0.96 ± 0.02) or lymphoblastoid cell lines (0.97 ± 0.02). This suggests that breaks in resting cells are dynamic — whether they are reformed and rebroken at the same or different sites awaits further investigation. Greer and Kaplan [12] report that 5 mM 3-methoxybenzamide causes an increase in the breaks of murine lymphocytes both resting and Con A activated. However this was only evident after 11 h incubation under conditions giving 4—10% cell death (the methoxy derivative is more toxic than the amino) and so its significance is unclear. Also contrary to these authors [9], I can find no change in strand breaks after 2 h incubation of resting murine or human lymphocytes with nicotinamide (A.P. Johnstone, unpublished).

To allow the lymphocyte data to be viewed in the context of changes from quiescence to growth in general, fibroblasts were also studied. Amino- and methoxybenzamide had no effect on either the normal proliferation of these cells or the serum-induced switch back to proliferation of cells that had become quiescent following serum starvation (A.P. Johnstone, unpublished). Furthermore, no changes in DNA strand breaks were detected by nucleoid analysis either upon cessation or restarting of growth (Table 1). These differences between lymphocytes and fibroblasts rule out the possibility that the phenomena are merely consequences of changes from quiescence to growth and support the concept that they are involved in eukaryotic differ-entiation, possibly related to changes in gene exression [1–4, 13].

Acknowledgments. I thank Gwyn Williams and Farzin Farzaneh for much helpful discussion, David Darling for technical assistance, Mike Neuberger for the gift of c-myc probe and Prof. Kaplan for showing me preprints of their work. Supported by grants from the Wellcome Trust and the MRC.

References

1. Farzaneh F, Shall S, Zalin R (1980) ADP-ribosylations and cell differentiation. In: Smulson M, Sugimura T (eds) Novel ADP-ribosylations of regulatory enzymes and proteins. Elsevier, Amsterdam, p 217
2. Farzaneh F, Shall S, Brill D, Zalin R (1982) DNA strand breaks and ADP-ribosyltransferase activation during cell differentiation. Nature (London) 300:362–366
3. Johnstone AP, Williams GT (1982) Role of DNA breaks and ADP-ribosyltransferase activity in eukaryotic differentiation demonstrated in human lymphocytes. Nature (London) 300:368–370
4. Johnstone AP (1984) Rejoining of strand breaks is an early nuclear event during the stimula-tion of quiescent lymphocytes. Eur J Biochem 140:401–406
5. Milam KM, Cleaver JE (1984) Inhibitors of poly(ADP-ribose) synthesis: effects on other meta-bolic processes. Science 223:589–591
6. Lehmann AR, Kirk-Bell S, Shall S, Whish WJD (1974) The relationship between cell growth, macromolecular synthesis and poly ADP-ribose polymerase in lymphoid cells. Exp Cell Res 83:63–72
7. Berger NA, Adams JW, Sikorski GW, Petzold SJ, Shearer WT (1978) Synthesis of DNA and poly(ADP-ribose) in normal and CLL lymphocytes. J Clin Invest 62:111–118
8. Rochette-Egly C, Ittel ME, Bilen J, Mandel P (1980) Effect of nicotinamide on RNA and DNA synthesis and on poly (ADP-ribose) polymerase activity in normal and PHA stimulated human lymphocytes. FEBS Lett 120:7–11
9. Greer WL, Kaplan JG (in press) Regulation of repair of naturally occurring DNA strand breaks in lymphocytes. Biochem Biophys Res Commun
10. Johnstone AP, Thorpe R (1982) Immunochemistry in practice. Blackwell, Oxford, pp 87–89
11. Kelly K, Cochran BH, Stiles CD, Leder P (1983) Cell-specific regulation of the c-myc gene by lymphocyte mitogens and PDGF. Cell 35:603–610
12. Greer WL, Kaplan JG (in press) A decrease in the steady state level of DNA strand breaks as a factor in the regulation of lymphocyte proliferation. In: Growth and the cell cycle: molecular, cellular and developmental biology. Humana, New York
13. Williams GT, Johnstone AP (1983) ADP-ribosyltransferase, rearrangement of DNA, and cell differentiation. Biosci Rep 3:815–830
14. Halldorsson H, Gray DA, Shall S (1978) Poly(ADP-ribose) polymerase activity in nucleotide permeable cells. FEBS Lett 85:349–352
15. Neuberger MS, Calabi F (1983) Reciprocal chromosome translocation between c-myc and IgG2b genes. Nature (London) 305:240–243
16. Williams GT, Lau KMK, Coote JM, Johnstone AP (in press) NAD metabolism and mitogen stimulation of human lymphocytes. Exp Cell Res
17. Johnstone AP, Darling DC (1985) Some early events in the primary mitogenic stimulation of lymphocytes differ from later stimulation and other quiescence to growth activation systems. Immunol 55:685–692

Specific Inhibition of the Appearance of a T-Lymphocyte Surface Marker in Thymus Organ Cultures

INTISAR H. MIRZA, ERIC J. JENKINSON, and GWYN T. WILLIAMS[1]

T-cells are the lymphocytes responsible for cell-mediated immunity, rather than directly for antibody secretion. They are the only lymphocytes which pass through critical stages of their development in the thymus. The developmental pathway which produces immunocompetent T-lymphocytes from their precursor cells has been extensively studied in the thymus of the foetal mouse [1].

The aquisition of competence by developing T-lymphocytes is preceded by changes in cell surface composition. These changes include the appearance of surface markers in a defined sequence [2]. The plasma membrane proteins on which these markers are found presumably play some part either in the immune response or in the process of development itself.

Whatever the exact roles played by cell surface changes, the recognition of many of the markers concerned by monoclonal antibodies has provided a powerful approach to the analysis of development in this system. The observation that the same developmental programme, including the production of immunocompetent T-cells, is followed in organ cultures of foetal thymus has considerably helped these investigations [1]. Thymic lobes can be removed from foetal mice after 14 days of gestation and maintained for several days in vitro. The precursor cells of T-lymphocytes both differentiate and proliferate during this period. Proliferation can be monitored by including [^{3}H]-thymidine in hanging-drop cultures and measuring the incorporation into acid-insoluble material during the last 15 h of culture. This incorporation provides a sensitive index of the proliferation rate during the entire in vitro culture period. Since the cells go through about six cell cycles in culture, 20% inhibition of the rate of proliferation would produce a reduction of 70% to 80% in the amount of DNA synthesis detected in the final 15 h of the incubation.

Figure 1 shows the effect of 3-methoxybenzamide, a potent inhibitor of ADP-ribosyl transferase [3], and cyclosporin A, an inhibitor of T-lymphocyte activation [4] on the incorporation of [^{3}H]-thymidine by thymic lobes after 5 days in vitro. In both cases a small but statistically significant reduction in incorporation was observed, indicating that both compounds produced a limited inhibition of proliferation at the concentrations used. This is in sharp contrast to the strong inhibition of the activation of mature T-cells produced by these compounds [4, 5].

---

1 Anatomy Department, University of Birmingham Medical School, Vincent Drive, Birmingham B15 2TJ, Great Britain

ADP-Ribosylation of Proteins
(ed. by F.R. Althaus, H. Hilz, and S. Shall)
© Springer-Verlag Berlin Heidelberg 1985

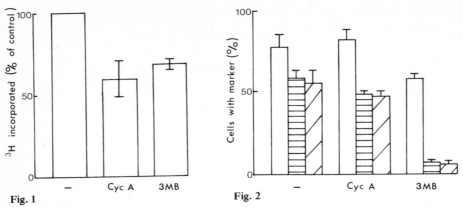

Fig. 1. Effect of 3-methoxybenzamide and cyclosporin A on thymidine incorporation by thymus organ cultures. Thymic lobes were removed from 14-day-old mouse embryos and cultured on the surface of a nucleopore filter on gelatin foam squares in Dulbecco's Modified Eagles's Medium supplemented with nonessential amino acids (Flow), 50 μM 2-mercaptoethanol and 10% foetal calf serum. After 5 days at 37°C, the lobes were transferred to hanging-drop cultures in Terasaki plates and [^{3}H]-thymidine was added to a final concentration of 2 μCi ml^{-1}. After a further 15 h at 37°C the thymic lobes were washed, disaggregated with trypsin-EDTA and harvested for determination of the level of incorporation of [^{3}H]-thymidine. In drug-treated cultures, either 1 μg ml^{-1} cyclosporin A or 2.5 mM 3-methoxybenzamide was included in the medium (both before and during the [^{3}H]-thymidine pulse). Means and SE of three experiments, each of which involved at least four thymuses, are shown

Fig. 2. Effect of 3-methoxybenzamide and cyclosporin A on Lyt-2 and Thy-1 expression by developing thymocytes. Thymus organ cultures with and without inhibitor were set up as described in the legend to Fig. 1. After 5 days of culture, the proportions of cells expressing Lyt-2 and Thy-1 were determined by immunofluorescence. Cells were stained with the monoclonal anti-Thy-1 antibody clone 30-H12 (Becton Dickinson) conjugated with biotin and the monoclonal anti-Lyt-2 antibody 53-6.7 (Becton Dickinson) conjugated with arsinilate. The second step reagents were fluorescein-conjugated avidin (which binds to biotin) and rhodamine-conjugated anti-arsinilate. Both markers could therefore be detected in the same preparation. Means and SE of three experiments are shown. *Clear bar:* % of cells expressing Thy-1. *Horizontally lined bar:* % of cells expressing Lyt-2. *Diagonally lined bar:* % of cells expressing Thy-1 and Lyt-2

The effects of cyclosporin A and 3-methoxybenzamide on the expression of the lymphocyte surface proteins Thy-1 and Lyt-2 were, however, clearly different (Fig. 2). Cyclosporin A had no effect on the distribution of either of the markers. 3-Methoxybenzamide also had only a limited effect on the proportion of cells expressing Thy-1, but dramatically reduced the fraction of cells expressing Lyt-2 from 52% to 7%. Since the inhibition of proliferation produced by the two compounds was similar, this could not have caused the specific inhibition of Lyt-2 expression produced by 3-methoxybenzamide.

The expression of antigens of the major histocompatibility complex by thymus stromal cells was not significantly affected by incubation with 3-methoxybenzamide, which again suggested that the inhibition of Lyt-2 expression was not due to nonspecific cytotoxic effects or to a block on total proteins synthesis.

These initial observations of the effect of 3-methoxybenzamide on Lyt-2 expression may be related to the specific inhibition of differentiation (including the expression of new genes) seen with myoblasts [6, 7], mature lymphocytes [5, 8, 9], hepatocytes in culture [10], *Trypanosoma cruzi* [11, 12], granulocyte-macrophage progenitors [13] and HL-60 myeloid leukemia cells (F. Farzaneh et al., pers. comm.). Our data therefore suggest that ADP-ribosyl transferase may be involved in the differentiation of T-lymphocyte precursors in the thymus, as appears to be the case for a wide range of eukaryotes [14]. The expression of Thy-1 in the absence of Lyt-2 may therefore be related to the earlier appearance of Thy-1 during development. The thymus cells of the 14 day foetus have advanced further in the process of expressing Thy-1 than in expressing Lyt-2. A proportion of the cells in this asynchronous population already express Thy-1 at day 14. 3-Methoxybenzamide may therefore inhibit a stage in the pathway leading to expression which has been passed for Thy-1 at day 14, but has not been passed for Lyt-2. Further investigation of the time-course of the susceptibility of these markers to inhibition should clarify this point.

The study of the suggested role of ADP-ribosyl transferase in thymic development may well prove valuable in the investigation of the molecular mechanisms of this particularly interesting stage of lymphocyte development. Thymus organ cultures provide an attractive experimental system for this investigation, since they allow the events which occur to be analysed in vitro.

Acknowledgments. We thank the MRC, the CRC and the Wellcome Trust for financial support.

References

1. Owen JJT, Jenkinson EJ (1981) Embryology of the lymphoid system. Prog Allergy 29:1–34
2. Owen JJT, Raff MC (1970) Studies on the differentiation of thymus-derived lymphocytes. J Exp Med 132:1216–1232
3. Purnell MR, Whish WJD (1980) Novel inhibitors of poly(ADP-ribose) synthetase. Biochem J 185:775–777
4. Britton S, Placios R (1982) Cyclosporin A – usefulness, risks and mechanisms of action. Immunol Rev 65:5
5. Johnstone AP, Williams GT (1982) Role of DNA breaks and ADP-ribosyl transferase activity in eukaryotic differentiation demonstrated in human lymphocytes. Nature (London) 300:368–370
6. Farzaneh F, Shall S, Zalin R (1980) DNA strand breaks and poly(ADP-ribose) polymerase activity during chick muscle differentiation. In: Smulson ME, Sugimura T (eds) Novel ADP-ribosylations of regulatory enzymes and proteins. Elsevier, Amsterdam, p 217
7. Farzaneh F, Zalin R, Brill D, Shall S (1982) DNA strand breaks and ADP-ribosyl transferase activation during cell differentiation. Nature (London) 300:362–366
8. Johnstone AP (1984) Rejoining of DNA strand breaks is an early nuclear event during the stimulation of quiescent lymphocytes. Eur J Biochem 140:401–406
9. Greer WL, Kaplan JG (1983) DNA strand breaks in murine lymphocytes: induction by purine and pyrimidine analogues. Biochem Biophys Res Commun 115:834–840

10. Althaus FR, Lawrence SD, He Y-Z, Sattler GL, Tsukada Y, Pitot H (1982) Effects of altered (ADP-ribose)$_n$ metabolism on expression of foetal functions by adult hepatocytes. Nature (London) 300:366–368
11. Williams GT (1983) *Trypanosoma cruzi:* inhibition of intracellular and extracellular differentiation by ADP-ribosyl transferase antagonists. Exp Parasitol 56:409–415
12. Williams GT (1984) Specific inhibition of the differentiation of *Trypanosoma cruzi.* J Cell Biol 99:79–82
13. Francis GE, Gray DA, Berney JJ, Wing MA, Guimares JET, Hoffbrand AV (1983) Role of ADP-ribosyl transferase in differentiation of granulocyte-macrophage progenitors to the macrophage lineage. Blood 62:1055–1062
14. Williams GT, Johnstone AP (1983) ADP-ribosyl transferase, re-arrangement of DNA, and cell differentiation. Biosci Rep 3:815–830

Transient Formation of DNA Strand Breaks During DMSO-Induced Differentiation of Promyelocytic Cell Line, HL-60

FARZIN FARZANEH[1], ROBERT A. LEBBY[2], DAVID BRILL[2], SYDNEY SHALL[2], JEAN-CLAUDE DAVID[3], and SYLVIE FÉON[3]

Introduction

We have previously shown that during the spontaneous induction of differentiation in primary avian myoblasts several hundred single-strand DNA breaks are formed in the genome [1]. This increase in the detectable number of DNA strand breaks is not due to a general deficiency in DNA repair mechanisms and γ-radiation induced breaks are as efficiently repaired in the post-mitotic nuclei of terminally differentiating myotubes as they are in the proliferating myoblasts. The formation of DNA strand breaks during myoblast differentiation is accompanied by an increase in ADPRT activity [1]. The nuclear ADPRT activity, which is totally dependent on the presence of DNA strand breaks, is required for the efficient ligation of DNA strand breaks [2]. Inhibition of this activity, either by enzyme inhibitors or substrate deprivation, blocks myoblast differentiation but not their proliferation [1]. Here, we report the transient formation of DNA strand breaks during the induced differentiation of the promyelocytic cell line, HL-60. Inhibition of ADPRT activity blocks the religation of these breaks which are formed in the face of a proficient DNA repair mechanism. ADPRT activity is also required for the excision repair of DNA strand breaks formed by either γ-irradiation or exposure to the mono-functional methylating agent dimethylsulphate (DMS).

DMSO Induction of Differentiation in HL-60 Cells

HL-60 cells are a human promyelocytic leukaemia cell line developed by Collins et al. [3], from a patient with acute promyelocytic leukaemia. The HL-60 cells in culture consists predominantly of promyelocytes (approximately 85%) and an smaller per-

1   Harris Birthright Research Center for Fetal Medicine, Department of Obstetrics and Gynaecology, Kings College School of Medicine, Denmark Hill, London SE5 8RX, Great Britain
2   Cell and Molecular Biology Laboratory, University of Sussex, Brighton BN1 9QG, Great Britain
3   Laboratoire de Biochimie du Development, Universite de Renne, LA CNRS 256, Campus de Beaulieu, 35402 Rennes, France

ADP-Ribosylation of Proteins
(ed. by F.R. Althaus, H. Hilz, and S. Shall)
© Springer-Verlag Berlin Heidelberg 1985

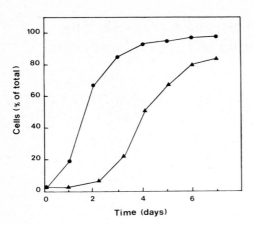

Fig. 1. DMSO induced differentiation of HL-60 cells. Nitro-blue-tetrazolium reduction (●), phagocytosis of opsonised latex particles (▲)

centage of myeoloblasts and myeolocytes, metamyelocytes and granulocytes. HL-60 cells can be induced to differentiate along either the granulocytic or macrophage lineages, depending on the choice of the inducing agent [4]. DMSO induces differentiation along the granulocytic lineage as judged by the morphological appearance, the ability to reduce nitro-blue-tetrazolium (NBT) and phagocytosis of latex particles coated with the human complement component c3b (Fig. 1).

DNA Repair and its Dependence on ADPRT Activity in HL-60 Cells

DNA strand breaks formed as a result of exposure of cells to low doses of either γ-radiation or DMS can be easily quantitated using the nucleoid gradients which contain a saturating concentration of ethidium bromide. The inclusion of ethidium bromide in the gradient has the dual advantage of both avoiding interference from changes in the superhelical structure of the DNA and the rapid and efficient detection of the position of the nucleoid band in the gradient by visualisation under UV light. This procedure can accurately measure DNA strand breaks formed by exposure to either 50 rad of γ-radiation (Fig. 2a) or 5 μM DMS (Fig. 2b).

There is no detectable difference in the rate of DNA repair between undifferentiated and differentiated HL-60 cells. DNA strand breaks formed by exposure of HL-60 cells to either 100 or 400 rad of γ-radiation is as efficiently repaired in both undifferentiated and differentiated cells (Fig. 3). Similarly, the rate of repair of DNA strand breaks formed by exposure to either 10 μM or 50 μM DMS is the same in both cells (Fig. 4).

The excision repair of both γ-radiation and DMS-induced breaks, in either undifferentiated or differentiated cells, can be blocked by the ADPRT inhibitors 3-aminobenzamide and 3-methoxybenzamide, but not by the acid analogues of these compounds (Fig. 5).

These observations demonstrate that there is no reduction in the DNA repair capacity of HL-60 cells after the induction of differentiation. Having established the pro-

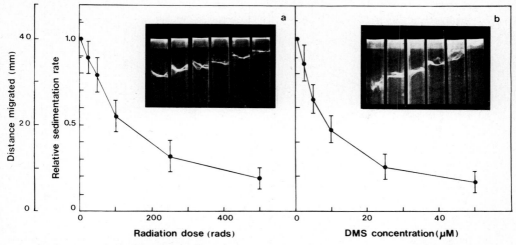

Fig. 2a,b. Estimation of DNA strand breaks by the nucleoid sedimentation procedure. **a** Sedimentation of nucleoids from HL-60 cells either exposed to γ-radiation; **b** treated with the indicated concentration of DMS for 15 min

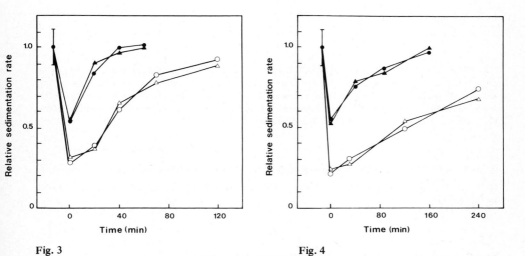

Fig. 3   **Fig. 4**

Fig. 3. Repair of DNA strand breaks induced by γ-radiation. Undifferentiated (●, ○) or differentiated (▲, △) cells were irradiated with 100 rad (*closed symbols*) or 400 rad (*open symbols*) of γ-radiation. Cells were then incubated at 37°C for the indicated times prior to being applied to the nucleoid gradients

Fig. 4. Repair of DNA strand breaks induced by DMS. Undifferentiated (●, ○) or differentiated (▲, △) cells were treated with either 10 μM (*closed symbols*) or 50 μM (*open symbols*) DMS for 15 min (defined as time 0 of repair). At indicated times cells were removed and their nucleoid sedimentation rate was determined

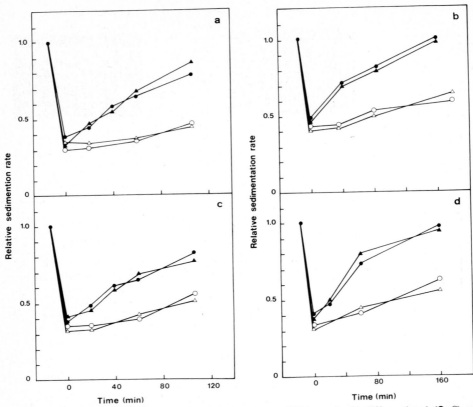

Fig. 5a–d. Effect of inhibition of ADPRT activity on DNA repair. Undifferentiated (●, ○) or differentiated (▲, △) HL-60 cells were either irradiated with 400 rad of γ-radiation (a, c) or treated with 10 μM DMS (b,d). Cells were then incubated in 5 mM 3-aminobenzamide (a, b *open symbols*), 5 mM 3-aminobenzoic acid (a, b *closed symbols*), 2 mM 3-methoxybenzamide (c, d *open symbols*), or 2 mM 3-methoxybenzoic acid (c, d *closed symbols*)

ficiency of DNA repair and its dependence on ADPRT activity, we examined the possible physiological formation of DNA strand breaks during the induced differentiation of HL-60 cells.

The Transient Formation of DNA Strand Breaks During HL-60 Differentiation

During the DMSO-induced differentiation of HL-60 cells along the granulocytic lineage DNA, strand breaks are formed. This is reflected in the reduced sedimentation rate of the nucleoids during the first 24 h of treatment with DMSO, as compared to the sedimentation rate of nucleoids from parallel uninduced cultures (Fig. 6). At this time

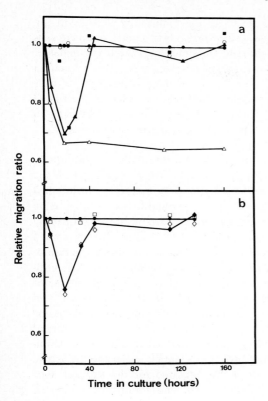

Fig. 6a,b. Transient formation of DNA strand breaks during DMSO-induced differentiation of HL-60 cells. Nucleoid sedimentation rate in uninduced control cultures (●) and in DMSO-induced cultures (▲, ■, ◆). a Control (○) or DMSO-treated cultures (△) in the presence of 2 mM 3-methoxybenzamide. b Control (□), and DMSO-treated cultures (◇) in the presence of 2 mM 3-methoxybenzoic acid

very few cells are capable of reducing nitro-blue-tetrazolium, which is one of the earliest markers of differentiation in these cells (see Fig. 1). These breaks are only transiently maintained and are religated during the next 24 h. Inhibition of ADPRT activity by 3-methoxybenzamide blocks the religation of these physiologically formed DNA strand breaks. The addition of the same concentration of 3-methocybenzamide to uninduced cells does not alter the sedimentation rate of the nucleoids. 3-methoxybenzoic acid does not affect either the formation of removal of these breaks in the DMSO-treated cells and has no effect on the sedimentation rate of the nucleoids from uninduced cultures.

Discussion

The data presented here demonstrate that the induction of differentiation by DMSO in the human promyelocytic leukaemia cells, HL-60, along the granulocytic lineage does not retard the ability of these cells to repair DNA lesions induced by exposure to either ionising radiation or treatment with the mono-functional methylating agent, DMS. This is consistent with our previous observation that post-mitotic primary muscle cell nuclei undergoing terminal differentiation are as efficient as their proliferating precursor myoblasts in repairing DNA damage induced by γ-radiation [1]. The present studies also demonstrate that in both proliferating, undifferentiated cells and in the post-mitotic cells induced to differentiate along the granulocytic lineage, ADPRT activity is required for the efficient repair of DNA damage induced by either DMS or γ-radiation. This is an interesting observation, because Kanai et al. [5] have reported a decrease in both the basal and potential activity of ADPRT after the DMSO induction of differentiation in HL-60 cells. However, these authors do detect an eight-fold increase in the cellular level of poly(ADP-ribose) in the HL-60 cells induced to differentiate by DMSO [5].

The studies reported here also demonstrate the formation of DNA strand breaks during the DMSO-induced differentiation of HL-60 cells. These breaks, which are formed despite a proficient DNA repair mechanism, are only transiently maintained and then religated. The ligation of these physiologically formed breaks is blocked by the inhibition of ADPRT activity, as is the ligation of DNA strand breaks formed by exposure to DMS or γ-irradiation. The formation of these DNA strand breaks is not a passive consequence of the DMSP treatment but the induction of differentiation in these cells. In mouse lymphoma cells, L1210, which can not be induced to differentiate, DMSO treatment does not result in the formation of DNA strand breaks (data not shown). DNA strand breaks are also formed during the cellular differentiation in mouse erythroleukaemic cells, induced by a number of different agents, including DMSO [6, 7]. Resting human peripheral blood lymphocytes contain DNA strand breaks which are ligated following their mitogen-stimulated activation [8, 9]. Inhibition of ADPRT activity blocks the ligation of these DNA strand breaks and the activation of lymphocytes [8, 9]. ADPRT activity is also required for the expression of foetal functions in cultured adult rat hepatocytes [10] and the differentiation of primary bone marrow granulocyte-macrophage progenitor cells (CFU-gm) [11], primary avian myoblast [1] and *Trypanosoma cruzi* [12]. At present it is not clear whether this obligatory requirement for ADPRT activity is restricted only to the nuclear enzyme or whether the cytoplasmic enzyme is also involved. However, the inhibition of ADPRT activity does not affect cellular proliferation in any of these cells. This suggests that ADPRT activity is required for the expression of new genes only and not for the control of gene expression during the cell cycle. Therefore, DNA strand break formation and rejoining, modulated by ADPRT activity, may be involved in the developmental control of cellular differentiation. This may be due to the possible involvement of gene amplification, mobility in the genome and/or regional chromatin relaxation mediated by the action of topoisomerases in the developmental regulation of gene expression.

References

1. Farzaneh F, Zalin R, Brill D, Shall S (1982) DNA strand breaks and ADP-ribosyl transferase activation during cell differentiation. Nature (London) 300:362–366
2. Durkacz WD, Omidiji O, Gray DA, Shall S (1980) (ADP-ribose)$_n$ participates in DNA excision repair. Nature (London) 283:593–596

3. Collins SJ, Ruscetti FW, Gallagher RE, Gallo RC (1979) Normal functional characteristics of cultivated human promyelocytic leukemia cells (HL60) after induction of differentiation by dimethylsulfoxide. J Exp Med 149:969–974

4. Abraham J, Rovera J (1981) In: Baserga R (ed) Handbook of experimental pharmacology, vol 57. Springer, Berlin Heidelberg New York, p 405

5. Kanai M, Miwa M, Kondo T, Tanaka Y, Nakayasu M, Sugimura T (1982) Involvement of poly-(ADP-ribose) metabolism in induction of differentiation of HL-60 promyelocytic leukemia cells. Biochem Biophys Res Commun 105:404–411

6. Terada M, Nudel U, Fibach E, Rifkind RA, Marks PA (1978) Changes in DNA associated with induction of differentiation by dimethyl sulfoxide in murine erythroleukemia cells. Cancer Res 38:835–840

7. Scher W, Friend C (1978) Breakage of DNA and alterations in folded genomes by inducers of differentiation in Friend erythroleukemic cells. Cancer Res 38:841–849

8. Johnstone AP, Williams GT (1982) Role of DNA breaks and ADP-ribosyl transferase activity in eukaryotic differentiation demonstrated in human lymphocytes. Nature (London) 300: 368–370

9. Greer WL, Kaplan JG (1983) DNA strand breaks in murine lymphocytes: induction by purine and pyrimidine analogues. Biochem Biophys Res Commun 115:834–840

10. Althaus FR, Lawrence SD, He Y-Z, Sattler GL, Tsukada Y, Pitot HC (1982) Effects of altered (ADP-ribose)$_n$ metabolism on expression of fetal functions by adult hepatocytes. Nature (London) 300:366–368

11. Francis GE, Gray DA, Berney JJ, Wing MA, Guimaraes JET, Hoffbrand AV (1983) Role of ADP-ribosyl transferase in differentiation of human granulocyte-macrophage progenitors to the macrophage lineage. Blood 62:1055–1062

12. Williams GT (1983) *Trypanosoma cruzi*: Inhibition by ADP-ribosyl transferase antagonists of intracellular and extracellular differentiation. Exp Parasitol 56:409–415

Differentiation of a Myoblast Cell Line and Poly(ADP-Ribosyl)ation

HAIM HACHAM and RUTH BEN-ISHAI[1]

1. Introduction

Myogenesis in vivo and in vitro involves conversion of cycling and undifferentiated myoblasts to terminally differentiated and noncycling myotubes [1, 2]. The sequence of events associated with differentiation of muscle cells in culture has been well defined and involves: myoblast proliferation to confluency, cessation of DNA synthesis with irreversible withdrawal of myoblasts from the cell cycle, plasma membrane fusion and formation of multinucleated myotubes, and induction of muscle specific proteins and isoenzymes (e.g., myosin, α-actin, α-tropomysin, creatine phosphokinase, etc.) at the time of onset of fusion [3—5].

At present the events triggering and the mechanism controlling muscle differentiation are not well understood; they may include changes in chromatin structure, gene amplification, and specific DNA and/or protein modification. One such modification, namely, poly(ADP-ribosyl)ation of nuclear proteins has recently attracted much attention and several studies have indicated that the enzyme poly(ADP-ribose) synthetase plays a role in cellular differentiation [6—9]. This enzyme is a nuclear chromatin-associated protein which catalyzes covalent modification of both histone and non-histone protein acceptors (for reviews see [10—13]). The synthetase is activated by DNA strand breaks and it has been suggested that DNA fragmentation and the consequent increase in poly(ADP-ribose) activity are obligatory events for chick muscle differentiation [6].

The purpose of this investigation has been to test whether myogenesis is accompanied by, or dependent upon, modification of specific nuclear proteins by poly(ADP-ribose).

1   Department of Biology, Technion – Israel Institute of Technology, Haifa 32000, Israel

ADP-Ribosylation of Proteins
(ed. by F.R. Althaus, H. Hilz, and S. Shall)
© Springer-Verlag Berlin Heidelberg 1985

2. Results

2.1 The Myogenic Cell Line System

The isolation of established lines of myoblasts which retain the ability to fuse and form myotubes has provided a very useful system for studying muscle differentiation [14]. These cells display the same characteristic morphological and biochemical changes as freshly explanted muscle cells in culture and offer the advantage that pure populations of myoblasts and myotubes can be obtained.

We have examined the involvement of poly(ADP-ribose) in myogenesis using a highly myogenic clone (E63) which had been isolated by Kaufman et al. from the L8 rat skeletal muscle cell line [15]. Usually 7×10^5 myoblasts were plated in 100 mm dishes and grown in DME medium containing 10% horse serum. Under these conditions proliferation of myoblasts slows down on day 4, and cultures become confluent with replication being essentially turned off on day 5. Fusion and synthesis of muscle specific proteins start on day 6–7 and increases until day 10–12 when > 90% of cell nuclei are in multinucleated myotubes.

2.1.1 Poly(ADP-Ribose) Synthetase Inhibitors

Preliminary experiments using permeabilized E63 myoblasts have shown that as observed for chick muscle cells [6], poly(ADP-ribose) synthetase activity increases threefold during myoblast fusion. E63 myoblasts also behave similarly to the primary chick myoblasts in their response to inhibitors of poly(ADP-ribose) synthetase. As monitored by cell fusion, nicotinamide, and 3-aminobenzamide at concentrations of 10 mM almost completely inhibit differentiation. At this concentration the inhibitors have no apparent adverse effect on E63 myoblast proliferation. Inhibition of fusion was observed if either the inhibitor was present continuously, or if it was added up to day 5 after the initial plating of myoblasts (shown for nicotinamide in Fig. 1). Addition of nicotinamide at days 4, 5, and 5 1/2 reduced fusion compared with control cultures, to the extent of 80, 85, and 30%, respectively. These results indicate that highest sensitivity to the inhibitor occurs prior to fusion when mononucleated cells are undergoing or have undergone their last presumptive replication cycle. It is noteworthy that inhibition of cell fusion is reversible (Fig. 1F) since myoblasts prevented from undergoing fusion by prolonged exposure to nicotinamide remain differentiation competent and undergo fusion upon removal of inhibitor.

2.1.1.1 Poly(ADP-Ribose) Modified Proteins. Poly(ADP-ribose) appears to be involved as a modifier of chromatin-associated proteins in several cellular processes (reviewed in [12]). To determine the role of poly(ADP-ribose) in myogenesis we attempted to identify the acceptor proteins which undergo this modification, and to define the specificity of such modification.

Poly(ADP-ribose) modified proteins were isolated from E63 cells at different stages of myogenesis. Isolation and analysis of these proteins was performed essentially as described by Adamietz et al. [16] using boronate cellulose chromatography and per-

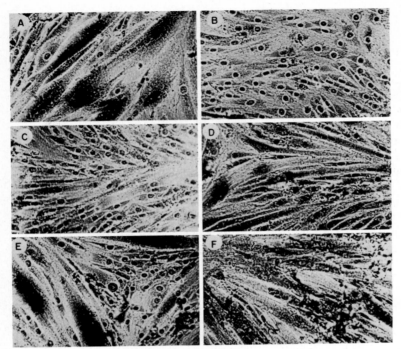

Fig. 1A–F. Effect of nicotinamide on cell fusion and myotube formation. A–E E63 myoblasts were exposed to nicotinamide (10 m*M*) and examined microscopically for fusion at day 11 in culture. **A** Control without nicotinamide; **B–E** nicotinamide added at day 3, 5, 5 1/2, and 6, respectively. **F** Cells exposed to nicotinamide from day 3–8 were examined for fusion 5 days after inhibitor had been removed

forming electrophoresis in 8% polyacrylamide gels at pH 6 in the presence of 4.5 *M* urea [17]. Proteins were detected on the gels by silver stain [18]. Figure 2 shows electrophoretic separation of the boronate cellulose-bound proteins isolated from E63 cells maintained for 4, 5, 6, and 10 days in culture. It is evident that several proteins become modified and that changes in the pattern of poly(ADP-ribose) modified proteins accompany myogenesis.

The most notable changes observed involve a protein that comigrates with the 116 kD marker; this modified protein is present in the 4 and 5 day cultures, but absent at the time of onset of fusion (6 days) and in multinucleated myotubes (10 days). Concomitantly with the disappearance of the 116 kD protein on day 6 a slower migrating component, which is present in prefusion cultures (< 200 kD in Fig. 3), appears to increase progressively in molecular mass and abundance. This protein may represent an acceptor which is modified during differentiation by ADP-ribose chains of greater length. Other notable changes include a very broad band at the position of approximately 54–70 kD which increases in amount during differentiation. Alkaline treatment, which releases poly(ADP-ribose) residues from the conjugates [16] showed, however, that this band is contaminated with a non-ADP ribosylated protein.

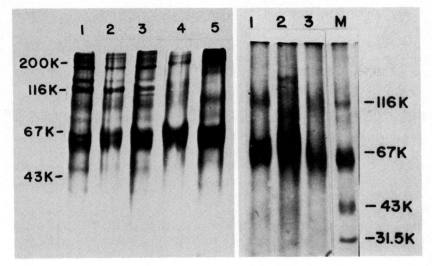

Fig. 2   Fig. 3

Fig. 2. Electrophoretic analysis of poly(ADP-ribose) modified proteins of E63 cells at different stages of myogenesis. Proteins of boronate cellulose-bound fraction isolated from E63 muscle cells maintained in culture for (*lane 1*) 4 days; (*lanes 2 and 3*) 5 days; (*lane 4*) 6 days; and (*lane 5*) 10 days

Fig. 3. Comparison of poly(ADP-ribosyl)ated proteins isolated from E63 muscle cells and a non-fusing variant (Nf-1). (*Lane 1*) prefusion E63 myoblasts (5 days); (*lane 2*) differentiated E63 cells (8 days); (*lane 3*) Nf-1 cells maintained for 8 days in culture

It may be asked whether the changes in the pattern of poly(ADP-ribose) modified proteins are specific for myoblast differentiation. To answer this question we tested a myoblast clone that had lost its capacity to fuse. This clone (Nf-1) was isolated from the E63 myoblast cell line [19]. Cultures of Nf-1 were maintained for 8 days and were analyzed for poly(ADP-ribose) modified proteins. Figure 3 shows that the pattern of modified proteins isolated from the nonfusing variant is similar to that of prefusion E63 myoblasts (5 days). The 116 kD modified protein is present in both cultures, whereas the changes in poly(ADP-ribose) acceptor(s) accompanying differentiation are not observed. We have recently observed that the 116 kD modified protein is also formed if differentiation of E63 myoblasts is inhibited by DMSO or UV light.

3. Discussion

The present study represents an attempt to characterize the proteins undergoing poly-(ADP-ribose) modification in association with myogenesis. Although we have not yet performed analysis of poly(ADP-ribose) conjugates of histones which are the main acceptors in many systems [20, 21], several interesting observations have emerged from our study.

During the limited period in which prefusion myoblasts are sensitive to inhibitors of poly(ADP-ribose) synthetase, two significant changes involving modification of acceptor proteins were observed. A modified protein of 116 kD disappears and a slower migrating component becomes increasingly modified and abundant. That the disappearance of the poly(ADP-ribose)modified 116 kD protein accompanies differentiation is also indicated by our results with a nonfusing myoblast variant.

It is feasible that the 116 kD conjugate represents an automodified form of the poly(ADP-ribose) synthetase. Automodification of this enzyme has been observed in vitro and it has been reported that such modification decreases enzyme activity [22]. The presence of an automodified form of the enzyme in prefusion myoblasts could account for the low poly(ADP-ribose) synthetase activity of these cells. On the other hand, enhanced enzyme activity concurrent with the onset of fusion [6, 7] could be directly related to lack of such automodification.

The events triggering the observed changes in the pattern of poly(ADP-ribose) modified proteins could result from an alteration in the structure of chromatin during myogenesis. Such alteration may affect the accessibility of nuclear acceptors and shift the equilibrium from modification of the 116 kD protein to that of "differentiation associated" acceptors of poly(ADP-ribose). Identification of the poly(ADP-ribose) modified acceptors will help elucidate their role in myogenesis.

Acknowledgments. We are indebted to Drs. D. Yaffe and S.J. Kaufman for providing cell lines. This research was supported in part by Technion J. and A. Taub Biological Research Fund.

References

1. Okazaki K, Holtzer H (1966) Mogenesis, fusion, myosin synthesis and the mitotic cycle. Proc Natl Acad Sci USA 56:1484–1488
2. Yaffe D (1974) Developmental changes preceding cell fusion during muscle differentiation in vitro. Exp Cell Res 66:33–43
3. Merlie JP, Buckingham ME, Whalen RG (1977) Molecular aspects of myogenesis. Curr Top Dev Biol 77:61–114
4. Nadal-Ginard B (1978) Commitment, fusion and biochemical differentiation of a myogenic cell line in the absence of DNA synthesis. Cell 15:855–864
5. Shani M, Zevin Sonkin D, Saxel O, Carmon Y, Katcoff D, Nudel U, Yaffe D (1981) The correlation between the synthesis of skeletal muscle actin, myosin heavy chain, and myosin light chain and the accumulation of corresponding mRNA sequences during myogenesis. Dev Biol 86:483–492
6. Farzaneh F, Zalin R, Brill D, Shall S (1982) DNA strand breaks and ADP-ribosyl transferase activation during cell differentiation. Nature (London) 300:362–366
7. Cherney BW, Midura RJ, Caplan AI (1982) Poly (ADP-ribose) and the differentiation of embryonic tissue. In: Hayaishi O, Ueda K (eds) ADP-ribosylation reactions, biology and medicine. Academic Press, London New York, pp 389–404
8. Johnstone AP, Williams GT (1982) Role of DNA breaks and ADP-ribosyltransferase activity in eukaryotic differentiation demonstrated in human lymphocytes. Nature (London) 300:368–370
9. Althaus FR, Lawrence SD, He YZ, Sattler GL, Tsukada Y, Pitot HC (1982) Effect of altered (ADP-ribose)n metabolism on expression of fetal functions by adult hepatocytes. Nature (London) 300:366–368

10. Sugimura T (1973) Poly(adenosine diphosphate ribose). Prog Nucleic Acid Res Mol Biol 13: 127–151
11. Hilz H, Stone PR (1976) Poly(ADP-ribose) and ADP-ribosylation of proteins. Rev Physiol Biochem Pharmacol 76:1–58
12. Mandel P, Okazaki H, Niedergang C (1982) Poly(adenosine diphosphate ribose). Prog Nucleic Acid Res Mol Biol 27:1–51
13. Hayaishi O, Ueda K (1977) Poly(ADP-ribose) and ADP-ribosylation of proteins. Annu Rev Biochem 46:95–116
14. Yaffe D (1968) Retention of differentiation potentialities during prolonged cultivation of myogenic cells. Proc Natl Acad Sci USA 61:477–483
15. Kaufman SJ, Parks CM, Bohn J, Faiman LE (1980) Transformation is an alternative to normal skeletal muscle development. Exp Cell Res 125:333–349
16. Adamietz P, Klapproth K, Hilz H (1979) Isolation and partial characterization of the ADP-ribosylated nuclear proteins from Ehrlich ascites tumor cells. Biochem Biophys Res Commun 91:1232–1238
17. Holtlund J, Kristensen T, Ostvold AC, Laland SG (1983) ADP-ribosylation in permeable HeLa S3 cells. Eur J Biochem 130:47–51
18. Porro M, Viti S, Antoni G, Saletti M (1982) Ultrasensitive silver stain for the detection of proteins in polyacrylamide gels and immunoprecipitates on agarose gels. Anal Biochem 127: 316–321
19. Hacham H (1981) Effect of UV irradiation on muscle cell differentiation in vitro. Thesis, Technion-Israel Inst Technol, Haifa
20. Wong M, Kanai Y, Miwa M, Bustin M, Smulson M (1983) Immunological evidence for the in vivo occurrence of a crosslinked complex of poly(ADP-ribosylated) histone H1. Proc Natl Acad Sci USA 80:205–209
21. Kreimeyer A,Wielckens K, Adamietz P, Hilz H (1984) DNA Repair associated ADP-ribosylation in vivo. Modification of histone H1 differs from that of the principal acceptor proteins. J Biol Chem 259:890–896
22. Kawaichi M, Ueda K, Hayaishi O (1981) Multiple autopoly(ADP-ribosyl)ation of rat liver poly(ADP-ribose) synthetase. Mode of modification and properties of automodified synthetase. J Biol Chem 256:9483–9489

Inhibitors of Poly(ADP-Ribose) Polymerase Prevent Friend Cell Differentiation

TIM BRAC and KANEY EBISUZAKI[1]

1. Introduction

Friend et al. [1] made the striking observation that dimethyl sulfoxide (DMSO)-treated Friend erythroleukemia cells (FEL) differentiate in vitro. Subsequently, a number of other chemicals, including butyric acid, hypoxanthine and hexamethylene bis-acetamide were shown to induce Friend cells [2]. These inducers appear to remove a block in the differentiation process but the mechanisms involved are unknown. The finding that benzamide and nicotinamide induced Friend cells [3, 4] suggested that since both compounds were inhibitors of poly(ADP-ribose) polymerase, poly(ADP-ribosylation) may have a role in the differentiation process. Furthermore, since poly(ADP-ribose) polymerase requires DNA strand breaks for activity [5], these observations implicated DNA strand breaks in FEL differentiation.

The possible involvement of DNA strand breaks and poly(ADP-ribosylation) reactions has been indicated in other differentiating cell types such as myoblasts and lymphocytes [6—8] as well as in DNA repair [9]. However, the existence of DNA strand breaks in differentiating Friend cells [10, 11] is controversial since recent studies have shown that induction does not result in detectable strand breaks [12—15].

Our studies have been concerned with the effect of inhibitors of poly(ADP-ribose) polymerase on the differentiation of Friend cells. We have also investigated the question of DNA strand breaks, particularly as it relates to poly(ADP-ribose) polymerase inhibitors and DNA repair.

2. Materials and Methods

FEL clone 745A was cultured at $37°C$ in Iscove's modified Dulbecco medium supplemented with 10% fetal calf serum (Gibco), penicillin (50 U ml^{-1}) and streptomycin (0.5 μg ml^{-1}) in a humidified atmosphere with 5% CO_2. Logarithmically growing cells, seeded at 2×10^5 ml^{-1}, doubled in approx. 11 h.

---

1 Cancer Research Laboratory, University of Western Ontario, London, Ontario, Canada N6A 5B7

ADP-Ribosylation of Proteins
(ed. by F.R. Althaus, H. Hilz, and S. Shall)
© Springer-Verlag Berlin Heidelberg 1985

Benzidine staining for hemoglobin was assayed by mixing cells with 1/10 volume of 0.2% benzidine dihydrochloride in 0.05 M acetic acid and 0.0012% hydrogen peroxide. Benzidine-positive cells appeared bright blue and at least 200 cells were counted for each analysis.

Hemoglobin was extracted [16] and estimated spectrophotometrically using bovine hemoglobins as a standard [17].

Logarithmically growing Friend cells were labeled with [^{14}C]-thymidine (0.03 μCi ml^{-1}) for 16 h, then diluted into fresh medium containing test chemicals. Subsequently these cells were removed at various times following the treatment and analyzed for DNA strand breaks using the alkaline elution method [18].

3. Results

Measurement of FEL differentiation by benzidine staining [17] showed that 3-amino-benzamide (3AB) inhibited spontaneous and DMSO-induced differentiation at concentrations that did not significantly inhibit cell growth (Table 1). Under all of the test conditions 3AB and 3-methoxybenzamide (3MB) did not induce differentiation. At concentrations of 3AB that partially inhibited cell growth (20 mM), differentiation was completely inhibited (Table 1).

Since these results appear to contradict those reported previously [3, 4], we re-examined the effects of benzamide on Friend cell differentiation. As shown previously [3] at higher concentrations, benzamide was a weak inducer (Table 2). However, at concentrations which did not inhibit cell growth, benzamide inhibited induction of FEL differentiation (Table 2).

These results indicate that some inhibitors of poly(ADP-ribose) polymerase also function as inhibitors of differentiation of Friend cells. Since some models of differentiation have invoked DNA rearrangements [8] and since DNA strand breaks are required for poly(ADP-ribose) polymerase activity [5], we examined the relationship of inducers of differentiation and poly(ADP-ribose) polymerase inhibitors for DNA

Table 1. Inhibitors of poly(ADP-ribose) polymerase prevent Friend cell differentiation[a]

DMSO (%)	Inhibitor	(mM)	% Benzidine +
—	—		1
—	3AB	10	0.01
1.5	3AB	0	82
1.5	3AB	2.5	31
1.5	3AB	5.0	16
1.5	3AB	10	7
1.5	3AB	20	0.01
1.5	3MB	5	9

[a] Friend cell differentiation was examined by benzidine staining 5 days after culturing of the cells with inducers and inhibitors

Table 2. Benzamide can inhibit and induce Friend cell differentiation[a]

DMSO (%)	Benzamide (mM)	% Benzidine +
–	–	1
–	5	1
–	20	12
1.5	–	77
1.5	5	42

[a] The analysis is described in Table 1

strand breaks. DMSO did not induce DNA strand breaks after the first 24 h of induction although a 24 h exposure resulted in a twentyfold increase in benzidine-positive cells (Fig. 1). Analysis of DNA strand breaks by alkaline elution, showed no difference between controls and DMSO-treated cells at 0, 2, 4, 6, 8 and 12 h (data not shown). Addition of 10 mM 3AB together with DMSO (Fig. 1) or the addition of 3AB separately did not cause an increase in DNA strand breaks (data not shown). Sodium butyrate and N'-methylnicotinamide did not induce single-strand breaks in DNA at 12 or 24 h of incubation (data not shown). Inducers of FEL differentiation and inhibitors of poly(ADP-ribose) polymerase did not cause detectable DNA strand breaks under the experimental conditions used in these studies.

If poly(ADP-ribose) polymerase was involved in differentiation, would activation of the polymerase by DNA damaging agents alter the course of differentiation? The alkylating agent, dimethyl sulfate, caused DNA single-strand breaks in Friend cells which were repaired (Fig. 2). Addition of 10 mM 3AB potentiated the DNA damage at each time-point examined. To confirm the observations that dimethyl sulfate increased the number of benzidine-positive cells (about 3% compared to 1% for controls), hemoglobin was also measured spectrophotometrically from cell extracts.

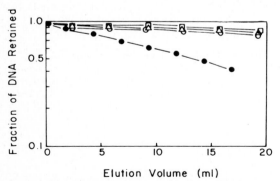

Elution Volume (ml)

Fig. 1. Dimethyl sulfoxide does not induce DNA single-strand breaks in Friend cells. Friend cells (10⁶), prelabeled with [^{14}C]-thymidine, were treated with 1.5% DMSO 24 h plus (△) and minus 3AB (□) and analyzed by alkaline elution. Control cells had no treatment (○), and 100 rad of gamma rays from ^{137}Cesium (●). The elution profiles of Friend cell DNA treated DMSO for 0, 2, 4, 6, 8 and 12 h were not different from the controls

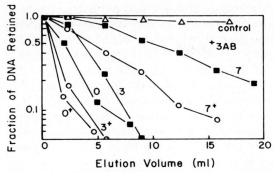

Fig. 2. Dimethyl sulfate causes DNA single-strand breaks in Friend cells that are potentiated by 3AB. [^{14}C]-thymidine labeled Friend cells, treated with 50 μM dimethyl sulfate for 30 min, were centrifuged and resuspended in fresh media with or without 3AB. Samples were removed at 0, 3 and 7 h after treatment and analyzed by alkaline elution. *Numbers* refer to h and + to presence of 3AB

Table 3. Dimethyl sulfate is a weak inducer of Friend cell differentiation[a]

Treatment	μg Hemoglobin/10^7 cells
DMS	7.2
DMS + 10 mM 3AB	3.4
10 mM 3AB	1.9
1.5% DMSO	45.0
1.5% DMSO + 10 mM 3AB	11.3

[a] Friend cells were treated with 50 μM dimethyl sulfate (DMS) for 30 min ± 10 mM 3AB. Parallel cultures were induced with 1.5% DMSO and 10 mM 3AB. After 5 days the cells were counted, lysed and centrifuged. The supernatants were analyzed for hemoglobin content. Uninduced cells contained 3.8 μg hemoglobin 10^{-7} cells

Induction of differentiation measured by analysis of hemoglobin showed that dimethyl sulfate was a weak inducer (Table 3). Addition of 3AB together with dimethyl sulfate inhibited FEL differentiation.

4. Discussion

ADP-ribosylation reactions may be involved in DNA repair and differentiation [6—9]. However, the role of poly(ADP-ribose) polymerase has been often inferred through the use of inhibitors such as 3AB. In most studies using inhibitors of poly(ADP-ribose) polymerase it was assumed that the inhibitors function in vivo as they do in vitro, however, the possibility of side effects [19] suggest caution in these interpretations.

For example, since benzamide and nicotinamide induce Friend cell differentiation, it has been suggested that induction resulted from the inhibition of poly(ADP-ribose) polymerase [3, 4]. However, these investigators noted that the inhibition of poly(ADP-ribose) polymerase was not a prerequisite of inducers since N'-methylnicotinamide, a strong inducer, was not an inhibitor of this enzyme in vitro [3]. We have found that 3AB inhibited FEL differentiation while benzamide had dual effects, inducing at high concentrations and blocking differentiation at low concentrations (see Results).

These observations could be understood if two different reactions were involved in prevention and induction of differentiation. Benzamide at high concentrations might function as an inducer of Friend cell differentiation based on its polar-planar structure [20]. Inducers are typically most active at high concentrations. At lower concentrations, the primary effect of benzamide might be through its inhibition of poly(ADP-ribose) polymerase [21]. The dual effect of induction and inhibition of differentiation is a property of other FEL inducers such as bromodeoxyuridine [22].

Inhibitors of poly(ADP-ribose) polymerase prevent differentiation in other cells [6, 7]. The previous reports showing that benzamide was an inducer of FEL differentiation suggested that poly(ADP-ribose) polymerase might have a different type of role in Friend cells. However, our results show that Friend cell differentiation is not exceptional and that common mechanisms may be involved in differentiation of diverse cell types.

If poly(ADP-ribose) polymerase was involved in the differentiation of Friend cells, then the role of DNA strand breaks must be examined since strand breaks modulate enzyme activity [5]. In a provocative model for differentiation, it has been suggested that DNA transposition reactions accompanied by DNA strand breaks are involved in gene activation [8]. In turn, these DNA strand breaks might trigger ADP-ribosylation reactions. We have found that treatment of Friend cells with dimethyl sulfate, which causes strand breaks, weakly induces FEL differentiation. However, the addition of 3AB blocked the differentiation process despite the increase in strand breaks. Therefore, DNA strand breaks by themselves do not stimulate differentiation but they may contribute to the differentiation process by activating poly(ADP-ribose) polymerase.

Several studies have shown that inducers of Friend cells increased DNA strand breaks when assayed by sucrose density gradient centrifugation [10, 11]. However, strand breaks were not detected when alternate analytical methods were used [12–15]. Our study shows that the treatment of Friend cells with DMSO, sodium butyrate or N'-methylnicotinamide separately or with 3AB did not induce DNA strand breaks. If DNA strand breaks are involved in differentiation of Friend cells, these breaks are either very transient or few in number.

In summary, we suggest that poly(ADP-ribosylation) reactions are needed for differentiation of Friend cells. This requirement might reflect the need for poly(ADP-ribosylation) reactions for modulating DNA-protein interactions by altering histones or HMG proteins [23] or by affecting enzymes such as DNA toposiomerase I [24]. The consequence resulting from an interference of poly(ADP-ribosylation) reactions may be far-reaching, affecting such diverse functions as differentiation and DNA repair.

Acknowledgments. We thank the NCI and MRC of Canada for financial support and D. Marsh for preparing the figures.

References

1. Friend C, Scher W, Holland JG, Sato T (1971) Hemoglobin synthesis in murine virus-induced leukemic cells in vitro: stimulation of erythroid differentiation by dimethyl sulfoxide. Proc Natl Acad Sci USA 68:378–382
2. Marks PA, Rifkind RA (1978) Erythroleukemic differentiation. Annu Rev Biochem 47:419–498
3. Terada M, Fujiki H, Marks PA, Sugimura T (1979) Induction of erythroid differentiation of murine erythroleukemic cells by nicotinamide and related compounds. Proc Natl Acad Sci USA 76:6411–6414
4. Morioka K, Tanaka K, Nokuo T, Ishizawa M, Ono T (1979) Erythroid differentiation and poly-(ADP-ribose) synthesis in Friend leukemia cells. Gann 70:37–46
5. Benjamin RC, Gill DM (1980) Poly (ADP-ribose) synthesis in vitro programmed by damaged DNA. A comparison of DNA molecules containing different types of strand breaks. J Biol Chem 255:10502–10508
6. Farzaneh F, Zalin R, Brill D, Shall S (1982) DNA strand breaks and ADP-ribosyl transferase activation during cell differentiation. Nature (London) 300:362–366
7. Johnstone AP, Williams GT (1982) Role of DNA breaks and ADP-ribosyl transferase activity in eukariotic differentiation demonstrated in human lymphocytes. Nature (London) 300:368–370
8. Williams GT, Johnstone AP (1983) ADP-ribosyl transferase, rearrangement of DNA, and cell differentiation. Biosci Rep 3:815–830
9. Shall S (1982) ADP-ribose in DNA repair. In: Hayaishi O, Ueda K (eds) ADP-ribosylation reactions in biology and medicine. Academic Press, London New York, p 478
10. Scher W, Friend C (1978) Breakage of DNA and alterations in folded genomes by inducers of differentiation in Friend erythroleukemic cells. Cancer Res 38:841–849
11. Terada M, Nudel U, Fibach E, Rifkind RA, Marks PA (1978) Changes in DNA associated with induction of erythroid differentiation by dimethyl sulfoxide in mouse erythroleukemic cells. Cancer Res 38:835–840
12. Pantazis P, Erickson LC, Kohn K (1981) Preservation of DNA integrity in human and mouse leukemic cells induced to terminally differentiate by chemical agents. Dev Biol 86:55–60
13. Wintersberger E, Mudrak I (1982) Butyrate does not induce single strand breaks in Friend erythroleukemic cells or in 3T6 mouse fibroblasts. FEBS Lett 138:218–220
14. Pulito VL, Miller DL, Sassa S, Yamane T (1983) DNA fragments in Friend erythroleukemia cells induced by dimethyl sulfoxide. Proc Natl Acad Sci USA 80:5912–5915
15. Sugiura M, Fram R, Munroe D, Kufe D (1984) DNA strand scission and ADP-ribosyltransferase activity during murine erythroleukemia cell differentiation. Dev Biol 104:484–488
16. Gopalakrishnan TV, French Anderson W (1979) Mouse erythroleukemia cells. In: Jakoby WB, Pastan IH (eds) Cell culture. Methods in enzymology, vol 58. Academic Press, London New York, pp 506–511
17. Conscience JF, Miller RA, Henry J, Ruddle FH (1977) Acetylcholinesterase, carbonic anhydrase and catalase activity in Friend erythroleukemic cells, non-erythroid mouse cell lines and their somatic hybrids. Exp Cell Res 105:401–412
18. Kohn KW, Ewig RAG, Erickson LC, Zwelling LA (1981) Measurements of strand breaks and crosslinks in DNA by alkaline elution. In: Friedberg E, Hanawalt P (eds) DNA repair: a laboratory manual of research procedures. Marcel Dekker, New York, p 379
19. Milam KM, Cleaver JE (1984) Inhibitors of poly(ADP-ribose) synthesis: effect on other metabolic processes. Science 223:589–591
20. Reuben RC, Khanna P, Gazitt Y, Breslow R, Rifkind RA, Marks PA (1978) Inducers of erythroleukemic differentiation. J Biol Chem 253:4214–4218
21. Purnell MR, Whish WJD (1980) Novel inhibitors of poly (ADP-ribose) synthetase. Biochem J 185:775–777

22. Bilello JA, Gauri KK, Kuhne J, Warnecke G, Koch G (1982) Induction and inhibition of Friend erythroleukemic cell differentiation by pyrimidine analogs: analysis of the requirement for intracellular accumulation and incorporation into DNA. Mol Cell Biol 2:1020–1024
23. Tanuma S, Johnston LD, Johnston GS (1983) ADP-ribosylation of chromosomal proteins and mouse mammary tumor virus gene expression. J Biol Chem 258:15731–15735
24. Ferro AM, Higgin NP, Olivera BM (1983) Poly (ADP-ribosylation) of a DNA topoisomerase. J Biol Chem 258:6000–6003

Poly(ADP)-Ribosylation of Nuclear Proteins in the Mouse Testis

ENZO LEONE†[1], PIERA QUESADA[1], MARIA R. FARAONE MENNELLA[1],
BENEDETTA FARINA[1], MARIA MALANGA[1], and ROY JONES[2]

Abbreviations:

ADP	adenosine diphosphate
NAD	nicotinamide adenine dinucleotide
M_r	molecular weight
HMG	high mobility group
LMG	low mobility group
PCA	perchloric acid
PMSF	phenylmethylsulfonyl fluoride

Introduction

Poly(ADP-ribosylation) of nuclear proteins is one of a number of post-translational events which has been demonstrated in eukaryotic cells. The reaction is catalysed by poly(ADP-ribose) synthetase which transfers ADP-ribose from NAD to a suitable acceptor protein. Histone H1 and HMG proteins 1, 2, 14 and 17 have been reported as acceptors in a variety of tissues ranging from trout testis [1] to mammary carcinoma cells [2]. Poly(ADP-ribosylation) of HMG proteins is of particular interest because of the reported association of these proteins with actively transcribed genes [3]. Functionally, poly(ADP-ribosylation) has been implicated in a variety of regulatory events such as DNA synthesis [4], DNA excision repair [5, 1], gene expression [6] and cell differentiation [7].

We have reported previously [8] that in mouse testis non-histone proteins are labelled to a higher specific activity than histones following an i.p. injection of [14C]-ribose and [3H]-adenine, and that the radioactivity is covalently bound to protein in the form of ADP-ribose. This was in contrast to the situation in liver where histones incorporated more radiolabel than non-histone proteins. We have recently extended these observations on ADP-ribosylation of nuclear proteins in testis [9] and we report

1 Dipartimento di Chimica Organica e Biologica, Universita di Napoli, Via Mezzocannone 16, 80134 Napoli, Italy
2 AFRC Institute of Animal Physiology, Animal Research Station, 307 Huntingdon Road, Cambridge CB3 0JQ, Great Britain

ADP-Ribosylation of Proteins
(ed. by F.R. Althaus, H. Hilz, and S. Shall)
© Springer-Verlag Berlin Heidelberg 1985

here that a group of nuclear proteins, originally defined as low mobility group proteins (LMG proteins) [10], can also act as acceptors for poly(ADP-ribose) and that modification of these proteins may be important during spermatogenesis for the production of spermatozoa with intact and competent DNA.

Labelling, Extraction and Analysis of Acceptor Proteins for ADP-Ribose from Testis and Liver Nuclei: In Vivo Experiments

Adult male mice (CFLP strain) were anaesthetized with diethyl ether and injected intraperitoneally (for liver) with a mixture of 25 μCi of [^{14}C]-ribose and 50 μCi of [^{3}H]-adenine (Amersham International), or intratesticularly with 5 μCi [^{14}C]-ribose + 5 μCi [^{3}H]-adenine. After 1 h (for liver) or 2 h (for testis) tissues were removed, nuclei prepared [11] and extracted three times by homogenization in ice-cold 5% (v/v) perchloric acid (PCA). The supernatant was removed and the insoluble pellet was re-extracted three times in 0.25 N-HCl [12]. PCA- and HCl-extracted proteins were chromatographed on ion-exchange columns as described previously [9] and as outlined in the Flow Sheet. Protein fractions were analysed by electrophoresis on polyacrylamide slab gels at pH 2.9 [10, 13] or on polyacrylamide gels containing SDS and 2-mercaptoethanol [9]. Proteins judged to be > 90% pure on SDS polyacrylamide gels, were hydrolysed in 6 M HCl at 110°C for 20 or 70 h and released amino acids analysed on a Beckman (model 119) amino acid analyser. Amino acid analyses was also performed on proteins electroeluted from gel slices [14].

To estimate the length of the poly(ADP-ribose) chain, labelled proteins were digested with snake venom phosphodiesterase (Sigma) and acid soluble products chromatographed on a AG1-X2 resin (Bio-Rad) column [9].

Labelling, Extraction and Analysis of Acceptor Proteins for ADP-Ribose from Testis and Liver: In Vitro Experiments

Nuclei from 2 g tissue were prepared by the method of Utakoji [1], and incubated for 30 min at 20°C in 2 ml of 0.25 M buffered sucrose containing 10 mM Tris/HCl pH 8.2, 10 mM MgCl$_2$, 5 mM 2-mercaptoethanol, 5 mM NaF, 50 mM NaCl, 0.5 mM PMSF and 2 μCi [^{14}C]-NAD [1]. Nuclei were washed two times in cold buffer to remove free label and extracted three times in ice-cold PCA. After centrifuging, the pellet was re-extracted with 0.25 N-HCl and soluble proteins in the PCA, and HCl supernatants analysed on 20% acid urea polyacrylamide gels at pH 2.9 [13]. Gels were either stained directly with 0.05% Coomassie Blue R in methanol:acetic acid:water (40:7:53 by vol) or cut into 5 mm slices which were then dissolved in 30% H$_2$O$_2$ at 50°C for 16 h followed by counting in 5 ml Packard scintillator 299.

Flow Sheet. Outline of the procedure for purification of ADP-ribosylated non-histone acceptor proteins from testis nuclei labelled in vivo. For details see [9]

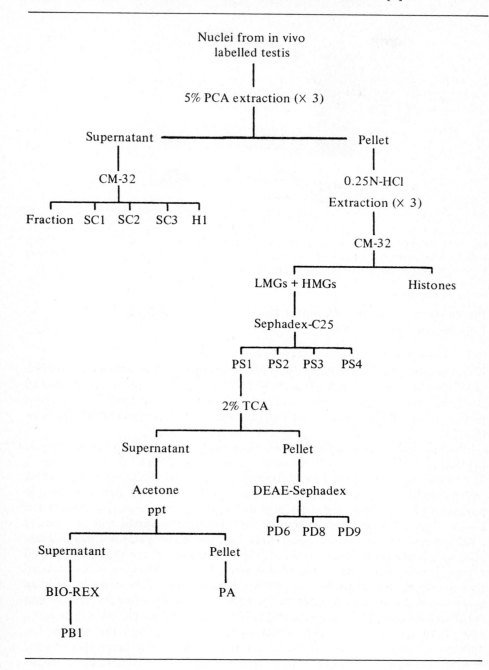

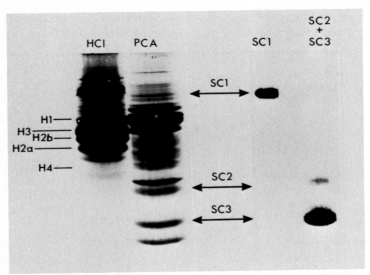

Fig. 1. Electrophoresis on a 15% polyacrylamide slab gel at pH 2.9 [10] of HCl- and PCA-soluble proteins from mouse testis. The mobilities of proteins SC1, SC2 and SC3 are shown for comparison as are a mixture of histone standards from calf thymus

Purification of Nuclear Acceptor Proteins for ADP-Ribose Labelled in Vivo

Electrophoresis at pH 2.9 of PCA-soluble proteins extracted from testis nuclei revealed, in addition to histone H1, a wide variety of LMG- and HMG-like proteins as defined by Goodwin [10] (Fig. 1). The HCl-soluble fraction on the other hand was comprised principally of histones H3, H2a, H2b and H4 and was enriched in LMG-like proteins (Fig. 1).

Chromatography of labelled PCA-soluble proteins from testis on CM-cellulose 32 [9] yielded four major labelled protein peaks designated SCl (1.3×10^6 cpm mg^{-1}), SC2 (93000 cpm mg^{-1}), SC3 (42000 cpm mg^{-1}) and SC4 (62000 cpm mg^{-1}). Peak SC4 consisted principally of histone H1 but SC1, SC2 and SC3 were composed of proteins with low (SC1) and high (SC2 and SC3) mobility on polyacrylamide gels pH 2.9 (Fig. 1). Amino acid analysis of these fractions suggested that SC1 was an LMG-like protein (35% acidic amino acids, 7% basic amino acids) and that SC2 and SC3 were HMG-like proteins (Table 1). Chromatography of the PCA-insoluble HCl extracted proteins on CM-32 revealed that approximately 90% of the total radioactivity was associated with non-histone proteins. Fractionation of these labelled proteins according to the scheme outlined in the Flow Sheet yielded five purified proteins designated PA (100 000 cpm mg^{-1}), PB1 (170 000 cpm mg^{-1}), PD6 (3000 cpm mg^{-1}), PD8 (4200 cpm mg^{-1}) and PD9 (3800 cpm mg^{-1}). All of these proteins had low mobility on polyacrylamide gels at pH 2.9 (Fig. 2) and a low proportion of basic amino acids (range 8.1% to 12.5%; Table 1), both characteristics of LMG-like proteins.

Table 1. Amino acid composition of ADP-ribosylated non-histone acceptor proteins purified from mouse testis nuclei[a]

| | Composition (mol/100 mol) | | | | | | | |
	SC1	SC2	SC3	PB1	PA	PD6	PD8	PD9
LYS	4.6	17.1	17.1	4.7	5.8	4.6	4.0	5.0
HIS	1.6	1.8	1.5	1.7	1.4	2.2	2.0	2.1
ARG	1.3	4.1	4.6	2.2	5.3	1.3	1.8	1.4
ASN	12.0	10.6	9.5	13.8	11.9	9.2	10.2	11.5
THR	5.4	5.2	7.3	5.2	4.7	1.8	3.7	2.9
SER	10.8	8.7	7.0	10.9	11.0	12.2	12.5	8.4
GLN	23.3	11.1	14.5	11.6	10.6	17.9	17.0	16.4
PRO	2.6	2.0	2.1	ND	ND	ND	ND	ND
GLY	16.0	12.3	13.1	17.5	14.5	24.0	24.5	23.0
ALA	9.8	9.1	5.1	10.4	13.3	9.2	7.4	9.1
1/2 CYS	ND	ND	ND	ND	ND	ND	ND	ND
VAL	6.1	4.8	5.5	5.8	5.7	4.5	4.7	6.0
MET	0.9	1.0	1.6	1.3	1.3	1.5	0.6	1.2
ILE	1.9	4.0	3.2	3.4	2.6	2.8	2.8	3.4
LEU	3.1	7.0	4.6	6.5	5.7	4.5	4.8	6.0
TYR	0.6	1.3	0.9	2.1	1.9	2.5	1.5	1.2
PHE	1.1	1.7	2.4	2.7	2.3	1.7	2.0	1.9
LYS/ARG	3.5	4.2	3.7	2.1	1.1	3.5	2.2	3.5
Basic AA%	7.5	23.1	23.2	8.6	12.5	8.1	7.8	8.5
Acidic AA%	35.3	21.7	24.0	25.4	23.0	27.0	27.9	27.9
Acidic/basic	4.7	0.9	1.0	2.9	1.8	3.3	3.5	3.3
$10^{-3} \times M_r$	16.0	10.5	11.5	29.0	68.0	58.0	61.0	60.0

[a] Non-histone acceptor proteins for ADP-ribose were extracted and purified as described in the Flow Sheet. Values for aspartic acid and glutamic acid include asparagine and glutamine, respectively. Abbreviation used: ND = non-detectable

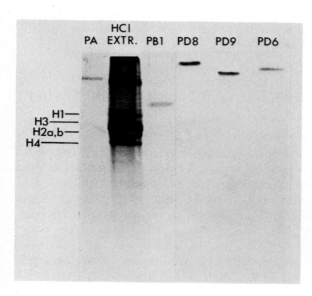

Fig. 2. Electrophoresis on a 15% polyacrylamide slab gel at pH 2.9 [10] of purified proteins PA, PB1, PD6, PD8 and PD9. The mobilities of histone standards are shown for reference

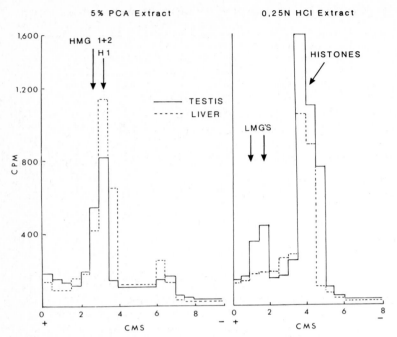

Fig. 3. Distribution of radioactivity in gel slices after electrophoresis at pH 2.9 [10] of PCA- and HCl-soluble proteins from testis and liver nuclei incubated in vitro with [^{14}C]-NAD (as described in the text)

Digestion of purified labelled proteins with snake venom phosphodiesterase consistently yielded two major labelled products which co-chromatographed with authentic 5'-AMP and ADP-ribose standards on AG1-X2 (formate form) exchange columns [9]. From the relative proportions of radioactivity recovered in these fractions the ADP-ribose chain length was estimated to be 4–6 repeating units.

Analysis of Nuclear Acceptor Proteins for ADP-Ribose Labelled in Vitro

The distribution of radioactivity in proteins extracted with PCA and HCl from nuclei incubated with [^{14}C]-NAD and analysed on polyacrylamide gels at pH 2.9 is shown in Fig. 3. In PCA extracts of both testis and liver nuclei most of the radioactivity was associated with the region of the gel containing histone H1 and HMG 1 and 2. Similarly, histones were labelled predominantly in the HCl extracts but it was also found in testis that proteins with low mobility incorporated significant levels of radioactivity. No radioactivity was associated with these proteins in HCl extracts of liver nuclei (Fig. 3).

A comparison of HCl-soluble proteins from testis and liver on polyacrylamide gels at pH 2.9 showed that in testis the labelled proteins with low mobility comprised a

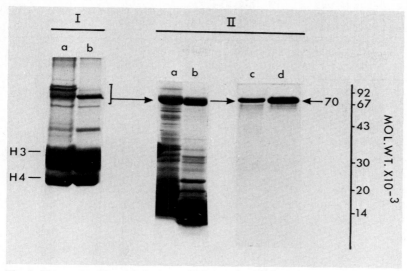

Fig. 4. Electrophoresis on polyacrylamide gels of HCl-soluble proteins from testis (*a*) and liver (*b*) at pH 2.9 (*I*) or on reducing SDS gels (*II*). The purity of the 70,000 M_r proteins from testis (*c*) and liver (*d*) are shown after elution from gel slices

group of about three proteins (Fig. 4). These proteins all had a M_r of about 70,000 when analysed on reducing SDS polyacrylamide gels. A similar group of low mobility proteins were present in HCl extracts of liver nuclei and these also had a M_r of about 70,000. The 70,000 M_r proteins were purified by electroeluting them from gel slices cut out of preparative SDS gels (Fig. 4). Amino acid analysis showed the 70,000 M_r proteins to have a relatively low content of basic amino acids (13–14%) and a high proportion of acidic residues (20–21%), characteristics of LMG-like proteins (Table 2). Protein PA described previously is probably one of the group of 70,000 M_r proteins as its composition and M_r are very similar (Table 2). A preliminary investigation into HCl extracts of different organs from mice suggest that varying amounts (as judged by the intensity of staining with Coomassie Blue on SDS gels) are present in all tissues (results not shown).

Discussion

The experiments described in this paper indicate that in mouse testis nuclei non-histone proteins, particularly LMG protein, are ADP-ribosylated to relatively high specific activity. This conclusion is based on experiments performed in vivo and in vitro and which, in addition, suggest that there are significant differences between tissues in the type of nuclear acceptor proteins for ADP-ribose. Since quantitatively the LMG proteins are minor components of the PCA and HCl extracts these results demonstrate the potential for specific proteins to be ADP-ribosylated.

Table 2. Amino acid composition of the 70,000 M_r LMG-like proteins from testis and liver compared with protein PA (Table 1)

| | Composition (mol/100 mol) | | | | |
	Testis	Liver	PA	C14[a]	Calf LMGs[b]
LYS	7.46	7.28	5.8	7.9	7.9
HIS	2.20	2.56	1.4	1.3	2.4
ARG	4.43	3.37	5.3	4.6	6.2
ASN	8.43	8.85	11.9	11.5	9.7
THR	5.41	6.05	4.7	6.8	5.2
SER	5.31	5.52	11.0	7.2	6.3
GLN	11.86	12.60	10.6	13.5	13.5
PRO	4.1	3.9	ND	4.1	4.0
GLY	25.27	22.8	14.5	8.7	8.0
ALA	8.74	9.4	13.3	8.0	8.1
1/2 CYS	Traces	Traces	–	1.5	0.7
VAL	5.17	5.37	5.1	5.1	6.4
MET	1.24	1.10	1.3	2.1	0.8
ILE	2.53	2.03	2.6	4.7	4.9
LEU	8.10	8.95	5.7	7.4	9.0
TYR	0.60	0.42	1.9	2.2	2.6
PHE	3.22	3.66	2.3	3.3	3.5
LYS/ARG	1.68	2.16	1.1	1.72	1.27
Basic AA%	14.11	13.21	12.5	13.8	16.5
Acidic AA%	20.29	21.45	23.0	25.0	23.2
Acidic/basic	1.43	1.6	1.8	1.8	1.4
$10^{-3} \times M_r$	70.0	70.0	68.0	70.0	–

[a] C14 from [17]
[b] Total calf thymus LMG proteins from [10]

Our classification of proteins SC1, PB1, PA, PD6, PD8, PD9 and the 70,000 M_r proteins eluted from gels of testis and liver as LMG-like proteins is based on their poor ability to penetrate 20% polyacrylamide gels at pH 2.9 and their low content of basic amino acids. The composition of these proteins compares favourably with the analysis of total LMG proteins extracted from calf thymus (Table 2) [10]. We consider it unlikely that the LMG-like proteins we have described here are degradation products of histones or other non-histone proteins. First, they have lower mobilities on polyacrylamide gels at pH 2.9 than HMG proteins. Proteinase digestion products of HMG 1 and 2, for example, have higher mobilities than the parent molecules [15]. Second, the LMG proteins described here generally have higher M_r than HMG proteins or histones. Third, PCA extraction is considered to be highly effective in inhibiting degradation of nuclear proteins [16]. In many respects protein PA resembles protein C-14 from Novikoff hepatoma ascites cells (Table 2). Protein C-14 was found to bind avidly to DNA and to stimulate RNA synthesis [17], and it is interesting to speculate that protein PA may have similar activity in mouse testis.

At present it is not possible to ascribe a specific function to ADP-ribosylation of LMG-like proteins in the testis. One obvious difference between testis and liver is that the former tissue shows a relatively high rate of cell division. A role for poly(ADP-ribosylation) in DNA synthesis and cell replication is suggested by several pieces of evidence [18] and by recent investigations into the effects of inhibitors of poly(ADP-ribose) synthetase on DNA excision repair [5, 19]. Seydin and Kistler [20] reported a close association between the levels of HMG 2 and proliferative activity in various tissues but did not investigate if HMG 2 was ADP-ribosylated or not. The cellular localization of the LMG-like proteins in mouse testis is not known but the fact that poly-(ADP-ribose) synthetase is present in isolated seminiferous tubules at levels close to those in whole testis (P. Quesada and R. Jones, unpubl. results) suggests that the process may have a role during spermatogenesis. It is known that during spermiogenesis there is a sequential synthesis and replacement of histones and testis-specific proteins with protamines and that this results in the formation of spermatids with a highly condensed nucleus [21]. It is not known how these protein transitions are regulated although they must be controlled very carefully and efficiently in order to produce large numbers of spermatozoa with intact and competent DNA. It will be important in future work to investigate if poly(ADP-ribosylation) is associated with any of those nucleoprotein transitions.

Acknowledgments. This paper is dedicated to the memory of Prof. E. Leone who died suddenly in June 1984. His guidance and friendship is greatly missed by all his colleagues.

Part of this work was done with the financial assistance of NATO research grant 1850. We thank Miss Giuseppina Travaglione for help with the amino acid analysis.

References

1. Levy-Wilson B (1981) ADP-ribosylation of trout testis chromosomal proteins: Distribution of ADP-ribosylated proteins among DNase 1-sensitive and -resistant chromatin domains. Arch Biochem Biophys 208:528–534
2. Tanuma S, Johnson LD, Johnson GS (1983) ADP-ribosylation of chromosomal proteins and mouse mammary tumor virus gene expression. J Biol Chem 258:15371–15375
3. Weisbrod S (1982) Active chromatin. Nature (London) 297:289–295
4. Burzio L, Koide SS (1973) Activation of the template activity of isolated rat liver nuclei for DNA synthesis and its inhibition by NAD. Biochem Biophys Res Commun 53:572–579
5. Durkacz BW, Omidiji O, Gray DA, Shall S (1980) (ADP-ribose)n participates in DNA excision repair. Nature (London) 283:593–596
6. Muller WEC, Zahn RK (1976) Poly ADP-ribosylation of DNA-dependent RNA polymerase I from Quail oviduct. Dependence on progesterone stimulation. Mol Cell Biochem 12:147–158
7. Farzaneh F, Zalin R, Brill D, Shall S (1982) DNA strand breaks and ADP-ribosyl transferase activation during cell differentiation. Nature (London) 300:362–366
8. Faraone-Mennella MR, Quesada P, Farina B, Leone E, Jones R (1983) ADP-ribosylation of nuclear proteins in mouse testis. Biochem J 205:245–248
9. Faraone-Mennella MR, Quesada P, Farina B, Leone E, Jones R (1984) Purification of non-histone acceptor proteins for ADP ribose from mouse testis nuclei. Biochem J 221:223–233
10. Goodwin GH, Sanders C, Johns EW (1973) A new group of chromatin-associated proteins with a high content of acidic and basic amino acids. Eur J Biochem 38:14–19
11. Utakoji T, Muramatsu M, Sugano H (1968) Isolation of pachytene nuclei from the Syrian hamster testis. Exp Cell Res 53:447–458
12. Ueda K, Omachi A, Kawaichi M, Hayaishi O (1975) Natural occurrence of poly(ADP-ribosyl) histones in rat liver. Proc Natl Acad Sci USA 72:205–209

13. Panyim S, Chakley R (1969) High resolution acrylamide gel electrophoresis of histones. Arch Biochem Biophys 130:337–346
14. Levy-Wilson B, Connor W, Dixon GH (1979) A subset of trout testis nucleosomes enriched in transcribed DNA sequences contains high mobility group proteins as major structural components. J Biol Chem 254:609–620
15. Cary PD, Turner CH, Mayes E, Crane-Robinson C (1983) Conformation and domain structure of the non-histone chromosomal proteins, HMG 1 and 2. Eur J Biochem 131:367–374
16. Minaga T, Romaschin AD, Kirsten E, Kun E (1979) The *in vivo* distribution of immunoreactive larger than tetrameric polyadenosine diphosphoribose in histone and non-histone protein fractions of rat liver. J Biol Chem 254:9663–9668
17. James GT, Yeoman LC, Matsui S, Goldberg AH, Busch H (1977) Isolation and characterization of nonhistone chromosomal protein C-14 which stimulates RNA synthesis. Biochemistry 16: 2384–2389
18. Mandel P, Okazaki H, Niedergang C (1982) Poly(adenosine diphosphate ribose). Prog Nucleic Acid Res Mol Biol 37:1–51
19. Nudka N, Shall S (1980) 5-Methylnicotinamide resistant variant of mouse lymphoma L1210 cells. Biochem Biophys Res Commun 96:997–1002
20. Seyedin SH, Kistler WS (1979) Levels of chromosomal protein high mobility group 2 parallel the proliferative activity of testis, skeletal muscle and other organs. J Biol Chem 254:11264–11271
21. Meistrich ML, Trostle PK, Brock WA (1981) Association of nucleoprotein transitions with chromatin changes during rat spermatogenesis. In: Jagiello G, Vogel HJ (eds) Bioregulators of reproduction. Academic Press, London New York, pp 151–165

ADP-Ribosylation and Neoplastic Transformation in Vitro and in Vivo

A Role for Poly(ADP-Ribose) in Radiogenic and Chemically-Induced Malignant Transformation and Mutagenesis

CARMIA BOREK[1], AUGUSTINUS ONG[1], XI-CANG GUO[2,3],
and JAMES E. CLEAVER[2]

Introduction

3-aminobenzamide (3AB) is a nicotinamide analog which was first introduced as an inhibitor of poly(ADP-ribose) synthesis [29]. 3AB has a multitude of effects on cells previously exposed to DNA damaging agents, which have been taken to indicate a role for poly(ADP-ribose) synthesis in DNA repair [1–3, 9, 10, 12, 13, 15–17, 22–27, 31–33]. Other side effects suggest, however, that 3AB may not be as specific as supposed [9, 10, 14, 18, 21]. 3AB also inhibits malignant transformation by radiation and methylating agents in human, mouse, and hamster cells [5–7, 19]. We have now compared the effects of low concentrations of 3AB on mutagenesis and transformation, in view of the supposed similarities in these processes.

Methods

Cell Culture. Chinese hamster ovary (CHO) cells were used for mutagenetic studies. These cells were grown in Eagle's minimum essential medium with 10% fetal calf serum, in which they had a population doubling time of 10 h.

Syrian hamster embryo and mouse C3H 10T1/2 cells were used for transformation assays. Hamster cells were prepared and grown in Dulbecco's modified Eagle's medium supplemented with 10% fetal calf serum, penicillin (50 U ml^{-1}), and streptomycin (50 μg ml^{-1}; GIBCO), as previously described [4].

Mouse embryo fibroblast cells C3H 10T1/2 (clone 8) were grown in Eagle's basal medium containing 10% heat-inactivated fetal calf serum, penicillin (50 U ml^{-1}), and streptomycin (50 μg ml^{-1}) [4, 30].

---

1 Department of Radiology, College of Physicians and Surgeons, Columbia University, New York, NY 10032, USA
2 Laboratory of Radiobiology and Environmental Health, University of California, San Francisco, CA 94143, USA
3 Department of Radiological Medicine, Suzhou Medical College, Suzhou, People's Republic of China

ADP-Ribosylation of Proteins
(ed. by F.R. Althaus, H. Hilz, and S. Shall)
© Springer-Verlag Berlin Heidelberg 1985

Irradiation and Alkylation Conditions. X-ray doses of 3–6 Gy were delivered from a Phillip's RT100 X-ray machine (200 kVp 15 mA with a nominal half value layer of 0.35 cm copper). Cultures were irradiated with an incident dose rate of 1.3 J m^{-2} s^{-1} UV light (254 nm). Cultures were exposed to various concentrations of N-methyl-N'-nitro-N-nitrosoguanidine (MNNG) (Aldrich Chemical Co., Milwaukee, WI), MMS (ICN Pharmaceuticals, Plainview, NY), ethyl methanesulfonate (EMS) (ICN Pharmaceuticals), or ethyl nitrosourea (ENU) (a gift of B. Singer, University of California, Berkeley), in tissue culture medium for 60 min to 4 h.

Cell Survival Determinations. Immediately after exposure to X-rays, UV light, or alkylating agents, cells were either allowed to grow with or without 3AB (1 mM) for mutation or transformation expression, or diluted to low density and inoculated into Petri dishes for colony formation. Colonies were left to grow for 8 days (CHO cells), 10 days (Syrian hamster ovary cells), or 6 weeks (C3H 10T1/2 cells) before fixation.

Mutation Frequency Determinations. Cultures were allowed to grow for 3 to 8 days for expression of ouabain or 6-thioguanine resistance, respectively. 3AB (1 mM) was included during the first 3 days of growth. For mutant selection, cultures were trypsinized and cells were resuspended and inoculated at a range of densities in normal medium and in medium containing 0.06 mM 6-thioguanine or 3 mM ouabain, as described previously [8], with due care to avoid artifacts from cell crowding [11]. Mutation frequencies were calculated from the ratio of the plating efficiency in 6-thioguanine or ouabain to that in normal medium.

Cell Transformation Assays. Cells were irradiated or exposed to alkylating agents 24 h after plating in Petri dishes. Cultures were then maintained with or without 3AB (1 mM) with weekly changes of medium for a 10-day incubation period for the hamster cells and 6 weeks for the 10T1/2 cells, and then fixed and stained with Giemsa. Scoring transformed cells was carried out as described previously [4, 30].

Results

Survival and Mutagenesis from Radiation in CHO Cells. 3AB (1 mM) had no significant effect on cell survival or the yield of 6-thioguanine-resistant mutations after irradiation with X-rays (Tables 1 and 2). Ouabain mutations were not investigated because of the extremely low yield of ouabain-resistant mutants from X-rays [8].

3AB (1 mM) slightly increased the toxicity and reduced mutagenesis from UV light. The reduction in D_0 was approximately 20% (Table 1). The rate of mutagenesis per J m^{-2} to ouabain or 6-thioguanine resistance was reduced about 30% by 3AB.

Survival and Mutagenesis from MMS. 3AB (1 mM) caused a large increase in the toxicity of MMS, which was expressed predominantly by elimination of the shoulder of the survival curve (Table 1). The extrapolation number was reduced from 18.0 to 1.0, but the D_0 was not changed significantly. The mutation frequencies to ouabain resistance

Table 1. Survival parameters of Chinese hamster ovary cells grown with or without 3-aminobenzamide[a]

Agent	3AB	n	D_0
X-rays	0	1.2	2.16 Gy
	1 mM	1.2	2.16 Gy
UV light	0	1.5	5.8 J m^{-2}
	1 mM	1.5	4.8 J m^{-2}
MMS	0	18.0	0.83 mM
	1 mM	1.0	0.92 mM

[a] Absolute plating efficiencies were 47.9% with no 3AB, 42.9% (1 mM 3AB), and 33.8% (5 mM 3AB)

Table 2. Mutation frequencies in Chinese hamster ovary cells grown with or without 3-aminobenzamide (1 mM) during the expression period

Agent	3AB	Mutation frequencies	
		Ouabain ($\times 10^6$)	6-thioguanine ($\times 10^5$)
X-rays	0	–	4.35 Gy
	1 mM	–	5.13 Gy
UV light	0	3.97 ± 0.35 J^{-1} m^{-2}	5.52 ± 0.21 J^{-1} m^{-2}
	1 mM	2.73 ± 0.34 J^{-1} m^{-2}	4.08 ± 0.40 J^{-1} m^{-2}
MMS	0	60.40 ± 10.6 mM^{-1}	17.00 ± 3.8 mM^{-1}
	1 mM	56.20 ± 0.2 mM^{-1}	33.50 ± 10.0 mM^{-1}

were unaffected by 3AB, but mutagenesis to 6-thioguanine resistance was approximately doubled by 3AB.

Survival and Transformation of Syrian Hamster and C3H 10T1/2 Mouse Fibroblasts. The survival of cells exposed to each agent in these experiments was in the range of 80 to 100% (Tables 3 and 4). Transformation by X-rays, UV light, MNNG, MMS, ENU, and EMS was in the range of 10^{-4} to 10^{-3} with 3AB alone having no toxicity or transforming ability. Addition of 3 AB after exposure resulted in contrasting effects on transformation by radiations and methylating agents as compared to ethylating agents. Transformation by radiations, MNNG, and MMS was inhibited, whereas transformation by ENU and EMS was increased (Tables 3 and 4). The effect was small with ENU, but much larger with EMS.

Discussion

The effects of 3AB on transformation frequencies were much larger and qualitatively different from its effects on mutagenesis. Mutations from radiation were scarcely

Table 3. Transformation of mouse C3H 10T1/2 cells irradiated with X-rays or UV light or exposed to alkylating agents for 4 h and grown in 3-aminobenzamide (1 mM) for 7 days starting immediately after exposure

Damaging agent	3AB	Surviving fraction[a] (%)	Transformation frequency ($\times 10^{-4}$)
None	0	100	0
	1 mM	96.0 ± 7.0	0
X-rays (4 Gy)	0	70.0 ± 11.0	10.3 ± 2.1
	1 mM	75.0 ± 11.0	2.83 ± 1.15
UV (8.6 J m^{-2})	0	68.0 ± 13.0	9.97 ± 2.17
	1 mM	75.0 ± 11.0	2.83 ± 1.15
MMS (0.15 mM)	0	89.0 ± 1.0	5.64 ± 1.2
	1 mM	84.5 ± 2.5	0.19 ± 0.9
MNNG (3.4 μM)	0	83.0 ± 2.0	18.4 ± 1.2
	1 mM	91.0 ± 6.0	3.88 ± 1.5
ENU (1 mM)	0	92.5 ± 2.5	2.72 ± 1.4
	1	91.0 ± 2.0	5.30 ± 1.8
EMS (1 mM)	0	88.0 ± 8.0	2.58 ± 1.5
	1	90.0 ± 5.0	9.10 ± 3.3

[a] Plating efficiency of 15.0 ± 0.2% was normalized to 100% for unexposed cells

Table 4. Transformation of Syrian hamster embryo cells exposed to X-rays or MNNG for 4 h and grown in 3-aminobenzamide for 7 days starting immediately after exposure

Damaging agent	3AB	Surviving fraction[a] (%)	Transformation frequency ($\times 10^{-3}$)
None	0	100	0
	1 mM	96.0 ± 8.0	0
X-rays (3 Gy)	0	66.0 ± 13.0	11.2 ± 0.10
	1 mM	59.0 ± 11.0	1.2 ± 0.09
MNNG (3.4 μM)	0	55.0 ± 14.0	8.8 ± 0.17
	1 mM	49.0 ± 13.0	1.9 ± 0.09

[a] Plating efficiency of 4.8 ± 0.3% was normalized to 100% for unexposed cells

affected, whereas transformation was inhibited. Mutation from MMS was increased at the 6-thioguanine resistant locus, but not at the ouabain locus, but transformation was inhibited. This is similar to the increased mutagenesis observed in previous experiments using EMS (W.F. Morgan, pers. comm.). Only for ethylating agents was mutagenesis and transformation affected in a similar way.

In these experiments, 3AB was added only after exposures were complete. This prevented interaction of 3AB with the production of DNA damage and concentrated its action on possible changes in gene activation during growth immediately after exposure, when poly(ADP-ribose) synthesis is stimulated [5, 10, 19].

In previous investigations [5, 10, 21] we studied the effect of 3AB on other cellular end points, such as DNA repair (excision of N-7-guanine and N-3-adenine, strand breaks, and repair replication) and de novo purine metabolism. We concluded that the inhibition of transformation could not be mediated by effects of 3AB on any step of DNA repair or on purine metabolism at the low concentration we used. Instead, it appeared likely that some specific change in poly(ADP-ribosylation) of nuclear proteins that affects gene expression or amplification was the most probable explanation. It is interesting to note that some tumor virus gene products show increased expression when poly(ADP-ribose) synthesis is inhibited [34]. Also, the product of the myc oncogene is a nuclear protein that could be subject to regulation by poly(ADP-ribosylation).

The ethylating agents responded differently from most other agents with respect to the involvement of poly(ADP-ribose) synthesis in transformation. For these agents, prevention of poly(ADP-ribose) synthesis enhanced transformation (Table 3). It is perhaps significant for interpretation of our results that Durrant et al. [13], using another inhibitor of poly(ADP-ribose) synthesis, nicotinamide, observed a small inhibition of 0^6-alkyltransferase. It is possible that for ethylating agents alone, under our conditions, 3AB exerts some of its effects by inhibiting the 0^6 alkyl transferase pathway [28]. Inhibition of removal of the 0^6 ethyl group could cause the observed increases of both transformation and mutagenesis ([7, 20] W.F. Morgan, pers. comm.) from ethylating agents.

Our results consequently indicate that 3AB has contrasting effects on mutagenesis and transformation when radiation or methylating agents are used as damaging agents, but similar effects when ethylating agents are used. Transformation by radiation and methylating agents may, therefore, involve nonmutagenic events, such as gene amplification or changed gene expression,whereas transformation by ethylating agents more closely resembles mutagenic events.

These results highlight a significant role for poly(ADP-ribose) synthesis in malignant transformation. This role is complex, and precise attention to scheduling and concentrations of 3AB is required to identify mechanisms of action as distinct from nonspecific effects. Our results demonstrate the existence of unique metabolic processes involved in transformation that can be modulated by changes in cell physiology, and an in vivo investigation of these processes is currently under way.

Acknowledgments. This work was supported by Grant CA 12536-11 from the National Cancer Institute (C.B.), by the U.S. Department of Energy, contract no. DE-AMO3-76-SF01012 (J.E.C.), and by the People's Republic of China (G. X.-C.). We are grateful to Dr. W.F. Morgan for permission to cite his unpublished observations and to Mrs. Susan Brekhus and Ms. Mary McKenney for assistance in preparing the manuscript.

References

1. Althaus FR, Lawrence SD, Sattler GL, Pitot HC (1982) ADP-ribosyltransferase activity in cultured hepatocytes. Interactions with DNA repair. J Biol Chem 257:5528–5535
2. Berger NA, Sikorski GW (1980) Nicotinamide stimulates repair of DNA damage in human lymphocytes. Biochem Biophys Res Commun 95:67–72
3. Berger NA, Sikorski GW, Petzold SJ, Kurohara KK (1979) Association of poly(adenosine diphosphoribose) synthesis with DNA damage and repair in normal human lymphocytes. J Clin Invest 63:1164–1171
4. Borek C, Miller C, Pain C, Troll W (1979) Conditions for inhibiting and enhancing effects of the protease inhibitor antipain on X-ray induced neoplastic transformation in hamster and mouse cells. Proc Natl Acad Sci USA 76:1800–1803
5. Borek C, Morgan WF, Ong A, Cleaver JE (1984) Inhibition of malignant transformation in vitro by inhibitors of poly(ADP-ribose) synthesis. Proc Natl Acad Sci USA 81:243–247
6. Borek C, Ong A, Morgan WF, Cleaver JE (1984) Inhibition of X-ray- and ultraviolet light-induced transformation in vitro by modifiers of poly(ADP-ribose) synthesis. Radiat Res 99:219–227
7. Borek C, Ong A, Cleaver JE (1984) Methylating and ethylating carcinogens have different requirements for poly(ADP-ribose) synthesis during malignant transformation. Carcinogenesis 5:1573–1576
8. Cleaver JE (1977) Induction of thioguanine- and ouabain-resistant mutants and single-strand breaks in the DNA of Chinese hamster ovary cells by ^{3}H-thymidine. Genetics 87:129–138
9. Cleaver JE (1984) Differential toxicity of 3-aminobenzamide to wild-type and 6-thioguanine resistant Chinese hamster cells by interference with pathways of purine biosynthesis. Muat Res 131:123–127
10. Cleaver JE, Bodell WJ, Morgan WF, Zelle B (1983) Differences in the regulation by poly(ADP-ribose) of repair of DNA damage from alkylating agents and ultraviolet light according to cell type. J Biol Chem 258:9059–9068
11. Cox RP, Krauss MR, Balis ME, Dancis J (1970) Evidence for transfer of enzyme product as the basis of metabolic cooperation between tissue culture fibroblasts of Lesch-Hyhan disease and normal cells. Proc Natl Acad Sci USA 67:1573–1579
12. Durkacz BW, Omidiji O, Gray DA, Shall S (1980) (ADP-ribose) participants in DNA excision repair. Nature (London) 283:593–596
13. Durrant LG, Margison GP, Boyle JM (1981) Effects of 5-methylnicotinamide on mouse L1210 exposed to N-methyl-N'-nitrosourea: mutation induction, formation and removal of methylation products in DNA and unscheduled synthesis. Carcinogenesis 2:1013–1017
14. Grunfeld C, Shigenaga JK (in press) Nicotinamide and other inhibitors of ADP-ribosylation block deoxyglucose uptake in cultured cells. Biochem Biophys Res Commun
15. Hayaishi O, Ueda K (1977) Poly(ADP-ribose) and ADP-ribosylation of proteins. Annu Rev Biochem 46:95–116
16. Hori T-A (1981) High incidence of sister chromatid exchanges and chromatid interchanges in the conditions of lowered activity of poly(ADP-ribose) polymerase. Biochem Biophys Res Commun 102:38–45
17. James MR, Lehmann AR (1982) Role of poly(adenosine diphosphate ribose) in deoxyribonucleic acid repair in human fibroblasts. Biochemistry 21:4007–4013
18. Johnson GS (1981) Benzamide and its derivatives inhibit nicotinamide methylation as well as ADP-ribosylation. Biochem Int 2:611–617
19. Kun E, Kirsten E, Milo GE, Kurian P, Kumari HL (1983) Cell cycle dependent intervention by benzamide of carcinogen-induced neoplastic transformation and in vitro poly(ADP-ribosyl)-ation of nuclear proteins in human fibroblasts. Proc Natl Acad Sci USA 80:7219–7223
20. Lubet RA, McGarvill JT, Putnam DL, Schwartz JL, Schechtman LM (1980) Effect of 3-aminobenzamide on induction of toxicity and transformation by ethylmethane sulfonate and methyl cholanthenein BALB/3T3 cells. Carcinogenesis 5:459–462

21. Milam K, Cleaver JE (1984) Inhibitors of poly(ADP-ribose) synthesis: effect on other metabolic processes. Science 233:589–591
22. Miwa M, Kinai M, Kondo T, Hoshino H, Isihara K, Sugimura T (1981) Inhibitors of poly(ADP-ribose) polymerase enhance unscheduled DNA synthesis in human peripheral lymphocytes. Biochem Biophys Res Commun 100:463–470
23. Morgan WF, Cleaver JE (1982) 3-aminobenzamide synergistically increases sister chromatid exchanges in cells exposed to methyl methane sulfonate but not to ultraviolet light. Mutat Res 104:361–366
24. Morgan WF, Cleaver JE (1983) Effects of 3-aminobenzamide on the rate of ligation during repair of alkylated DNA in human fibroblasts. Cancer Res 43:3104–3107
25. Natarajan AT, Scukas I, van Zeeland AA (1981) Contribution of incorporated 5-bromodeoxyuridine in DNA to the frequencies of sister-chromatid exchanges induced by inhibitors of poly-(ADP-ribose)-polymerase. Mutat Res 84:125–132
26. Nduka N, Skodmore CJ, Shall S (1980) The enhancement of cytotoxicity of N-methyl-N'-Nitro-N-nitrosourea and of γ-irradiation by inhibitors of poly(ADP-ribose) polymerase. Eur J Biochem 105:525–530
27. Oikawa A, Tohda H, Kanai M, Miwa M, Sugimura T (1980) Inhibitors of poly(adenosine diphosphate ribose) polymerase induce sister chromatid exchanges. Biochem Biophys Res Commun 97:1311–1316
28. Pegg AE (1983) Akylation and subsequent repair of DNA after exposure to dimethylnitrosamine and related carcinogens. In: Hodgson E, Bend JR, Philpot RM (eds) Reviews in biochemical toxicology, vol 5. Elsevier/North Holland Miomed Press, Amsterdam New York, p 83
29. Purnell MR, Wish WJD (1980) Novel inhibitors of poly(ADP-ribose) synthetase. Biochem J 185:775–780
30. Reznikoff CA, Bertram JS, Brankow DW, Heidelberger C (1973) Quantitative and qualitative studies of chemical transformation of cloned C3H mouse embryo cells sensitive to postconfluence inhibition. Cancer Res 33:3239–3249
31. Sims JL, Sikorski GW, Catino DM, Berger SJ, Berger NA (1982) Poly(adenosine diphosphoribose) polymerase inhibitors stimulate unscheduled deoxyribonucleic acid synthesis in normal human lymphocytes. Biochemistry 21:1813–1821
32. Sugimura T, Miwa M, Saito H, Kanai Y, Ikejima M, Terada M, Yamada M, Utakoji T (1980) Studies of nuclear ADP-ribosylation. Adv Enzyme Regul 18:195–220
33. Sugimura T, Miwa M (1983) Poly(ADP-ribose) and cancer research. Carcinogenesis 4:1503–1506
34. Tanuma S, Johnson LD, Johnson GS (1983) ADP-ribosylation of chromosomal proteins and mouse mammary tumor virus gene expression. J Biol Chem 258:15371–15375

Effects of 3-Aminobenzamide on the Induction of Toxicity and Transformation by Various Chemicals in BALB/3T3 Cl A 31-1 Cells

RONALD A. LUBET[1], JOHN T. McCARVILL[1], and JEFFREY L. SCHWARTZ[2]

Introduction

Poly(ADP-ribosyl) synthetase, a nuclear-associated enzyme transfers ADP-ribose from NAD^+ to a variety of proteins, including itself, specific histones, and a number of acidic proteins in the nucleus. Employing relatively specific inhibitors of poly(ADP-ribosyl) synthetase, it has been demonstrated that the ribosylation process is involved in differentiation of various cells, proliferation, and in DNA repair processes [1, 2].

Within the context of DNA repair processes, it has been shown that inhibitor of poly(ADP-ribosyl) synthetase enhances the effects of a variety of compounds when measuring cellular cytotoxicity [1, 2], induction of sister chromatid exchanges [3], chromosome aberation [3], or induction of single-strand breaks [1]. The effects of inhibitors of poly(ADP-ribosyl) synthetase on carcinogenesis is more controversial. Although the in vivo data seems most amenable to the interpretation that inhibitors of poly(ADP-ribose) exhibit cocarcinogenic and/or promoting effects [4, 5], the in vitro data is conflicting. Thus, two different laboratories have observed profound inhibition of chemically-induced transformation following treatment with inhibitors of poly(ADP-ribose) synthetase [6, 7], while other laboratories have observed marked enhancement of X-ray [8] or short-chain alkylating agent [8, 9], induced transformation in Balb 3T3 cells.

Materials and Methods

Cells and Medium. BALB/3T3 clone A31-1 cells were obtained from Dr. T. Kakunaga (NCI, NIH, Bethesda, MD) and were maintained as previously described [9].

Transformation Assay. The transformation assay, which is a minor modification of that of Kakunaga [10] was briefly as follows. BALB/3T3 clone A31-1 cells were dis-

---

1 Microbiological Associates, Inc., Bethesda, MD 20816, USA
2 Harvard School of Public Health, Boston, MA, USA

ADP-Ribosylation of Proteins
(ed. by F.R. Althaus, H. Hilz, and S. Shall)
© Springer-Verlag Berlin Heidelberg 1985

Table 1. Effects of 3-aminobenzamide on the induction of transformation by methylcholanthrene or aflatoxin B_1 in BALB 3T3 Cl 31-1 cells

Treatment[a]	3AB	RCE	Foci/treated plate		
μM (μg ml^{-1})	mM (μg ml^{-1})		Type II	Type III	TF × 10^{-4}
Acetone control	0	100	0	1	0.17
Acetone control	3 (408)	98	1	1	0.35
MCA 3.7 (1)	0	28	8	16	12.00
MCA 3.7 (1)	3 (408)	43	8	14	7.40
MCA 1.9 (0.5)	0	54	4	10	4.60
MCA 1.9 (0.5)	3 (408)	54	3	9	4.00
AFB$_1$ 3.8 (1.2)	0	29	3	3	3.58
AFB$_1$ 3.8 (1.2)	3 (408)	34	2	7	4.23
AFB$_1$ 1.9 (0.6)	0	56	1	2	0.67
AFB$_1$ 1.9 (0.6)	3 (408)	71	4	4	1.84

[a] Cells were exposed simultaneously to methylcholanthrene, aflatoxin B_1 and 3-aminobenzamide for a 24 h period

pensed such that 15 transformation plates were seeded with 1×10^4 cells/dish and three cytotoxicity plates seeded with 250 cells/dish were established per condition. After an overnight (16–20 h) incubation at 37°C in a humidified atmosphere of 5% CO_2 in air, cells were exposed in complete medium containing the specified concentrations of 3AB and/or EMS, MMS, MNNG, AFB$_1$, or MCA for the indicated times at 37°C. Thus, 3AB was in contact with the target cells only during the period of chemical exposure. Dishes were washed at the indicated times then replenished with 4 ml complete medium, which was changed twice weekly for the duration of the assay. Toxicity and transformation were determined as previously described [9].

Table 2. The effect of varying concentrations of 3AB on the induction of transformation by EMS in BALB 3T3 Cl A 31-1 cells

Treatment[a]	3AB concentration		Foci/treated dishes		
mM (μg ml^{-1})	mM (μg ml^{-1})	RCE	Type II	Type III	TF × 10^{-4}
Acetone control	0	100	2	0	0.28
Acetone control	3 (408)	98	1	1	0.14
EMS 0.6 (75)	0	80	0	4	0.68
EMS 0.6 (75)	3 (408)	6	10	19	68.3
EMS 0.6 (75)	1 (136)	13	2	13	12.02
EMS 0.6 (75)	0.33 (45)	53	6	10	4.76
EMS 0.6 (75)	0.11 (15)	67	3	7	2.35

[a] Cells were exposed simultaneously to EMS and the indicated concentration of 3AB for 24 h

Table 3. The effects of varying lengths of exposure to 3AB on the induction of transformation by MNNG in BALB 3T3 Cl A 31-1 cells

| Treatment[a] | 3AB | Time of exposure | RCE | Foci/treated dishes | | TF × 10^{-4} |
| | | | | Type II | Type III | |
μM (μg ml^{-1})	mM (μg ml^{-1})					
Acetone	0	–	100	1	1	0.28
Acetone	1.5 (204)	24 h	102	0	0	<0.14
MNNG 2 (0.3)	0	4 h	66	2	4	1.02
MNNG 2 (0.3)	1.5 (204)	4 h	38	5	10	5.83
MNNG 2 (0.3)	0	24 h	79	3	8	1.86
MNNG 2 (0.3)	1.5 (204)	24 h	6	13	26	93.0
MNNG 2 (0.3)	0	72 h	76	1	3	0.75
MNNG 2 (0.3)	1.5 (204)	72 h	10	13	32	57.4

[a] Cells were exposed simultaneously to MNNG and 3-aminobenzamide for the indicated periods of time

Results

When cells were exposed to 3AB (1 or 3 mM) together with aflatoxin B_1, or methylcholanthrene, limited effects on toxicity or transformation were observed (Table 1).

When cells were exposed simultaneously to EMS plus varying doses of 3AB a dose-dependent increase in toxicity and transformation frequency was observed (Table 2).

When cells were exposed simultaneously to the short-lived methylating agent (MNNG) and 3AB for varying lengths of time, one observed a significant enhancement following a 4 h treatment. However, a much more striking albeit similar effect was observed following longer treatment (24 or 72 h) (Table 3).

Discussion

We had previously shown that 3AB could enhance the toxicity and transformation caused by short-chain alkylating agents in Balb 3T3 cells [8, 9]. To confirm and expand those results, we exposed cells to EMS plus varying nontoxic doses of 3AB for 24 h. The higher doses of 3AB caused a striking dose-dependent increase in EMS-induced toxicity and transformation. Interestingly we still observed increases in transformation frequencies at doses of 3AB (0.11 or 0.33 mM). These doses are considerably lower than those typically employed to induce SCE or inhibit repair of strand breaks.

We also examined the effect of duration of exposure to 3AB on the enhancement we observed. Following treatment with alkylating agents, target molecules for poly-(ADP-ribosylation) are both ribosylated and deribosylated quite rapidly. If the biologic effects of 3AB were quite rapid one might expect 4 h of exposure to yield as great an effect as 24 or 72 h, on the other hand, if one supposed a slow cumulative effect one might have expected an increasing effect over the full 72 h period. To test this hypo-thesis, we tested cells with MNNG and 3AB for 4, 24, or 72 h. MNNG was chosen because it is a relatively short-lived alkylating agent and most DNA binding resulting from this compound occurs within 150 min. Therefore, we felt that all three experi-mental conditions would achieve a similar effective exposure to MNNG. We found that while a 4 h exposure to 3AB caused significant enhancement there was still greater enhancement following a 24 or 72 h exposure. These results would seem com-patible with the view that 3AB inhibits the repair of DNA damage and that this damage is "fixed" as the cells go through S-phase. However, most of the cells go through S-phase and "fix damage" within a 24 h period, thereby obviating any advan-tage to a more extended exposure (i.e., 72 h). Although the specific nature of the fixed damage is still undetermined, previous work by Kasid et al. [8] shows that foci induced by a combination of X-rays and benzamide display dominant transforming genes, when transfected into NIH 3T3 cells. In contrast to the enhanced toxicity and transformation observed with EMS and MNNG, we observed minimal effects on AFB_1 or MCA-induced transformation.

We have shown that the ability of 3AB to enhance transformation is (a) chemically dependent; (b) dose dependent; and (c) time dependent.

References

1. James MR, Lehrmann AR (1982) Role of poly(adenosine diphosphate ribose) in DNA repair in human fibroblasts. Biochemistry 21:4008–4013
2. Nduka N, Skidmore CJ, Shall S (1980) The enhancement of cytotoxicity of N-methyl-N-nitrosourea and of gamma radiation by inhibitors of poly(ADP-ribose) polymerase. Eur J Biochem 105:525–530
3. Morgan WF, Cleaver JE (1982) 3-Aminobenzamide synergistically increases sister-chromatid exchanges in cells exposed to methyl methanesulfonate but not to ultraviolet light. Mutat Res 104:361–366
4. Kazumi T, Yoshino G, Baba S (1980) Pancreatic islet cell tumors found in rats given alloxan and nicotinamide. Endocrinol Jpn 27:387–393
5. Takahashi S, Makae D et al. (1984) Enhancement of DEN initiation of liver carcinogens by inhibitors of NAD + ADP ribosyl transferase in the rat. Carcinogenesis 5:901–906
6. Borek C, Ong A, Cleaver JE (1984) Inhibition of radiogenic and chemical oncogenesis by inhibitors of poly (ADP-ribose). Proc Natl Acad Sci USA 81:273–277
7. Kun E, Kirsten E, Milo GE, Kurian P, Kunari HI (1983) Cell cycle-dependent intervention by benzamide of carcinogen induced neoplastic transformation and in vitro poly(ADP-ribosyla-tion) of nuclear proteins in human fibroblasts. Proc Natl Acad Sci USA 80:7219–7223
8. Kasid U, Stefanik D, Lubet RA, Dritschilo A, Smulson M (in press) Relationship between poly ADP-ribosylation and DNA sequence alterations as measured by transformation. Carcino-genesis
9. Lubet RA, McCarvill JT, Putman DL, Schwartz JL, Schechtman LM (1984) Effect of 3-amino-benzamide on the induction of toxicity and transformation by EMS and methylcholanthrene in BALB 3T3 cells. Carcinogenesis 5:459–462
10. Kakunaga T (1973) A quantitative system for assay of malignant transformation by chemical carcinogens using a clone derived from BALB/3T3. Int J Cancer 12:463–473

Specific Binding of Benzamide to DNA:
Possible Correlation to Antitransforming Activity

ERNEST KUN and PAL I. BAUER[1]

Introduction

Experimental evidence obtained in at least two laboratories [2, 3] has demonstrated that benzamide-related reagents, known to act in vitro as inhibitors of poly(ADP-ribose) polymerase [4], when applied at specifically nontoxic concentrations to various cells in culture prevent cellular transformation in vitro initiated by an ultimate carcinogen [2], X-ray, or UV light [3, 5]. That these cellular transformations can lead to neoplasia was supported by recent experiments with 3T3 (NIH) cells which, if transformed, are known to produce lethal tumors in nude mice [6]. A seemingly paradox observation has also been reported [2] — and will be discussed in this symposium, i.e., that relatively large doses of the same polymerase inhibitors, if administered in vivo can enhance the formation of precancerous lesions induced by chemical carcinogens. It appears, therefore, that, depending on dosage, either prevention of transformation or synergism to carcinogens can be achieved by benzamides. We have studied biochemical and molecular structural requirements for the antitransforming propensity of over 15 different molecular species and found that the enzyme inhibitory property is irrelevant to the slow (t1/2 = 5 to 10 h) and cell cycle specific antitransforming biological effect [7]. Two types of nuclear binding sites were identified: (1) a low affinity, enzyme inhibitory binding site and (2) a high affinity antitransformation related binding site [7]. Several powerful antitransforming drugs do not possess affinity for site 1. Therefore, the dual affinity of benzamides to both sites introduces an ambiguity that requires clarification in molecular terms. It is predictable that antitransformers like benzamide, with significant poly(ADP-ribose) polymerase inhibitory potency (site 1), are undesirable from the standpoint of potential antitransforming drug design, because at higher doses inhibition of the polymerase will abolish the desired biological response, in fact, the results will be a cooperation with the carcinogens.

We have observed [2] that radioactive benzamide accumulates in the nuclei of cells that are exposed to 1 mM extracellularly applied drug, and the intranuclear concentration of benzamide corresponding to prevention of transformation, if applied for 10 h in early S phase, is not more than 4 to 8 μM. In the present report we describe experiments

---

1 Department of Pharmacology, Biochemistry and Biophysics, and the Cardiovascular Research Institute, The University of California, San Francisco, CA 94143, USA

ADP-Ribosylation of Proteins
(ed. by F.R. Althaus, H. Hilz, and S. Shall)
© Springer-Verlag Berlin Heidelberg 1985

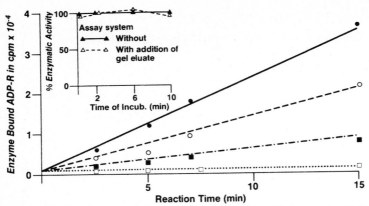

Fig. 1. The binding of coenzymic DNA to benzamide affinity matrix. *Ordinate:* auto-poly(ADP-ribosylation) of the enzyme; *abscissa:* time of enzymic reaction. ●——● Recovery of free coenzymic DNA after incubation with matrix (30 min 25°) containing *no* covalently-bound benzamide (control 1). —○—○— 0.2 ml gel, —■——■— 0.4 ml gel. Recovery of free coenzymic DNA after incubation (30 min 25°) with increasing amounts of benzamide-affinity matrix. —□——□— System containing no DNA (control 20; *inset* shows the absence of inhibitory component in the gel eluate

demonstrating an intranuclear association of benzamide not to the enzyme protein, but preferentially with its coenzymic DNA [8]. The quantities and kinetics of DNA binding agree with the conditions required for antitransformation in cell cultures.

Experimental Procedures

Isolation of calf thymus poly(ADP-ribose) polymerase (at least 95% homogeneous, specific activity = 1500 nmol ADP-ribose incorporated mg protein in 1 min) and co-enzymic DNA were carried out using published methods [8]. The synthesis of 0-amino-benzamide agarose affinity matrix, DNA-cellulose, the method of iodination of the enzyme protein (with [^{125}I]), the preparation of the Bolton-Hunter reagent adduct of 0-aminobenzamide and its iodination [^{125}I] have been described in detail [1].

Results

Benzamide covalently bound to agarose matrix (linker at the ortoposition) does not bind poly(ADP-ribose) polymerase (data not shown). However, the binding to this affinity matrix of the coenzymic DNA of poly(ADP-ribose) polymerase is readily demonstrable (Fig. 1). The free coenzymic DNA was determined by poly(ADP-ribose) polymerase, an experiment based on the linear rate of enzyme-autopoly(ADP-ribosyla-tion) that is directly proportional to the concentration of free coenzymic DNA in the

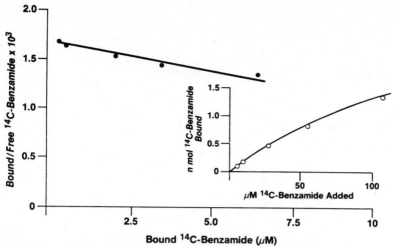

Fig. 2. Association of [^{14}C]-benzamide with calf thymus DNA, determined by equilibrium dialysis, illustrated as a Scatchard plot. Dialysis was carried out for 40 h at 4°C. *Inset:* the relationship between binding as a function of benzamide concentration

system. The coenzymic DNA was incubated for 30 min at 25° with 0, and increasing quantities of benzamide affinity gel, then the gel was spun down and remaining DNA was determined in the supernatant fluid with poly(ADP-ribose) polymerase [1]. The gel without the benzamide did not bind coenzymic DNA. The association of coenzymic DNA with calf thymus DNA was also determined by equilibrium dialysis (40 h at 4°) as shown in Fig. 2. Benzoic acid did not bind to DNA. The calculated binding was 2.7 μM benzamide/μg DNA, a value which is in good agreement with the quantity of intranuclear benzamide found in transformation resistant cells. The t1/2 of benzamide binding to DNA was 8 to 10 h, in agreement with the biological t1/2 for antitransforming activity. We have tested the influence of the benzamide-DNA association on the binding of [^{125}I]-labeled enzyme protein to DNA-cellulose, and found no effect. The DNA binding of benzamide was also shown with isolated liver nuclei as illustrated in Fig. 3. Digestion of DNA in liver nuclei decreased and abolished the binding of the benzamide ligand ([^{125}I]-labeled BH adduct of 0-aminobenzamide) as demonstrated in Table 1, illustrating the specific role of DNA in benzamide binding in nuclear systems. Preferential association of benzamide with the coenzymic DNA vs thymus DNA is summarized in Table 2. It is noteworthy that the benzamide affinity column can isolate the coenzymic DNA species from unfractionated calf thymus DNA.

Conclusions

Several lines of evidence point to the existence of certain nuclear sites [11] that may be involved in the control of multigene expressions, as would be expected to be the

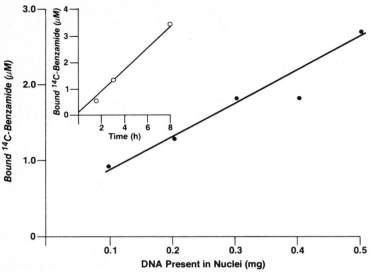

Fig. 3. The association of benzamide with DNA present in liver nuclei, (t = 2.5 h). The *main plot* illustrates a DNA dependence, whereas the *inset* the dependence on the time of incubation (nuclear DNA = 0.2 mg)

Table 1. Correlation between DNA content of liver nuclei prior to and after digestion with DNAse I with the binding of the $[^{125}I]$-labeled adduct of the Bolton-Hunter reagent and 0-aminobenzamide

	Time of incubation with DNAse I (min)	DNA content (μg DNA mg^{-1} protein)	$[^{125}I]$-ligand binding (pmol mg^{-1} protein)	% Decrease in binding
1.	0	345	981	0
2.	10	242	392	60
3.	30	76	0	100

Table 2. Preferential association of $[^{125}I]$-BH-benzamide with coenzymic DNA

DNA species	Ligand binding:pmol/100 μg DNA
Calf thymus DNA	16
Coenzymic DNA	60

molecular basis of regulation of phenotypic cellular expression. Inhibitory sites of poly(ADP-ribose) polymerase enzyme are unlikely to be participatory in this complex regulation, but selective DNA binding sites could fulfill these biological requirements and we are presently comparing this reaction as a possible common denominator for a variety of antitransforming molecules. The cell killing effect of certain antibiotics is known to be related to their mode of binding to certain DNA sequences [9]. Therefore, it seems plausible to assume that less drastic biological modifications, as presumed to operate in the control of cellular phenotypes by drugs, could be initiated by selective binding to DNA domains. The common link to the poly(ADP-ribose) system of a variety of antitransformers may be sought in their preferential binding to the coenzymic DNA of the enzyme which may in turn be related to the inducibility of this enzyme [10].

Acknowledgments. This work was supported by Grant F49620-81-C-0007 from the Office of Scientific Research of the United States Air Force and by Grant HL-27317 from the United States Public Health Service. P.I.B. is a visiting scientist from the Semmelweiss University, 2nd Institute of Biochemistry, Budapest and E.K. is a recipient of the Research Career Award of the U.S. Public Health Service.

References

1. Bauer PI, Hakam A, Kun E (in press) The association of benzamide with DNA
2. Kun E, Kirsten E, Milo GE, Kurian P, Kumari HL (1983) Cell cycle-dependent intervention by benzamide of carcinogen induced neoplastic transformation and in vitro poly(ADP-ribosyl)-ation of nuclear proteins in human fibroblasts. Proc Natl Acad Sci USA 80:7219–7223
3. Borek C, Morgan WF, Ong A, Cleaver JE (1984) Inhibition of malignant transformation in vitro by inhibitors of poly(ADP-ribose) synthesis. Proc Natl Acad Sci USA 81:243–247
4. Purnell MR, Wish WJD (1980) Novel inhibitors of poly(ADP-ribose) synthetase. Biochem J 185:775–777
5. Milo GE, D'Ambrosio S, Kun E (unpublished)
6. Milo GE, Kun E (unpublished)
7. Milo GE, Kurian P, Kirsten E, Kun E (1985) Inhibition of carcinogen induced cellular transformation of human fibroblast by drugs that interact with the poly(ADP-ribose) polymerase system. FEBS Lett 179(2):332–336
8. Yoshihara K, Hashida T, Tanaka Y, Ohgushi H, Yoshihara H, Kamiya T (1978) Bovine thymus poly(adenosine-diphosphate ribose) polymerase. J Biol Chem 253:6459–6466
9. Waring MJ (1981) DNA modification and cancer. Annu Rev Biochem 50:149–192
10. Griffin MJ, Kirsten E, Caribelli R, Palakodety RB, McLick J, Kun E (1984) The in vivo effect of benzamide and pheno barbital on liver enzymes: poly(ADP-ribose) polymerase, cytochrom P-450 styrene oxide hydrolase, cholesterol oxide hydrolase, glutathione S-transferase and UDP-glucuronyl transferase. Biochem Biophys Res Commun 122:770–775
11. Stark GR, Wahl GM (1984) Gene amplification. Annu Rev Biochem 53:447–491

Enhancement by m-Aminobenzamide of Methylazoxymethanol Acetate-Induced Hepatoma of the Small Fish "Medaka" (Oryzias latipes)

MASANAO MIWA, TAKATOSHI ISHIKAWA, TOMOKO KONDO, SHOZO TAKAYAMA, and TAKASHI SUGIMURA[1]

Introduction

A poly(ADP-ribosyl)ation reaction is suggested to be involved in DNA repair, mutagenesis, cell transformation, and cell differentiation [1-3].

We have been interested in the effect of poly(ADP-ribosyl)ation on chemical carcinogenesis in vivo. The system used was the induction of hepatocarcinogenesis by methylazoxymethanol acetate (MAM acetate) in the small fish "Medaka", *Oryzias latipes*. Medaka was selected as a suitable model animal for the following reasons: It is easy to breed, adapts to a wide range of temperatures, is easy to sexually differentiate, has a short reproductive cycle, and is available as a pure strain. In addition, continuous administration to Medaka of the drug is possible by simply dissolving it in tank water.

Results

Medaka were treated with MAM acetate at concentrations of $60\ \mu M$ for 1 h at $25°C$ and after being washed several times in freshly distilled water they were transferred to tanks containing distilled water with or without 13 mM m-aminobenzamide (m-AB). This m-AB treatment was performed for 5 days, and then the m-AB was removed from the tank. Medaka were reared for 2 months after treatment with MAM acetate and were then sacrificed. The numbers of Medaka-bearing hepatoma were counted and are summarized in Table 1 [4].

There was a significant enhancement of the incidence of MAM acetate-induced hepatoma in Medaka by treatment with m-AB. The m-AB alone did not show a carcinogenic effect. m-Aminobenzoic acid, which is an analogue of m-AB, but is not inhibitory to poly(ADP-ribose)polymerase, did not show enhancement of MAM acetate-induced hepatoma.

---

1 Virology Division, National Cancer Center Research Institute, 1-1, 5-chome, Tsukiji, Chuo-ku, Tokyo 104 and
Experimental Pathology, Cancer Institute, 37-1, 1-chome, Kami-Ikebukuro, Toshima-ku, Tokyo, Japan

ADP-Ribosylation of Proteins
(ed. by F.R. Althaus, H. Hilz, and S. Shall)
© Springer-Verlag Berlin Heidelberg 1985

Table 1. Enhancement by m-aminobenzamide of MAM acetate-induced hepatocarcinogenesis

Treatment MAM acetate	m-AB	Fish with tumor per fish examined	Incidence (%)
$60 \mu M$, 1 h	–	4/20	20
–	13 mM, 5d	0/29	0
$60 \mu M$, 1 h	13 mM, 5d	18/21	86

When m-AB treatment was given from day 5 to day 10, the enhancement by m-AB of hepatoma formation was less significant. Therefore, the effect of m-AB is thought to be closely coupled to the cellular events occuring in the early period after the MAM acetate treatment.

To better understand the involvement of NAD and poly(ADP-ribose) metabolism during the early stage of MAM acetate treatment NAD metabolism was studied.

The NAD level of the liver of Medaka treated with MAM acetate, with or without m-AB treatment, was measured using a sensitive method established in our laboratory.

As shown in Fig. 1, treatment with $60 \sim 180 \mu M$ MAM acetate gradually decreased the NAD level of the liver and the amount of this decrease was dependent on the MAM acetate concentration, with the lowest level being achieved at 7 h after treatment with MAM acetate. The decrease of NAD was greatly inhibited by administration of m-AB. m-AB itself had some effect on increasing the NAD level even in the absence of MAM acetate treatment. The decreased level of NAD gradually recovered

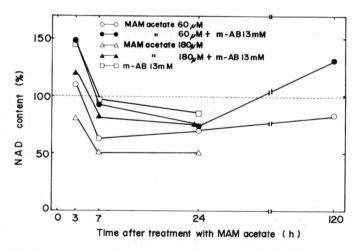

Fig. 1. NAD levels of the liver of Medaka treated with MAM acetate with or without m-AB. The liver of Medaka was removed at the indicated times, and each liver was homogenized with 0.5 N perchloric acid containing [^{14}C]-NAD as a tracer. The supernatant after centrifugation was neutralized and subjected to high performance liquid chromatography. The amount of NAD was calculated by the peak height with reference to the standard and was normalized by the recovery of [^{14}C]-NAD. The concentration of NAD was measured per mg of protein, and the value of the liver of untreated Medaka, was taken as 100 %, to which the concentration of NAD was expressed

482 M. Miwa et al.

during the 5 day period. These results were in accordance with the data obtained in the cultured cell system which showed that DNA damaging alkylating agents decreased the cellular NAD level and this decrease was inhibited by m-AB [5].

Next we studied DNA repair by the number of nicks in DNA of the liver after MAM acetate treatment. The length of single-stranded DNA was analyzed by agarose gel electrophoresis under alkaline conditions [6].

Treatment with MAM acetate increased the number of nicks and caused a decrease in the length of single-stranded DNA, 2 h after the MAM acetate treatment. By 24 h after MAM acetate treatment, a significant portion of the nicked DNA was repaired and identified as a smear near the original band. However, when m-AB was added after treatment with MAM acetate the length of the single-stranded DNA did not recover to any great extent even at 24 h after MAM acetate treatment (data not shown).

Discussion

These results suggest the involvement of NAD and probably poly(ADP-ribose) metabolism in the initial step of liver carcinogenesis by MAM acetate. It should be clarified whether the inhibition by m-AB of poly(ADP-ribosyl)ation simply shortens the latency of MAM acetate-induced hepatoma formation or if it affects some other aspects to cause pathologically or molecularly different kinds of tumors. Our data are consistent with those of Konishi et al. that m-AB enhances the formation of preneoplastic foci in the liver of rat treated with diethylnitrosamine [7]. It is of interest that m-AB and benzamide were reported to decrease the transformation of cultured cells in vitro [8,9]. However, the mechanism of carcinogenesis may involve multisteps, which consist of gene mutation, rearrangement, amplification, and translocation. Further works will clarify the key step in carcinogenesis which involves poly(ADP-ribosyl)ation.

References

1. Sugimura T, Miwa M (1983) Poly(ADP-ribose) and cancer research. Carcinogenesis 4:1503–1506
2. Hayaishi O, Ueda K (eds) (1982) ADP-ribosylation reaction, biology and medicine. Academic Press, London New York
3. Miwa M, Hayaishi O, Shall S, Smulson M, Sugimura T (eds) (1983) ADP-ribosylation, DNA repair and cancer. Proc 13th Int Symp Princess Takamatsu Cancer Res Fund, Tokyo, 1982. Jpn Sci Soc Press, Tokyo
4. Ishikawa T, Prince Masahito, Aoki K, Takayama S (1983) Inhibitor of poly(ADP-ribose) polymerase enhances hepatocarcinogenesis in small aquarium fish, medaka. Proc 42nd Annu Meet Jpn Cancer Assoc, p 43
5. Shall S (1983) ADP-ribosylation, DNA repair, cell differentiation and cancer. In: Miwa M, Hayaishi O, Shall S, Smulson M, Sugimura T (eds) ADP ribosylation, DNA repair and cancer. Proc 13th Int Symp Princess Takamatsu Cancer Res Fund, Tokyo, 1982. Jpn Sci Soc Press, Tokyo, pp 3-25

6. Michael W, Martha NS, Williams S (1977) Analysis of restriction fragments of T7 DNA and determination of molecular weight by electrophoresis in neutral and alkaline gels. J Mol Biol 110:119–146
7. Takahashi S, Ohnishi T, Denda A, Konishi T (1982) Enhancing effect of 3-aminobenzamide on induction of γ-glutamyl transpeptidase positive foci in rat liver. Chem Biol Interact 39: 363–368
8. Kun E, Kirsten E, Milo G, Kurian P, Kumari H (1983) Cell cycle-dependent intervention by benzamide of carcinogen induced neoplastic transformation and in vitro poly(ADP-ribosyl) ation of nuclear protein in human fibroblasts. Proc Natl Acad Sci USA 80:7219–7233
9. Borek C, Morgan W, Ong A, Cleaver J (1984) Inhibition of malignant transformation in vitro by inhibitors of poly(ADP-ribose)synthesis. Proc Natl Acad Sci USA 81:243–247

Influence of Dietary Nicotinamide on N-Nitrosodimethylamine Carcinogenesis in Rats

EDWARD G. MILLER[1]

Abbreviations:

DMN   N-nitrosodimethylamine
ADP-ribose   adenosine 5'-diphosphoribose

In the late 1970's as evidence started to mount [1,2] linking poly(ADP-ribose) polymerase to DNA repair, this laboratory started to explore the possible clinical significance of this nuclear enzyme. A number of papers had already been published concerning the effects of carcinogens on the metabolism of nicotinamide, NAD, and poly(ADP-ribose). These results can be summarized as follows:

1. Carcinogens can lower the cellular levels of NAD [3].
2. Carcinogens can increase the urinary excretion of 1-methylnicotinamide [4].
3. Carcinogens can stimulate the activity of poly(ADP-ribose) polymerase [5,6].

Another piece of evidence connecting carcinogens to nicotinamide is found in the studies of Warwick and Harington [7]. Their research on the epidemiology and etiology of esophageal cancer concentrated on the Xhosa tribe in the Transkei of South Africa. The incidence of carcinoma of the esophagus is extremely high in this population; another problem is pellagra. The authors presented evidence linking the two problems and suggested that the dominance of maize as the monostaple crop might be increasing the susceptibility of these people to an environmental carcinogen.

All of this data has been combined into a working hypothesis that links nicotinamide, NAD, and poly(ADP-ribose) polymerase to carcinogenesis. This hypothesis can be used to make several predictions. One, if the level of nicotinamide in the diet is below normal, then the concentration of NAD in the tissues will be reduced. This, in turn, will inhibit the ability of poly(ADP-ribose) polymerase to respond to damage in the DNA. This slowing down of the repair process should magnify the effects of ultimate carcinogens and accelerate carcinogenesis. The second prediction is the opposite of the first. If the level of nicotinamide in the diet is above normal, then this will increase the concentration of NAD in the tissues, enhance the ability of poly

---

1 Department of Biochemistry, Baylor College of Dentistry, 3302 Gaston Avenue, Dallas, TX 75246, USA

ADP-Ribosylation of Proteins
(ed. by F.R. Althaus, H. Hilz, and S. Shall)
© Springer-Verlag Berlin Heidelberg 1985

(ADP-ribose) polymerase to respond to carcinogen-induced damage, and inhibit tumorigenesis. This work represents our first attempt to test this working hypothesis.

One of the purposes of this experiment was to reproduce within a laboratory setting some of the nutritional problems that are found in the Transkei. With the help of Dr. R. Rose at Teklad Test Diets three special rat chows were formulated. Each diet was deficient in protein (the protein content was 5.5 % instead of 20 %) and the only source of protein was a vitamin-free casein hydrolysate. A second change that was common to the three diets was in the content of carbohydrate. This was increased from 67 % which is considered normal to 81.5 %. Other ingredients, except for vitamin B_3 (niacin and nicotinamide), were kept at normal levels. The content of vitamin B_3 varied from 0 to 50 to 500 mg of nicotinamide kg^{-1} of food. The nicotinamide-deficient diet was given to animals in groups designated L or LC. Fifty mg of nicotinamide kg^{-1} of food is considered normal and this chow was given to rats in the group labeled N. The diet containing a ten fold excess of the vitamin was given to the animals in the H group.

Figure 1 gives the experimental design. As can be seen the rats were on the special diets for 5 weeks. Before and after this period of time all of the animals were fed Purina Rat Chow. The rats were given 10 days to adjust to the new diets and following this interval of time the animals in the experimental groups (L, N, and H) were given 12 injections of N-nitrosodimethylamine, DMN. The injections took 18 days to complete and were given i.p. Following the treatment with the proximate carcinogen the rats remained on the diets for one more week. During the last 3 weeks of the treatment some of the animals from the experimental groups were sacrificed. The livers and kidneys were excised and analyzed for NAD-NADH and NADP-NADPH [8].

Following the treatment the animals were monitored on a daily basis. Whenever a rat died it was autopsied. At the end of 600 days the experiment was terminated. At this time the remaining animals were sacrificed and autopsied.

The severity of the special diets was readily apparent. Six of the animals died during the treatment and all of the rats lost weight. At the start of the 5 week period,

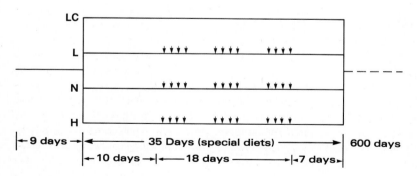

↑ injections of DMN (dose--5mg/kg)

Fig. 1. Experimental design highlighting time relationships during the treatment with special diets and carcinogens

the average weight for the animals in each of the groups was 160 g. At the end of the treatment, the averages for the four groups ranged from 105-108 g. Overall the weight loss patterns were nearly identical. This result suggests that the loss in weight was due to the restricted protein content of the diets and the poor quality of the protein that was present. Through the use of this data it was calculated that the average total intake of DMN for the rats in the experimental groups ranged from 6.8-6.9 mg. All of the symptoms of extreme malnutrition disappeared after the rats were returned to a normal diet.

Table 1 gives the data for the NAD-NADH and NADP-NADPH analyses. From Table 1 it is apparent that the treatment with special diets and DMN had a profound effect on the hepatic concentrations of these coenzymes. The values for the L, N, and H groups were 45-65 % lower than the corresponding values for the control group (animals fed Purina Rat Chow). In the kidney the results were less dramatic. The only numbers that differed significantly ($p < 0.02$) from the results for the control group were the NAD-NADH values for the L group.

Table 1. Variations in the renal and hepatic concentrations of NAD-NADH and NADP-NADPH during the treatment with DMN and special diets. (Reprinted with permission from Miller and Burns [9])

Group	Mean NAD-NADH[a] (mg g^{-1} tissue) Liver	Kidney	Mean NADP-NADPH (mg g^{-1} tissue) Liver	Kidney
Control[b]	0.53 ± 0.02^c	$0.45 \pm 0.03_d$	$0.27 \pm 0.03_d$	0.110 ± 0.01
L	$0.20 \pm 0.02^{d,e}$	0.35 ± 0.01^d	0.11 ± 0.01^d	0.092 ± 0.004
N	0.26 ± 0.02^d	0.38 ± 0.01	0.12 ± 0.01^d	0.096 ± 0.003
H	0.30 ± 0.01^d	0.39 ± 0.02	0.11 ± 0.01^d	0.094 ± 0.006

[a] The data in the table gives the means for seven separate determinations for NAD-NADH and NADP-NADPH. In each determination one of the animals from each of the groups was sacrificed. The liver and kidneys were removed and assayed for NAD-NADH and NADP-NADPH. Duplicate samples were taken for each assay

[b] Rats receiving Purina Rat Chow

[c] Mean ± SE

[d] Significantly less than the corresponding values for the control group

[e] Significantly less than the corresponding value for the H group

Figure 2 gives the cumulative probability of death due to renal tumors for the animals in the three experimental groups. After a lag of approximately 30 weeks the rate of attrition from these mesenchymal tumors increased. Analysis of the data from week 30 to the end of the experiment showed that there were significant differences ($p < 0.025$) in the life expectancies of the animals in the L group, 44.1 ± 1.6 weeks (SE), and the animals in the H group, 51.9 ± 2.8 weeks. The data for the N group (47.9 ± 3.1 weeks) was not statistically significant. None of the animals in the LC group developed renal tumors and the average life expectancy for the control group was 82.9 weeks.

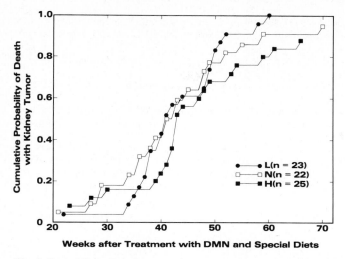

Fig. 2. Probability of death with kidney tumor in rats treated with DMN. The cumulative probability of death was calculated according to the method of Saffiotti et al. [10]. Reprinted with permission from Miller and Burns [9]

Taken together the results showed that the rate of DMN-induced tumorigenesis was inversely related to the concentration of nicotinamide in the special diets. The reason for this effect is not known, but the data in Table 1 suggested that the level of NAD-NADH in the tissues might be related to the differences in renal carcinogenesis. If such a relationship does exist, then this would increase the possibility that changes in the activity of poly(ADP-ribose) polymerase might be the underlying metabolic reason for these dietary effects. The data neither supports nor rejects this premise, but evidence from other laboratories has shown that DMN can affect the metabolism of nicotinamide [4], NAD [3], and poly(ADP-ribose) [11].

References

1. Miller EG (1975) Stimulation of nuclear poly(adenosine diphosphate-ribose) polymerase activity from HeLa cells by endonucleases. Biochem Biophys Acta 395:191–200
2. Smulson ME, Schein P, Mullins DW, Sudhakar S (1977) A putative role for nicotinamide adenine dinucleotide-promoted nuclear protein modification in the antitumor activity of N-methyl-N-nitrosourea. Cancer Res 37:3006–3012
3. Schein PS (1969) 1-Methyl-1-nitrosourea and dialkylnitrosamine depression of nicotinamide adenine dinucleotide. Cancer Res 29:1226–1232
4. Chu BCF, Lawley PD (1975) Increase urinary excretion of nucleic acid and nicotinamide derivatives by rats after treatment with alkylating agents. Chem Biol Interact 10:333–338
5. Berger NA, Sikorski GW, Petzold SJ, Kurohara KK (1979) Association of poly(adenosine diphosphoribose) synthesis with DNA damage and repair in normal human lymphocytes. J Clin Invest 63:1164–1171
6. Jacobson MK, Levi V, Juarez-Salinas H, Barton RA, Jacobson EL (1980) Effect of carcinogenic N-alkyl-N-nitroso compounds on nicotinamide adenine dinucleotide metabolism. Cancer Res 40:1797–1802

7. Warwick GP, Harington JS (1973) Some aspects on the epidemiology and etiology of esophageal cancer with particular emphasis on the Transkei, South Africa. Adv Cancer Res 17: 81–229
8. Nisselbaum JS, Green S (1969) A simple ultramicro method for determination of pyridine nucleotides in tissues. Anal Biochem 27:212–217
9. Miller EG, Burns H (1984) N-nitrosodimethylamine carcinogenesis in nicotinamide-deficient rats. Cancer Res 44:1478–1482
10. Saffiotti U, Montesano R, Sellakumar AR, Cefis R, Kaufman DG (1972) Respiratory tract carcinogenesis in hamsters induced by different numbers of administrations of benzo(a)pyrene and ferric oxide. Cancer Res 32:1073–1081
11. Romaschin AD, Kirsten E. Jackowski G, Kun E (1981) Quantitative isolation of oligo- and polyadenosine-diphosphoribosylated proteins by affinity chromatography from livers of normal and dimethylnitrosamine-treated Syrian hamsters. J Biol Chem 256:7800–7805

Involvement of Poly(ADP-Ribosylation) Reactions in Liver Carcinogenesis of Diethylnitrosamine in Rats

YOICHI KONISHI[1], SEIICHI TAKAHASHI[1], DAI NAKAE[1], KAZUHIKO UCHIDA[1], YOHKO EMI[1], KAZUMI SHIRAIWA[1], and TAKEO OHNISHI[2]

Abbreviations:

DEN	diethylnitrosamine
ABA	3-aminobenzamide
γ-GTP	γ-glutamyltranspeptidase
PB	phenobarbital
NAD	nicotinamide adenine dinucleotide
PH	partial hepatectomy
G-6-Pase	glucose-6-phosphatase
ATPase	adenosine triphosphatase

Introduction

Our investigations of the possible involvement of poly(ADP-ribosylation) reaction in liver carcinogenesis were initiated by the evidence indicating that poly(ADP-ribose) participates in DNA damage and repair [1,2] since induction of DNA lesions by carcinogens and their repair have been considered to be closely related to carcinogenesis [3-6]. Previously, we reported enhancement of diethylnitrosamine (DEN) initiation of liver carcinogenesis by inhibitors of NAD^+ ADP-ribosyl transferase in rats [7-9]. However, evidence for the direct involvement of the poly(ADP-ribosylation) reaction in the induction of liver carcinomas was lacking. In this paper, we describe a disturbance in NAD^+ ADP-ribosyl transferase-dependent formation of homopolymers of repeating ADP-ribose units from NAD and possible inhibition of repair of DEN damaged DNA by ABA. The foci initiated by DEN and ABA were found to be capable of development into hepatocellular carcinomas without promotion by phenobarbital (PB).

1 Department of Oncological Pathology, Cancer Center, Nara Medical College, 840 Shijo-cho, Kashihara, Nara 634, Japan
2 Department of Biology, Nara Medical College, 840 Shijo-cho, Kashihara, Nara 634, Japan

ADP-Ribosylation of Proteins
(ed. by F.R. Althaus, H. Hilz, and S. Shall)
© Springer-Verlag Berlin Heidelberg 1985

Effects of DEN and/or ABA on Rat Liver NAD Level and Poly(ADP-Ribose)

Cellular NAD levels were measured by the method described by Bernofsky and Swan [10] and studied by the indirect immunofluorescence technique established by Ikai et al. [11].

Male Fischer 344 rats received various doses of DEN and/or 600 mg kg^{-1} body wt. of ABA prior to measurement of NAD. DEN was dissolved in saline (4 mg ml^{-1}) and ABA in DMSO (300 mg ml^{-1}) and given intraperitoneally. The dose-response effect of DEN on NAD levels 12 h after administration is shown in Fig. 1a. DEN depleted dose-dependently the NAD level. Figure 1b represents the NAD level as a function of time after a 200 mg kg^{-1} body wt. injection of DEN. The depletion of NAD reached a maximum at 12 h and returned to the control level 48 h after DEN administration. The effect of 600 mg kg^{-1} body wt. ABA on NAD level as a function of time after the administration is shown in Fig. 2a. The NAD level was increased after 4 h

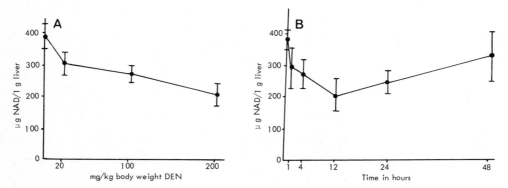

Fig. 1. A Dose-response effect of DEN on NAD level 12 h after the administration in rat liver. **B** NAD level as a function of time after administration of 200 mg kg^{-1} body wt. DEN in rat liver

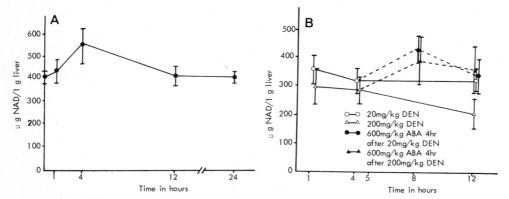

Fig. 2. A NAD level as a function of time after administration of 600 mg kg^{-1} body wt. ABA in rat liver. **B** Preventive effect of ABA on depletion of NAD level by DEN in rat liver

and returned to the control level after 12 h remaining constant thereafter. The preventive effect of ABA on depletion of AND level by DEN is shown in Fig. 2b. ABA administered 4 h after 20 or 200 mg kg^{-1} body wt. DEN, clearly prevented depletion of NAD. For indirect immunohistochemical investigation of poly(ADP-ribose), purified antibody against poly(ADP-ribose) was kindly supplied by Dr. K. Ueda, Department of Medical Chemistry, Kyoto University, Kyoto, Japan. Conspicuous intranuclear immunofluorescence of poly (ADP-ribose) was observed in the livers of control rats treated with DMSO, while it was decreased 1 h after administration of 600 mg kg^{-1} body wt. ABA. The decrease in intranuclear immunofluorescence pf poly (ADP-ribose) is presumably the result of decreased activity of NAD$^+$ ADP-ribosyl transferase due to ABA inhibition.

Effects of Partial Hepatectomy (PH) Performed Before, Simultaneously, and After DEN and ABA Administration on the Induction of γ-GTP Positive Foci in Rat Liver

It was reported that the initiation of chemical carcinogenesis requires cell proliferation [12] and liver regeneration induced by PH is effective in fixing carcinogen-induced DNA damage if PH is performed before DNA repair can be completed [13].

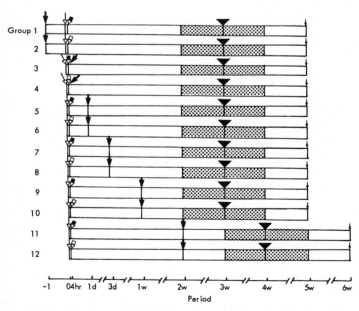

Fig. 3. Experimental design for effect of timing of PH on the induction of γ-GTP positive foci in rat liver. ➔ Partial hepatectomy (PH); ⊳ ABA 300 mg kg^{-1} i.p.; ▼ CCl$_4$ 1ml kg^{-1} i.p.; ☐ Basal diet; ➝ DEN 20 mg kg^{-1} i.p.; ⊳ DMSO 2ml kg^{-1} i.p.; ▦ 0.02 % AAF diet; ➝ Sacrifice

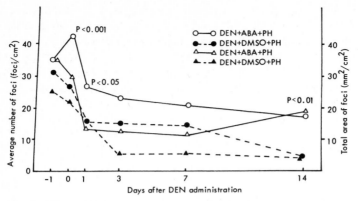

Fig. 4. Effect of PH on numbers and area of γ-GTP positive foci induced in rat liver initiated by DEN and/or ABA

The protocol followed in this experiment was developed by Cayama et al. [14] in which PH was performed at various time intervals as shown in Fig. 3. This protocol allows a determination of the time-dependent effect of ABA on the repair of DEN-induced DNA lesions since it is known that the process of repair is generally achieved within 48 h [5]. These results (Fig. 4) indicate that PH performed 14 days after DEN plus DMSO was ineffective, while after DEN plus ABA induction of γ-GTP positive foci remained enhanced, suggesting that ABA had, in both the short- and long-term, partially inhibited the repair of DNA damaged by DEN under these conditions.

Enzymatic Characterization of Foci Induced by DEN and ABA

Enzymes, such as γ-GTP, glucose-6-phosphate (G-6-Pase), and canalicular adenosine triphosphatase (ATPase) have been generally used as markers for assaying initiated cell populations in liver carcinogenesis [15]. In order to confirm the essential similarity of foci induced by DEN plus ABA to those observed after DEN alone, a comparison of histochemical characteristics was made. The foci induced by DEN plus PH, typically showing positive for γ-GTP and periodic acid Schiff stain and negative for G-6-Pase and ATPase.

Evolution of Foci Induced by DEN plus ABA to Hepatoma

It was reported that DEN-initiated enzyme-altered foci develop to form hepatocellular carcinomas [16]. However, it was reported that the vast majority of foci and hyperplastic nodules disappear after removal of the animals from the carcinogenic diet [17]. In the present series of experiment, we administered PB as a promoter continuously after initiation and induction of γ-GTP positive foci with selection pressure (Fig. 5) which is frequently used for detection of promoters [18]. The results are shown in Table 1. Hepatocellular carcinomas developed after initiation by DEN plus PH and DEN plus ABA with or without subsequent promotion by PB. Whereas with respect to tumor incidence, effect of PB could not be discerned after initiation with DEN plus PH (compare groups 4 and 5) and the promotion of lesions initiated by DEN plus ABA was observed (compare groups 6 and 7).

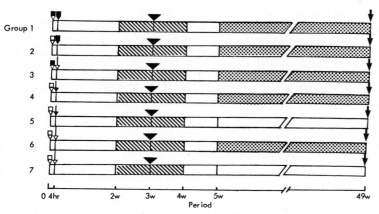

Fig. 5. Experimental design for liver carcinogenesis inititated by DEN and/or ABA and PH.
⊃ DEN 20 mg kg^{-1} i.p.; ➤ DMSO 2ml kg^{-1} i.p.; ▨ 0.02 % AAF diet; → ABA 600 mg kg^{-1} i.p.; ▼ CCl$_4$ 1ml kg^{-1} i.p.; ▨ 0.05 % PB diet; ➤ Saline 2ml kg^{-1} i.p.; → Partial hepatectomy (PH); ☐ Basal diet; ⊃ Sacrifice

Table 1. Incidence of hepatomas in the livers of rats treated with DEN and/or ABA and PH with or without subsequent PB treatment

Group	Treatment	Effective No. of rats[a]	Body weight (g)[b] Initial	Final	Liver weight (g)[b]	Incidence of hepatoma (%)
1	Saline + DMSO and PB	12	166 ± 9	459 ± 24	14.1 ± 1	0/12 (0)
2	DEN + DMSO and PB	9	164 ± 9	429 ± 38	17.0 ± 6	9/9 (100)
3	Saline + ABA and PB	15	157 ± 9	439 ± 23	13.1 ± 2	0/15 (0)
4	DEN + PH and PB	12	148 ± 4	403 ± 38	23.7 ± 7	12/12 (100)
5	DEN + PH and BD	16	153 ± 9	373 ± 46	11.8 ± 4	15/16 (95)
6	DEN + ABA and PB	11	156 ± 7	422 ± 23	13.7 ± 2	9/11 (82)
7	DEN + ABA and BD	9	164 ± 10	410 ± 13	11.0 ± 2	5/9 3 (56)

[a] Based on histological examinations
[b] g ± SD

The present results strongly support our previous findings that ABA enhanced the numbers of γ-GTP positive foci initiated by DEN in rat liver [7-9]. Independent of subsequent promotion DEN plus ABA initiated foci were capable of development through hyperplastic nodules to form hepatocellular carcinomas. It was recently found that ABA enhanced hepatocarcinogenesis of methylazoxymethanol acetate in a small fish, Medaka (*Oryzias latipes*) (T. Ishikawa et al., unpubl.). Our present findings clearly offer support for the view that DNA lesions are necessary for initiation of liver carcinogenesis and that the repair process, including changes in levels of NAD and NAD^+ ADP-ribosyl transferase, are intimately involved in neoplasia in the rat liver. Possible mechanism of involvement of poly(ADP-ribosylation) reaction in irreversible initiation of liver carcinogenesis by DEN are schematically shown in Fig. 6. Carcinogenesis is regarded as a multistep process initiated by DNA damage, gene mutation, gene rearrangement of gene transformation, and ending with final tumor development [2]. To clarify the biological function of the poly(ADP-ribosylation) process and its contribution to the development of hepatomas, detailed studies of its involvement in each of the multiple steps leading to expression of the cancer cell phenotype should be made in the future.

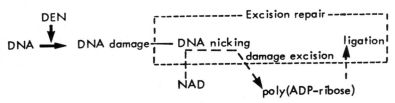

Fig. 6. Possible involvement of the poly(ADP-ribosylation) reaction in irreversible initiation of liver carcinogenesis by DEN

Acknowledgement. The investigation was supported in part by Grants-in-Aid for Cancer Research from the Ministry of Education, Science and Culture (401538, 401057, 56010070, 57010069, 58010072, and 58010083) of Japan.

References

1. Shall S (1982) ADP-ribose in DNA repair. In: Hayaishi O, Ueda K (eds) ADP-ribosylation reactions. Academic Press, London New York, p 478
2. Sugimura T, Miwa M (1983) Poly(ADP-ribose) and cancer research. Carcinogenesis 12:1503–1506
3. Lieberman MW, Dipple A (1972) Removal of bound carcinogen during DNA repair in non-dividing human lymphocytes. Cancer Res 32:1855–1860
4. Cleaver JE, Bootsma DA (1975) Xeroderma pigmentosum: biochemical and genetic characteristics. Annu Rev Genet 9:19–38
5. Sarma DSR, Rajalakshmi S, Farber E (1975) Chemical carcinogenesis: Interaction of carcinogen with nucleic acid. In: Becker FF (ed) Cancer 1. Plenum Press, New York London, p 235

6. Kondo S (1976) Misrepair model for mutagenesis and carcinogenesis. In: Magee PH et al. (eds) Fundamentals in prevention of cancer. Univ Tokyo Press, Tokyo, p 417

7. Takahashi S, Ohnishi T, Denda A, Konishi Y (1982) Enhancing effect of 3-aminobenzamide on induction of γ-glutamyltranspeptidase positive foci in rat liver. Chem-Biol Interact 39: 363–368

8. Konishi Y, Takahashi S, Ohnishi T, Denda A, Mikami S, Emi Y, Nakae D (1983) Possible role of poly(ADP-ribose) polymerase on the early stage of liver carcinogenesis by diethylnitrosamine in rats. In: Miwa M et al. (eds) ADP-ribosylation, DNA repair and cancer. Jpn Sci Soc Press, Tokyo/VNU Science Press, Utrecht, p 309

9. Takahashi S, Nakae D, Yokose Y, Emi Y, Denda A, Mikami S, Ohnishi T, Konishi Y (1984) Enhancement of DEN initiation transferase in rats. Carcinogenesis 5:901–906

10. Bernofsky C, Swan M (1973) An improved cycling assay for nicotinamide adenine dinucleotide. Anal Biochem 53:452–458

11. Ikai K, Ueda K, Hayaishi O (1980) Immunohistochemical demonstration of poly(adenosine diphosphate-ribose) in nuclei of various rat tissue. J Histochem Cytochem 28:670–676

12. Farber E, Cameron R (1980) The sequential analysis of cancer development. Adv Cancer Res 31:125–226

13. Ishikawa T, Takayama S, Kitagawa T (1980) Correlation between time of partial hepatectomy after a single treatment with diethylnitrosamine and induction of adenosinetriphosphatase-defficient islands in rat liver. Cancer Res 40:4261–4262

14. Cayama E, Tsuda H, Sarma DSR, Farber E (1978) Initiation of chemical carcinogenesis requires cell proliferation. Nature (London) 275:60–62

15. Pitot HC, Barsness L, Kitagawa T (1978) Stage in process of hepatocarcinogenesis in rat liver. In: Slage TJ, Sivak A, Boutwell RK (eds) Mechanisms of tumor promotion and carcinogenesis. Raven Press, New York, p 433

16. Scherer E, Emmelot P (1975) Kinetics of induction and growth of precancerous liver-cell foci, and liver tumor formation by diethylnitrosamine in the rat. Eur J Cancer 11:689–701

17. Farber E (1976) On the pathogenesis of experimental hepatocellular carcinoma. In: Okuda K, Peters RL (eds) Hepatocellular carcinoma. Wiley, New York, p 1

18. Ito N, Tatematsu M, Nakanishi K, Hasegawa R, Takano T, Imaida K, Ogiso T (1980) The effects of various chemicals on the development of hyperplastic liver nodules in hepatectomized rats treated with N-nitrosodiethylamine of N-2-fluorenylacetamide. Gann 71:832–842

Immunofluorescent Staining of Poly(ADP-Ribose) in Situ in Acetone-Fixed HeLa Cells: Its Usefulness for Clinical and Basic Sciences

YOSHIYUKI KANAI[1] and SEI-ICHI TANUMA[2]

1. Introduction

Since the discovery of naturally-occuring antibodies against poly(ADP-ribose) in patients with systemic lupus erythematosus (SLE) [1], a number of studies on anti-poly(ADP-ribose) in SLE patients have been reported [2–4]. Morrow et al. stressed that levels of anti-poly(ADP-ribose) antibodies correlated with clinical activity of SLE patients better than double-stranded (ds)DNA [3]. Very recently, Clayton et al. have reported that the measurement of anti-poly(ADP-ribose) antibodies is more specific and less sensitive than the anti-ds(DNA) binding assay [4]. However, wide application of the measurement of anti-poly(ADP-ribose) antibodies to the diagnosis of rheumatic diseases seems to be limited because of the shortage of available poly(ADP-ribose) antigen. Unlike DNA, poly(ADP-ribose) must be synthesized from NAD by nuclear enzyme(s) [5] and purified with a relatively low yield [6]. Therefore, we have attempted to overcome this disadvantage by establishing a fluorescent antibody-staining method, for detecting naturally-occuring antibodies against poly(ADP-ribose). Previously, we have shown that poly(ADP-ribose) could be synthesized from endogeneous NAD even in acetone-fixed HeLa cells during the incubation with antibody [7]. In the present experiments, the fluorescent staining patterns of poly(ADP-ribose) shown by rabbit antibody, changed when the cells were preincubated with NAD. The staining pattern before incubation with NAD was homogeneous with unstained spots, whereas after incubation it became homogeneous with intense fluorescence. This indicated that poly(ADP-ribose) was distributed evenly in acetone-fixed HeLa cells after incubation with NAD. Based upon these findings we have tested a number of serum samples from SLE patients who had high antibody titers against poly(ADP-ribose), for the fluorescence by this method.

1   Department of Molecular Oncology, Institute of Medical Science, University of Tokyo, Minato-ku, Tokyo 108, Japan
2   Faculty of Pharmaceutical Sciences, Teikyo University, Kanagawa 199-01, Japan

ADP-Ribosylation of Proteins
(ed. by F.R. Althaus, H. Hilz, and S. Shall)
© Springer-Verlag Berlin Heidelberg 1985

2. Materials and Methods

2.1 Cell Culture and Synchronization of Growth

HeLa S3 cells were maintained in monolayer culture at 37°C in Eagle's medium/10 % calf serum. Growth was synchronized by exposing the cells to hydroxyures (1 mM) for 16 h [7].

2.2 NAD Treatment of Cells

Cells grown on coverslips were fixed in cold acetone (-10°C). Then 100 μl of an solution (1 mM in 82.5 mM Tris-HCl, pH 8.0, 10 mM MgCl$_2$, 50 mM KCl, 3.3 mM 2-mercaptoethanol) was applied to the cells which were then incubated in a moist chamber. After incubation, the coverslips were rinsed four times with phosphate buffered saline (PBS) and once with distilled water. The cells with or without treatment were quickly stored in a freezer (-20°C).

2.3 Antisera, Human Serum Samples, and Measurement of Antinucleic Acid Antibodies

Antisera against poly(ADP-ribose) were raised in rabbits [8] and affinity-purified antibodies from γ-globulin were obtained [9]. We routinely examined antibodies against poly(ADP-ribose), dsDNA or single-stranded DNA (ssDNA) by an enzyme-linked immunosorbent assay (ELISA) [10]. Nine serum samples from SLE patients with high titers against all or any of these nucleic acids were selected for fluorescent antibody-staining for poly(ADP-ribose).

2.4 Fluorescent Antibody-Staining of HeLa Cells

The procedure for fluorescent antibody staining has been described previously [7]. For the indirect immunofluorescent staining, fluorescein isothiocyanate (FITC)-labeled antihuman IgG (Fc fragment specific) goat antibody (Cappel) was used as the second antibody. Human serum samples were diluted 1:20 in PBS, and affinity-purified primary antibody was used at 20 μg ml^{-1}, the FITC-labeled antihuman IgG and the FITC-labeled antirabbit IgG goat antibody (Cappel) were diluted 1:40. The photographs were taken with a microscope equipped with an Olympus epiillumination fluorescent module (Model BH2-RFK). The color print film used was Ektachrome 400 (Kodak).

3. Results and Discussion

3.1 Increase in the Specific Binding Activity of an Affinity-Purified Antibody

The specific activity of affinity-purified anti-poly(ADP-ribose) antibody was ten times greater than that of the γ-globulin fraction; the specific activity was defined as the amount of protein required to give 1 A_{405} unit (antibody titer) under certain conditions. The poly(ADP-ribose) binding specificity of the rabbit antibody [8] was confirmed by both direct and competitive ELISA (data not shown).

3.2 Change of the Immunofluorescent Staining Pattern of Poly(ADP-Ribose) After Incubation with NAD

The staining patterns (by the direct immunofluorescent antibody technique) of poly-(ADP-ribose) before incubation with NAD was "homogeneous", and nucleoli were unstained [7]. After incubation with NAD, the unstained spots (nucleoli) disappeared and the pattern of staining became totally homogeneous with more intense fluorescent (not shown), indicating that poly(ADP-ribose) polymerase distributed evenly in HeLa cell nuclei. This change in the distribution of poly(ADP-ribose) and the increase in its density could be useful parameters for detecting anti-poly(ADP-ribose) antibodies amongst a number of autoantibodies reactive with other nuclear antigens, in the sera of patients with rheumatic diseases.

Ikai et al. introduced the NAD incubation system with the indirect immunofluorescent antibody staining of poly(ADP-ribose) to effectively visualize the formation of poly(ADP-ribose) in situ in methanol-fixed tissue sections in various mammalian organs [11]. Furthermore, they reported that the immunoflurescent staining by anti-poly(ADP-ribose) was prominent in the marginal area rather than the central region of nuclei [11]. They confirmed that the staining pattern of nuclei by anti-poly(ADP-ribose) antibody was very similar to that by anti-poly (ADP-ribose) polymerase antibody [12]. These experiments suggested that immunoreactive poly(ADP-ribose) could be synthesized in situ from exogeneous NAD by poly(ADP-ribose) polymerase [12].

As a medical application of poly(ADP-ribose) immunofluorescence, Ikai et al. reported that myeloblasts in peripheral blood as well as in bone marrow in patients with acute myeloblastic leukemia (AML) showed intense fluorescence of poly(ADP-ribose) in nuclei and that this was the case for blastic crisis of chronic myelocytic leukemia (CML) [13].

In our experiments using acetone-fixed HeLa cells as the substrate, the pattern of immunofluorescence of poly(ADP-ribose) was so-called homogeneous, irrespective of preincubation with NAD, with intense fluorescence after incubation with NAD. Thus, acetone-fixed HeLa cells after NAD incubation are of practical value for the screening of the presence of anti-poly(ADP-ribose) antibodies in the sera of patients with rheumatic diseases. The preparation of acetone-fixed HeLa cells is easy and the poly-(ADP-ribose) is distributed evenly in the cells particularly after incubation with NAD and behaves just like the immobilized poly(ADP-ribose) antigen as in the case of ELISA [10].

Table 1. Patterns of indirect immunofluorescent staining by human antibodies in HeLa cells before and after incubation with NAD[a]

Subject		Before	After	Antibody titers (A_{405} Unit) against:		
				Poly(ADP-ribose)	dsDNA	ssDNA
SLE No.	21	Peripheral	Homogeneous	0.248	0.198	0.097
	22	Peripheral and homogeneous	Homogeneous	0.217	0.315	0.175
	30	Fine homogeneous	—[b]	0.054	0.060	0
	61	Speckled	—	0.199	0.184	0
	109	Patchy and homogeneous	—	0.263	0.079	0.032
	110	Fine homogeneous	Homogeneous	0.256	0.172	0.067
	131	Fine homogeneous	Homogeneous	0.281	0.154	0.182
	136	Speckled	Homogeneous	0.405	0.083	0,074
	150	Speckled	—	0.187	0.061	0.043
PSS No.	10	Speckled	Speckled	ND[c]	ND	ND

[a] Normal values of anti-poly(ADP-ribose), anti-dsDNA, and anti-ssDNA titers were less than 0.03 A_{405} U

[b] Diminution of fluorescence

[c] Not determined

3.3 Indirect Immunofluorescent Staining Pattern by Human Serum Antibodies

Of eight serum samples from SLE patients with relatively high antibody titers against poly(ADP-ribose), five samples gave a homogeneous staining pattern after incubation with NAD. The patterns of immunofluorescent staining observed in nine serum samples of SLE patients and one serum sample of a PSS patient are shown in Table 1. Two samples (Nos. 21 and 22) with relatively high titers against dsDNA showed a peripheral/homogeneous pattern (Fig. 1A,C) which is considered to be associated with the presence of anti-DNA antibodies [14]; it was recently confirmed by a monoclonal antibody that the peripheral pattern was in fact associated with the presence of anti-DNA antibody [15]. After incubation with NAD the fluorescent pattern given by these two serum samples became homogeneous (Fig. 1B,D) as observed by the indirect immunofluorescent staining using affinity-purified rabbit anti-poly-(ADP-ribose) antibodies. This change may have been due to the effective interactions of antibodies with newly formed poly(ADP-ribose) in situ. Previously, we found that poly(ADP-ribose) could be synthesized from endogenous NAD during incubation with antibody [7]. However, taking account of the serological diversity of SLE patients [16] who have a number of different classes of antibodies even against a certain kind of antigen, the poly(ADP-ribose) synthesized in situ from endogenous NAD could not cover the full interaction with naturally occurring antibodies against poly(ADP-ribose). A serum sample of one patient with PSS showed a speckled pattern of immunofluorescence, irrespective of incubation with NAD (Table 1). Therefore, this serum seems not to contain anti-poly(ADP-ribose) antibodies and could serve as a negative control.

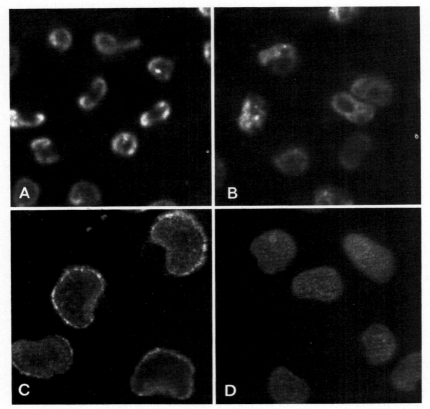

Fig. 1A–D. Indirect immunofluorescent staining of acetone-fixed HeLa cells before and after incubation with NAD. A Sample No. 21, before incubation with NAD; B No. 21, after incubation with NAD; C No. 22, before incubation with NAD; D No. 22, after incubation with NAD. Magnification: **A,B** × 200; **C,D** × 400

It is well known that antihistone antibodies give homogeneous immunofluorescent patterns, but the incidence of antihistone antibodies is relatively low in patients with rheumatic diseases [17, 18]. Therefore, in a few cases, it would be necessary to differentiate antihistone antibodies from anti-poly(ADP-ribose) antibodies when a homogeneous staining pattern is observed before incubation with NAD. In the present experiments, samples showing typical homogeneous patterns were only observed after incubation with NAD. Whether the homogeneous pattern is due to antihistone or anti-poly(ADP-ribose) could be determined by an absorption experiment with a histone antigen. Therefore, if a sample gives a strongly homogeneous pattern, it is relatively easy to differentiate anti-poly(ADP-ribose) antibodies from antihistone antibodies. One case (sample No. 110) with diminished immunofluorescence after incubation with NAD, although it showed a high titer against poly(ADP-ribose), remains to be studied as an exception.

Thus, acetone-fixed HeLa cells that are incubated with NAD could be of practical use as substrates for screening for the presence of anti-poly(ADP-ribose) antibodies by the indirect immunofluorescent antibody technique.

Acknowledgment. This work was supported in part by Grants-in Aid for Sceintific Research from the Ministry of Health and Welfare, Japan.

References

1. Kanai Y, Kawaminai Y, Miwa M, Matushima T, Sugimura T, Moroi Y, Yokohari R (1977) Naturally-occurring antibodies to poly(ADP-ribose) in patients with systemic lupus erythematosus. Nature (London) 265:175–177
2. Okolie EE, Shall S (1979) The significance of antibodies to poly(adenosine diphosphate-ribose) in systemic erythematosus. Clin Exp Immunol 36:151–164
3. Morrow WJ, Isenberg DA, Parry HF, Shen L, Okolie LL, Farzaneh F, Shall S, Snaith ML (1982) Studies on autoantibodies to poly(adenosine diphosphate-ribose) in SLE and other autoimmune diseases. Ann Rheum Dis 41:396–402
4. Clayton A-L, Bernstein RM, Tavassoli M, Shall S, Bunn C, Hughes GRV, Chantler SM (1984) Measurement of antibody to poly(adenosine diphosphate-ribose): its diagnostic value in systemic lupus erythematosus. Clin Exp Immunol 56:263–271
5. Sugimura T (1973) Poly(adenosine diphosphate ribose). Prog Nucleic Acid Res Mol Biol 12: 127–151
6. Kanai Y, Kawamitsu H, Tanaka M, Matsushima T, Miwa M (1980) A novel method for purification of poly(ADP-ribose). J Biochem (Tokyo) 88:917–920
7. Kanai Y, Tanuma S, Sugimura T (1981) Immunofluorescent staining of poly(ADP-ribose) in situ in HeLa cell chromosome in the M phase. Proc Natl Acad Sci USA 78:2801–2904
8. Kanai Y, Miwa M, Matsushima T, Sugimura T (1974) Studies on anti-poly(adenosine diphosphate ribose) antibody. Biochem Biophys Res Commun 59:300–306
9. Kanai Y, Fujiwara M (1985) Naturally occurring antibodies to poly(ADP-ribose) in autoimmune MRL/Mp-lpr/lpr mice. Clin Exp Immunol 59:132–138
10. Kanai Y, Tauchi M, Aotsuka S, Yokohari R (1982) A simple and rapid microenzyme-linked immunosorbent assay for antibodies to poly(ADP-ribose) in systemic lupus erythematosus. J Immunol Methods 53:355–365
11. Ikai E, Ueda K, Hayaishi O (1980) Immunohistochemical demonstration of poly(adenosine diphosphate-ribose) in nuclei of various rat tissues. J Histochem Cytochem 28:670–767
12. Ikai K, Ueda K, Hayaishi O (1982) Immunohistochemistry of poly(ADP-ribose). In: Hayaishi O, Ueda K (eds) ADP-ribosylation reactions. Academic Press, London New York, pp 339–360
13. Ikai K, Ueda K, Fukushima M, Nakamura T, Hayaishi O (1980) Poly(ADP-ribose) synthesis, a marker of granulocyte differentiation. Proc Natl Acad Sci USA 77:3682–3685
14. Tan EM, Rothfeld NF (1978) Systemic lupus erythematosus. In: Samter M (ed) Immunological diseases. Little Brown, Boston, pp 1038–1060
15. Lerner EA, Lerner MR, Janeway CA, Steiz JA (1981) Monoclonal antibodies to nucleic acid-containing cellular constituents: Probes for molecular biology and autoimmune diseases. Proc Natl Acad Sci USA 78:2737–2741
16. Hahn BH (1980) Systemic lupus erythematosus. In: Parker CW (ed) Clinical immunology. Saunders WB, pp 583–631
17. Tan EM, Robinson J, Robitaill P (1976) Studies on antibodies to histones by immunoflurescence. Scand J Immunol 5:811–818
18. Fishbein E, Alarcon-Segovia D, Vega JM (1979) Antibodies to histones in systemic lupus erythematosus. Clin Exp Immunol 36:145–150

Diadenosine Tetraphosphate: A New Target for Cancer Chemotherapy

THOMAS ALDERSON[1]

Introduction

Many antimetabolic drugs already used in the treatment of cancer are analogues of the normal nucleic acid bases, such as 5-fluorouracil, 6-mercaptopurine, the 2- and 8-halo-adenines, and the adenine and cytosine arabinonucleosides. These analogues and others are generally believed to act by the inhibition of nucleic acid synthesis and metabolism, and/or the incorporation of the analogue into DNA and/or RNA. Some of these analogues have sufficient selectivity against tumour cells to make them useful anti-tumour agents [1]. But the search for more tumour-specific antimetabolites, which are also less toxic for the organism, still remains a major task for chemotherapy.

The recent establishment of a link between ribonucleotide metabolism, specifically the synthesis of the cellular metabolite, diadenosine tetraphosphate, and its relationships with adenosine diphosphate-ribosylation reactions in the cell, presents a significant new target.

An example of one adenosine analogue (and other related adenosine analogues) which may now be considered to act through this pathway will be described.

Formaldehyde-Induced Mutagenesis

My own interest in nucleic acid base-analogues began during my investigation into the mechanism of mutagenesis induced by formaldehyde in *Drosophila* larvae.

In the larval feeding method, formaldehyde is added to the agar-based culture medium on which the larvae develop. My original observation identifying the basis for its mode of action was that formaldehyde exhibits a mutagenic effect only when a source of (yeast) RNA is present in the culture medium [2]. Further analysis showed that the requirement for RNA was due solely to a need for its adenylic acid component.

An essential role for the N^6-amino group of adenosine or adenylic acid was demonstrated when it was found that the corresponding hypoxanthine derivatives, inosine

---

1 Ludwig Institute for Cancer Research, MRC Centre, University Clinical School, Cambridge, Great Britain

ADP-Ribosylation of Proteins
(ed. by F.R. Althaus, H. Hilz, and S. Shall)
© Springer-Verlag Berlin Heidelberg 1985

and inosinic acid, which carry an hydroxyl group in place of the N^6-amino group, were unable to mediate formaldehyde mutagenesis [4].

It was therefore concluded that formaldehyde reacts in the culture medium to produce a mutagenic base-analogue(s). Free formaldehyde itself injected into larvae shows no mutagenic effects [6], which indicates that an initial in vitro reaction of formadehyde with either adenosine or adenylic acid is necessary for its mutagenicity. Formaldehyde-treated RNA, from which unreacted formaldehyde is removed by dialysis, shows a strong mutagenic effect [7].

The primary reaction of formaldehyde with the N^6-amino groups of adenosine or adenylic acid is to form a labile hydroxymethyl grouping ($HN\text{-}CH_2\text{-}OH$), although a secondary reaction of the hydroxymethyl group can produce a stable methylene-bridged dimer on its condensation with the unreacted N^6-amino group of a corresponding molecule ($HN\text{-}CH_2\text{-}NH$). It is now established [5] that formaldehyde reacts with either the free bases, adenine or guanine, or their nucleosides and nucleotides, to hydroxymethylate their extranuclear amino groups, which, by a slow secondary reaction under suitable conditions, generate methylene-bridged dimers of either adenine-adenine, adenine-guanine, or guanine-guanine.

The yield of N^6-hydroxymethyl adenosine or adenylic acid after addition of formaldehyde to the culture medium (at $60°C$) is unknown, since a proportion of the formaldehyde will be effectively lost by reaction with many other food components. The formation of the methylene bis-derivatives will be much less, and probably nonexistent, since their slow and limited formation (7% of adenylic acid residues) over several days under optimal conditons [5], now makes it unlikely that they will be formed in time under the conditions used for the majority of the mutagenesis experiments. Moreover, methylene bis-adenosine, unlike the corresponding adenylic acid dimer, precipitates out of solution as a very insoluble product [5], which would severely limit its effective concentration; yet adenosine and adenylic acid are equally effective in mediating formaldehyde mutagenesis [3].

It appears that the in vitro reaction must be in the form of the N^6-hydroxymethyl product, and this may well be the active mutagen. However, subsequent in vivo cross-linking could still form dimeric products, particularly when two appropriate molecules, one N^6-hydroxymethylated and the other not, come into a stereochemically appropriate relationship for methylene-bridge formation.

Genetical Effects Induced

By the larval feeding method, formaldehyde induces all types of genetical effects known to be produced by X-rays: dominant and recessive lethals, visibles, small and large deficiencies, duplications, inversions, translocations, and gynandromorphs [8].

Studies at the molecular level, after gene cloning and the DNA sequence analysis of four alcohol dehydrogenase negative mutant alleles formaldehyde-induced by the larval feeding method, have shown [9] that all four *Drosophila* mutants have a small deletion within the gene, ranging in size from 3—34 base pairs. Two of these lie within a 65 base pair intervening sequence, and are accompanied by other aberrations; one has a duplication accompanying a 6 base pair deletion, whilst the other has a base pair substitution at the 3' end of the deletion.

Formaldehyde administered by larval feeding therefore induces a wide variety of major genetic effects, which contrasts with the simple base pair changes largely found to be induced by nucleic acid base-analogues in prokaryotes. Several mutagenic purine and pyrimidine base-analogues of this latter type have been tested on *Drosophila* larvae, but found not to induce mutation [10]: namely, 2-aminopurine; 2,6-diaminopurine; 5-bromodeoxyuridine; and 5-bromodeoxycytidine.

A New Class of Adenine Ribonucleoside Analogue

The actual pathway for the mutagenic activity of the N^6-hydroxymethyl derivatives of adenosine and adenylic acid has proved difficult to establish.

It was first assumed [3] that the analogues were effective by incorporation into DNA after conversion to their deoxyribose forms. However, it was later found that only the ribose derivative is effective [11]; the deoxyribose derivatives of adenosine and adenylic acid do not mediate formaldehyde mutagenesis when present in the culture medium.

It was therefore considered [12] that the analogues may produce a mutagenic effect by incorporation into RNA. In this case, the transcription of genetic information from chromosomal DNA to an RNA intermediate, and thence to newly synthesised DNA by a reverse transcriptional process, would seem necessary to explain this type of mechanism.

However, the recent identification in mammalian cells of a regulatory role for the cellular metabolite, diadenosine tetraphosphate (Ap4A), and its relationship with ADP-ribosylation reactions via the enzyme poly(ADP-ribose) polymerase [27], now suggest that the analogues may act by their utilisation in Ap4A synthesis. The utilisation of the N^6-hydroxymethyl derivatives of adenosine and adenylic acid in Ap4A synthesis, where one or both of the adenosine residues of Ap4A are N^6-substituted, may thus produce an antimetabolite(s) which interferes with Ap4A-mediated events. The N^6-hydroxymethyl derivates may consequently be required to be in the ribose form, since Ap4A is a ribose derivative; this would explain why the deoxyribose derivatives of adenosine and adenylic acid are not mediators of formaldehyde mutagenesis.

Also the involvement of poly(ADP-ribose) polymerase in DNA excision repair [13], and its indicated regulatory role for DNA ligase activity [14], suggest that the antimetabolic analogues may produce their mutagenic effects by interference with DNA repair.

A new class of nucleic acid base-analogue represented by N^6-hydroxymethyl adenosine, therefore appears to have been identified whose mutagenic mode of action may be produced by effects on DNA repair.

Cell Stage Restriction

A most important feature of formaldehyde mutagenesis by the larval feeding method is the striking germ cell stage sensitivity shown for its activity in male *Drosophila* larvae. The larval testis is populated initially by spermatogonia, which after their final division, develop into primary spermatocytes. This latter stage, the auxocyte stage, has a protracted period of growth and nuclear preparation prior to meiosis before it eventually divides to produce secondary spermatocytes in the late pupa. It is found [15] that the mutagenic activity of formaldehyde is entirely restricted to the larval auxocyte stage; spermatogonia are not affected.

The (male) larval auxocyte stage therefore appears to have some particular attributes of metabolism which allow this mechanism to operate. It might be that the spermatogonia are dividing so rapidly that the mutagenic analogue(s) has no time to be effective, whereas in the auxocyte stage it has.

Alternatively, the pre-meiotic auxocyte cell may be a stage of differentiation where some mobility of the genome is taking place. The appearance of single-strand breaks in mammalian cells during differentiation, which is not due to a general deficiency of DNA repair, has been reported [16]. If this is occurring in the auxocyte stage, it would make it a prime target for effects on Ap4A-mediated DNA repair, since poly(ADP-ribose) polymerase is activated by DNA strand breaks.

Ap4A and Methylene Bis-Adenosine Diphosphate

The circumstantial evidence linking the activity of the N^6-hydroxymethyl derivatives of adenosine and adenylic acid with effects on DNA repair, has some indirect support, at least for the case of their methylene bis-derivatives.

It has been claimed by one group of workers [17], though with scant experimental details, that methylene bis-ADP competes in vitro with Ap4A at a 1:1 ratio for its binding site on DNA polymerase α; ADP itself competes only at a hundred-fold excess with Ap4A, and methylene bis-AMP is inactive. Also, using in vitro systems [17] of either purified DNA polymerase α and activated DNA, or a complex cell lysate, methylene bis-ADP strongly inhibited DNA synthesis, whereas ADP and methylene bis-AMP again had no effect. Further studies [22] in vivo on mouse 3T3 and SV40 transformed 3T3 fibroblasts claim that methylene bis-(5')AMP, methylene bis-adenosine and methylene bis-(2' or 3')AMP inhibit DNA synthesis. However, methylene bis-adenosine is a very insoluble compound, and in order to be taken up by mammalian cells, the 5' or 2' or 3' methylene bis-AMP's need to be dephosphorylated to their adenosine derivatives by extracellular phosphorylases.

In contrast, it has been reported elsewhere [18] that methylene bis-adenylic acid is inactive on mammalian cells because of its extracellular dephosphorylation to the insoluble methylene bis-adenosine. This observation also contrasts with the further claim [17] by the previous group that methylene bis-adenosine inhibits DNA synthesis and proliferation of tissue culture cells.

These two studies, though contradictory at the cellular level, nonetheless indicate that appropriate methylene bis-nucleotides have an antagonistic effect on Ap4A, at least in the in vitro studies.

Anti-Tumour Effects of Purine Ribonucleotide Analogues

On the basis of my mutation studies, experiments were carried out to test the anti-tumour effects of the formaldehyde-induced base-analogue(s) in the hope that a difference for its activity between normal and tumour cells may be exploited. These experiments were initiated on the assumption that the methylene bis-ribonucleotides were the active products.

When testing for effects on mammalian cells, it is necessary first to remove free formaldehyde because of its cytotoxicity. Thus, experiments were carried out with formaldehyde-treated RNA from which unbound formaldehyde was removed by dialysis [12]. The RNA was treated with formaldehyde under conditions where the formation of methylene-bridged dimers of A⁻A, A⁻G and G⁻G had been shown to involve 15 to 19% of the purine bases in RNA [5]. A proportion of the remaining purine bases may be either unmodified or hydroxymethylated, even after dialysis.

It was therefore hoped to retain sufficient modified residues in the RNA to detect an effect after cellular uptake of the RNA by pinocytosis, and its subsequent intracellular degradation into its constituent nucleotides.

A strong cytotoxic effect was observed [12] in vitro on a human epitheloid carcinoma (HELA) and a human epidermoid carcinoma (KB) cell line, but little or no effect was observed on the non-tumorous cell lines, Chinese hamster V79-4, and a freshly isolated rapidly growing human foetal fibroblast cell line.

A pronounced effect in vivo was also observed after injection of either formaldehyde-treated RNA or polyadenylic acid into the peritoneal cavity of the mouse where the murine leukaemia cell line (P388) was growing as an ascitic tumour [12].

The in vitro studies have been confirmed by others [19] for formaldehyde-treated RNA and extended to its alkaline hydrolysate. An approximately ten-fold greater cytotoxicity was produced by the hydrolysed RNA against a range of human tumour cell lines than against non-tumorous human cell lines.

In both these treatments, the products formed are the A⁻A, A⁻G, and G⁻G dimers as well as unmodified and N-hydroxymethylated nucleotides. In contrast to the form-aldehyde-treated RNA, the products in the hydrolysate need to be dephosphorylated extracellularly before being taken up by the cells. The effect produced by the hydrolysate may therefore be from the soluble A⁻G and G⁻G dimers and from any N-hydroxymethyl nucleotides. The A⁻A dimer is insoluble when converted to its nucleoside, and the pyrimidine nucleotide, cytidylic acid, is not involved in dimer formation [5].

These experiments in themselves do not allow specific identification of the nucleotide analogue(s) responsible for the cytotoxic effect, nor do they exclude the N-hydroxymethyl derivatives as active products.

It has been reported [20] that a mixture of A⁻A and A⁻G dimers, or the individual G⁻G dimer, produce effects on the mitotic activity of a human amniotic cell line, and a transient effect on the reproduction of the RNA viruses, polio virus and vesicular stomatitis virus, grown on these cells. These effects may be attributable to the utilisation of the dimers in RNA synthesis of the cells and the viruses. A similar effect on RNA synthesis may also be produced in experiments [20] demonstrating a cytotoxic effect of the G⁻G and A⁻G dimers on a human adenocarcinoma cell line. In both reports, a high concentration of dimers (0.6 mM) was used.

The mechanism mediating the cytotoxicity of purine analogues is usually thought to result from their incorporation into DNA. However, the purine analogue, 6-thio-guanine, can produce cytotoxic effects by its incorporation into DNA or RNA. Thus, pretreatment of the L1210 murine leukaemia cell line with methotrexate prior to 6-thioguanine results in a marked enhancement of the cytotoxic potency of 6-thio-guanine which is found to be associated with an increase in its incorporation into RNA rather than DNA [21]. By contrast, the simultaneous administration of methotrexate and 6-thioguanine antagonises 6-thioguanine toxicity.

In conclusion therefore, it is possible that the A⁻G and G⁻G dimers contribute to the cytotoxic effects produced by their utilisation in RNA synthesis, but that the major cytotoxic effect is due to the N-hydroxymethyl derivatives. If the dimers are producing an effect, then the N-hydroxymethyl derivatives may exhibit more specificity towards tumour cells than so far seems indicated. Later experiments with formaldehyde-treated RNA, under conditions where negligible dimer formation takes place [5], still show a strong cytotoxic effect (T. Alderson, unpubl. data), as is also the case for its mutagenic activity in *Drosophila* larvae [7]. This implicates the N-hydroxymethyl derivatives as the active products.

N⁶-Hydroxymethyl Adenosine

In the reaction of formaldehyde with adenosine, the formation of insoluble methylene bis-adenosine is dependent on the concentration of formaldehyde, as seen from the symmetrical curve produced when the amount of methylene bis-adenosine is plotted against the logarithm of the initial concentration of formaldehyde [5]. The maximum production of methylene bis-adenosine occurs at a concentration of formaldehyde which converts about 50% of adenosine to the hydroxymethyl derivative during the primary reaction. This latter condition is achieved in practice at a molar ratio of about 4.6:1 [5]; and, as the concentration of formaldehyde increases, the amount of dimer produced is progressively reduced to nil at a molar ratio of formaldehyde:adenosine of about 20:1, when the majority of the adenosine molecules are hydroxymethylated, and cross-linked products cannot form.

Taking advantage of this latter reaction condition, unreacted formaldehyde was separated from the monohydroxymethylated adenosine adduct by quickly passing the mixture through a P-2 Biogel column, and then collecting the reacted (and un-reacted) adenosine as a frozen front (as in [22]); the product was then lyophilised.

Strong cytotoxic effects were observed against several tumour cell lines in vitro at low concentrations (< 1 µg ml^{-1}), where free formaldehyde was not detected in the culture medium (< 1 ppm) by the Schryver test [23] for free formaldehyde. Similar preparations of deoxyadenosine and guanosine showed no effects at this and higher concentrations (T. Alderson, unpubl. data).

The Need for A Prodrug of N⁶-Hydroxymethyl Adenosine

No progress has been made so far in testing N⁶-hydroxymethyl adenosine in vivo on mammalian tumour systems, because of the unstable nature of the hydroxymethyl group (breakdown into adenosine and formaldehyde would be expected to be complete before it reached the target cells). However, recent work on the reaction of formaldehyde with nucleic acid bases has shown that the hydroxymethyl group can be stabilised to some extent.

Thus, reacting formaldehyde with adenosine, cytidine, or guanosine in ethanol solution produces the corresponding N-ethoxymethyl derivatives [H-N-CH$_2$O(C$_2$H$_5$)] [24], or reacting first with formaldehyde and then with bisulphite, forms the N-sulphomethyl derivatives (H-N-CH$_2$SO$_3^-$) [25]. Such derivatives are shown to be in a much more stable form. But the nature of these stabilising groups does not make them suitable for constructing drugs by latentiation; that is, the chemical modification of a biologically active compound to form a new compound or prodrug, which upon in vivo enzymatic attack, will liberate the parent compound [1].

Work is now in progress to produce an appropriate prodrug of N⁶-hydroxymethyl adenosine.

Anti-Tumour Effects of Other N⁶-Substituted Adenosines

One group of workers has synthesised a range of N⁶-substituted adenosines in the hope that they may show anti-tumour effects; a significant proportion of them was found to have anti-tumour activity both in vitro and in vivo.

Those of interest in the present context are short chain N⁶-substituted adenosine analogues, namely, the allyl, -CH$_2$CH-CH$_2$; isopropyl, -CH(CH$_3$)$_2$; propargyl, -CH$_2$C≡H; and 2-(methylallyl), -CH$_2$C(CH$_3$)=CH$_2$ [26]. All show cytotoxic effects in a number of in vitro and in vivo tumour cell systems. The authors state that "an interesting finding was that these compounds had no cytotoxic effect on a leucocyte cell line in vitro which had originated from normal cells"; this observation was not pursued further by them.

These N⁶-substituted adenosines have not been tested for mutagenic effects; and if their mechanism of action is similar to N⁶-hydroxymethyl adenosine, it would require the use of the *Drosophila* larval feeding system to detect activity. It is also not known whether their deoxyribose derivatives are inactive, as is the case with N⁶-hydroxymethyl deoxyadenosine, or if they are incorporated into RNA or DNA.

Mechanism of Action of N⁶-Substituted Adenosines

The tumour studies generally indicate that N⁶-substituted adenosines exhibit specificity for tumour cells. By analogy with the mechanism proposed for the mutagenic activity of N^6-hydroxymethyl adenosine in formaldehyde-induced mutagenesis in *Drosophila* larvae, it is likely that their mechanism of action for cytotoxicity is also mediated via Ap4A.

The utilisation of N^6-hydroxymethyl adenosine, or the other N^6-substituted adenosine molecules, in Ap4A synthesis, where one (or both) adenosine residues of Ap4A are N^6-substituted, may produce antimetabolites which interfere with Ap4A-mediated events.

The N^6-hydroxymethyl group may possibly cross-link the N^6-amino groups of Ap4A by a methylene-bridge in vivo, but it seems improbable that the other N^6-substituted adenosines can cross-link in this way.

In both cases, however, an antimetabolic analogue of Ap4A may be generated. Cytotoxic effects then would presumably result from influences on the ADP-ribosylation reactions of proteins, and/or DNA repair, and/or cellular metabolism, to which tumour cells may be particularly susceptible.

References

1. Montgomery JA (1982) Cancer chemotherapy: Conger synthesis. J Med Chem 23:1063–1067
2. Alderson T (1960) Significance of RNA in the mechanism of formaldehyde mutagenesis. Nature (London) 182:508–510
3. Alderson T (1960) The mechanism of formaldehyde mutagenesis. The uniqueness of adenylic acid in the mediation of the mutagenic activity of formaldehyde. Nature (London) 187:485–489
4. Alderson T (1961) Mechanism of mutagenesis induced by formaldehyde. The essential role of the 6-amino group of adenylic acid (or adenosine) in the mediation of the mutagenic activity of formaldehyde. Nature (London) 191:251–253
5. Feldman M Ya (1967) Reaction of formaldehyde with nucleotides and RNA. Biochem Biophys Acta 49:20–34
6. Sobels FH (1954) Injection of formaldehyde into *Drosophila* larvae. Drosophila Inf Serv 28:156–157
7. Alderson T (1964) The mechanism of formaldehyde-induced mutagenesis. The monohydroxymethylation reaction of formaldehyde with adenylic acid as the necessary and sufficient condition for the mediation of the mutagenic activity of formaldehyde. Mutat Res 1:77–85
8. Auerbach C, Moser H (1953) An analysis of the mutagenic action of formaldehyde food. II The mutagenic potentialities of the treatment. Z Vererbungsl 85:547–563
9. Benyajati C, Place AR, Sofer W (1983) Formaldehyde mutagenesis in *Drosophila*. Molecular analysis of ADH-negative mutants. Mutat Res 111:1–8
10. Khan AH, Alderson T (1968) Studies of the mutagenic activity of nucleic acid base-analogues in *Drosophila*. Mutat Res 5:155–161
11. Alderson T (1967) Induction of genetically recombinant chromosomes in the absence of induced mutation. Nature (London) 215:1281–1283
12. Alderson T (1973) Chemotherapy for an elective effect on mammalian tumour cells. Nature (London) New Biol 244:3–6
13. Durkacz BW, Omidiji O, Gray DA, Shall S (1980) (ADP-ribose)$_n$ participates in DNA excision repair. Nature (London) 283:593–596
14. Creissen D, Shall S (1982) Regulation of DNA ligase activity by poly(ADP-ribose). Nature (London) 296:271–272
15. Auerbach C (1953) Sensitivity of *Drosophila* germ cells to mutagens. Heredity 6 (Suppl): 247–257

16. Farzaneh F, Zalin R, Brill D, Shall S (1983) DNA strand breaks and ADP-ribosyl transferase activation during cell differentiation. Nature (London) 300:362–366

17. Grummt F, Waltl G, Jantzen HM, Hamprecht K, Huebscher U, Kuenzle CC (1979) Diadenosine tetraphosphate (Ap4A) – a ligand of DNA polymerase α and trigger of replication. In: Regulation of macromolecular synthesis by low molecular weight mediators. Academic Press, London New York

18. Feldman M Ya (1977) Mechanism of anti-tumour and cytotoxic activity of formaldehyde-treated RNA. Abstr 11th FEBS Meet, Copenhagen, and personal communication

19. Feldman M Ya, Balabanova H, Bachrach U, Pyshnov M (1977) Effect of hydrolysed formaldehyde-treated RNA on neoplastic and normal human cells. Cancer Res 37:501–506

20. Feldman M Ya, Zalmanson ES, Mikhailova LN (1971) Mechanism of action of products of reactions of formaldehyde with nucleotides (methylene bis-dinucleotides) on the cell and virus reproduction (in Russian). Mol Biol 5:847–857

21. Armstrong RD, Very R, Snyder P, Cadman E (1982) Enhancement of 6-thioguanine cytotoxic activity with methotrexate. Biochem Biophys Res Commun 109:595–601

22. McGhee JD, von Hippel PH (1975) Formaldehyde as a probe of DNA structure. I Reaction with exocyclic amino groups of DNA bases. Biochemistry 14:1281–1296

23. Walker JF (1964) Formaldehyde, 3rd edn. Reinhold, New York

24. Bridson PK, Jiricny J, Kemal O, Reese CB (1980) Reactions between ribonucleoside derivatives and formaldehyde in ethanol solution. J Chem Soc Chem Commun 208–209

25. Hyatsu H, Yamashita Y, Yui S, Yamagata Y, Tomita K, Negishi K (1980) N-sulfomethylation of guanine, adenine and cytidine with formaldehyde-bisulfite. A selective modification of guanine in DNA. Nucleic Acids Res 10:6281–6293

26. Fleysher MH, Bernacki RJ, Bullard GA (1980) Some short-chain N^6-substituted adenosine analogues with antitumour properties. J Med Chem 23:1448–1452

27. Alderson T (1985) Formaldehyde-induced mutagenesis: a novel mechanism for its action. Mutat Res 154:101–110

Mono(ADP-Ribosylation) Reactions

Characterization of NAD: Arginine Mono(ADP-Ribosyl)-Transferases in Turkey Erythrocytes: Determinants of Substrate Specificity

JOEL MOSS, ROBERT E. WEST, Jr., JAMES C. OSBORNE, Jr.,
and RODNEY L. LEVINE[1]

Introduction

Mono(ADP-ribosylation) is catalyzed by transferases identified in viruses, bacteria, and animal cells [1]. Its function has thus far been clearly defined only for certain bacterial toxins that exert their effects on animal tissues by catalyzing the mono(ADP-ribosylation) of critical cellular proteins [1—5]. One of these toxins, choleragen (cholera toxin), an NAD:arginine mono(ADP-ribosyl)transferase, causes the activation of the hormone-sensitive adenylate cyclase from animal tissues by ADP-ribosylating a guanine nucleotide-binding stimulatory protein termed G_s [5]. In vitro, choleragen also catalyzes the ADP-ribosylation of several proteins not related to the cyclase system as well as low molecular weight guanidino compounds, such as the amino acid arginine [6—8]. Animal tissues contain NAD:arginine (ADP-ribosyl)transferases that catalyze reactions similar to those of choleragen [7, 9—11].

Identification, Localization, and Characterization of NAD:Arginine Mono(ADP-Ribosyl)Transferases in Turkey Erythrocytes

Prior studies disclosed the presence of two NAD:arginine mono(ADP)ribosyltransferases, designated A and B, in turkey erythrocyte cytosol, both of which were purified to homogeneity [9, 12]. Transferase A is a 28,000 dalton enzyme that is stimulated by chaotropic salts, histones, phospholipids, and certain detergents [12—16], all of which appear to convert the enzyme from an inactive high molecular weight species to an active protomeric form. Transferase B is a 32,000 dalton enzyme that is inhibited by salt and unaffected by histones [9]. In addition to its different physical, chromatographic, and regulatory properties, transferase B differs from A in substrate specificity; the former uses NADP as well as NAD as a substrate, while the latter does not ([17], D.A. Yost, J. Moss, 1983, unpubl. observations).

Recently, the spectrum of mono(ADP-ribosyl)transferase activities in turkey erythrocytes was expanded with the identification of membrane-associated (trans-

---

1   National Heart, Lung, and Blood Institute, National Institutes of Health, Bethesda, MD 20205, USA

ADP-Ribosylation of Proteins
(ed. by F.R. Althaus, H. Hilz, and S. Shall)
© Springer-Verlag Berlin Heidelberg 1985

ferase C) and nuclear (transferase D) enzymes ([18], R.E. West Jr., J. Moss, 1983, unpubl. observations). The two enzymes were extracted at high ionic strength from an exhaustively washed turkey erythrocyte particulate fraction and purified 60,000-fold. Transferase C eluted from a TSK 2000 silica gel sizing column between transferases A and B ($M_r \sim 30,000$) [18]. It was distinguished by a relative insensitivity to salt or histone ($<$ twofold stimulation), unique chromatographic properties, and a K_m for NAD lower than those of transferases A and B. The nuclear enzyme, also a mono-(ADP-ribosyl)transferase, was stimulated more than tenfold by histone or salt. Its K_{av} on a TSK 2000 column was slightly less than that of transferase A (R.E. West Jr., J. Moss, 1984, unpubl. observations). The presence of a number of ADP-ribosyltransferases exhibiting differences in localization and properties may be related to a diversity of function for this family of enzymes.

Effects on Catalytic Activity of Escherichia coli Glutamine Synthetase Caused by Site-Specific Transferase A-Catalyzed ADP-Ribosylation

Specific arginine residues appear to be critical for the function of some proteins. With ovine brain glutamine synthetase, chemical modification of a single arginine residue results in a loss of enzymatic activity [19]. The glutamine synthetases from ovine brain, chicken heart, and E. coli were ADP-ribose acceptors in the erythrocyte transferase A-catalyzed reaction ([20], J. Moss, S.J. Stanley, R.L. Levine, 1984, unpubl. observations). In each case, transfer of approximately one ADP-ribose moiety per subunit resulted in the loss of the ability of the synthetase to catalyze the Mg^{2+}-dependent formation of glutamine from glutamate. This stoichiometry is consistent with a site-specific reaction. Denaturation of the synthetase increased the extent of ADP-ribosylation and thus appeared to enhance nonspecifically the availability of arginine residues on the protein. It is evident from the selectivity of the ADP-ribosylation reaction that protein conformation is critical to site reactivity. Using the E. coli synthetase, it was observed that ADP-ribosylation and loss of the Mg^{2+}-dependent activity was inhibited by the glutamine synthetase substrates MgATP and glutamate. Whereas ADP-ribosylation decreased the Mg^{2+}-dependent activity, it had a significantly lesser effect on the reaction when Mg^{2+} was replaced with Mn^{2+}; since Mg^{2+} is presumed to be the physiologically important cation, ADP-ribosylation causes a functional inactivation. The selective inhibitory effect of ADP-ribosylation on Mg^{2+}-dependent activity parallels the findings obtained when a tyrosine residue in E. coli glutamine synthetase is enzymatically adenylylated, a covalent modification thought to be of physiological significance [21, 22]. Chemical modification of other residues on the synthetase caused similar changes in the catalytic properties of the enzyme [23]. Thus, different types of enzymatic and chemical modification on different amino acids appear to have similar effects on enzyme function.

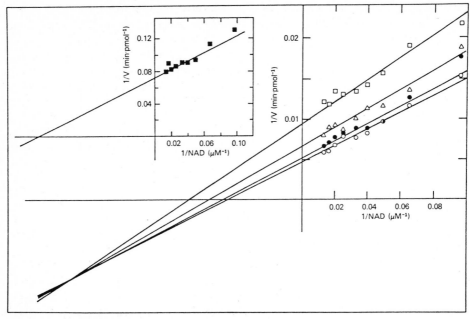

Fig. 1. Effect of agmatine and NAD on the activity of NAD:arginine ADP-ribosyltransferase plotted with NAD as the variable substrate at different fixed concentrations of agmatine. Assays contained 50 mM potassium phosphate (pH 7.0), 1 mg ml^{-1} ovalbumin, and the following concentrations of agmatine: $\square$ 0.5 mM; $\triangle$ 1 mM; $\bullet$ 2 mM; $\bigcirc$ 3 mM. Reactions were performed at 30°C and initiated with erythrocyte transferase A (9.1 ng/assay). *Inset:* Double reciprocal plot of the activity of transferase A in the absence of acceptor. *Solid lines* are the theoretical fit of the data to the random rapid equilibrium sequential mechanism described in the text

Kinetic Studies with NAD:Arginine ADP-Ribosyltransferase A

The kinetic mechanism for the ADP-ribosylation of low molecular weight guanidino compounds and proteins by the cytosolic transferase A was investigated. Two possible mechanisms were considered.

ordered   random

E + NAD $\rightleftharpoons$ E · NAD   E + NAD $\rightleftharpoons$ E · NAD

         +   +   +

   Acceptor   Acceptor   Acceptor

      $\big\updownarrow$ $\big\updownarrow$ $\big\updownarrow$

E · NAD · Acceptor   E · Acceptor + NAD $\rightleftharpoons$ E · Acceptor · NAD

Kinetic analysis revealed that the Michaelis constant for NAD increased with increasing concentrations of the ADP-ribose acceptor agmatine (Fig. 1). This result is inconsistent

Table 1. Kinetic constants for NAD:arginine ADP-ribosyltransferase A[a]

Parameter	Value ± SE
K_m (NAD)	7 ± 0.7 μM
K_m (Agmatine)	260 ± 20 μM
α	3.8 ± 0.3
V_{max}	31 μmol min^{-1} mg^{-1}
β	0.05

[a] Summary of kinetic parameters for erythrocyte transferase A using the random rapid equilibrium sequential mechanism described in the text. α is the factor by which binding of one substrate changes the affinity for the second substrate and β is the ratio of maximal velocities of glycohydrolase to transferase reactions

with an ordered mechanism; if NAD must bind to the enzyme prior to agmatine, then binding of agmatine cannot decrease the apparent affinity for NAD.

Detailed studies with agmatine as the acceptor in the presence and absence of the product inhibitor nicotinamide (data not shown) were in agreement with the sequential rapid equilibrium, random mechanism shown below (kinetic parameters are summarized in Table 1).

$$
\begin{array}{ccccc}
 & K_{NAD} & & \beta K_3 & \\
E + NAD & \rightleftharpoons & E \cdot NAD & \longrightarrow & \text{hydrolysis} \\
+ & & + & & \\
B & & B & & B = \text{ADP-ribose acceptor} \\
K_B \updownarrow & & \updownarrow \alpha K_B & & \beta = \text{ratio of hydrolysis to transfer} \\
 & & & K_3 & \\
E \cdot B + NAD & \rightleftharpoons & E \cdot B \cdot NAD & \longrightarrow & \text{transfer} \\
 & \alpha K_{NAD} & & &
\end{array}
$$

Under the framework of this mechanism, binding of one substrate changes the affinity for the second substrate by a factor α.

With agmatine, a low molecular weight arginine derivative, as the acceptor, site-site interactions between donor and acceptor are negative ($\alpha > 1$) and binding of one substrate lowers the affinity for the second substrate (Fig. 1, Table 1). With the larger protein substrates arginine-rich histone and lysozyme, binding of one substrate did not alter the affinity for the second substrate ($\alpha = 1$).

These results and those of the glutamine synthetase studies are compatible with the hypothesis that protein conformation (1) is a critical determinant of the availability of specific arginine residues for modification and (2) affects the affinity of the transferase for NAD. It might be predicted that a favored substrate would enhance the affinity for NAD and thus perhaps promote the reaction.

Summary

A family of NAD:arginine mono(ADP-ribosyl)transferases present in turkey erythrocytes exhibited different physical, regulatory, and kinetic properties. The transferases were localized to the cytosolic, membrane, and nuclear fractions. With *Escherichia coli* glutamine synthetase as an acceptor protein, ADP-ribosylation by the histone- and salt-activated cytosolic transferase was site-specific. The transfer of approximately one ADP-ribose moiety per protomer inhibited the Mg^{2+}-dependent synthetic activity of the synthetase to a greater extent than the Mn^{2+}-dependent reaction. ADP-ribosylation of a critical arginine thus converted glutamine synthetase to a form with catalytic properties similar to that of the adenylylated enzyme in which a single tyrosine is modified. Site-specificity of the ADP-ribosylation reaction was lost following thermal denaturation of the synthetase. Kinetic studies on the action of the histone- and salt-activated ADP-ribosyltransferase were most consistent with a rapid equilibrium random sequential mechanism. Either substrate (i.e., ADP-ribose donor or acceptor) may bind to the free enzyme and, through donor-acceptor site-site interactions, alter affinity for the second substrate. With agmatine as acceptor, negative site-site interactions resulted in a 70% decrease in affinity for the second substrate; negative site-site interactions were not observed with lysozyme or arginine-rich histone as the acceptor substrate. These studies are consistent with the hypothesis that the topology of the ADP-ribose acceptor may be a critical factor in determining the affinity of the ADP-ribosyltransferase for NAD and in the reactivity of specific arginine residues on the acceptor substrate.

References

1. Vaughan M, Moss J (1981) Mono(ADP-ribosyl)transferases and their effects on cellular metabolism. Curr Top Cell Regul 20:205–246
2. Pappenheimer AMJr (1977) Diphtheria toxin. Annu Rev Biochem 46:69–94
3. Iglewski BH, Kabat D (1975) NAD-dependent inhibition of protein synthesis by *Pseudomonas aeruginosa* toxin. Proc Natl Acad Sci USA 72:2284–2288
4. Katada T, Ui M (1982) ADP ribosylation of the specific membrane protein of C6 cells by islet-activating protein associated with modification of adenylate cyclase activity. J Biol Chem 257:7210–7216
5. Gilman AG (1984) Guanine nucleotide-binding regulatory proteins and dual control of adenylate cyclase. J Clin Invest 73:1–4
6. Moss J, Vaughan M (1977) Mechanism of action of choleragen. Evidence for ADP-ribosyltransferase activity with arginine as an acceptor. J Biol Chem 252:2455–2457
7. Moss J, Vaughan M (1978) Isolation of an avian erythrocyte protein possessing ADP-ribosyltransferase activity and capable of activating adenylate cyclase. Proc Natl Acad Sci USA 75:3621–3624
8. Watkins PA. Moss J, Vaughan M (1980) Effects of GTP on choleragen-catalyzed ADP ribosylation of membrane and soluble proteins. J Biol Chem 255:3959–3963
9. Yost DA, Moss J (1983) Amino acid-specific ADP-ribosylation. Evidence for two distinct NAD:arginine ADP-ribosyltransferases in turkey erythrocytes. J Biol Chem 258:4926–4929
10. Soman G, Mickelson JR, Louis CF, Graves DJ (1984) NAD:guanidino group specific mono ADP-ribosyltransferase activity in skeletal muscle. Biochem Biophys Res Commun 120:973–980

11. Tanigawa Y, Tsuchiya M, Imai Y, Shimoyama M (1984) ADP-ribosyltransferase from hen liver nuclei. Purification and characterization. J Biol Chem 259:2022–2029
12. Moss J, Stanley SJ, Watkins PA (1980) Isolation and properties of an NAD- and guanidine-dependent ADP-ribosyltransferase from turkey erythrocytes. J Biol Chem 255:5838–5840
13. Moss J, Stanley SJ (1981) Histone-dependent and histone-independent forms of an ADP-ribosyltransferase from human and turkey erythrocytes. Proc Natl Acad Sci USA 78:4809–4812
14. Moss J, Stanley SJ, Osborne JCJr (1981) Effect of self-association on activity of an ADP-ribosyltransferase from turkey erythrocytes. Conversion of inactive oligomers to active protomers by chaotropic salts. J Biol Chem 256:11452–11456
15. Moss J, Stanley SJ, Osborne JCJr (1982) Activation of an NAD:arginine ADP-ribosyltransferase by histone. J Biol Chem 257:1660–1663
16. Moss J, Osborne JCJr, Stanley SJ (1984) Activation of an erythrocyte NAD:arginine ADP-ribosyltransferase by lysolecithin and nonionic and zwitterionic detergents. Biochemistry 23:1353–1357
17. Moss J, Stanley SJ, Oppenheimer NJ (1979) Substrate specificity and partial purification of a stereospecific NAD- and guanidine-dependent ADP-ribosyltransferase from avian erythrocytes. J Biol Chem 254:8891–8894
18. West REJr, Moss J (1984) NAD:arginine mono-ADP-ribosyltransferases in turkey erythrocytes: Characterization of a membrane-associated transferase different from the cytosolic enzymes. Fed Proc 43:1786
19. Powers SG, Riordan JF (1975) Functional arginyl residues as ATP binding sites of glutamine synthetase and carbamyl phosphate synthetase. Proc Natl Acad Sci USA 72:2616–2620
20. Moss J, Watkins PA, Stanley SJ, Purnell MR, Kidwell WR (1984) Inactivation of glutamine synthetases by an NAD:arginine ADP-ribosyltransferase. J Biol Chem 259:5100–5104
21. Kingdon HS, Shapiro BM, Stadtman ER (1967) Regulation of glutamine synthetase, VIII ATP: Glutamine synthetase adenylyltransferase, an enzyme that catalyzes alterations in the regulatory properties of glutamine synthetase. Proc Natl Acad Sci USA 58:1703–1710
22. Stadtman ER, Ginsburg A, Ciardi JE, Yeh J, Hennig SB, Shapiro BM (1970) Multiple molecular forms of glutamine synthetase produced by enzyme catalyzed adenylylation and deadenylylation reactions. Adv Enzyme Regul 8:99–118
23. Cimino F, Anderson WB, Stadtman ER (1970) Ability of nonenzymic nitration or acetylation of *E. coli* glutamine synthetase to produce effects analogous to enzymic adenylylation. Proc Natl Acad Sci USA 66:564–571

Mono(ADP-Ribosyl)ation and Phospho(ADP-Ribosylation)Reactions in Eukaryotic Cells

HELMUTH HILZ, ROBERT KOCH, ANDREAS KREIMEYER,
PETER ADAMIETZ, and MYRON K. JACOBSON[1]

Introduction

The existence in vivo of mono(ADPR) protein conjugates has been described by this and other laboratories [1, 2]. Also, a subfractionation of endogenous mono(ADPR) conjugates was introduced by Bredehorst and colleagues who observed that the amounts of conjugates susceptible and resistant to 0.5 M neutral NH_2OH showed considerable variation, depending on tissue, status of differentiation, and rate of cell proliferation [3]. Several enzymes and enzymic activities [4—8] — besides the toxins — as well as nonenzymic formation of acid-insoluble ADPR conjugates [9] — have been described. All these reactions could contribute to the formation of endogenous ADPR protein conjugates, but for none of these mono(ADP-ribosyl)ation reactions a physiological function has been established so far.

In order to analyze individual mono(ADPR) proteins we first set out to purify the acceptor(s) of the mitochondrial system which according to the subcellular distribution in liver represent a major fraction of this type of conjugate [10]. In the course of these studies evidence for a nonenzymic formation of mitochondrial ADPR protein conjugates accumulated [11].

Nonenzymic ADP-Ribosylation and Phospho(ADP-Ribosylation) of Specific Acceptor Proteins

Nonenzymic formation of acid-insoluble ADPR protein conjugates with free poly-(ADPR) has been described first by Kun and colleagues [12]. They pertained to unspecific reactions involving basic groups like lysine or arginine residues and, therefore, were of general unspecificity. In the following section nonenzymic ADP-ribosylation as well as phospho(ADP-ribosylation) reactions are described that exhibit a high degree of preference with respect to the acceptors [11].

---

1 Institut für Physiologische Chemie, Universität Hamburg, Martinistr. 52, 2000 Hamburg 20, FRG

ADP-Ribosylation of Proteins
(ed. by F.R. Althaus, H. Hilz, and S. Shall)
© Springer-Verlag Berlin Heidelberg 1985

Mitochondrial ADP-Ribosylation

Mitochondrial ADP-ribosylation as described by Kun et al. [2, 4] involves the apparent transfer of ADPR residues from NAD to a 50 kD polypeptide. Yet another ADP-ribosylation reaction was reported by Richter et al. [13, 14]. They described the modification of a 30 kD polypeptide in submitochondrial particles when incubated with labeled NAD. This reaction was brought into connection with Ca^{++} efflux from mitochondria as induced by treatment of mitochondria with organic peroxides. Such treatment was known to result in the concomitant degradation of NAD and NADP to ADPR and P-ADPR, respectively [15]. During our attempts to characterize these mitochondrial systems, evidence accumulated that nonenzymic ADP-ribosylation is involved. The following observations support this interpretation:

1. NAD is rapidly converted to ADPR by mitochondrial extracts and by SMP.
2. Free ADPR is incorporated into the TCA-insoluble fraction at an equal or higher rate than NAD, depending on pH.
3. NIH, an inhibitor of NAD glycohydrolase blocks incorporation of adenine equivalents from NAD, not from ADPR.
4. Separation of NAD glycohydrolase from rat liver mitochondrial extracts eliminates incorporation from NAD, not of ADPR.
5. NADase added to SMP increased incorporation from NAD to the level of that of ADPR.
6. Trapping of free ADPR by NH_2OH, semicarbazide or arginine methylester also reduces or eliminates formation of conjugates not only from ADPR, but also from NAD. In addition: AMP and especially ATP are highly effective inhibitors of both reactions.
7. In rat liver mitochondrial extracts, the same 50–55 kD polypeptide reacts with ADPR and NAD, while in SMP from bovine heart and rat liver a 30 kD polypeptide is the acceptor for both "substrates", NAD and ADPR.
8. Chemical stability of the 30 kD conjugate is different from other "enzymic" and "nonenzymic" ADPR protein conjugates.

The preferential reaction of ADPR with two mitochondrial proteins indicated that these acceptors contain specific sites responsible for the formation of the nonenzymic ADPR conjugates. And these sites should not correspond to the abundant basic lysine or arginine residues found in practically all proteins. A comparison of the reaction of ADPR with poly(Lys) and poly(Arg) with that involving the mitochondrial acceptors, revealed important differences with respect to pH-dependency, stability towards 3 M NH_2OH, and sensitivity towards picrylsulfonate. In fact, it can be shown that the mitochondrial conjugates exhibit chemical features not shown by any other enzymic or nonenzymic ADPR conjugates, with the possible exception of nonenzymic ADPR-poly(His) conjugates [16].

Table 1. Formation of ADPR- and P-ADPR-protein conjugates in sub-mitochondrial particles from bovine heart. SMP were incubated with 250 μM "substrate" at pH 6.5 under conditions described elsewhere [11]

"Substrate"	Adenine equivalents incorporated	
	(pmol $\pm$ SD)	+ NH_2OH
NAD	8.11 $\pm$ 0.31	0.78 $\pm$ 0.11
ADPR	7.99 $\pm$ 0.10	0.34 $\pm$ 0.15
NADP	15.43 $\pm$ 1.70	1.08 $\pm$ 0.08
P-ADPR	15.98 $\pm$ 1.15	0.99 $\pm$ 0.17

Nonenzymic ADP-Ribosylation in Plasma Membranes

In rat liver plasma membranes enzymic hydrolysis of NAD leads to free ADPR that reacts rapidly and with preference with several polypeptides. The pattern obtained with NAD and ADPR is practically identical. However, membrane-bound adenylate cyclase activity, which is subject to stimulation by cholera toxin [17], is not signif-icantly influenced by nonenzymic ADP-ribosylation of the membrane preparation.

From the data presented, we conclude that apparent ADPR transferase activities in mitochondria and in plasma membranes are simulated completely or to a large extent by a NAD glycohydrolase action on NAD followed by a nonenzymic yet highly pref-erential reaction of free ADPR with specific acceptors.

Phospho(ADP-Ribosylation): A New Type of Covalent Modification in Submitochondrial Particles from Rat Liver?

The reaction of free ADPR with the mitochondrial acceptors can account — within the limits of the method — for the incorporation of ADPR equivalents from NAD. In this case, not only ADPR should react, but also the 2′-phospho(ADPR) (P-ADPR) which can be generated by NAD(P) glycohydrolase acting on NADP. Indeed, when SMP from bovine heart were incubated with 2′-^{33}phosphate-labeled NADP, the same incor-poration of P-ADPR residues into the acid-insoluble fraction was obtained as when free 2′-^{33}phosphate-labeled P-ADPR was used (Table 1). Both reactions surpassed those of NAD or ADPR. Furthermore, incorporation of [^{3}H, ^{33}P]P-ADPR equivalents from labeled NADP was largely eliminated by heating the SMP preparation, but fully restored by the addition of brain NAD(P) glycohydrolase.

Quite different, however, was the situation when SMP from rat liver was analyzed (Table 2). Here, incorporation of P-ADPR equivalents from 2′-phosphate-labeled NADP clearly surpassed that from free P-ADPR, which in turn was again significantly higher than incorporation from NAD or ADPR. An eight times higher incorporation of phospho(ADPR) equivalents from NADP over that of free P-ADPR was also seen in

Table 2. ADP-ribosylation and phospho(ADP-ribosylation) of proteins in submitochondrial particles of rat liver. SMP were incubated with 50 μM "substrate" at pH 6.5. Further details are described in [11]

"Substrate"	Adenine equivalents incorporated (pmol ± SD)
[^{3}H]NAD	1.09 ± 0.17
[^{3}H]ADPR	1.26 ± 0.27
[^{3}H, ^{33}P]NADP	8.76 ± 0.68
[^{3}H, ^{33}P]P-ADPR	2.46 ± 0.48

a kinetic study. Furthermore, the increased incorporation of activated P-ADPR in NADP seemed to involve a new acceptor protein. This reaction may, therefore, represent the first demonstration of enzymic phospho(ADP-ribosylation) as a new type of covalent modification of proteins. However, the observation that a major part of the increased incorporation of NADP still pertained to the same 31 kD and 50–55 kD polypeptides that served as acceptors of nonenzymic modification by free ADPR or P-ADPR, and the finding that incorporation of adenine equivalents from all the precursors (including NADP) was completely inhibited by 0.1 M NH$_2$OH, calls for a cautious interpretation of these preliminary data.[1]

Nuclear Mono(ADP-Ribosyl)ation in Alkylated Hepatoma Cells

In hepatoma cells treated with dimethyl sulfate, the increase in protein-bound monomeric ADPR residues surpasses polymeric ADPR residues by a factor of 3–4, most of which is associated with histones H1 and H2B [18, 19]. Histone H1 was not a significant acceptor for the highly stimulated formation of poly(ADPR) chains in vivo, while histone H2B carried a large fraction of both mono- and poly(ADPR) residues [18, 19]. We could also show that in alkylated cells mono(ADPR) residues were linked to these histones by NH$_2$OH-resistant bonds (> 90%). Several routes could have led to these NH$_2$OH-resistant conjugates: (1) Either they were the "remnants" of a relatively small fraction of poly(ADPR) chains that were linked by NH$_2$OH-resistant bonds, the remnants being formed in the course of poly(ADPR) turnover, (2) or they were synthesized under the catalytic action of a specific nuclear mono(ADPR) transferase like the one described recently by Tanigawa et al. [7], (3) or they might have been products of nonenzymic ADP-ribosylation, since large amounts of free ADPR (an equivalent of 60% of the cellular NAD) must have been formed within 30 min by the highly stimulated turnover of poly- and mono(ADPR) residues under the conditions applied [20].

---

1 Recent data indicate that the reaction represents a phospho adenylation catalyzed by a NADP-specific transferase located mainly in microsomes

To test these possibilities, the histone H1 fraction was purified from alkylated cells and subjected to the stability treatments described above. Chemical stability of the endogenous material was different in all aspects from ADPR-histone H1 formed in vitro by nonenzymic reaction and from glu- or arg-linked ADPR conjugates. From our data, we conclude that in alkylated hepatoma cells, mono(ADP-ribosyl) groups in H1 are neither linked by an ester group, nor via arginine residues. They also do not represent products of nonenzymic reaction.

We have also isolated *histone H2B* as another major acceptor of the alkylation-induced mono(ADPR) group [21]. Here again, the same stabilities and sensitivities were seen as with the ADP-ribosyl histone H1. These two polypeptides carry more than 50% of the mono(ADPR) residues induced by DMS treatment. Since the increase of these ADPR groups surpasses the augmentation of polymeric ADPR residues by a factor of 3, while turnover is significantly slower, it appears that in alkylated cells the function of nuclear mono(ADP-ribosyl)ation is different from that of the modification by poly(ADPR). In this respect, it is interesting to note that the mono(ADPR) groups in histone H1 and H2B are exclusively located at those nonglobular extensions that mediate a strong interaction with DNA (cf. Kreimeyer, Adamietz, Hilz, this volume).

So far, no specific mono(ADPR) transferase was found which could be responsible for the ADP-ribosylation of histone H1 and H2B in response to alkylation. This certainly does not exclude the existence of such an enzyme. However, the possibility of an intramolecular shift of mono(ADPR) or poly(ADPR) residues from an energy-rich ester linkage to a more stable type of bond must also be taken into consideration.

Differentiation-Associated Mono(ADP-Ribosyl)ation

The alkylation-induced changes had demonstrated that important acceptors like histone H1 may become nearly exclusively mono(ADP-ribosyl)ated. Mono(ADP-ribosyl)ation of proteins also appears to be of importance in certain differntiation processes. When F9 teratocarcinoma cells were incubated with low concentrations of retinoic acid, they differentiated into endoderm cells [22]. Under these conditions, mono(ADPR) conjugates increased considerably. The increase was time dependent, and it was mainly due to a fraction that was neither carboxyl- nor arginine-linked, nor did it correspond to nonenzymic conjugates. However, further experiments are required to determine the nature of these conjugates.

Tumor Promotors and ADP-Ribosylation

In spite of the notion that tumor promotion does not involve mutation of the genome, this process must nevertheless be able to markedly influence nuclear processes like cell proliferation and differentiation. In certain systems, promotors like 12-O-tetradecanoyl-phorbol-13-acetate (TPA) or benzoylperoxide (BPO) can induce DNA fragmentation [23]. This may, however, not be a general phenomenon. When 3T3 cells are incubated with TPA (up to 50 mg ml^{-1}) for 3 h no significant change in nucleoid sedimentation

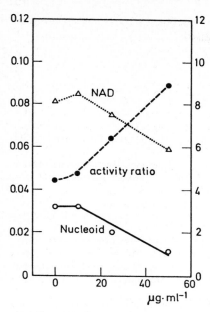

Fig. 1. Benzoylperoxide-induced activation of poly(ADPR) polymerase in 3T3 cells. Cells were incubated for 3 h with increasing concentrations of benzoylperoxide and analyzed for DNA fragmentation (nucleoid sedimentation). NAD content, and poly(ADPR) polymerase activity in permeabilized cells ± DNase

or activation of poly(ADPR) polymerase in permeabilized cells was seen. Benzoylperoxide, on the other hand, induced a concentration-dependent fragmentation of DNA (nucleoid sedimentation), a depletion of NAD, and an activation of poly(ADPR) polymerase (Fig. 1).

This appears to be an early and perhaps an unspecific effect of BPO not related to tumor promotion. In F9 cells, benzoylperoxide provoked time-dependent changes in mono(ADP-ribosyl) conjugates that were comparable to the differentiation-associated alterations seen with retinoic acid.

Conclusion

Neither transfer of single ADPR residues by specific eukaryotic transferases nor the existence of other than ester-type linkages in ADPR protein conjugates were recognized for a number of years [24]. Now, there is increasing evidence that the eukaryotic cell possesses a rather elaborate instrumentarium of ADP-ribosylation reactions to modulate important processes. The use of mono(ADPR) groups for the covalent modification of protein provides a tool that may prove to be as versatile as poly(ADP-ribosyl)ation. In addition to the systems discussed here (alkylated cells, differentiation, tumor promotors), mono(ADP-ribosyl)ation may be involved also in the modulation of protein synthesis [8], in the response to heat shock (M. Jacobson, personal communication), and in glutamine synthesis [25]. Nonenzymic mono(ADP-ribosyl)ation reactions must also be taken into consideration whenever the functions of active (P)ADP-ribose as present in the pyridine nucleotides are studied. Finally, preliminary data may indicate that an additional way to modulate protein functions could be provided by phospho-(ADP-ribosylation) reactions.

Acknowledgments. This work was supported by the Deutsche Forschungsgemeinschaft and by the Alexander von Humboldt-Stiftung.

References

1. Bredehorst R, Goebel M, Renzi F, Kittler M, Klapproth K, Hilz H (1979) Intrinsic ADP-ribose transferase activity versus levels of mono(ADP-ribose) protein conjugates in proliferating Ehrlich ascites tumor cells. Hoppe-Seyler's Z Physiol Chem 360:1737−1743
2. Hilz H, Bredehorst R, Adamietz P, Wielckens K (1982) Subfractions and subcellular distribution of mono(ADP-ribosyl) proteins in eukaryotic cells. In: ADP-ribosylation reactions, chap 11. Academic Press, London New York
3. Bredehorst R, Wielckens K, Gartemann A, Lengyel H, Klapproth K, Hilz H (1978) Two different types of bonds linking single ADP-ribose residues covalently to protein. Eur J Biochem 92:129−135
4. Kun E, Zimber PH, Chang ACY, Puschendorf B, Grunicke H (1975) Macromolecular enzymatic product of NAD$^+$ in liver mitochondria. Proc Natl Acad Sci USA 72:1436−1440
5. Moss J, Vaughan M (1980) In: Smulson ME, Sugimura T (eds) Novel ADP-ribosylation of regulatory enzymes and proteins. Elsevier/North Holland, Amsterdam New York, pp 391−399
6. Moss J, Stanley SJ (1981) Histone-dependent and histone-independent forms of an ADP-ribosyltransferase from human and turkey erythrocytes. Proc Natl Acad Sci USA 78:4809−4812
7. Tanigawa Y, Tsuchiya M, Imai Y, Shimoyama M (1984) ADP-ribosyltransferase from hen liver nuclei. J Biol Chem 259:2022−2029
8. Lee H, Iglewski WJ (1984) Cellular ADP-ribosyltransferase with the same mechanism of action as diphtheria toxin and pseudomonas toxin A. Proc Natl Acad Sci USA 81:2703−2707
9. Kun E, Chang AC, Sharma MC, Ferro AM, Nitecki D (1976) Covalent modification of proteins by metabolites of NAD$^+$. Proc Natl Acad Sci USA 73:3131−3135
10. Adamietz P, Wielckens K, Bredehorst R, Lengyel H, Hilz H (1981) Subcellular distribution of mono(ADP-ribose) protein conjugates in rat liver. Biochem Biophys Res Commun 101:96−103
11. Hilz H, Koch R, Fanick W, Klapproth K, Adamietz P (1984) Nonenzymic ADP-ribosylation of specific mitochondrial polypeptides. Proc Natl Acad Sci USA 81:3929−3933
12. Kun E, Kirsten E, Piper WN (1979) Stabilization of mitochondrial functions with digitonin. Method Enzymol 55:115−118
13. Hofstetter W, Mühlebach T, Lötscher H, Winterhalter K, Richter C (1981) ATP prevents both hydroperoxide-induced hydrolysis of pyridine nucleotides and release of calcium in rat liver mitochondria. Eur J Biochem 117:361−367
14. Richter C, Winterhalter K, Baumhüter S, Lötscher H, Moser B (1983) ADP-ribosylation in inner membrane of rat liver mitochondria. Proc Natl Acad Sci USA 80:3188−3192
15. Lötscher HR, Winterhalter KH, Carafoli E, Richter C (1980) Hydroperoxide-induced loss of pyridine nucleotides and release of calcium from rat liver mitochondria. Proc Natl Acad Sci USA 76:4340−4344
16. Koch R, Jacobson M, Hilz H (in preparation) Stability of various ADPR polypeptide conjugates to chemical treatment
17. Gill DM, Meren R (1978) ADP-ribosylation of membrane proteins catalyzed by cholera toxin: Basis of the activation of adenylate cyclase. Proc Natl Acad Sci USA 75:3050−3054
18. Kreimeyer A, Wielckens K, Adamietz P, Hilz H (1984) DNA repair-associated ADP-ribosylation in vivo. J Biol Chem 259:890−896
19. Adamietz P, Rudolph A (1984) ADP-ribosylation of nuclear proteins in vivo. J Biol Chem 259:6841−6846
20. Wielckens K, Schmidt A, George E, Bredehorst R, Hilz H (1982) DNA fragmentation and NAD depletion. J Biol Chem 257:12872−12877

21. Kreimeyer A, Adamietz P, Hilz H (1985) Mono(ADPR) histone conjugates induced by alkylation of hepatoma cells. Biol Chem Hoppe Seyler 366:537–544
22. Strickland S, Smith KK, Marotti KR (1980) Hormonal induction of differentiation of teratocarcinoma stem cells: Generation of partial endoderm by retionic acid and dibutyryl cAMP. Cell 21:347
23. Birnboim HC (1982) DNA strand breakage in human leukocytes exposed to a tumor promoter, phorbol myristate acetate. Science 215:1247–1249
24. Hayaishi O, Ueda K (1977) Poly(ADP-ribose) and ADP-ribosylation to proteins. Annu Rev Biochem 46:95–116
25. Moss J, Watkins PA, Stanley SJ, Purnell MR, Kidwell WR (1984) Inactivation of glutamine synthetase by an NAD:arginine ADP-ribosyltransferase. JBC 259:5100–5104

Mono(ADP-Ribosylation) of Proteins at Arginine in Vivo

MYRON K. JACOBSON[1], D. MICHAEL PAYNE[1], KELLY P. SMITH[1],
MARIA ELENA CARDENAS[1], JOEL MOSS[2], and ELAINE L. JACOBSON[3]

Mono(ADP-Ribosyl) Transferases

Mono(ADP-ribosyl)transferases are present in both prokaryotic and eukaryotic cells and in bacterial viruses [1]. While the physiological roles of some of the prokaryotic transferases are fairly well understood, the function of the eukaryotic transferases is unknown. We describe here recent studies designed to aid our understanding of endogenous mono(ADP-ribosylation) reactions in eukaryotic cells.

Proteins Modified by ADP-Ribose in Vivo

Different mono(ADP-ribosyl)transferases are known to catalyze the formation of N-glycosylic linkages to arginine, asparagine, and modified histidine (diphthamide) residues in acceptor proteins [1]. In addition, mono(ADP-ribose) residues bound to proteins via carboxylate esters can be generated by the combined action of poly(ADP-ribose) polymerase and poly(ADP-ribose) glycohydrolase [2]. Transferases isolated from eukaryotic cells thus far specifically modify arginine residues. Thus, we have focused our efforts on the detection of proteins in vivo that are modified by ADP-ribosylation at arginine. The methodology we have developed for this purpose has been described in detail elsewhere [3]. A key factor in the detection of proteins modified at arginine was the determination of conditions for the selective release of ADP-ribose from arginine. Figure 1 shows the stability of the known protein-(ADP-ribose) linkages at neutral pH in the presence and absence of hydroxylamine. Carboxylate esters are released at neutral pH in the absence of hydroxylamine, while arginine linkages require hydroxylamine for release. Asparagine and histidine (diphthamide) linkages are stable under both conditions. Utilizing these conditions, we have examined

1 Department of Biochemistry, North Texas State University/Texas College of Osteopathic Medicine, Denton, TX 76203, USA
2 Laboratory of Cellular Metabolism, National Heart, Lung, and Blood Institute, National Institutes of Health, Bethesda, MD 20205, USA
3 Department of Biology, Texas Woman's University, Denton, TX 76204, USA

ADP-Ribosylation of Proteins
(ed. by F.R. Althaus, H. Hilz, and S. Shall)
© Springer-Verlag Berlin Heidelberg 1985

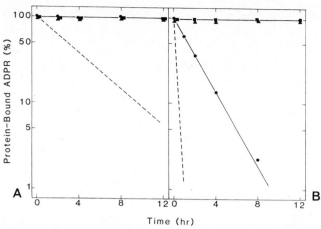

Fig. 1A,B. Stability of different (ADP-ribose)-protein linkages. Samples of [^{14}C]ADP-ribosylated histone (●), [^{32}P]ADP-ribosylated EF-2 (■), or [^{32}P]ADP-ribosylated transducin (▲) (containing ADP-ribosylated arginine, diphthamide, and asparagine, respectively) were added to liver extract and incubated at pH 7.0, 37°C in the absence (A) or presence (B) of 1 M NH$_2$OH. The amount of ADP-ribose remaining bound to protein was determined after removal of free ADP-ribose by column centrifugation. The *dashed line* in each panel represents the stability of the (ADP-ribose)-glutamate linkage

Table 1. Estimated levels of NAD and ADP-ribose residues in adult rat liver

ADP-ribose moieties as	Amount detected	
	pmol mg^{-1} protein	Relative values
NAD$^+$ + NADH[a]	3,110	1.0
"Arginine-linked" mono(ADP-ribose)	31.8	0.01
"Glutamate-linked" mono(ADP-ribose)	9.7	0.003
Poly(ADP-ribose)	0.083	0.00003

[a] Calculated from the data of Bredehorst et al. (1980) Hoppe-Seyler's Z Physiol Chem 361:559–562

total adult rat liver proteins for the presence of linkages with characteristics of carboxylate ester- and arginine-bound ADP-ribose (Table 1). Of the residues detected, approximately 80% had the characteristics of the arginine linkage and 20% the characteristics of the carboxylate ester linkage. The total amount of protein-bound monomeric ADP-ribose represented about 1% that of the total NAD pool; however, these were present in about 400-fold excess over polymeric ADP-ribose residues. In order to characterize the size distribution of rat liver proteins ADP-ribosylated at arginine, proteins were fractionated by size exclusion HPLC prior to analysis (Fig. 2). The results show a major peak of ADP-ribose in the mass range of 40 to 50 kD.

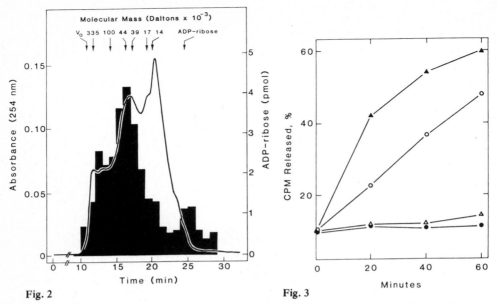

Fig. 2

Fig. 3

Fig. 2. Fractionation of rat liver proteins modified by ADP-ribosylation at arginine. The acid-insoluble fraction from rat liver was dissolved in a buffer containing 6 M guanidinium chloride, adjusted to pH 7.0, and incubated at 37°C for 4 h. An aliquot was then subjected to column centrifugation to eliminate noncovalently bound nucleotides. Aliquots of the protein fraction were fractionated by size exclusion HPLC using two columns in series (Bio-Sil TSK 250 plus TSK 400, 300 mm × 7.5 mm i.d., each) preceded by a guard column (Bio-Sil TSK 125, 75 mm × 7.5 mm i.d.). Fractions of 1 ml each were collected and either analyzed for protein or incubated with neutral hydroxylamine (1 M, 37°C, 12 h) and analyzed for ADP-ribose [3]

Fig. 3. Release of radioactivity from [^{14}C]ADP-ribosylated histone. ○ Standard incubation; ● boiled control (5 min); ▲ standard incubation plus 5 mM MgCl$_2$; △ boiled control plus 5 mM MgCl$_2$

Processing of Protein-Bound Mono(ADP-Ribose)

We have searched for enzymatic activities that will utilize as substrates ADP-ribosyl residues bound to protein at arginine. This search was facilitated by the preparation of NAD$^+$ radiolabeled in the nicotinamide-proximal ribose, which was then employed to modify a mixture of histones using purified turkey erythrocyte mono(ADP-ribosyl)-transferase [4]. Figure 3 shows the result obtained when these substrates were incubated with lysates of cultured mouse cells (SVt2). Activity catalyzing the release of radiolabel was observed (standard incubation), and the addition of MgCl$_2$ stimulated this release. The nature of the products was examined by paper chromatography. In the standard incubation, radiolabel comigrated with ADP-ribose and a second compound closely related to ADP-ribose. The stimulation of release in the presence of MgCl$_2$ was coincident with the appearance of a large amount of radiolabel that migrated with ribose 5-phosphate. This result (and additional data not shown) suggested the presence of a two-step processing pathway involving the sequential release of 5′-AMP followed by ribose 5-phosphate. In order to detect an activity that would release ribose 5-phosphate from proteins, we prepared a radiolabeled (5-phosphoribosyl)histone substrate. Cell extracts also catalyzed the removal of ribose 5-phosphate from this substrate.

Discussion

We have detected significant amounts of ADP-ribose residues bound to proteins in vivo in which the chemical linkages appear to be "arginine-like." While the actual identification of these linkages as arginine has not been confirmed, these studies make it higly likely that the proteins detected represent target proteins for endogenous mono(ADP-ribosyl) transferases. Most of the ADP-ribose residues detected were bound to proteins in the mass range of 40–50 kD. In this regard, it is of interest that the known target proteins of cholera and pertussis toxins are two specific G-proteins (Gs and Gi) of the adenylate cyclase complex, with masses of 45 and 39 kD, respectively. It has previously been suggested that these components of the adenylate cyclase complex may also represent target proteins for endogenous transferases [5]. We are currently isolating and identifying the proteins in this size class that are ADP-ribosylated.

If endogenous ADP-ribosylation of arginine residues represents a regulatory post-translational modification of proteins analogous to phosphorylation, one would predict that the modification should be reversible. We have detected activities in cell lysates which process proteins modified at arginine. This suggests that mono(ADP-ribosylation) is indeed a reversible modification. Our data also suggest that there may be multiple pathways of processing of ADP-ribose residues. The option of removing ADP-ribose intact or via a multistep process adds interesting potential regulatory capabilities to mono(ADP-ribosylation) reactions.

Acknowledgments. This work was supported in part by NIH grant CA23994, the North Texas State University Faculty Research Fund, and the Texas Ladies Auxillary to the Veterans of Foreign Wars. D.M.P. was a fellow of the Samuel Roberts Noble Foundation.

References

1. Vaughan M, Moss J (1981) Mono(ADP-ribosyl)transferases and their effects on cellular metabolism. Curr Top Cell Regul 20:205–246
2. Hayaishi O, Ueda K (1982) Poly- and mono(ADP-ribosyl)ation reactions: Their significance in molecular biology. In: Hayaishi O, Ueda K (eds) ADP-ribosylation reactions. Academic Press, London New York, pp 3–16
3. Jacobson MK, Payne DM, Juarez-Salinas H, Alvarez-Gonzalez R, Sims JL, Jacobson EL (1983) Determination of in vivo levels of polymeric and monomeric ADP-ribose by fluorescence methods. In: Wold F, Moldave K (eds) vol 106. Academic Press, London New York, pp 483–494
4. Moss J, Yost DA, Stanley SJ (1983) Amino acid-specific ADP-ribosylation. J Biol Chem 258:6466–6470
5. DeWolf MKS, Vitti P, Ambesi-Impiomboto FS, Kohn LD (1981) Thyroid membrane ADP-ribosyltransferase activity. J Biol Chem 256:12287–12296

Calcium Transport and Mono(ADP-Ribosylation) in Mitochondria

CHRISTOPH RICHTER, BALZ FREI, and JÖRG SCHLEGEL[1]

Abbreviations:

$\Delta\psi$ Mitochondrial transmembrane electrical potential, negative inside
SDS-PAGE   Sodium dodecyl sulfate polyacrylamide gel electrophoresis
SMP   Submitochondrial particles

1. The Importance of Cellular Ca^{2+} Homeostasis

The intracellular distribution of Ca^{2+} and, in particular, the level of cytosolic free Ca^{2+} are known to be extremely important in the regulation of many cellular processes, including metabolism, motility, exocytosis, membrane transport, and cell division [1, 2]. In many cells, such as hepatocytes, the cytosolic free Ca^{2+} concentration is thought to be controlled through the concerted action of Ca^{2+} transport systems located in the membranes of mitochondria [3], endoplasmatic reticulum [4], and plasmalemma [5].

2. Ca^{2+} Movements Across the Inner Mitochondrial Membrane

Since its discovery in 1962 by Vasington and Murphy [6] the mitochondrial Ca^{2+} transport system has been studied extensively. The molecular mechanism of Ca^{2+} uptake, and particularly that of Ca^{2+} release, are still not fully understood. However, there is now general consensus that Ca^{2+} uptake and release occur via two distinct transport systems.

2.1 Uptake of Ca^{2+} by Mitochondria

Addition of Ca^{2+} to energized mitochondria results in stimulation of respiration, extrusion of protons, and uptake of Ca^{2+} into mitochondria. The uptake is a purely

---

1   Laboratorium für Biochemie I, Eidgenössische Technische Hochschule, Universitätsstraße 16, CH−8092 Zürich, Switzerland

ADP-Ribosylation of Proteins
(ed. by F.R. Althaus, H. Hilz, and S. Shall)
© Springer-Verlag Berlin Heidelberg 1985

electrogenic process, driven by the electrical component ($\Delta\psi$, membrane potential) of the total proton-motive force. Ca^{2+} is transported with two positive charges via a so far unidentified carrier ("uniport"). The electrophoretic uniport has a high affinity to Ca^{2+} with a K_m between about 5 and 30 μM, depending on the tissue origin of mitochondria and the composition of the medium. In the absence of inorganic phosphate, the maximal uptake rate varies between 2 and 10 nmol mg^{-1} protein and s^{-1} in liver and heart mitochondria, respectively. Ca^{2+} uptake can be blocked by the specific inhibitor ruthenium red.

2.2 Release of Ca^{2+} from Mitochondria

Clearly, mitochondria cannot take up Ca^{2+} in unlimited amounts, and in order to prevent calcification of mitochondria there must be a process of Ca^{2+} release at a rate comparable to that of Ca^{2+} uptake. Many compounds are known to induce Ca^{2+} release from mitochondria. With a few exceptions, they cannot be of physiological relevance since they induce release by de-energizing mitochondria. Examples are uncouplers, respiratory inhibitors, and substrates that cause nonspecific damage to mitochondria. An important criterion for the identification of a physiologically important Ca^{2+} release agent is, therefore, intactness of mitochondria during release induced by this agent. In mitochondria of heart, brain, and some other tissues a ruthenium red-insensitive Ca^{2+} release can be induced by Na^+ [8]. The release is probably electroneutral and takes place in the presence of a high $\Delta\psi$. A significant Na^+-induced Ca^{2+} release from liver mitochondria is not observed. It is known, however, that in liver mitochondria Ca^{2+} is released in exchange with protons [3]. Until recently, very little was known about the molecular properties or the regulation of the Ca^{2+} release pathways. The finding that during Ca^{2+} release from intact liver mitochondria intramitochondrial pyridine nucleotides are hydrolyzed, and that nearly exclusively one protein in the inner mitochondrial membrane can be ADP-ribosylated upon NAD^+ hydrolysis (see below) opens new and exciting perspectives.

3. Ca^{2+} Release from Rat Liver Mitochondria

3.1 Oxidation and Hydrolysis of Pyridine Nucleotides Parallels Ca^{2+} Release

3.1.1 Releasing Compounds

Release of Ca^{2+} from intact rat liver mitochondria can be induced by oxidation of mitochondrial pyridine nucleotides. This was first shown by the group of Lehninger [8] who oxidized mitochondrial pyridine nucleotides with acetoacetate or oxaloacetate at the level of the citric acid cycle. Our group [9] used hydroperoxides, known to be produced by the respiratory chain of mitochondria and to be linked to pyridine nucleotides by glutathione peroxidase and reductase, to induce Ca^{2+} release and oxidation of pyridine nucleotides. Orrenius and co-workers employed [10] menadione (2-methyl-1, 4-naphthoquinone) to induce pyridine nucleotide oxidation and Ca^{2+} release. The latter groups also reported intramitochondrial hydrolysis of pyridine nucleotides during Ca^{2+} release.

3.1.2 Qualitative and Quantitative Aspects

Oxidation alone of pyridine nucleotides is not sufficient to induce Ca^{2+} release. In the presence of ATP, the hydroperoxide-induced pyridine nucleotide oxidation is even accelerated, yet pyridine nucleotide hydrolysis and Ca^{2+} release are inhibited [11]. Similar observations were made during the menadione-induced Ca^{2+} release [10]. When liver mitochondria are treated with N-ethyl maleimide to lower intramitochondrial glutathione, both oxidation of pyridine nucleotides and Ca^{2+} release are inhibited (S. Baumhüter, C. Richter, unpubl.). Finally, both pyridine nucleotide hydrolysis and Ca^{2+} release show the same sigmoidal dependence on the mitochondrial Ca^{2+} load [15]. Thus, there is clear, albeit circumstantial, evidence that pyridine nucleotide hydrolysis and Ca^{2+} release are functionally related. The link between the two processes may be protein ADP-ribosylation.

4. Protein ADP-Ribosylation in the Inner Mitochondrial Membrane

4.1 NAD$^+$ Glycohydrolase Activity

Intramitochondrial hydrolysis of pyridine nucleotides and release of nicotinamide from mitochondria exposed to hydroperoxide [9] suggested the existence of an intra-mitochondrial NAD$^+$ glycohydrolase. Subsequent studies [12] located the enzyme on the inner side of the inner mitochondrial membrane. Activity was also present in the matrix fraction of mitochondria, but release from the inner membrane into the matrix fraction during isolation of the membrane was not ruled out. The activity in inner mitochondrial membranes (submitochondrial particles, SMP) is inhibitable by ATP [11]. In addition, ATP-sensitive covalent modification of SMP concomitant with enzymatic hydrolysis of NAD$^+$ was indicated.

The activity of the enzyme in SMP was further investigated [13]. It is moderately inhibited by ATP and seemingly strongly by nicotinamide. Later it was found (U. Heimgartner, C. Richter 1984, unpubl.) that in the presence of nicotinamide the enzyme catalyzes a transglycosidation reaction. Based on this finding, the inhibition of the enzyme by nicotinamide is only slight. It was also shown [13] that the enzyme is not inhibited by arginine-blocking reagents and not stimulated by arginine methyl ester.

The NAD$^+$ glycohydrolase was isolated and purified [14] from SMP to apparent homogeneity as judged by sodium dodecyl sulfate polyacrylamide gel electrophoresis (SDS-PAGE) and showed a mol.wt. of about 64,000. The purified enzyme had highest activity with NAD$^+$, moderate activity with NADP$^+$, and was inactive with NAD(P)H.

With a slightly modified purification procedure (J. Schlegel, C. Richter 1984, unpubl.) we were recently able to purify two proteins with NAD$^+$ glycohydrolase activity to homogeneity upon SDS-PAGE with mol.wt. of about 32,000 and 64,000, respectively. Whether the two proteins represent monomeric and dimeric forms of the same enzyme remains to be seen.

4.2 Protein Mono(ADP-Ribosylation)

Incubation of SMP with NAD^+ labeled at the adenine moiety leads to a time-dependent incorporation of radioactivity into acid-precipitable material [13]. Maximal incorporation is achieved within about 60 min. With NAD^+ labeled at the NMN-ribose moiety (kindly provided by K. Ueda) the efficiency of incorporation is doubled and amounts to about 220 pmol mg^{-1} protein (B. Frei, C. Richter 1984, unpubl.). With NAD^+ labeled at the nicotinamide moiety incorporation is not observed.

When labeled SMP were analyzed on SDS-PAGE, radioactivity was found almost exclusively in the region of proteins of mol.wt. of about 32,000 [13]. Treatment of the labeled protein with snake venom phosphodiesterase liberated predominantly 5′-AMP, indicating modification of the protein by mono(ADP-ribose).

The mono(ADP-ribose) residue modifying the protein turns over rapidly. This was indicated by two lines of evidence [13]. First, when SMP were incubated with [^{3}H]-NAD^+ and, after attaining a steady state level of modification as judged by incorporation of tritium, were then supplied with [^{14}C]-NAD^+, there was loss of tritium and simultaneously incorporation of [^{14}C] into the protein. Secondly, when ATP was added to SMP, mono(ADP-ribose) was lost from the protein within seconds. Whether the removal of mono(ADP-ribose) from the protein is enzyme-catalyzed remains to be seen. The bond between mono(ADP-ribose) and the 32,000 mol.wt. protein in SMP is acid stable, alkali sensitive, and labile in neutral hydroxyl amine. Formation of it is inhibited by arginine blocking reagents [13].

As mentioned above, we have recently isolated two NAD^+ glycohydrolase activities from SMP. Interestingly, the enzyme of mol.wt. of about 32,000 undergoes a time-dependent auto(ADP-ribosylation) when incubated with 1 mM NAD^+ (J. Schlegel, C. Richter 1984, unpubl.). It is, therefore, conceivable that the isolated protein and the protein labeled in SMP are identical.

5. Conclusions and Outlook

Ca^{2+} release from intact rat liver mitochondria is induced by compounds that oxidize intramitochondrial pyridine nucleotides. Oxidation alone of pyridine nucleotides is not sufficient to cause Ca^{2+} release. Release is, however, observed when oxidized pyridine nucleotides are hydrolyzed at the N-glycosidic bond linking ADP-ribose and nicotinamide. Both pyridine nucleotide hydrolysis and Ca^{2+} release are very similarly dependent on the intramitochondrial Ca^{2+} load.

When inner mitochondrial membranes hydrolyze NAD^+ a protein with mol.wt. of about 32,000 is modified by mono(ADP-ribose). The modification turns over rapidly. A NAD^+ glycohydrolase can be isolated from SMP. Its mol.wt. as judged by SDS-PAGE is about 32,000. In the presence of NAD^+ the enzyme undergoes auto(ADP-ribosylation).

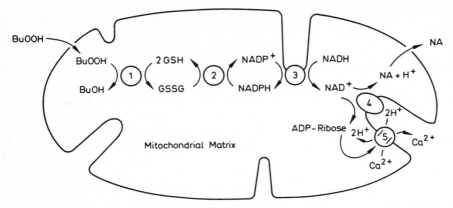

Fig. 1. Hydroperoxide-induced release of Ca^{2+} from rat liver mitochondria. _1_ Glutathione peroxidase; _2_ glutathione reductase; _3_ energy-linked pyridine nucleotide transhydrogenase; _4_ NAD^+ glycohydrolase; _5_ Ca^{2+}/H^+ antiporter

Based on our studies on the hydroperoxide-induced release of Ca^{2+} from rat liver mitochondria a close correlation between protein mono(ADP-ribosylation) in the inner mitochondrial membrane and Ca^{2+} release is apparent. Whether this correlation represents a regulatory mechanism of Ca^{2+} release is presently not proven. It seems, however, justified to propose a regulation of a physiologically relevant Ca^{2+} release pathway by mono(ADP-ribosylation). Our results and the proposed mechanism of the hydroperoxide-induced release of Ca^{2+} from rat liver mitochondria are summarized in Fig. 1.

The release of Ca^{2+} from intact rat liver mitochondria induced by hydroperoxide has been investigated in detail in vitro. Since hydroperoxides are physiological reactants of the mitochondrial respiratory chain, the question naturally arises whether such a release might also be operative in vivo. In this context it is worth noting that the amounts of hydroperoxide used in the above experiments are close to those that occur under physiological or pathological conditions. Furthermore, the quantities of $NAD(P)^+$, important coenzymes in many biochemical reactions, that are lost in mitochondria due to enzymatic hydrolysis during Ca^{2+} release can be quite large. At present, very little is known about the biosynthesis of mitochondrial pyridine nucleotides. Exciting new experiments are clearly ahead of us.

Acknowledgments. The work performed in the author's laboratory was supported by the Swiss National Science Foundation, grants No. 3.119-0.77, 3.699-0.80, and 3.503-0.83.

References

1. Rasmussen H, Goodman DBP (1977) Relationships between calcium and cyclic nucleotides in cell activation. Physiol Rev 57:421−509
2. Cheung WY (1980) Calmodulin plays a pivotal role in cellular regulation. Science 207:19−27
3. Fiskum G, Lehninger AL (1980) The mechanisms and regulation of mitochondrial Ca^{2+} transport. Fed Proc 39:2432−2436
4. Moore L, Chen T, Knapp HR, Landon EJ (1975) Energy-dependent calcium sequestration activity in rat liver microsomes. J Biol Chem 250:4562−4568
5. Van Rossum GDV (1970) Net movements of calcium and magnesium in slices of rat liver. J Gen Physiol 55:18−32
6. Vasington FD, Murphy JV (1962) Ca^{2+} uptake by rat kidney mitochondria and its dependence on respiration and phosphorylation. J Biol Chem 237:2670−2677
7. Crompton M, Capano M, Carafoli E (1976) The sodium-induced efflux of calcium from rat heart mitochondria. Eur J Biochem 69:453−462
8. Lehninger AL, Vercesi A, Bababunmi EA (1978) Regulation of Ca^{2+} release from mitochondria by the oxidation-reduction state of pyridine nucleotides. Proc Natl Acad Sci USA 75:1690−1694
9. Lötscher HR, Winterhalter KH, Carafoli E, Richter C (1979) Hydroperoxides can modulate the redox state of pyridine nucleotides and the calcium balance in rat liver mitochondria. Proc Natl Acad Sci USA 76:4340−4344
10. Bellomo G, Jewell SA, Orrenius S (1982) The metabolism of menadione impairs the ability of rat liver mitochondria to take up and retain calcium. J Biol Chem 257:11558−11562
11. Hofstetter W, Mühlebach T, Lötscher HR, Winterhalter KH, Richter C (1981) ATP prevents both hydroperoxide-induced hydrolysis of pyridine nucleotides and release of calcium in rat liver mitochondria. Eur J Biochem 117:361−367
12. Lötscher HR, Winterhalter KH, Carafoli E, Richter C (1980) Hydroperoxide-induced loss of pyridine nucleotides and release of calcium from rat liver mitochondria. J Biol Chem 255:9325−9330
13. Richter C, Winterhalter KH, Baumhüter S, Lötscher HR, Moser B (1983) ADP-ribosylation in inner membrane of rat liver mitochondria. Proc Natl Acad Sci USA 80:3188−3192
14. Moser B, Winterhalter KH, Richter C (1983) Purification and properties of a mitochondrial NAD^{+} glycohydrolase. Arch Biochem Biophys 224:358−364
15. Frei B, Winterhalter KH, Richter C (1985) Quantitative and mechanistic aspects of the hydroperoxide-induced release of Ca^{2+} from rat liver mitochondria. Eur J Biochem 149:633−639

A Cellular Mono(ADP-Ribosyl)Transferase which Modifies the Diphthamide Residue of Elongation Factor-2

WALLACE J. IGLEWSKI, HERBERT LEE, and PATRICK MULLER[1]

Abbreviations:

EF-2   elongation factor 2
NAD   nicotinamide adenine dinucleotide
DTT   dithiothreitol
pyBHK   polyoma virus transformed baby hamster kidney cells

1. Introduction

ADP-ribosylation of eukaryotic elongation factor-2 (EF-2) is recognized as the essential biochemical event in the intoxication of cells by both diphtheria toxin and *Pseudomonas* toxin A. Both toxins catalyze the transfer of the adenosine 5'-diphosphate ribosyl (ADPR) moiety of nicotinamide adenine dinucleotide (NAD) onto EF-2 [1, 2]. The resultant ADPR-EF-2 complex is inactive in protein synthesis. When EF-2 becomes rate limiting, protein synthesis ceases and the intoxicated cell dies.

Diphtheria toxin and *Pseudomonas* toxin A demonstrate a remarkable specificity in ADP-ribosylating a single amino acid in EF-2. This is a modified histidine residue, 2-[3-carboxyamide-3-(trimethylammonio)propyl] histidine, termed diphthamide [3]. The ability of EF-2 from extracts of all eukaryotic cells to be ADP-ribosylated by diphtheria toxin fragment A or *Pseudomonas* toxin A suggests that the diphthamide acceptor site has been conserved throughout eukaryotic evolution [4, 5]. Since enzymatically active toxin can not gain entrance into many of these cell types, it seems unreasonable that diphthamide has been conserved only for the microbial intoxication of cells. It is more likely that diphthamide is an acceptor for an ADP-ribosylation reaction catalyzed by a cellular enzyme which would modify the structure and presumably the function of EF-2. In fact, a cellular enzyme has been isolated from polyoma virus transformed baby hamster kidney (pyBHK) cells and from beef liver which also ADP-ribosylates EF-2. This report summarizes recent studies on the isolation and characterization of this enzyme and presents additional data on the immunological characterization of the ADP-ribosyltransferase from pyBHK cells.

---

1   Department of Microbiology and Immunology, Oregon Health Sciences University, 3181 SW Sam Jackson Park Road, Portland, OR 97201, USA

ADP-Ribosylation of Proteins
(ed. by F.R. Althaus, H. Hilz, and S. Shall)
© Springer-Verlag Berlin Heidelberg 1985

2. Production of EF-2 Preparations
Containing the Cellular Transferase Activity

EF-2 was isolated from cell extracts prepared from pyBHK-induced tumors in hamsters or from beef liver [6]. Tissue was homogenized in a Waring blender with Littlefields medium and a postmitochondrial fraction was obtained following centrifugation. A protein fraction was precipitated by adding ammonium sulfate to 80% saturation. EF-2 was purified from the ammonium sulfate precipitate by successive chromatography on DEAE-cellulose, Sephadex DEAE-A50, Sephacryl S-200, and phosphocellulose. The purified EF-2 exhibited one major band on sodium dodecyl sulfate (SDS)-polyacrylamide gel electrophoresis with an apparent M_r of 93,000.

3. Detection of Transferase Activity

During the characterization of EF-2 preparation from pyBHK cells, it was noted that there was a transfer of the $[^{14}C]$-adenosine moiety from NAD to a trichloroacetic acid precipitable form during incubation in a Tris-HCl buffered reaction mixture containing no toxin, only the EF-2 preparation and [adenosine-^{14}C]-NAD [7]. The transfer of label was time dependent, reaching a maximum level at 80 to 160 min of incubation. Adding diphtheria toxin fragment A to the reaction resulted in a more rapid transfer of label to an acid precipitable form, but the maximum amount of label transferred in the two reactions was similar. The ability to transfer $[^{14}C]$-adenosine from NAD to an acid precipitable form suggested that the EF-2 preparations contained a novel transferase activity. This activity was probably a contaminant in the EF-2 preparations since the majority, but not all, of the EF-2 preparations examined contained the transferase activity.

4. Characterization of the pyBHK ADP-Ribosyltransferase
Catalyzed Reaction

The $[^{14}C]$-adenosine labeled product formed in the cellular transferase catalyzed reaction was analyzed by SDS-polyacrylamide gel electrophoresis and the radioactive product was visualized by autoradiography. Essentially all of the $[^{14}C]$-adenosine was transferred to EF-2 [7]. Digestion of the $[^{14}C]$-adenosine labeled EF-2 with snake venom phosphodiesterase produced $[^{14}C]$-AMP indicating that ADPR was being transferred to EF-2 from $[^{14}C]$-NAD by the cellular ADP-ribosyltransferase as monomeric units. In addition, the forward reaction catalyzed by pyBHK ADP-ribosyltransferase was able to be reversed by diphtheria toxin fragment A with the production of $[^{14}C]$-NAD from $[^{14}C]$-adenosine labeled EF-2. These results imply that the ADPR-EF-2 bonds formed by diphtheria toxin fragment A and pyBHK (ADP-ribosyl)transferase are very similar and suggest that the cellular transferase also ADP-ribosylates the diphthamide residue of EF-2.

Table 1. Reconstitution of a reticulocyte cell-free protein synthesizing system by addition of pyBHK EF-2 or pyBHK EF-2 ADP-ribosylated by pyBHK ADP-ribosyltransferase or by diphtheria toxin fragment A

Incubation time (min)	[^{3}H]-Leucine incorporated into protein (CPM)			
	Without EF-2	EF-2	EF-2 ADP-ribosylated by cellular transferase	EF-2 ADP-ribosylated by fragment A
10	1,552	12,575	2,132	1,878
20	4,750	22,837	5,458	6,514
30	5,972	39,827	8,457	7,981

ADP-ribosylation of EF-2 by diphtheria toxin fragment A results in the inability of ADPR-EF-2 to function in protein synthesis [4]. An inactivation of EF-2 should also occur if pyBHK ADP-ribosyltransferase is modifying EF-2 in a manner similar to that catalyzed by fragment A. To test this possibility, a rabbit reticulocyte cell-free protein synthesizing system deficient in functional EF-2 was prepared as previously described [2]. The system was then reconstituted with either 55 μg ml^{-1} purified pyBHK EF-2, EF-2 which was ADP-ribosylated by the cellular transferase or EF-2 which was ADP-ribosylated by diphtheria toxin fragment A (Table 1). Although the cell-free protein synthesizing system could be effectively reconstituted with pyBHK EF-2, no stimulation of proteins synthesis was obtained by adding either EF-2 ADP-ribosylated by the cellular transferase or by fragment A. Thus, the two transferases have identical effects on the function of EF-2 in protein synthesis confirming that they modify EF-2 in the same manner.

Although fragment A of diphtheria toxin and pyBHK ADP-ribosyltransferase share a number of properties, they can clearly be distinguished [7]. Diphtheria toxin fragment A antibody neutralizes the enzyme activity of this bacterial toxin, but has no effect on pyBHK ADP-ribosyltransferase. Secondly, charcoal adsorbed cytoplasmic extracts of pyBHK cells inhibit the cellular transferase activity, but have no effect on fragment A enzyme activity. The inhibitory activity of the cytoplasmic extracts is eliminated by boiling, suggesting a heat labile component is involved in the inhibition. Finally, the cellular transferase is inhibited by histamine, while diphtheria toxin fragment A and *Pseudomonas* toxin A activities are not affected.

5. ADP-Ribosyltransferase from Beef Liver

Since pyBHK cells are both virus infected and a transformed cell line, it was of interest to determine whether the cellular transferase also could be isolated from normal tissue. Beef liver EF-2 preparations containing the transferase activity were isolated and characterized by the same methods used to study the pyBHK ADP-ribosyltransferase [8]. Both eukaryotic enzymes transfer [adenosine-^{14}C] ADP-ribose from NAD to EF-2 [7, 8]. Snake venom phosphodiesterase digestion of the [^{14}C]-labeled beef

Table 2. Effect of *Pseudomonas* toxin A and pyBHK ADP-ribosyltransferase antisera on transferase activities

Transferase	ADP-ribosyltransferase activity[a] (% of control)[b]	
	Anti-*Pseudomonas* toxin A serum	Anti-pyBHK ADP-ribosyltransferase serum
Pseudomonas toxin A	3%	100%
pyBHK ADP-ribosyltransferase	2%	20%
Diphtheria toxin fragment A	NT[c]	100%

[a] Toxin containing reactions were incubated for 20 min and pyBHK ADP-ribosyltransferase containing reactions were incubated for 160 min at 22°C to obtain maximum transfer of ADP-ribose to EF-2

[b] The activity of the transferase treated with antiserum as a percent of the activity of the transferase treated with normal serum

[c] NT = not tested; *Pseudomonas* toxin A antisera does not neutralize diphtheria toxin fragment A [10]

liver EF-2 yields [^{14}C]-AMP. In addition, the forward transferase reaction catalyzed by beef liver ADP-ribosyltransferase is reversible by excess diphtheria toxin fragment A, with the formation of [^{14}C]-NAD, indicating that both enzymes modify the same site on EF-2. The isolation of similar ADP-ribosyltransferases from different tissues of two species of animals suggests the enzyme may be ubiquitous among eukaryotic cells.

6. Immunological Studies on the pyBHK ADP-Ribosyltransferase

6.1 The Effect of Antisera on the Enzyme Activity of Pseudomonas Toxin A, Fragment A of Diphtheria Toxin, and pyBHK ADP-Ribosyltransferase

Previous studies have shown that diphtheria fragment A antibody has no effect on the enzyme activity of pyBHK ADP-ribosyltransferase [7]. However, antisera to other ADP-ribosyltransferases which modify EF-2 were not examined. *Pseudomonas* toxin A antibody was kindly provided by B.H. Iglewski. In addition, an antisera was produced by hyperimmunizing BALB/c mice with a preparation of pyBHK EF-2 containing the endogenous ADP-ribosyltransferase activity. The resulting sera was called pyBHK ADP-ribosyltransferase antisera since our earlier studies indicated that the EF-2 component of the immunogen was not antigenic (unpublished observations). *Pseudomonas* toxin A, diphtheria toxin fragment A or a pyBHK EF-2 preparation containing endogenous transferase were incubated with the two antisera or with nonimmune control sera for 5 min at 37°C (Table 2). The preincubation mixtures were added to our standard reaction mixture and assayed for transferase activity as previously described [7]. *Pseudomonas* toxin A antibody inhibits both *Pseudomonas* toxin A and pyBHK ADP-ribosyltransferase activities. However, pyBHK ADP-ribosyltransferase antibody inhibits the cellular transferase, but has no effect on the activity of *Pseudomonas* toxin A. The data suggest that *Pseudomonas* toxin A antibody recognizes determinants common to

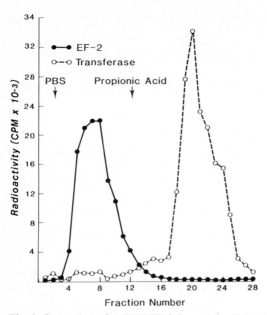

Fig. 1. Separation of pyBHK EF-2 from the cellular ADP-ribosyltransferase by immunoabsorbant chromatography. A 50 μg sample of an EF-2 preparation containing the cellular transferase activity was applied to a 250 μl column of Sepharose 4B resin coupled to *Pseudomonas* toxin A antibody. Unbound sample was washed from the resin with phosphate-buffered saline containing 1 m*M* DTT (PBS). Protein bound to the immunoabsorbant was dissociated and eluted in 1 *M* propionic acid containing 1 m*M* DTT. Acid-containing fractions were immediately neutralized with 2 *M* Tris-Base. A portion of each fraction was assayed for EF-2 (●—●) by the transfer of [^{14}C]-adenosine from NAD to the EF-2 in the presence of diphtheria toxin fragment A [6]. A second portion of each fraction was assayed for pyBHK ADP-ribosyltransferase activity (○—○) by the transfer of [^{14}C]-adenosine from NAD to EF-2 in our standard cellular transferase reaction mixture [7] which was supplemented with 4 μg of a pyBHK EF-2 preparation which lacked its own endogenous transferase activity. Acid precipitable radioactivity was detected using a liquid scintillation counter

both transferases, whereas the cellular transferase antibody recognizes a different determinant of the pyBHK transferase which is not shared by *Pseudomonas* toxin A.

6.2 Immunoabsorbant Chromatography

Since *Pseudomonas* toxin A antibody appears to bind to the pyBHK transferase, the antibody was used in immunoabsorbant chromatography to separate the transferase from EF-2. *Pseudomonas* toxin A antibody (300 μg) which had been affinity purified on a column of *Pseudomonas* toxin A was coupled to 0.4 ml of swollen CNBr-activated Sepharose 4B and packed in a glass tuberculin syringe. A pyBHK EF-2 preparation (50 μg) containing the endogenous transferase was applied to the column. Nonbound protein was washed from the resin with phosphate-buffered saline (PBS). Protein bound to the immunoabsorbant was eluted with propionic acid and immediately neutralized in fraction tubes containing 2 *M* Tris-Base. The EF-2 component of the preparation did not bind to the immunoabsorbant and eluted in PBS (Fig. 1). In contrast, the pyBHK ADP-ribosyltransferase bound to the antibody conjugated resin and was removed by washing the resin with 1 M propionic acid. Thus, the cellular ADP-ribosyltransferase is a contaminant in the pyBHK EF-2 preparations and can be separated from its substrate by immunoabsorbant chromatography.

6.3 Western Blot Analysis

Since the results of the immunoabsorbant chromatography demonstrate that the cellular ADP-ribosyltransferase binds to the resin, Western blot analysis [9] was used to identify the antigens which bound to the *Pseudomonas* toxin A antibody or pyBHK ADP-ribosyltransferase antibody coupled to the immunoabsorbant resins. Preparations of EF-2 containing the transferase activity were chromatographed on the two types of immunoabsorbant columns. Material eluting in propionic acid was concentrated, subjected to electrophoresis on a SDS-polyacrylamide gel, and blotted onto a nitrocellulose sheet. Purified *Pseudomonas* toxin A was also subjected to electrophoresis in the gel as a positive control. The sheet was sectioned and reacted with either sheep anti-*Pseudomonas* toxin A antibody or with mouse anti-pyBHK ADP-ribosyltransferase antibody. The antibody binding bands on the nitrocellulose sheet were detected by reacting with peroxidase conjugated rabbit anti-sheep or peroxidase conjugated goat anti-mouse immunoglobulins. The position of the bound peroxidase was detected with HRP Color Development Reagent (Bio-Rad). As shown in Fig. 2, *Pseudomonas* toxin A bound the *Pseudomonas* toxin A antibody, but toxin A did not react with the pyBHK ADP-ribosyltransferase antibody, supporting the data on neutralization of transferase activities presented in Table 2. Both the *Pseudomonas* toxin A antibody and the pyBHK ADP-ribosyltransferase antibody reacted with two pyBHK proteins which have with an M_r of approx. 100,000 and 80,000. Since both antisera react with the cellular transferase and neutralize its enzyme activity, it is likely that either the 80,000 or 100,000 proteins represent the cellular transferase. However, the data do not distinguish which protein is the active form of the enzyme or whether the 80,000 protein is a cleavage product of the 100,000 protein. Nevertheless, the immunoreactive proteins are larger in size than diphtheria toxin (62,000) and *Pseudomonas* toxin A (66,000) and clearly distinguish the cellular transferase from the bacterial transferases which modify EF-2.

7. Summary

An ADP-ribosyltransferase was found in EF-2 preparations from pyBHK cells and beef liver. This eukaryotic cellular enzyme transfers [^{14}C]-adenosine from NAD to EF-2. Digestion of the [^{14}C]-adenosine labeled EF-2 product of the cellular transferase reaction with snake venom phosphodiesterase yielded [^{14}C]-AMP, indicating the enzyme is a mono(ADP-ribosyl)transferase. The forward ADP-ribosylation reaction catalyzed by the pyBHK or beef liver enzyme is reversed by diphtheria toxin fragment A, yield-

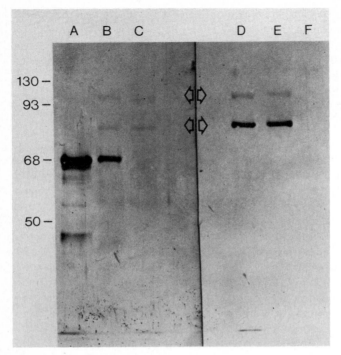

Fig. 2. Western blot analysis of protein bound to immunoabsorbant columns. PyBHK EF-2 preparations containing cellular ADP-ribosyltransferase activity were chromatographed on Sepharose 4B
coupled to *Pseudomonas* toxin A antibody or pyBHK ADP-ribosyltransferase antibody. After
washing unbound material from the resin with PBS, the bound protein was eluted from the columns
with 1 *M* propionic acid, concentrated under vacuum and subjected to electrophoresis on a 7.5%
SDS-polyacrylamide slab gel. In addition, *Pseudomonas* toxin A standards were subjected to
electrophoresis on the gel. Following electroblotting of the proteins from the gel to a nitrocellulose sheet, the sheet was sectioned and reacted with *Pseudomonas* toxin A antibody (*A−C*) or
with pyBHK ADP-ribosyltransferase antibody (*D−F*). *Pseudomonas* toxin A (*A*); protein from
Pseudomonas toxin A antibody coupled immunoabsorbant (*B*); protein from pyBHK ADP-ribosyltransferase antibody coupled immunoabsorbant (*C, D*); protein from *Pseudomonas* toxin A anti-
body-coupled immunoabsorbant (*E*); *Pseudomonas* toxin A (*F*). The *numbers* represent $M_r \times 10^{-3}$
of the mol.wt. standards

ing [^{14}C]-NAD. ADP-ribosylation of EF-2 by fragment A or by the cellular transferase
results in an inactivation of EF-2 in protein synthesis. The results indicate that the
eukaryotic transferase is a mono(ADP-ribosyl)transferase which ADP-ribosylates the
same diphthamide residue of EF-2 as does fragment A and *Pseudomonas* toxin A.
However, the eukaryotic transferase can be distinguished from fragment A and *Pseudomonas* toxin A by the inhibition of the pyBHK transferase with cytoplasmic extracts
of pyBHK cells or histamine. The cellular transferase is immunologically distinct from
fragment A, but cross-reacts with *Pseudomonas* toxin A antibody. In contrast, antibody to pyBHK transferase neutralizes the enzyme activity of the eukaryotic enzyme,
but has no effect on the activity of either *Pseudomonas* toxin A or fragment A of

diphtheria toxin. Immunoabsorbant chromatography of pyBHK EF-2 preparations on *Pseudomonas* toxin A antibody coupled to Sepharose 4B separates EF-2 from the cellular transferase which binds to the resin and can subsequently be eluted in an enzymatically active form. Western blot analysis of proteins from pyBHK EF-2 preparations which were purified on either *Pseudomonas* toxin A antibody or pyBHK ADP-ribosyltransferase antibody immunoabsorbants demonstrate both an 80,000 and a 100,000 mol.wt. protein which bind to both types of antibody, suggesting one or both of these proteins may represent the cellular ADP-ribosyltransferase.

Acknowledgments. This work was supported in part by the American Cancer Society, Oregon Division and the Collins Medical Trust and GM 34142.

References

1. Honjo T, Nishizuka Y, Hayaishi O, Kato I (1968) Diphtheria toxin-dependent adenosine diphosphate-ribosylation of aminoacyltransferase II and inhibition of protein synthesis. J Biol Chem 243:3553–3555
2. Iglewski BH, Liu PV, Kabat D (1977) Mechanism of action of *Pseudomonas aeruginosa* exotoxin A: Adenosine diphosphate-ribosylation of mammalian elongation factor 2 in vitro and in vivo. Infect Immun 15:138–144
3. Van Ness BG, Howard JB, Bodley JL (1980) ADP-ribosylation of elongation factor 2 by diphtheria toxin. J Biol Chem 255:10710–10716
4. Collier RJ (1975) Diphtheria toxin: Mode of action and structure. Bacteriol Rev 39:54–85
5. Pappenheimer AM Jr (1977) Diphtheria toxin. Annu Rev Biochem 46:69–94
6. Iglewski WJ, Lee H (1983) Purification and properties of an altered form of elongation factor 2 from mutant cells resistant to intoxication by diphtheria toxin. Eur J Biochem 134:237–240
7. Lee H, Iglewski WJ (1984) Cellular ADP-ribosyltransferase with the same mechanism of action as diphtheria toxin and *Pseudomonas* toxin A. Proc Natl Acad Sci USA 81:2703–2707
8. Iglewski WJ, Lee H, Muller P (1984) ADP-ribosyltransferase from beef liver which ADP-ribosylates elongation factor-2. FEBS Lett 173:113–118
9. Nicas T, Iglewski BH (1984) Isolation and characterization of transposon-induced mutants of *Pseudomonas aeruginosa* deficient in production of exoenzyme S. Infect Immun 45:470–474
10. Iglewski BH, Kabat D (1975) NAD-dependent inhibition of protein synthesis by *Pseudomonas aeruginosa* toxin. Proc Natl Acad Sci USA 72:2283–2288

Photochemical Transfer of the Nicotinamide Moiety of NAD to a Specific Residue in the Catalytic Center of Diphtheria Toxin Fragment A

STEPHEN F. CARROLL[1], NORMAN J. OPPENHEIMER[2],
THOMAS M. MARSCHNER[2], JAMES A. McCLOSKEY[3], PAMELA F. CRAIN[3],
and R. JOHN COLLIER[1]

1. Introduction

Photolabeling with enzyme substrates, effectors, or photolabile analogs thereof is one of the most useful means of identifying active site residues within the primary structures of enzymes. We have recently applied this method to the study of the NAD-binding sites of two mono(ADP-ribosyl) transferases, diphtheria toxin (DT) and exotoxin A (PT) from *Pseudomonas aeruginosa*. Both toxins (after appropriate activation steps) transfer ADP-ribose from NAD to elongation factor 2, and as a side reaction, catalyze the hydrolysis of NAD to ADP-ribose, nicotinamide, and a proton.

Earlier we characterized the enzymic and ligand-binding properties of DT fragment A, a M_r 21,164, enzymically active proteolytic fragment of the toxin [1]. The fragment was shown to have a single NAD site (K_d = 8.3 μM) with subsites for both aromatic moieties of the dinucleotide. Subsequently, we observed UV-induced transfer of label from NAD to fragment A, apparently mediated by photoexcitation of either or both of the nitrogenous bases of the dinucleotide [2]. Recently, we reported that label within the nicotinamide moiety of NAD, but not that within other moieties, is transferred to the protein with remarkably high efficiency and specificity [3]. Here, we summarize these results and more recent data regarding the structure of the photoproduct.

2. Results

2.1 Photolabeling of Proteins with NAD Labeled at Various Positions

Figure 1 shows results obtained when fragment A was irradiated under a low-pressure mercury lamp in the presence of three different preparations of NAD, which were

---

1 Department of Microbiology and Molecular Genetics and Shipley Institute of Medicine, Harvard Medical School, 25 Shattuck Street, Boston, MA 02115, USA
2 Department of Pharmaceutical Chemistry, University of California, San Francisco, CA 94143, USA
3 Departments of Medicinal Chemistry and Biochemistry, University of Utah, Salt Lake City, UT 84112, USA

ADP-Ribosylation of Proteins
(ed. by F.R. Althaus, H. Hilz, and S. Shall)
© Springer-Verlag Berlin Heidelberg 1985

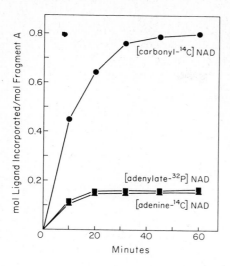

Fig. 1. Photolabeling of fragment A in the presence of three preparations of NAD radiolabeled in different positions. Reaction mixtures containing 50 mM Tris-HCl at pH 7.2, fragment A (20 μM), and NAD (40 μM) radiolabeled in the nicotinamide (●), the adenylate phosphate (■), or the adenine (▲) moiety were irradiated at 0°C under a low pressure mercury lamp (predominantly 253.7 nm). Aliquots were removed at intervals and denatured in guanidine hydrochloride, and the trichloroacetic acid-precipitable radioactivity was measured. (Reproduced from [3], with permission)

labeled in the adenine moiety, the adenylate phosphate, or the carbonyl group of the nicotinamide moiety [3]. With [carbonyl-^{14}C]NAD the rate and extent of incorporation of label into trichloroacetic acid-insoluble material (0.67–0.93 mol/mol) were much greater than with adenine- or phosphate-labeled dinucleotide (0.13–0.18 mol/mol). Similar results were observed when the nicked form of whole DT was substituted for fragment A; and the incorporated label migrated with the fragment A moiety of the toxin during electrophoresis under reducing conditions in SDS-polyacrylamide gels. The efficiency of labeling was very low when NAD was replaced with radioactive nicotinamide, either in the presence or absence of unlabeled ADP-ribose.

Photolabeling of a limited array of other proteins with [carbonyl-^{14}C]NAD was tested under identical conditions. Table 1 shows that besides fragment A, the ApUp-free form of DT, CRM-45, and activated *Pseudomonas* toxin incorporated label to much greater extents than did other proteins tested. The amounts of label incorporated by intact DT and by activated *Pseudomonas* toxin were virtually identical, suggesting similarities in the nicotinamide subsites of these two toxins.

2.2 Photolabeled Fragment A is Enzymically Inactive

After UV irradiation in the presence of [carbonyl-^{14}C]NAD and removal of unreacted dinucleotide, radiolabeled fragment A was separated from unlabeled fragment A by affinity chromatography on NAD-agarose. Labeled material, which was unretarded, was found to be inactive in either the NAD-glycohydrolase or the ADP-ribosyltransferase assay [3]. Unlabeled material bound to the column and was recovered by elution with 10 mM NAD. Although the unlabeled material had undergone some damage from irradiation [2], it nonetheless retained at least 50% of the activity of the non-irradiated control.

Table 1. UV-induced incorporation of label from [carbonyl-^{14}C]NAD into various proteins[a]. (Reproduced from [3], with permission)

Sample	mol % Labeling[b]
DT fragment A	74
DT, nucleotide-free	39
DT, nucleotide-bound	12
CRM-45	36
CRM-197	5
Exotoxin A	5
Activated exotoxin A[c]	33
Alcohol dehydrogenase	6
Lactate dehydrogenase	5
Malate dehydrogenase	6
Ovalbumin	3

[a] Each sample of protein (20 μM) was irradiated at 0°C in 50 mM Tris-HCl, pH 7.2, containing 40 μM [carbonyl-^{14}C]NAD. After 30 min, trichloroacetic acid-precipitable radioactivity was measured
[b] (mol label incorporated/mol protein) × 100
[c] Exotoxin A was activated by incubation with 4 M urea and 15 mM dithiothreitol, and desalted on Sephadex G-50 immediately before the experiment

2.3 Position of the Nicotinamide Label Within the Primary Structure of Fragment A

Fragment A photolabeled in the pesence of [carbonyl-^{14}C]NAD was digested with cyanogen bromide, and the products were fractionated on Sephadex G-75. Most of the label migrated with a single peak (of four). This peak contained peptide CNBr-3, which corresponded to residues 116–178 of fragment A.

When labeled material from the CNBr-3 peak was further digested with thermolysin and analyzed by reverse-phase HPLC at pH 2.0, a majority of the label was again associated with a single peak eluting at 45 min (Fig. 2). Hydrolysis of the major peak with acid, followed by reaction of the products with dansyl chloride gave three derivatives: Dns-Val, Dns-Tyr, and an unidentified residue migrating between Dns-Thr and Dns-Ser on polyamide thin-layer plates. Valine was identified as the amino-terminal residue and tyrosine as the carboxyl-terminal. By comparison of these results with the known primary structure of fragment A, the radiolabeled peptide eluting at 45 min on HPLC was identified as a derivative of the tripeptide Val-Glu-Tyr, corresponding to residues 147–149 of fragment A. The other labeled thermolytic peptides were resolved by HPLC and identified as partial digestion products containing the derivatized Val-Glu-Tyr tripeptide.

When the major labeled thermolytic peptide was sequenced by the dansyl-Edman and Edman methods, it was shown to have the sequence Val-X-Tyr, and 99% of the label was released at the second position (X), corresponding to a derivative of Glu-148. Analysis of a labeled chymotryptic peptide (residues 141–149) gave the expected sequence Ala-Glu-Gly-Ser-Ser-Ser-Val-X-Tyr, and most of the associated radioactivity

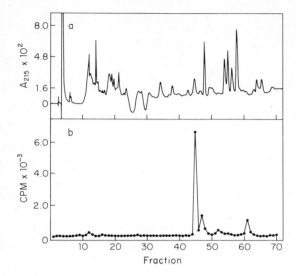

Fig. 2. High performance liquid chromatography of peptides from thermolytic digestion of photolabeled CNBr-3. Fragment A was photolabeled with [carbonyl-^{14}C]NAD and digested with CNBr, and the products were chromatographed on Sephadex G-75. The major radiolabeled peak (CNBr-3) was digested with 2% thermolysin and the products were chromatographed on an Alltech C_{18} reverse-phase column. Peptides were eluted with a linear 180 min gradient of 0–60% acetonitrile in 0.1% trifluoroacetic acid. (Reproduced from [3], with permission)

was again released at position 148. We concluded that most or all of the radioisotope incorporated from [carbonyl-^{14}C]NAD was attached to position 148 of fragment A.

2.4 Nature of the Photoadduct

A well-defined peak with λ_{max} about 260 nm was obtained when the absorbance difference spectrum was determined between the photolabeled fragment A and fragment A irradiated in the absence of NAD. Also, radiolabeled thermolytic or chemotryptic peptides showed an absorbance peak at 264 nm with a millimolar absorption coefficient of 6.1 mM^{-1} cm^{-1} at 260 nm. Inasmuch as tyrosine should contribute little to the absorbance at this wavelength, these results suggested that the nicotinamide ring, or an aromatic derivative thereof, was transferred to fragment A during the photolabeling reaction. Consistent with this conclusion was the finding that [nicotinamide-4-^{3}H]NAD was as efficient a donor of label as [carbonyl-^{14}C]NAD in the photolabeling reaction.

We also found that boronate affinity gels failed to retain nicotinamide-photolabeled fragment A or its peptides. In contrast, radiolabeled peptides from fragment A irradiated in the presence of phosphate- or adenine-labeled NAD bound quantitatively to such resins and could be eluted at pH 5.0. Since boronate gels bind compounds containing unblocked vicinal hydroxyls, we concluded that neither of the ribosyl moieties of NAD was present (at least in unmodified form) in the nicotinamide-labeled photoproduct.

Definitive evidence regarding the chemical structure of the photoadduct came from mass spectrometric and nuclear magnetic resonance (NMR) measurements. The labeled tripeptide, Val-X-Tyr (residues 147–149), was isolated from fragment A irradiated in the presence of either unlabeled NAD or NAD containing [^{13}C] or [^{15}N] within the nicotinamide moiety. Mass spectra of these peptides were taken following fast atom

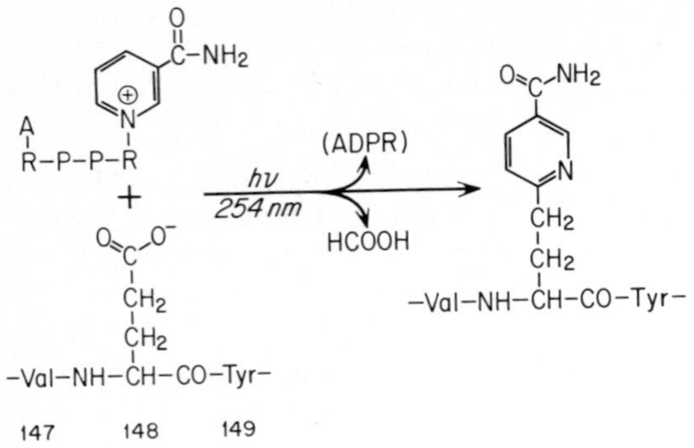

Fig. 3. Phototransfer of the nicotinamide moiety of NAD to residue 148 of DT fragment A

bombardment. By this method the composition of the tripeptide photoproduct was rigorously established as $C_{24}H_{32}N_5O_6$. This composition represents a net loss of a single oxygen atom and addition of $C_5H_4N_2$, compared with unmodified Val-Glu-Tyr. This is equivalent to decarboxylation with addition of nicotinamide.

The method of Sethi et al. [4] was employed to determine the active hydrogen content of the photoproduct. The results showed unequivocally that there are eight exchangeable hydrogens. This is consistent with structures in which the nicotinamide ring is linked to the decarboxylated residue via a ring carbon, and inconsistent with structures containing the nicotinamide moiety linked via its amide nitrogen (the latter giving seven exchangeable hydrogens).

NMR spectra of the photolabeled tripeptide showed that only the nicotinamide and tripeptide moieties were present; there was no indication of the presence of sugar or other moieties of NAD. The peptide portion of the spectrum showed the expected peaks, except that the β-methylene protons of glutamate were deshielded by about 0.5 ppm relative to the corresponding protons of an unmodified side chain in the same environment.

The chemical shifts of the nicotinamide moiety were consistent with the presence of an alkyl substitution on the ring, but not a direct bond to oxygen or nitrogen. Among the aromatic resonances of the nicotinamide, one of the protons was missing. From splitting patterns, this was determined to be either on the number four or the number six position of the ring. On the basis of chemical shifts it was determined that it was most likely that the proton in position six was absent. This would also be consistent with our finding that label from [nicotinamide-4-³H]NAD was incorporated into fragment A as efficiently as that from [carbonyl-¹⁴C]NAD. To make definite assignments of these protons, labeled tripeptide was prepared from fragment A that had been irradiated in the presence of NAD containing [¹³C] at position four of the nicotinamide ring or [¹⁵N] in the ring nitrogen. The [¹H]-resonance tentatively assigned as position four showed a large one-bond protein-[¹³C] coupling (165 Hz). This indicated that the attachment site must be the number six ring carbon. The [¹⁵N] labeled tripeptide not only showed geminal coupling to the proton at position two, but also coupling to the β-methylene protons of the decarboxylated glutamate moiety. This implied a maximum of three bonds between these protons and the ring nitrogen.

Figure 3 shows the photoadduct structure established by these results. ADP-ribose and formic acid are indicated as the other reaction products, although no experimental data have been gathered to demonstrate that these are actually the products released. Linkage of the nicotinamide ring to the peptide via a carbon-carbon bond is consistent with other results showing that the photoproduct formed from fragment A in the presence of [cabonyl-^{14}C]NAD is stable to 0.5 N NaOH or 0.5 M hydroxylamine (pH 8.3) for at least several hours at room temperature. It is conceivable that hydrolysis of the nicotinamide-ribose linkage may be an enzymatic event following the photochemical attachment reaction.

3. Discussion

We have described an efficient UV-induced reaction in which the nicotinamide moiety of NAD is transferred to the γ-methylene group of Glu-148 of diphtheria toxin fragment A, with concomitant decarboxylation of that group. Efficient photolabeling of *Pseudomonas* exotoxin A has also been observed under the same conditions [3], and we are in the process of determining the site of attachment and the nature of the photoproduct formed with this toxin.

The fact that most or all of the label transferred from [carbonyl-^{14}C]NAD to fragment A during photolysis was attached to Glu-148 strongly suggests that this residue is within the NAD binding site of DT. This notion is also supported by results showing that those forms of DT and PT known to have accessible NAD sites (fragment A, CRM-45, the nucleotide-free form of DT, and activated PT) were efficiently photolabeled in the presence of [carbonyl-^{14}C]NAD, whereas forms with altered or blocked binding sites (CRM-197, the nucleotide-bound form of DT, and native PT) showed greatly reduced levels of incorporation.

Other data, relevant to a nearby residue, Trp-153, also support the notion that the nicotinamide moiety of bound NAD may be in contact with this region of the primary structure of the toxin. Binding of NAD to fragment A strongly quenches the intrinsic tryptophan fluorescence of the fragment and generates a weak absorbance peak (λ_{max} = 360 nm) [2]. Both phenomena depend upon an N-substituted nicotinamide ring and may result from a charge transfer complex between the positively charged nicotinamide moiety and an adjacent indole side chain. Chemical modification of Trp-Trp-153, which is one of only two tryptophans in fragment A, is known to block enzymic activity [5].

If the γ-methylene of Glu-148 is adjacent to nicotinamide ring position six of bound NAD, then the carboxyl group of this amino acid may be immediately adjacent to the nicotinamide:ribose bond ruptured during ADP-ribosylation. It is noteworthy, and may be relevant to the mechanism of catalysis, that the negatively charged carboxyl may be associated with the cationic quaternary nitrogen of the nicotinamide ring. The possibility that Glu-148 is of major importance in catalysis has recently been probed by site-directed mutagenesis [6]. Glu-148 in a cloned fragment of the toxin was converted to Asp. This substitution produced a mutated protein product that was inactive in ADP-ribosylating EF-2. The fact that substitution of a residue (Asp) with the same charge as Glu was sufficient to inactivate the protein suggests that the precise position of the terminal side-chain carboxyl must be crucial to catalysis and, hence, further supports the notion that Glu-148 may play a pivotal role in ADP-ribosylation.

We are now in the process of examining a wider range of proteins, including the cholera toxin family of ADP-ribosyltransferases and other ADP-ribosylating enzymes, for efficient and specific photolabeling in the presence of [carbonyl-^{14}C]NAD.

Acknowledgments. This work was supported by NIH grants Al-22021 (R.J.C.) from the National Institute of Allergy and infectious Diseases, CA-39217 (R.J.C.) from the National Cancer Institute, and GM-29812 (J.A.M.) and GM-22982 (N.J.O.) from the National Institute of General Medical Sciences. N.J.O. was also supported by Research Career Development Award CA-00587.

References

1. Kandel J, Collier RJ, Chung DW (1974) Interaction of fragment A from diphtheria toxin with nicotinamide adenine dinucleotide. Proc Natl Acad Sci USA 249:2088–2097
2. Carroll SF, Lory S, Collier RJ (1980) Ligand interactions of diphtheria toxin. III. Direct photochemical crosslinking of ATP and NAD to toxin. J Biol Chem 255:12020–12024
3. Carroll SF, Collier RJ (1984) NAD binding site of diphtheria toxin: Identification of a residue within the nicotinamide subsite by photochemical modification with NAD. Proc Natl Acad Sci USA 81:3307–3311
4. Sethi SK, Smith DL, McCloskey JA (1983) Determination of active hydrogen content by fast atom bombardment mass spectrometry following hydrogen-deuterium exchange. Biochem Biophys Res Commun 112:126–131
5. Michel A, Dirkx J (1977) Occurrence of tryptophan in the enzymically active site of diphtheria toxin fragment A. Biochim Biophys Acta 491:286–295
6. Tweten RK, Barbieri JT, Collier RJ (1985) J Biol Chem 260:10392–10394

ADP-Ribosylation of a Membrane Protein Catalyzed by Islet-Activating Protein, Pertussis Toxin

MICHIO UI, TOSHIAKI KATADA, and MAKOTO TAMURA[1]

Introduction

The mono(ADP-ribosyl)ation, the transfer of the ADP-ribose moiety of NAD to a macromolecule, was discovered by Hayaishi et al. in 1968 as the mechanism of the cytotoxic effect of diphtheria toxin [1]. The substrate of this toxin-catalyzed ADP-ribosylation is elongation factor-2. The same reaction is catalyzed by *Pseudomonas* toxin. The second bacterial toxin involved in mono(ADP-ribosyl)ation of mammalian cell proteins is cholera toxin, the substrate of which was identified as the guanine nucleotide-binding regulatory component of membrane adenylate cyclase in 1978 [2]. *E. coli* heat-labile enterotoxin is similar to cholera toxin in many respects.

The subject of this short review is islet-activating protein (IAP), an exotoxin produced by *Bordetella pertussis,* which was discovered by us as the third bacterial toxin catalyzing mono(ADP-ribosyl)ation of a mammalian cell membrane protein [3]. IAP was purified from the supernatant of the culture medium of *B. pertussis* [4, 5] as a factor responsible for the unique action of pertussis vaccine to reverse α_2-adrenergic inhibition [6], and to enhance β-adrenergic stimulation, of insulin secretion from pancreatic islets in vivo [7, 8] and in vitro [9, 10]. These actions of pertussis vaccine were reproduced by purified IAP and were readily explained by its action to reverse α_2-adrenergic inhibition of cyclic AMP generation in islet cells [11, 12] or membrane adenylate cyclase [13]; cyclic AMP was an intracellular key messenger involved in pancreatic insulin secretion. IAP-induced modification of the membrane receptor adenylate cyclase system was not restricted to pancreatic islet cells, but was observed ubiquitously in a variety of cell types [14]. The direct action of IAP on the cell-free membrane preparation was observed only when the reaction mixture was supplemented with NAD [15]. IAP catalyzed the transfer of the ADP-ribose moiety of added NAD to the membrane M_r = 41,000 protein [15, 16] which was later identified as the GTP-binding subunit of the guanine nucleotide regulatory protein (N_i) involved in receptor-mediated inhibition of adenylate cyclase [17–20], whereas the membrane protein serving as the substrate of cholera toxin-catalyzed ADP-ribosylation proved to be N_s mediating adenylate cyclase activation [20].

---

1 Department of Physiological Chemistry, Faculty of Pharmaceutical Sciences, Hokkaido University, Sapporo 060, Japan

ADP-Ribosylation of Proteins
(ed. by F.R. Althaus, H. Hilz, and S. Shall)
© Springer-Verlag Berlin Heidelberg 1985

Islet-Activating Protein (IAP) as One of the A-B Toxins

Purified IAP was an oligomeric protein with a M_r value of 117,000. It was a hexamer composed of five dissimilar subunits, S_1 (M_r = 28,000), S_2 (23,000), S_3 (22,000), S_4 (11,700), and S_5 (9,300). Exposure of IAP to 5 M urea at 4°C for 3–4 days gave four separate protein peaks upon the subsequent column chromatography with CM-Sepharose; the two were S_1 and S_5 and the other two peaks were dimers (D_1 and D_2). These two dimers were further separated by exposure to 8 M urea for 16 h followed by DEAE-Sepharose chromatography to the constituent subunits; D_1 to S_2 and S_4, and D_2 to S_3 and S_4. Thus, these five subunits were separated from each other and purified to homogeneity as revealed by a sharp single peak on polyacrylamide gel electrophoresis [21]. Based on the relative color intensity of the individual subunit stained on sodium dodecy sulfate (SDS)-polyacrylamide gel, the molar ratio of these subunits was calculated as 1:1:1:2:1 in the native IAP molecule.

The original IAP molecule was reconstituted with the use of these purified subunits as follows. Combinations of S_2 with S_4 and of S_3 with S_4 in 2 M urea afforded D_1 and D_2, respectively. No dimer was formed from any other combination. Combination of D_1 with D_2 failed to form a tetramer, but the further addition of S_5 to the mixture of D_1 and D_2 was, but the addition of S_1 was not, effective in forming a pentamer, which exhibited no islet-activating activity when injected into rats. The subunit structure and the biological activity of IAP were then perfectly recovered by further combination of S_1 with the pentamer [21].

These results of reconstitution experiments, together with those of separation experiments, clearly showed that the whole IAP molecule is of such a subunit assembly that S_1 is associated with a pentamer composed of two dimers connected to each other by means of the smallest subunit, S_5. Evidence was further provided for ready dissociation of S_1 from the residual pentamer under mild conditions [21]. The solution of IAP in 5 M urea, after being maintained at 4°C for 6 h, was applied to a column of haptoglobin-Sepharose. A single sharp peak of the protein that passed through the column was identified as S_1, while the pentamer was bound to the column and then eluted by 0.5 M NaCl/3 M KSCN again as a sharp single peak.

Reductive cleavage of disulfide bonds in the S_1 peptide by incubation with 5–10 mM dithiothreitol was essential for this peptide to display an enzymic activity to hydrolyze NAD in the absence of any cellular component [22]. Neither the native IAP nor S_1 as such was active in this regard. Thus, S_1 is the enzymatically active component of the toxin, which should be henceforth referred to as the A-promoter. The native IAP was as effective as the preactivated A-promoter in catalyzing ADP-ribosylation of the M_r = 41,000 protein during incubation of membranes from rat glioma C6 cells with NAD. This action of the native IAP was not enhanced by dithiothreitol, but was interfered with by oxidized glutathione added to the reaction mixture, indicating that certain processing enzyme(s), including oxidoreductase present in the membrane preparation, are responsible for liberation and activation of the A-promoter. The native IAP was, however, without effect on the membrane preparation from rat heart or pancreatic tissues [23], probably due to lack of the processing enzyme(s) in these membranes. Thus, processing enzyme(s) appear to be located in membranes in some cells and in cytosol in other cell types, although they are not identified as yet.

The pentamer was referred to as the B-oligomer, because experimental evidence has been provided as follows for binding of IAP via its pentamer moiety to the cell surface. The binding of the B-oligomer is an essential step for the A-protomer to enter the cells across the plasma membrane [24]. (1) The action of the native IAP on intact cells was prevented by the B-oligomer in a competitive manner. (2) The direct action of the native IAP on C6 cell membranes was antagonized by anti-A-protomer rabbit antibodies (polyclonal IgG), but was not by anti-B-oligomer rabbit antibodies, excluding a possibility that the B-oligomer moiety of IAP is required for direct interaction of the A-protomer with the membrane protein serving as the substrate for ADP-ribosylation. (3) An essential role of the B-oligomer in interaction of IAP with intact cells was shown by the interference with anti-B-oligomer rabbit antibodies (polyclonal IgG). (4) The native IAP was mitogenic when added to T-lymphocytes. This action of the native IAP was reproduced by the isolated B-oligomer by itself indicating that the B-oligomer moiety was bound to glycoproteins on the cell surface, though receptor proteins for the B-oligomer of IAP are not identified as yet.

Thus, IAP is, like diphtheria and cholera toxins, a protein toxin with an A-B structure. The holotoxin is bound to particular sites on the cell surface via its B-oligomer moiety as the first step of its interaction with mammalian cells. The A-protomer (or holotoxin itself) is then inserted to the plasma membrane traversing the lipid bilayer gradually. This slow process of the entrance of the toxin molecule is reflected in a definite lag time invariably preceding the onset of the action of IAP on intact cells as analyzed with rat pancreatic islets by kinetic and immunological approaches [25]. The A-protomer is finally activated by certain processing enzyme(s) inside the membrane to catalyze ADP-ribosylation of the target membrane protein with intracellular NAD as substrate.

The Guanine Nucleotide-Binding Regulatory Protein in Cell Membranes as the Specific Substrate of IAP-Catalyzed ADP-Ribosylation

The membrane receptor-adenylate cyclase system is composed of three major protein components; the receptor protein (R), the catalytic protein of adenylate cyclase (C), and the guanine nucleotide regulatory protein (N) that communicates between R and C in this biosignaling system. N is a GTP- or GDP-binding protein and exhibits GTP-hydrolyzing (GTPase) activity. The GDP-bound form of N is inactive as a communicator. Binding of an agonist to R (stimulation of R) displaces GDP in exchange for GTP on N, thereby converting it to its GTP-bound active form which is capable of direct interaction with C. Thus, the GDP-GTP exchange reaction "turns on" signal transduction. GTP is immediately hydrolyzed on N, recovering the original inactive state. Thus, the GTPase reaction is a "turnoff" reaction. Prolonged exposure of cells to cholera toxin raises the intracellular cyclic AMP level enormously as a result of ADP-ribosylation of the membrane protein with M_r = 45,000–52,000 [52]. The ADP-ribosylated N displayed low GTPase activity; the turnoff reaction is, thus, prevented and adenylate cyclase continues to be in its active state, even without stimulation of coupled receptors in cholera toxin-treated cells. Cholera toxin has been a useful tool for analysis of the mechanism, whereby N mediates receptor-coupled activation of adenylate cyclase.

In addition to the membrane receptors involved in activation of adenylate cyclase, there are some receptors stimulation of which results in inhibition, rather than activation, of the cyclase via N [26]. These inhibitory receptors include α-adrenergic (α_2), muscarinic cholinergic (M_2), opiate, dopamine (D_2), and adenosine (A_1) receptors. Exposure of rat pancreatic islet [11–13], rat heart [23, 27], rat [18], and hamster [20] adipocyte and neuroblastoma × glioma hybrid NG108-15 [19] cells to IAP abolished the decreases in the cyclic AMP content of the cells (or reversed membrane adenylate cyclase inhibition) as induced by stimulation of α_2-adrenergic [11–13, 20], muscarinic [19, 27], opiate [19], and adenosine-A_1 [18, 27] receptors. The degree of IAP-induced attenuation of receptor-mediated decreases of cyclic AMP (or adenylate cyclase inhibition) was well correlated with the degree of ADP-ribosylation of the membrane M_r = 41,000 protein occurring during exposure of cells to various concentrations of IAP [16, 18, 19]. The IAP substrate was also a GTP-binding protein; the ADP-ribosylated M_r = 41,000 protein was more resistant to tryptic digestion in the presence of GTP analogues than in their absence. It is, thus, established that the IAP substrate (M_r = 41,000) is the GTP-binding subunit of N_i, while the cholera toxin substrate (M_r = 45,000) is the GTP-binding subunit of N_s. The IAP-catalyzed ADP-ribosylation of the M_r = 41,000 protein abolishes the function of N_i to communicate between inhibitory receptors and the adenylate cyclase catalyst. Accordingly, IAP was effective in inhibiting receptor-mediated GTPase activity [28, 29] as well as receptor-mediated GTP-GDP exchange [20] only when both enzymic reactions were exclusively dependent on the function of N_i.

Following purification of N_s from rabbit liver [30], N_i was first purified from the same tissue and its properties and functions were studied extensively [31–34]. Both N_s and N_i are trimers, each of which is composed of an α-, a β-, and a γ-subunit. The α-subunit possesses a GTP-binding site and is ADP-ribosylated by cholera toxin (α_s with M_r = 45,000) or IAP (α_i with M_r = 41,000). The β- and γ-subunits are common to N_s and N_i. Receptor-mediated GTP-GDP exchange occurring on α_s in the presence of Mg^{2+} is spontaneously accompanied by dissociation of $\beta\gamma$ from GTP-bound α_s, which directly activates C. No further dissociation of $\beta\gamma$ occurs under physiological conditions. Receptor-mediated GTP-GDP exchange on α_i is also accompanied by its dissociation from $\beta\gamma$ in the presence of Mg^{2+}. Since N_i is much more than N_s in membranes, $\beta\gamma$ thus dissociated from GTP-bound α_i could favor the equilibrium, $\alpha_s\beta\gamma \rightleftharpoons \alpha_s + \beta\gamma$, to the left, thereby decreasing the amount of α_s that is capable of activating C. This would be a mechanism for N_i-mediated inhibition of adenylate cyclase. ADP-ribosylation of α_s by cholera toxin favors the above equilibrium to the right, whereas $\alpha_i\beta\gamma$ is not capable of coupling with receptors any longer after α_i is ADP-ribosylated by IAP. Thus, signal transduction via N_i is completely blocked by IAP-catalyzed ADP-ribosylation of the GTP-binding subunit.

An additional GTP-binding protein involved in signal transduction is transducin which is responsible for rhodopsin-mediated activation of cyclic GMP phosphodiesterase in the rod outer segment of retinal cells. Transducin is also a trimeric protein. Its α-subunit (M_r = 39,000) was ADP-ribosylated by either cholera toxin or IAP [35–37]. Transducin lost its function as a signal transducer after IAP-induced ADP-ribosylation of its α-subunit [35, 37].

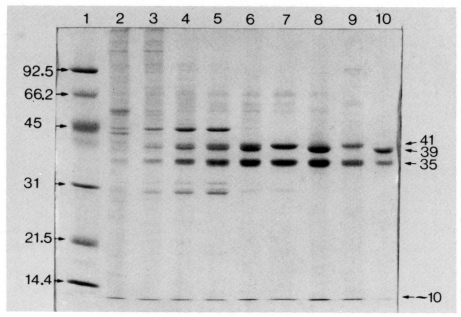

Fig. 1. SDS-polyacrylamide gel electrophoresis of IAP substrates. Electrophoretic gels (12.5% polyacrylamide) were stained with Coomassie blue. *Lane 1:* M_r markers; *lanes 2–6:* fractions obtained at purification steps of rat brain according to the procedure used for rabbit liver (see Table 1 in [31]) (*2* cholate extract; *3* DEAE-Sephacel; *4* AcA-34; *5* heptylamine-Sepharose; *6* DEAE-Sephacel II); *lane 7:* the first peak of Fig. 2A; *lane 8:* the second peak of Fig. 2A; *lane 9:* rabbit liver N_i; *lane 10:* bovine retinal transducin. $M_r \times 10^{-3}$ is shown by *arrows*

Based on a report by Sternweis and Robishaw [38] that there were two kinds of GTPγS-binding proteins ADP-ribosylated by IAP in bovine brain tissues, we have recently purified IAP substrate proteins from rat brain [39] according to the procedure originally adapted for the rabbit liver protein [31]. Fractions at purification steps were applied to SDS-polyacrylamide gel electrophoresis (Fig. 1, lanes 2–6). The protein at the final step of purification exhibited doublets for the α-subunit in addition to two single bands for the β- and γ-subunits. These two IAP substrates were then separated from each other by means of high performance liquid chromatography (HPLC) using a column of anion-exchange resin (Mono Q) as shown by two protein peaks in Fig. 2A. Both proteins were GTP-binding proteins (Fig. 2B, solid triangles) and were ADP-ribosylated by the preactivated A-protomer of IAP (Fig. 2B, solid circles). Both proteins inhibited adenylate cyclase when added to membrane preparations from human platelets (Fig. 3B, solid squares). The M_r values for the α-subunits of the first and the second peaks were 41,000 and 39,000, respectively (Fig. 1, lanes 7 and 8). The former was similar to liver N_i (Fig. 1, lane 9) and displayed N_i functions upon incubation with IAP-treated membranes, whereas the physiological function of the latter is still unknown, though it resembled bovine retinal transducin (Fig. 1, lane 10).

These two IAP substrates were individually incubated with GTPγS for the purpose of subunit dissociation (Fig. 3). In either case, the peak I (α-subunit) of HPLC gel

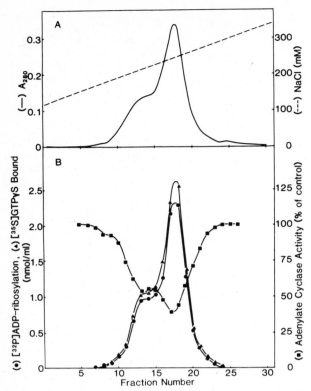

Fig. 2A,B. Further separation of two IAP substrates of rat brain by a Mono-Q column (HPLC). A Elution was performed with linear gradients of 0.1–0.35 M NaCl in 20 mM Tris-HCl (pH 7.5), 1 mM EDTA, 1 mM dithiothreitol, 0.1% Lubrol. B Each fraction was incubated with preactivated A-protomer of IAP and [α-^{32}P]NAD ($\bullet$), with [^{35}S]GTPγS ($\blacktriangle$), or with GTPγS and MgCl$_2$ ($\blacksquare$) at 30°C for 1 h. In the case of $\blacksquare$, an aliquot of the incubation mixture was further incubated with human platelet membranes and forskolin to measure adenylate cyclase activity of membranes

filtration was a GTP-binding protein (Fig. 3, solid triangles) and the peak II ($\beta\gamma$-sub-unit) displayed an activity to inhibit adenylate cyclase when added to human platelet membranes (Fig. 3, solid circles), in accordance with the inhibition mechanism proposed above. The isolated α-subunit of N$_i$, by itself, was not ADP-ribosylated by the preactivated A-protomer of IAP even in the presence of GTPγS (Fig. 4, plot 1); the ADP-ribose moiety of NAD was transferred to α only when $\beta\gamma$ was further added to the reaction mixture (Fig. 4, plot 2). The IAP-catalyzed ADP-ribosylation proceeded progressively as the amount of $\beta\gamma$ was increased until almost stoichiometric reaction occurred, i.e., 1 mol of ADP-ribose was transferred to 1 mol of α, when the equal amounts of α and $\beta\gamma$ were mixed (Fig. 4, plot 4). Thus, the $\alpha\beta\gamma$ trimer, or the inactive form of N$_i$, proved to be a real substrate for IAP-catalyzed ADP-ribosylation, though the site of ADP-ribosylation was located on its α-subunit moiety exclusively. Such was also the case with other IAP substrates, the branin trimer with an α-subunit of M_r = 39,000 and retinal transducin.

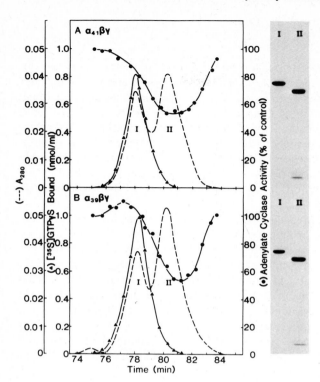

Fig. 3A,B. Dissociation of α_i from $\beta\gamma$ of the IAP substrate. The first (A) and the second (B) peaks shown in Fig. 2A were separately incubated with $[^{35}S]GTP\gamma S$, $MgCl_2$ at 30°C for 1 h and then applied to a TSK gel filtration column (HPLC). Proteins were eluted with 20 mM Hepes (pH 7.5), 1 mM EDTA, 1 mM dithiothreitol, 10 mM $MgCl_2$, 100 mM Na_2SO_4, 0.8% Na cholate. Adenylate cyclase activity (●) and $[^{35}S]GTP\gamma S$ binding (▲) were measured as in Fig. 2. SDS-polyacrylamide gel electrophoretic patterns of peak I and II proteins are shown on the *right-hand side*

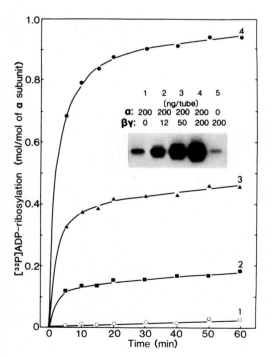

Fig. 4. ADP-ribosylation of α_i purified from rat brain. Separation of α_i from $\beta\gamma$ was achieved by incubating the rat brain N_i in the presence of Al^{3+}, F^-, and Mg^{2+}. They were combined at various ratios and incubated with 1 μM $[\alpha\text{-}^{32}P]NAD$ and 10 μg ml^{-1} of the preactivated A-protomer of IAP in 0.1 ml of 100 mM Tris-HCl (pH 8.0), 1 mM EDTA, 1 mM dithiothreitol, 10 mM thymidine. Aliquots were taken at the indicated time of incubation and applied to SDS-polyacrylamide gel electrophoresis as in [16]. The autoradiographic pattern of the M_r = 41,000 protein band (after 60 min incubation) is shown in *inset;* the radioactivity content of this band is plotted as a function of time. No radioactivity was detected in other protein bands under any condition. The amounts of α- and $\beta\gamma$-subunits used for *plots 1–4* are shown in the *inset*

Concluding Remarks

Properties of IAP, pertussis toxin, as an "enzyme-membrane transporter" complex catalyzing ADP-ribosylation of specific intramembrane proteins of intact cells, together with functions of these IAP substrate proteins as important mediators of membrane signal transduction, have been briefly described in this short review principally based on our recent experimental data. Pertussis toxin has been characterized by its diverse biological activities, including lymphocytosis-promoting, histamine-sensitizing, adjuvant, and hemagglutinating activities in addition to the islet-activating function. We have recently proposed dual mechanisms for these multiple functions based on our data concerning chemical modification of the lysine residues of the toxin molecule [40, 41].

Three IAP substrates have been discovered; N_i, transducin, and a transducin-like protein abundantly present in mammalian brain tissues. These IAP substrates are essential communicators between inhibitory receptors and adenylate cyclase (N_i) in a variety of cell types and between rhodopsin and cyclic GMP phosphodiesterase (transducin) in vertebrate retinal cells, though the real physiological function of the brain transducin-like protein remains to be disclosed. Several lines of evidence have recently been afforded for an additional role of the IAP-sensitive GTP-binding protein, probably N_i, in "Ca^{2+}-mobilizing" receptor-mediated biosignaling systems in guinea pig neutrophils [42, 43], rat mast cells [44–46], mouse 3T3 fibroblasts [47], and NG108-15 cells [48]. Thus, IAP will be a promising probe for future studies on mechanisms for receptor-mediated biosignaling. Several review articles have been issued in 1984 dealing with related problems [3, 29, 49–52].

Acknowledgment. This work was supported by research grants from the Scientific Research Fund of the Ministry of Education, Science, and Culture, Japan.

References

1. Hayaishi O, Ueda K (1977) Poly(ADP-ribose) and ADP-ribosylation of proteins. Annu Rev Biochem 46:95–116
2. Moss J, Vaughan M (1979) Activation of adenylate cyclase by choleragen. Annu Rev Biochem 48:581–600
3. Ui M (1984) Islet-activating protein, pertussis toxin: a probe for functions of the inhibitory guanine nucleotide regulatory component of adenylate cyclase. Trend Pharmacol Sci 5:277–279
4. Yajima M, Hosoda K, Kanbayashi Y, Nakamura T, Nogimori K, Nakase Y, Ui M (1978) Islet-activating protein in *Bordetella pertussis* that potentiates insulin secretory responses of rats. Purification and characterization. J Biochem (Tokyo) 83:295–303
5. Yajima M, Hosoda K, Kanbayashi Y, Nakamura T, Takahashi I, Ui M (1978) Biological properties of islet-activating protein purified from the culture medium of *Bordetella pertussis*. J Biochem (Tokyo) 83:305–312
6. Yamazaki S, Katada T, Ui M (1982) *Alpha*$_2$-adrenergic inhibition of insulin secretion via interference with cyclic AMP generation in rat pancreatic islets. Mol Pharmacol 21:648–653

7. Sumi T, Ui M (1975) Potentiation of the adrenergic β-receptor-mediated insulin secretion in pertussis-sensitized rats. Endocrinology 97:352−358

8. Katada T, Ui M (1976) Accelerated turnover of blood glucose in pertussis-sensitized rats due to combined actions of endogenous insulin and adrenergic β-stimulation. Biochim Biophys Acta 421:57−69

9. Katada T, Ui M (1977) Perfusion of the pancreas isolated from pertussis-sensitized rats: Potentiation of insulin secretory responses due to β-adrenergic stimulation. Endocrinology 101: 1247−1255

10. Katada T, Ui M (1979) Effect of in vivo pretreatment of rats with a new protein purified from *Bordetella pertussis* on in vitro secretion of insulin: Role of calcium. Endocrinology 104: 1822−1827

11. Katada T, Ui M (1979) Islet-activating protein. Enhanced insulin secretion and cyclic AMP accumulation in pancreatic islets due to activation of native calcium ionophores. J Biol Chem 254:469−479

12. Katada T, Ui M (1981) In vitro effects of islet-activating protein on cultured rat pancreatic islets. Enhancement of insulin secretion, cAMP accumulation and ^{45}Ca flux. J Biochem (Tokyo) 89:979−990

13. Katada T, Ui M (1981) Islet-activating protein. A modifier of receptor-mediated regulation of rat islet adenylate cyclase. J Biol Chem 256:8310−8317

14. Katada T, Amano T, Ui M (1982) Modulation by islet-activating protein of adenylate cyclase activity in C6 glioma cells. J Biol Chem 257:3739−3746

15. Katada T, Ui M (1982) Direct modification of the membrane adenylate cyclase system by islet-activating protein due to ADP-ribosylation of a membrane protein. Proc Natl Acad Sci USA 79:3129−3133

16. Katada T, Ui M (1982) ADP-ribosylation of the specific membrane protein of C6 cells by islet-activating protein associated with modification of adenylate cyclase activity. J Biol Chem 257: 7210−7216

17. Murayama T, Katada T, Ui M (1983) Guanine nucleotide activation and inhibition of adenylate cyclase as modified by islet-activating protein, pertussis toxin, in mouse 3T3 fibroblasts. Arch Biochem Biophys 221:381−390

18. Murayame T, Ui M (1983) Loss of the inhibitory function of the guanine nucleotide regulatory component of adenylate cyclase due to its ADP-ribosylation by islet-activating protein, pertussis toxin, in adipocyte membranes. J Biol Chem 258:3319−3326

19. Kurose H, Katada T, Amano T, Ui M (1983) Specific uncoupling by islet-activating protein, pertussis toxin, of negative signal transduction via α-adrenergic, cholinergic, and opiate receptors in neuroblastoma × glioma hybrid cells. J Biol Chem 258:4870−4875

20. Murayama T, Ui M (1984) [³H]GDP release from rat and hamster adipocyte membranes independently linked to receptors involved in activation or inhibition of adenylate cyclase. Differential susceptibility to two bacterial toxins. J Biol Chem 259:761−769

21. Tamura M, Nogimori K, Murai S, Yajima M, Ito K, Katada T, Ui M, Ishii S (1982) Subunit structure of islet-activating protein, pertussis toxin, in conformity with the A-B model. Biochemistry 21:5516−5522

22. Katada T, Tamura M, Ui M (1983) The A-protomer of islet-activating protein, pertussis toxin, as an active peptide catalyzing ADP-ribosylation of a membrane protein. Arch Biochem Biophys 224:290−298

23. Kurose H, Ui M (1983) Functional uncoupling of muscarinic receptors from adenylate cyclase in rat cardiac membranes by the active component of islet-activating protein, pertussis toxin. J Cycl Nucleotide Protein Phosphoryl Res 9:305−318

24. Tamura M, Nogimori K, Yajima M, Ase K, Ui M (1983) A role of the B-oligomer moiety of islet-activating protein, pertussis toxin, in development of the biological effects on intact cells. J Biol Chem 258:6756−6761

25. Katada T, Ui M (1980) Slow interaction of islet-activating protein with pancreatic islets during primary culture to cause reversal of α-adrenergic inhibition of insulin secretion. J Biol Chem 255:9580−9588

26. Rodbell M (1980) The role of hormone receptors and GTP-regulatory proteins in membrane transduction. Nature (London) 284:17−22

27. Hazeki O, Ui M (1981) Modification by islet-activating protein of receptor-mediated regulation of cyclic AMP accumulation in isolated rat heart cells. J Biol Chem 256:2856−2862

28. Burns DL, Hewlett EL, Moss J, Vaughan M (1983) Pertussis toxin inhibits enkephalin stimulation of GTPase of NG108-15 cells. J Biol Chem 258:1435−1438

29. Ui M, Katada T, Murayama T, Kurose H, Yajima M, Tamura M, Nakamura T, Nogimori K (1984) Islet-activating protein, pertussis toxin: A specific uncoupler of receptor-mediated inhibition of adenylate cyclase. In: Greengard P et al. (eds) Advances in cyclic nucleotide and protein phosphorylation research, vol 17. Raven Press, New York, pp 145−151

30. Smigel MD, Northup JK, Gilman AG (1982) Characteristics of the guanine nucleotide-binding regulatory component of adenylate cyclase. Recent Progr Horm Res 38:601−624

31. Bokoch GM, Katada T, Northup JK, Ui M, Gilman AG (1984) Purification and properties of the inhibitory guanine nucleotide-binding regulatory component of adenylate cyclase. J Biol Chem 259:3560−3567

32. Katada T, Koboch GM, Northup JK, Ui M, Gilman AG (1984) The inhibitory guanine nucleotide-binding regulatory component of adenylate cyclase. Properties and functions of the purified protein. J Biol Chem 259:3568−3577

33. Katada T, Northup JK, Bokoch GM, Ui M, Gilman AG (1984) The inhibitory guanine nucleotide-binding regulatory component of adenylate cyclase. Subunit dissociation and guanine nucleotide-dependent hormonal inhibition. J Biol Chem 259:3578−3585

34. Katada T, Bokoch GM, Smigel MD, Ui M, Gilman AG (1984) The inhibitory guanine nucleotide-binding regulatory component of adenylate cyclase. Subunit dissociation and the inhibition of adenylate cyclase in S49 lymphoma cyc⁻ and wild type membranes. J Biol Chem 259:3586−3595

35. Van Dop C, Yamanaka G, Steinberg F, Sekura RD, Manclark CR, Stryer L, Bourne HR (1984) ADP-ribosylation of transducin by pertussis toxin blocks the light-stimulated hydrolysis of GTP and cGMP in retinal photoreceptors. J Biol Chem 259:23−26

36. Manning DR, Fraser BA, Kahn RA, Gilman AG (1984) ADP-ribosylation of transducin by islet-activating protein. Identification of asparagine as the site of ADP-ribosylation. J Biol Chem 259:749−756

37. Watkins PA, Moss J, Burns DL, Hewlett EL, Vaughan M (1984) Inhibition of bovine rod outer segment GTPase by *Bordetella pertussis* toxin. J Biol Chem 259:1378−1381

38. Sternweis PC, Robishaw JD (1984) Isolation of two proteins with high affinity for guanine nucleotides from membranes of bovine brain. J Biol Chem 259:13806−13813

39. Katada T, Ui M (in preparation)

40. Nogimori K, Ito K, Tamura M, Satoh S, Ishii S, Ui M (1984) Chemical modification of islet-activating protein, pertussis toxin. Essential role of free amino groups in its lymphocytosis-promoting activity. Biochim Biophys Acta 801:220−231

41. Nogimori K, Tamura M, Yajima M, Ito K, Nakamura T, Kajikawa N, Maruyama Y, Ui M (1984) Dual mechanisms involved in development of diverse biological activities of islet-activating protein, pertussis toxin, as revealed by chemical modification of lysine residues in the toxin molecule. Biochim Biophys Acta 801:232−243

42. Okajima F, Ui M (1984) ADP-ribosylation of the specific membrane protein by islet-activating protein, pertussis toxin, associated with inhibition of a chemotactic peptide-induced arachidonate release in neutrophils. A possible role of the toxin substrate in Ca^{2+}-mobilizing biosignaling. J Biol Chem 259:13863−13871

43. Ohta H, Okajima F, Ui M (in preparation)

44. Nakamura T, Ui M (1983) Suppression of passive cutaneous anaphylaxis by pertussis toxin, an islet-activating protein, as a result of inhibition of histamine release from mast cells. Biochem Pharmacol 32:3435−3441

45. Nakamura T, Ui M (1984) Islet-activating protein, pertussis toxin, inhibits Ca^{2+}-induced and guanine nucleotide-dependent release of histamine and arachidonic acid from rat mast cells. FEBS Lett 173:414−418

46. Nakamura T, Ui M (1985) Simultaneous inhibitions of inositol phospholipid breakdown, arachidonic acid release and histamine secretion in mast cells by islet-activating protein, pertussis toxin. A possible involvement of the toxin-specific substrate in the Ca^{2+}-mobilizing receptor-mediated biosignaling system. J Biol Chem 260:3584–3593

47. Murayama T, Ui M (1985) Differential susceptibility to islet-activating protein, pertussis toxin, of multiple effects of thrombin on 3T3 fibroblasts including adenylate cyclase inhibition and arachidonic acid release. J Biol Chem 260:7226–7233

48. Kurose H, Ui M (1985) Dual pathways of receptor-mediated cyclic GMP generation of NG108-15 cells as differentiated by susceptibility to islet-activating protein, pertussis toxin. Arch Biochem Biophys 238:424–434

49. Ui M, Katada T, Murayama T, Kurose H (1984) Selective blockage by islet-activating protein, pertussis toxin, of negative signal transduction from receptors to adenylate cyclase. In: Kito S et al. (eds) Neurotransmitter receptors. Plenum Press, New York, pp 1–16

50. Ui M, Katada T, Murayama T, Nakamura T (1984) Islet-activating protein, pertussis toxin, as a probe for receptor-mediated signal transduction. In: Ebashi S et al. (eds) Calcium regulation in biological systems. Academic Press, London New York, pp 157–169

51. Ui M, Katada T, Murayama T, Kurose H (1984) Role of N protein in coupling of adrenergic receptors and adenylate cyclase. In: Proc 7th Int Congr Endocrinol. Elsevier, Amsterdam, pp 157–160

52. Ui M, Katada T, Murayama T, Kurose H (1984) Chemical modification of nucleotide regulatory proteins catalyzed by bacterial toxins. In: Proc IX Int Congr Pharmacol. Macmillan, London, pp 271–278

Why Does Cholera Toxin Need GTP to Act?

D. MICHAEL GILL and JENIFER COBURN[1]

Abbreviations:

CF   cytosolic factor
Gpp(NH)p   guanyl 5'-yl imidodiphosphate
MPU   3-chloromercuri-2 methoxy-propylurea

Cholera Toxin Requires GTP

Very soon after the discovery that NAD was required for cholera toxin to activate adenylate cyclase in pigeon erythrocyte membranes, it was recognized that a second nucleotide was also essential. At first the requirement seemed highly unspecific, being satisfied by a variety of sugar phosphates as well as by the majority of nucleoside diphosphates and triphosphates. However, as it became possible to simplify the test system, it became clear that these compounds were only effective by virtue of their abilities to generate GTP from endogenous GDP and GMP. Under conditions in which such change was prevented, notably by the removal of the magnesium ions essential for transphosphorylation reactions, only GTP proved to be effective: not ATP and not GDP. The closest contender is ITP, a nucleotide closely related to GTP that is commonly found to substitute for GTP. The amount of GTP required was most relevantly measured in the presence of an excess of ATP in order to regenerate the GTP split by the nonspecific nucleoside triphosphatases present. An alternative method was to use the nonhydrolyzable analog of GTP, Gpp(NH)p, (the usual alternative, GTPγS, is in fact slowly hydrolyzed to GTP by lysed cells). Both methods gave Ka values of 1 to 3 μM [1].

Adenylate Cyclase also Requires GTP

In the meantime much evidence had accumulated that GTP was essential for the activity of adenylate cyclase. This discovery was originally made by Rodbell et al. [2]

---

1 Department of Molecular Biology and Microbiology, Tufts University School of Medicine, 136 Harrison Avenue, Boston, MA 02111, USA

ADP-Ribosylation of Proteins
(ed. by F.R. Althaus, H. Hilz, and S. Shall)
© Springer-Verlag Berlin Heidelberg 1985

whose group also showed that GTPγS or Gpp(NH)p would substitute for GTP [3]. In all cases there was a short lag between the administration of the nucleotide and the attainment of the maximal cyclase activity. The seminal work of Cassel et al. [4] provided the attractive model that the cyclase activity was controlled by a GTP cycle. The cyclase system would hydrolyze to GDP a GTP molecule bound at a regulatory site and thereby turn off its ability to produce cyclic AMP. To restore activity the bound GDP has to be displaced and replaced by a new copy of GTP. The displacement of GDP is much accelerated by the presence of one of the cyclase-linked hormones, such as epinephrine, acting through the appropriate hormone receptor. The nonhydrolyzable analogs of GTP would remain indefinitely at the regulatory site and so would activate the cyclase without the periodic intervention of a hormone-receptor system.

The modern version of this model takes into account the structural data obtained from the isolation of cyclase and its component peptides. The part to which GTP binds is a regulatory protein called Ns. It has three subunits: α (M_r = 42,000 and sometimes also M_r = 47,000), β (M_r = 35,000), and γ (M_r = 10,000) [5, 6]. GTP binds to the α peptide and probably causes the dissociation of α(GTP) from the β and γ subunits. α(GTP) binds, and thereby activates, the catalytic subunit C. As above, inactivation consists of the hydrolysis of the GTP and the return of α to an association with β and γ rather than with C.

When loaded with GDP, Ns must also interact with occupied hormone receptors. Such Ns-receptor complexes can be detected by an increase in apparent receptor size and by the presence of ADP-ribosylated peptides (i.e., Nsα, see below) in the enlarged complex [7]. In the presence of GTP this complex dissociates. The net result is that Nsα acts as a shuttle, interacting alternately with hormone receptors and with the catalytic unit of cyclase and in effect informing the latter of the presence of the former.

Substrate of Cholera Toxin Binds GTP

As is now well known, cholera toxin turned out to ADP-ribosylate one particular site on the α peptide of Ns, thus, changing the properties of Ns in important ways. The change that probably results in the elevated cyclase activity is a reduced ability to hydrolyze GTP. GTP becomes almost as good at supporting cyclase activity as its nonhydrolyzable analogs and no hormone is required. Furthermore, the effect is long-lasting and cyclic AMP may continue to be made for hours and days after all extracellular toxin has been removed.

GTP of the Toxin and the GTP of the Substrate are Different

At first it seemed reasonable to interpret the requirement of the toxin for GTP as a need to have the substrate Ns in the GTP-liganded form. We supposed for a while [1] that the actual change catalyzed by cholera toxin was

Ns(GTP) → Ns(GTP)-ADPR

and this seemed such a reasonable postulate that it gained general acceptance. Nevertheless, it is not true. In retrospect, a cogent reason for discarding the hypothesis had always been that whereas GTP itself is an excellent supporter of the action of cholera toxin, it is so rapidly hydrolyzed on Ns that it supports hardly any cyclase activity. We managed to ignore this difficulty, however, until a second peculiarity was unearthed. We had shown that membranes could be soaked in GTPγS, washed free of unbound nucleotide, and made both to display adenylate cyclase activity and to act as substrates for cholera toxin. Presumably bound nucleotide allowed both functions:

Membrane + GTPγS → membrane(GTPγS).

However, when this experiment was repeated in the presence of GDPβS in addition to the GTPγS, membranes were obtained which could be ADP-ribosylated and yet had essentially no cyclase activity. With respect to cholera toxin they behaved as if they were responding to the GTPγS and yet their Ns site was clearly occupied by GDPβS. There had to be a second site for ADP-ribosylation. We called this second site "S" [8]. Presumably when membranes are soaked in GDPβS and GTPγS the two sites select different nucleotides:

Membrane + GDPβS + GTPγS → Ns(GDPβS) + S(GTPγS).

We are unable to ADP-ribosylate membranes after they have been incubated at $37°$ for extended periods even when supplied with fresh GTP and CF. This seems to be due to the decay of the S function. ADP-ribosylated membranes remain catalytically active (i.e., the ADP-ribose continues to activate cyclase) for many hours.

GTP of the Toxin is Influenced by a Cytosolic Factor

One of the major operational differences between S and Ns is that in erythrocytes and certain other tissues a ubiquitous soluble protein (cytosolic factor or CF) is required to bind a nucleotide to the S site efficiently. Only about 10% of the Ns of pigeon erythrocyte membranes can be ADP-ribosylated in the absence of CF. We had known of this soluble protein for several years before Enomoto proved that it was concerned only with the proper loading of the S site with a guanyl nucleotide [1]. Membranes so loaded with GTPγS or Gpp(NH)p could be washed and subsequently ADP-ribosylated without any further supply of nucleotide or of CF.

In some other tissues, including brain and the S49 lymphoma cell, neither we nor others can find any strong dependence on CF (although CF is present in the cytosols of these cells). Indeed, S49 and brain membranes can be ADP-ribosylated best if they have been preincubated with a guanyl nucleotide, but no CF is necessary during the preincubation. These membranes behave, therefore, like the 10% of erythrocyte membranes that are CF independent. Perhaps they already have CF permanently associated with their S sites.

Table 1. Conditions for binding and ADP-ribosylation

Affect guanyl nucleotide prebinding
    Mg required, EDTA blocks
    CF required in some tissues
    $37° > 25°$

Affect ADP-ribosylation
    Mg not required
    Sodium or potassium phosphate stimulate
    $25°-30° > 37°$
    Pretreatment of the membranes with isoproterenol plus GTPγS prevents subsequent ADP-ribosylation
    Inhibitors of NAD consumption (isoniazid, ADP-ribose, thymidine) may assist indirectly

---

Different Conditions for Loading S and for ADP-Ribosylation

We can now sort the environmental factors that influence the action of cholera toxin into those that affect guanyl nucleotide binding and those that affect ADP-ribosylation proper. Table 1 gives the current list. One point to emphasize is the effect of phosphate ions. Since its introduction by Moss [9], high phosphate has often been used to support ADP-ribosylation in those tissue fractions that do not respond to CF. We have recently found that phosphate stimulates ADP-ribosylation even in tissues that do require CF, but it is in no way a replacement for the CF. Phosphate stimulates the ADP-ribosylation itself, while the CF is required in the binding step.

Transfer of S Function from Membrane to Membrane

Recently, we have been able to transfer the S function from one membrane to another. We loaded S in the usual way by incubating membranes with GTPγS and CF; washed the membranes and made an extract in nonionic detergent solution. This extract, on addition to fresh membranes, conferred the ability for Ns to be ADP-ribosylated in the absence of further nucleotide or CF.

There were originally two possible interpretations of this result. One was that the detergent extract contained Ns of the original membranes in a state ready for ADP-ribosylation, and that this Ns was ADP-ribosylated after integration into the recipient membranes. This appears not to be the proper interpretation for two reasons. First, we have used brain membranes as the donors (these have Ns of $M_r = 42,000$ and $47,000$) and erythrocyte membranes as the recipients (their Ns is $M_r = 42,000$). We observed the ADP-ribosylation only of $M_r = 42,000$, in other words, only of the erythrocyte Ns. Secondly, we used the reagent 3-chloromercuri-2 methoxy-propylurea (MPU) to inactivate Ns and render it unfit for ADP-ribosylation. ADP-ribosylation was restricted by the pretreatment of recipient membranes with MPU, but not by pretreatment of the donor membranes. Thus, it seems most likely that the detergent extract contains

Table 2. Estimated activities of activated cholera toxin in catalyzing the ADP-ribosylation of various proteins[a]

Protein, concentration		V_{max} ADPR/enzyme/min	Reference
Ns, in situ	12 nM	38	Gill [11]
Ns, pure, solution,	40 nM	0.01	Schleifer et al. [12][b]
			Kahn and Gilman [13][c]
p21	50 nM	None detected	F. Houman, unpubl.
eIF2	200 nM	0.08	Cooper et al. [14]
Transducin α	250 nM	0.02	Abood et al. [15]
	300 nM	0.08	Cooper et al. [14]
Tubulin	6000 nM	0.003	Amir-Zaltsman et al. [16]

[a] All the proteins listed here bind GTP and this shows that GTP-binding proteins are not necessarily good substrates. The published data for each was converted to a V_{max} by assuming that the K_m for NAD was 3.6 mM. The values have not been adjusted for the different protein concentrations, which are given

[b] Calculated from their Fig. 1, assuming that each gel sample represented 500 cpm

[c] Calculated from their Fig. 3, assuming that the volume of the gel samples was 10 μl

S, or some derivative of S. If this is so, we have a powerful new assay for characterizing S and giving it some physical reality.

Are Other GTP-Binding Proteins Substrates for Cholera Toxin?

GTP is also required for the ADP-ribosylation of a number of "minor" proteins in membranes or other parts of cells. By the same reasoning as before, this requirement was taken to mean that the minor substrates were all GTP-binding proteins, but there is no longer any logical reason to believe this. In fact, it seems distinctly untrue; the provision, for example, of extra GTPγS boosts the labeling of the minor substrates in exactly the same manner as any other method of boosting the extent of ADP-ribosylation, such as by increasing the toxin concentration or the NAD concentration. Once again it is simplest to think of the guanyl nucleotide and its S site as agents which interact with the toxin rather than with the substrates of the toxin. In apparent ignorance of these facts, many workers have been led to ask whether known GTP-binding proteins would serve as substrates for cholera toxin (Table 2). Tubulin has been a favorite, with several groups reporting its ADP-ribosylation. Unfortunately, the V_{max} is so low that tubulin does not stand out from proteins that do *not* bind GTP. Likewise, eIF2, a GTP-binding initiation factor for protein synthesis, does serve as a substrate for cholera toxin, but its V_{max} also turns out to be trivial in relation to that of Ns. We, and Finkel et al. [10] have investigated p21, a viral GTP-binding oncogene product, and can find no evidence that it is ADP-ribosylated by cholera toxin even under our best conditions. Similarly, we find that elongation factor 2, another GTP-binding protein from the protein synthesizing system, is in no way special.

In sum, GTP-binding by the cholera toxin substrate does not seem to be important. As a class, GTP-binding proteins are no more likely than other proteins to serve as weak substrates for cholera toxin. The fact that the major substrate, Ns, happens to be a GTP-binding protein seems new to be merely an unfortunate coincidence that has distracted us from the real reason that cholera toxin needs GTP to work.

How is Ns Selected?

It is still appropriate to wonder what is special about Ns that makes it 1000-fold better than any other known substrate. The analogy with elongation factor 2 which has a unique amino acid residue at the ADP-ribosylation site appears inappropriate. On the contrary, Ns seems to be ADP-ribosylated at a regular arginine residue. We must look instead for something unique about the environment of the relevant arginine, and probably of the whole Nsα peptide, for even under optimal conditions the solubilized Ns behaves as a very poor substrate (Table 2).

We know of three specific ways of singling out the major from the minor substrates. The first, mentioned above, is to treat the membranes with an appropriate concentration of MPU, which specifically prevents the Ns from reacting, while hardly affecting the ADP-ribosylation of minor substrates in the same membranes. The second method is similar: it employs the protein modifying reagent diethyl pyrocarbonate which seems to specifically inactivate Ns. To a limited extent the reactivity is restored by hydroxylamine. The third maneuver is to pretreat membranes with both isoproterenol and GTPγS. This combination renders the Nsα irreversibly unreactive to cholera toxin, while it supports a very high adenylate cyclase activity. One likely hypothesis (suggested first by Murray Smigel, pers. comm.) is that the isoproterenol and GTPγS drive Ns into an irreversible condition, most probably one in which the link between Nsα and Nsβ is substantially weakened, perhaps even abolished, so that the two peptides physically dissociate. If so it would seem that Nsα requires a particular physical association with Nsβ in order to act as a respectable substrate.

References

1. Enomoto K, Gill DM (1980) Cholera toxin activation of adenylate cyclase. Roles of nucleoside triphosphates and a macromolecular factor in the ADP-ribosylation of the GTP-dependent regulatory component. J Biol Chem 255:1252–1258
2. Rodbell M, Birnbaumer L, Pohl SL, Krans HMJ (1971) The glucagon-sensitive adenyl cyclase system in plasma membranes of rat liver. J Biol Chem 246:1877–1882
3. Londos C, Salomon Y, Lin MC, Harwoos JP, Schramm M, Wolff J, Rodbell M (1974) 5′-Guanylylimidodiphosphate, a potent activator of adenylate cyclase systems in eukaryotic cells. Proc Natl Acad Sci USA
4. Cassel D, Levkovitz H, Selinger Z (1977) The regulatory GTPase cycle of turkey erythrocyte adenylate cyclase. J Cycl Nucleotide Res 3:393–406

 5. Northup JK, Sternweis PC, Smigel MD, Schleifer LS, Ross EM, Gilman AG (1980) Purification of the regulatory component of adenylate cyclase. Proc Natl Acad Sci USA 77:6516−6520
 6. Codina J, Hildebrandt JD, Sekura RD, Birnbaumer M, Bryan J, Manclark CR, Iyengar R, Birnbaumer L (1984) Ns and Ni, the stimulatory and regulatory components of adenylyl cyclase. J Biol Chem 259:5871−5886
 7. Limbird LE, Gill DM, Lefkowitz RJ (1980) Agonist-promoted coupling of the β adrenergic receptor with the guanine nucleotide regulatory protein of the adenylate cyclase system. Proc Natl Acad Sci USA 77:775−779
 8. Gill DM, Meren R (1983) A second guanyl nucleotide binding site associated with adenylate cyclase. Distinct nucleotides activate adenylate cyclase and permit ADP-ribosylation by cholera toxin. J Biol Chem 258:11908−11914
 9. Moss J, Manganiello VC, Vaughan M (1976) Hydrolysis of nicotinamide adenine dinucleotide by choleragen and its A protomer: Possible role in the activation of adenylate cyclase. Proc Natl Acad Sci USA 73:4424−4427
10. Finkel T, Der CJ, Cooper GM (1984) Activation of *ras* genes in human tumors does not affect localization, modification, or nucleotide binding properties of p21. Cell 37:151−158
11. Gill DM (1976) Multiple roles of erythrocyte supernatant in the activation of adenylate cyclase by *Vibrio cholerae* toxin in vitro. J Infect Dis 133:S55−S63
12. Schleifer LS, Kahn RA, Hanski E, Northup JK, Sternweis PC, Gilman AG (1982) Requirements for cholera toxin-dependent ADP-ribosylation of the purified regulatory component of adenylate cyclase. J Biol Chem 257:20−23
13. Kahn RA, Gilman AG (1984) ADP-ribosylation of Gs promotes the dissociation of its α and β subunits. J Biol Chem 259:6235−6240
14. Cooper DMF, Jagus R, Somers RL, Rodbell M (1981) Cholera toxin modifies diverse GTP-modulated regulatory proteins. Biochem Biophys Res Commun 101:1179−1185
15. Abood ME, Hurley JB, Pappone M-C, Bourne HR, Stryer L (1982) Functional homology between signal-coupling proteins. J Biol Chem 257:10540−10543
16. Amir-Zaltsman Y, Ezra E, Scherson T, Sutra A Littauer UZ, Salomon Y (1982) ADP-ribosylation of microtubule proteins as catalyzed by cholera toxin. EMBO J 1:181−186

Epilogue

MYRON K. JACOBSON[1]

At the close of the Seventh International Symposium on ADP-Ribosylation Reactions, it is perhaps appropriate that we reflect upon some of the progress that has occurred since the Sixth International Symposium held in 1982 in Tokyo. We have seen in this meeting continued progress along lines of investigation described in Tokyo in 1982 and we have seen new paths of investigation opened in the intervening period. We have also seen nagging problems which continue to limit progress in this field for which new solutions are required. In the following comments, I will present my own views with regard to each of these.

It should be noted that this international meeting has involved the participation of more junior investigators than any of the previous meetings. It is my feeling that these investigators have made an important contribution to this meeting and it is my hope that it will continue to be possible for them to participate in future meetings.

Mono(ADP-Ribosylation)

This meeting has shown us that ADP-ribosyl transfer catalyzed by the bacterial toxins continues to be the best understood area of ADP-ribosylation. We have seen in the presentations of Dr. Collier and Dr. Gill excellent examples of continued progress in this area. Cholera toxin and diphtheria toxin represent members of distinct families of ADP-ribosyltransferases that modify arginine and hypermodified histidine residues in their target proteins, respectively. The presentation of Dr. Ui has shown us that pertussis toxin represents yet a third type of transferase, one that modifies asparagine residues in acceptor proteins.

This meeting has also noted significant progress in our understanding of mono-(ADP-ribosylation) that is endogenous to eukaryotic cells. The presentation of Dr. Moss has shown us an increase to four in the number of known endogenous cellular mono-(ADP-ribosyl)transferases that target arginine residues in proteins. The presentation of Dr. Iglewski has provided the first evidence for a second type of endogenous trans-

1 Department of Biochemistry, North Texas State University, Texas College of Osteopathic Medicine, Denton, TX 76203, USA

ADP-Ribosylation of Proteins
(ed. by F.R. Althaus, H. Hilz, and S. Shall)
© Springer-Verlag Berlin Heidelberg 1985

ferase, one that modifies hypermodified histidine residues in a manner analogous to diphtheria toxin. An interesting question remaining relates to whether there are as yet undiscovered endogenous transferases that modify asparagine residues or even other amino acids. We should be reminded that one possible reason that most of the known transferases are arginine specific is that Dr. Moss and his co-workers have thoroughly searched for enzymes with this specificity. Certainly, the presentation of Dr. Kreimeyer raised the possibility that endogenous transferases with specificities other than those currently known may soon be detected. In that vein, the presentation of Dr. Hilz raised a note of caution with regard to the unambiguous detection of true mono(ADP-ribosyl)transferase activity. Clearly, his description of highly specific non-enzymatic ADP-ribosylation due to the presence of NAD glycohydrolase activity argues that inhibition of incorporation of ADP-ribose by nicotinamide is no longer sufficient to serve as a demonstration of mono(ADP-ribosyl)transferase activity in crude extracts.

Despite our increased knowledge of the enzymology of cellular transferases, many fundamental questions concerning endogenous mono(ADP-ribosylation) remain. At present, we still do not have a single clear cut example of a function for endogenous mono(ADP-ribosylation), although the postulation of Dr. Richter for a role in mito-chondrial calcium transport is intriguing. We have only very scanty information con-cerning the in vivo target proteins for endogenous mono(ADP-ribosyl)transferases. Still another question relates to whether mono(ADP-ribosylation) is a reversible post-translational protein modification analogous to phosphorylation or acetylation. Studies from the author's laboratory described at this meeting suggest that arginine-bound ADP-ribosylation may be a reversible modification. A major unanswered ques-tion concerns whether or not there is a connection between mono(ADP-ribose) metab-olism and poly(ADP-ribose) metabolism other than the sharing of a common substrate. At present, no direct link has been established, although the report in this meeting by Dr. Shimoyama of a nuclear mono(ADP-ribosyl)transferase still keeps this an inter-esting possibility. The continued lack of a clearly defined functional connection raises the question of whether or not studies of these two classes of ADP-ribosylation, both rapidly expanding research areas, should continue to coexist in international meetings of this type. Perhaps a risk-benefit analysis of separate vs combined meetings needs to be done in the near future.

Poly(ADP-Ribosylation)

We have seen in this meeting that inhibitors of poly(ADP-ribose) metabolism affect many biological responses. The most consistent effects are on cellular recovery from the cytotoxic effects of alkylating agents, sister chromatid exchanges, and cellular dif-ferentiation. The association of the effects of these inhibitors with many different conditions that involve changes in chromatin structure further strengthens the case for the involvement of poly(ADP-ribose) metabolism in fundamental types of chro-matin structural changes. However, we have also seen that our understanding of the details of poly(ADP-ribose) metabolism, its relationship to other chromatin components, and the functional consequences of these relationships remain difficult questions. Studies of poly(ADP-ribose) metabolism continue to be plagued with serious technical difficulties including: (1) the low steady state levels of polymer in vivo and it's rapid turnover; (2) the profound effects of polymer on the chromatographic and electrophoretic properties of the covalent acceptor proteins; (3) the instability of the linkages of polymer to protein at pH values necessary for boronate affinity chromatography, and the apparent heterogeneity of these linkages; (4) little knowledge of the complexity of the polymers; and (5) the lack of fruitful genetic approaches. We have heard a fine lecture by Dr. Koller concerning chromatin structure which also clearly pointed out that we have much yet to learn of chromatin structure, still another difficulty in relating poly(ADP-ribose) metabolism to changes in chromatin structure. Thus, it is clear that those of us in the field of poly(ADP-ribose) need to learn more of chromatin structure. We must also make those who primarily study chromatin structure aware of the technical problems associted with the study of poly(ADP-ribose) metabolism.

Despite this lament of continuing difficulties, interesting attempts to relate poly-(ADP-ribose) metabolism to chromatin structure have been described at this meeting. We have seen in the presentations of Dr. Althaus and Dr. Wielckens initial attempts to relate changes in poly(ADP-ribose) metabolism to changes in chromatin structure in intact cells by using DNA-binding dyes and nuclease probing. We have seen in the presentation of Dr. Adamietz the identification of covalent acceptor proteins in intact cells. The presentation of Dr. Alvarez-Gonzalez argued that noncovalent interactions of poly(ADP-ribose) with other components of chromatin are also likely to be important. Dr. Cerutti emphasized in his lecture the need to study well-defined regions of chromatin and we have seen in the presentations of Dr. Smulson and Dr. G. Johnson interesting initial attempts in this area. Thus, while we have far to go, approaches described in this meeting should allow us to circumvent many of the past difficulties in relating poly(ADP-ribose) metabolism to chromatin structure.

It is now well established that poly(ADP-ribose) metabolism is rapidly altered following DNA damage. Thus, this alteration is clearly coincident with the events of DNA excision repair. The present evidence is quite convincing that both poly(ADP-ribose) metabolism and DNA excision repair are necessary for cellular recovery from the cytotoxic effects of DNA-alkylating agents. However, whether or not these two processes are directly related remains a controversial question as evidenced by the intensity of the discussion on this point at this meeting. Dr. Shall in his opening lecture summarized the evidence for the involvement of ADP-ribosylation in DNA repair. Two major observations first made in Dr. Shall's laboratory are that inhibition of ADP-ribosylation following DNA damage results in an increased steady state number of DNA strand breaks and that the activation of DNA ligase II is prevented. These data have led Dr. Shall to propose that poly(ADP-ribosylation) is directly involved in DNA excision repair via effects on DNA ligation. In contrast, the presentations of Dr. Morgan and Dr. Wielckens argued that the increased number of DNA strand breaks observed in the absence of ADP-ribosylation are more likely due to effects on limiting incision rather than effects on ligation. Clearly, the mechanism by which an increased number of DNA strand breaks occurs requires more study. The elucidation of the mechanism by which DNA ligase II is activated by ADP-ribosylation will also be important to the clarification of this situation.

Past meetings have focused on the possibility that an increased number of DNA strand breaks provides the common factor by which inhibitors of ADP-ribosylation affect processes as diverse as the cytotoxicity of alkylating agents, sister chromatid exchanges, and cellular differentiation. The presentations of Dr. E. Jacobson and Dr. Zwanenburg have shown that no correlation between DNA strand breaks and the effects of inhibitors could be demonstrated on cytotoxicity or sister chromatic exchanges, respectively. Indeed, we may now need to seek other models to unify these diverse inhibitor effects. The observations of Dr. E. Jacobson, Dr. Smith, Dr. Wielckens, and Dr. Zwanenburg of abnormal cell cycle progression in the absence of ADP-ribosylation may provide a framework for new models. A final note on inhibitors: We have seen evidence at this meeting of the wide use of inhibitors of ADP-ribosylation. At present, it appears as though the selectivity of these inhibitors is reasonably good. However, the presentation of Dr. Kidwell has reminded us that the use of these inhibitors has certain inherent limitations to the study of these processes in intact cells. We must be acutely aware of these limitations. It is certainly possible that current work in this field may be overusing these inhibitors and by doing so we may be reworking a lot of old ground rather than breaking new ground. We also have a problem of the range of inhibitors. While we have a number of inhibitors that effectively inhibit poly(ADP-ribose) polymerase [and the mono(ADP-ribosyl)transferases], there is a clear need in this field for inhibitors of other reactions of poly(ADP-ribose) metabolism. We have no generally useful inhibitors of poly(ADP-ribose) glycohydrolase or ADP-ribosyl protein lyase. The presentation of Dr. Berger suggested that protease inhibitors may also prove useful for studying poly(ADP-ribose) metabolism.

We have seen a number of papers at this symposium concerned with a possible relationship between other biological endpoints related to DNA damage and ADP-ribosylation, namely, mutagenesis and malignant transformation. The data at present do not allow a firm consensus. However, the presentations of Dr. E. Jacobson, Dr. Konishi, and Dr. Miwa do argue that that poly(ADP-ribosylation) appears to be highly relevant to processes related to malignant transformation. Certainly, the involvement of a vitamin-derived molecule in these processes is of much interest in view of epidemiological data that conclude that nutritional factors are very important in human cancer.

A final comment concerning a serious limitation to progress in this field, the lack of genetic approaches. We have seen at this meeting two presentations which offer the exciting possibility of opening genetic approaches. The presentation of Dr. Ferro raised the interesting possibility that Drosophila and its powerful genetic system may be applied to this field and the presentation of Dr. Williams described an interesting finding that may provide mutant human cells in ADP-ribosyl protein lyase.

Subject Index